DEVELOPMENTS IN SEDIMENTOLOGY 55

CARBONATE RESERVOIRS
Porosity Evolution and Diagenesis in a Sequence Stratigraphic Framework

FURTHER TITLES IN THIS SERIES
VOLUMES 1-11, 13-15, 17, 21-25A, 27, 28, 31, 32 and 39 are out of print

12 R.G.C. BATHURST
CARBONATE SEDIMENTS AND THEIR DIAGENESIS
16 H.H. RIEKE III and G.V. CHILINGARIAN
COMPACTION OF ARGILLACEOUS SEDIMENTS
18A G.V. CHILINGARIAN and K.H. WOLF, Editors
COMPACTION OF COARSE-GRAINED SEDIMENTS, I
18B G.V. CHILINGARIAN and K.H. WOLF, Editors
COMPACTION OF COARSE-GRAINED SEDIMENTS, II
19 W. SCHARZACHER
SEDIMENTATION MODELS AND QUANTITATIVE STRATIGRAPHY
20 M.R. WALTER, Editor
STROMATOLITES
25B G. LARSEN and G.V. CHILINGAR, Editors
DIAGENESIS IN SEDIMENTS AND SEDIMENTARY ROCKS
26 T. SUDO and S. SHIMODA, Editors
CLAYS AND CLAY MINERALS OF JAPAN
29 P. TURNER
CONTINENTAL RED BEDS
30 J.R.L. ALLEN
SEDIMENTARY STRUCTURES
33 G.N. BATURIN
PHOSPHORITES ON THE SEA FLOOR
34 J.J. FRIPIAT, Editor
ADVANCED TECHNIQUES FOR CLAY MINERAL ANALYSIS
35 H. VAN OLPHEN and F. VENIALE, Editors
INTERNATIONAL CLAY CONFERENCE 1981
36 A. IIJIMA, J.R. HEIN and R. SIEVER, Editors
SILICEOUS DEPOSITS IN THE PACIFIC REGION
37 A. SINGER and E. GALAN, Editors
PALYGORSKITE-SEPIOLITE: OCCURRENCES, GENESIS AND USES
38 M.E. BROOKFIELD and T.S. AHLBRANDT, Editors
EOLIAN SEDIMENTS AND PROCESSES
40 B. VELDE
CLAY MINERALS—A PHYSICO-CHEMICAL EXPLANATION OF THEIR OCCURENCE
41 G.V. CHILINGARIAN and K.H. WOLF, Editors
DIAGENESIS, I
42 L.J. DOYLE and H.H. ROBERTS, Editors
CARBONATE-CLASTIC TRANSITIONS
43 G.V. CHILINGARIAN and K.H. WOLF, Editors
DIAGENESIS, II
44 C.E. WEAVER
CLAYS, MUDS, AND SHALES
45 G.S. ODIN, Editor
GREEN MARINE CLAYS
46 C.H. MOORE
CARBONATE DIAGENESIS AND POROSITY
47 K.H. WOLF and G.V. CHILINGARIAN, Editors
DIAGENESIS, III
48 J.W. MORSE and F.F. MACKENZIE
GEOCHEMISTRY OF SEDIMENTARY CARBONATES
49 K. BRODZIKOWSKI and A.J. VAN LOON
GLACIGENIC SEDIMENTS
50 J.L. MELVIN
EVAPORITES, PETROLEUM AND MINERAL RESOURCES
51 K.H. WOLF and G.V. CHILINGARIAN, Editors
DIAGENESIS, IV
52 W. SCHWARZACHER
CYCLOSTRATIGRAPHY AND THE MILANKOVITCH THEORY
53 G.M.E. PERILLO
GEOMORPHOLOGY AND SEDIMENTOLOGY OF ESTUARIES

DEVELOPMENTS IN SEDIMENTOLOGY 55

CARBONATE RESERVOIRS
Porosity Evolution and Diagenesis in a Sequence Stratigraphic Framework

CLYDE H. MOORE
Colorado School of Mines, Golden, CO, and Louisiana State University Baton Rouge, LA, USA

2001

ELSEVIER

Amsterdam - London - New York - Oxford - Paris - Shannon - Tokyo

ELSEVIER SCIENCE B.V.
Sara Burgerhartstraat 25
P.O. Box 211, 1000 AE Amsterdam, The Netherlands

© 2001 Elsevier Science B.V. All rights reserved.

This work is protected under copyright by Elsevier Science, and the following terms and conditions apply to its use:

Photocopying
Single photocopies of single chapters may be made for personal use as allowed by national copyright laws. Permission of the Publisher and payment of a fee is required for all other photocopying, including multiple or systematic copying, copying for advertising or promotional purposes, resale, and all forms of document delivery. Special rates are available for educational institutions that wish to make photocopies for non-profit educational classroom use.

Permissions may be sought directly from Elsevier Science Rights & Permissions Department, PO Box 800, Oxford OX5 1DX, UK; phone: (+44) 1865 843830, fax: (+44) 1865 853333, e-mail: permissions@elsevier.co.uk. You may also contact Rights & Permissions directly through Elsevier's home page (http://www.elsevier.nl), selecting first 'Customer Support', then 'General Information', then 'Permissions Query Form'.

In the USA, users may clear permissions and make payments through the Copyright Clearance Center, Inc., 222 Rosewood Drive, Danvers, MA 01923, USA; phone: (978) 7508400, fax: (978) 7504744, and in the UK through the Copyright Licensing Agency Rapid Clearance Service (CLARCS), 90 Tottenham Court Road, London W1P 0LP, UK; phone: (+44) 171 631 5555; fax: (+44) 171 631 5500. Other countries may have a local reprographic rights agency for payments.

Derivative Works
Tables of contents may be reproduced for internal circulation, but permission of Elsevier Science is required for external resale or distribution of such material.
Permission of the Publisher is required for all other derivative works, including compilations and translations.

Electronic Storage or Usage
Permission of the Publisher is required to store or use electronically any material contained in this work, including any chapter or part of a chapter.

Except as outlined above, no part of this work may be reproduced, stored in a retrieval system or transmitted in any form or by any means, electronic, mechanical, photocopying, recording or otherwise, without prior written permission of the Publisher. Address permissions requests to: Elsevier Science Rights & Permissions Department, at the mail, fax and e-mail addresses noted above.

Notice
No responsibility is assumed by the Publisher for any injury and/or damage to persons or property as a matter of products liability, negligence or otherwise, or from any use or operation of any methods, products, instructions or ideas contained in the material herein. Because of rapid advances in the medical sciences, in particular, independent verification of diagnoses and drug dosages should be made.

First edition 2001

Library of Congress Cataloging in Publication Data
A catalog record from the Library of Congress has been applied for.

ISBN hardbound: 0-444-50838-4
ISBN paperback: 0-444-50850-3

∞The paper used in this publication meets the requirements of ANSI/NISO Z39.48-1992 (Permanence of Paper).
Printed in The Netherlands.

This book is dedicated, with thanks, to Robin Bathurst, Bob Folk, and Bob Ginsburg. Your science, vision, scholarship, and teaching have impacted and guided all of us who deal with the world of carbonates.

PREFACE

In the decade since the publication of *Carbonate Diagenesis and Porosity*, I have given numerous public and in-house industry seminars under the auspices of Oil and Gas Consultants International (OGCI). These seminars were based on the carbonate diagenesis and porosity theme, using the first volume for course notes. During this period the emphasis of the seminars began to shift with the interests of the participants, guided by the needs of the industry and by new research directions being developed and applied by the carbonate community. As a result, the major theme of my seminars shifted from carbonate diagenesis to porosity evolution in carbonate reservoirs in a sea level-based stratigraphic framework. The title of the seminar was shortened to Carbonate Reservoirs and I began to consider a revision of the first book.

As I wrestled with the ultimate scope and direction of the project, it became obvious that I had to retain the diagenetic core of the first book because of the strong tie between porosity and diagenesis. Therefore, these sections have been up-dated and expanded to include important current research topics such as: microbial mediation of marine and meteoric diagenesis, deep marine diagenesis of mud mounds, climate driven diagenetic processes, expanded emphasis on karsting, tectonics and basin hydrology, temperate water carbonate diagenesis, basin-wide diagenetic trends and of course, new dolomitization models. The introductory section on the nature of carbonate porosity was significantly expanded and strengthened by consideration of Lucia's petrophysical-based carbonate porosity classification scheme.

The most significant changes in this new volume, however, center on consideration of the impact of climate and sea level cyclicity on facies development, diagenetic processes, and porosity evolution. The application of these core concepts of sequence stratigraphy to the study of porosity evolution and diagenesis in carbonate reservoirs has allowed us to develop a measure of predictability and to construct a series of rudimentary porosity/diagenesis models that should be useful to the exploration/production geologist in the future.

In addition, the discussion of burial diagenesis was strengthened by adding significant sections on the role of basin hydrologic history and the controls exerted by regional and global tectonics on the post-burial porosity and diagenetic evolution of carbonate rock sequences and reservoirs. The three diverse, detailed case histories presented in the last chapter were designed to reinforce the concepts and technology developed in the body of the text.

Finally, the aim of this book and the seminars upon which it is based, is to provide the working geologist, and the university graduate student with a reasonable overview of a complex, important, and above all fascinating subject, carbonate porosity evolution as driven by diagenesis.

The manuscript was read in its entirety by Emily Stoudt of Midland Texas. Her comments and suggestions were extremely helpful, were appreciated by the author, and strengthened the final product considerably. Clifford (Dupe) Duplechain produced all the figures in a timely fashion and with good humor. Lani Vigil read the manuscript too many times, served as resident grammarian and as "punctuation policeman." Lani's patience, steady support, and encouragement made the project possible and a pleasure. The core of this book was developed during my 30 year tenure at LSU. The Geology/GE Department of the Colorado School of Mines in Golden, Colorado is thanked for providing me with office space, a collegial atmosphere and logistical support. My editor, Femke Wallien of Amsterdam, is thanked for her patience and encouragement. The manuscript was produced camera ready, in house, on a Macintosh G-3 Powerbook.

Clyde H. Moore
Golden, Colorado
February 22, 2001

CONTENTS

PREFACE vii

CHAPTER 1. THE NATURE OF THE CARBONATE DEPOSITIONAL SYSTEM 1
  THE BASIC NATURE OF CARBONATE SEDIMENTS AND SEDIMENTATION 1
    Introduction 1
    Origin of carbonate sediments 1
    The reef: a unique depositional environment 3
    Unique biological control over the texture and fabric of carbonate sediments 4
    Carbonate grain composition 5
    Carbonate rock classification 6
    Efficiency of the carbonate factory and its impact on patterns of carbonate sedimentation 9
    Carbonate platform types and facies models 12
  SUMMARY 16

CHAPTER 2. CONCEPTS OF SEQUENCE STRATIGRAPHY AS APPLIED TO CARBONATE DEPOSITIONAL SYSTEMS 19
  INTRODUCTION 19
  SEQUENCE STRATIGRAPHY 19
    Eustasy, tectonics and sedimentation: the basic accommodation model 19
    Hierarchy of Stratigraphic Cycles 25
    Introduction to Carbonate Sequence Stratigraphic Models 25
    The Ramp Sequence Stratigraphic Model 25
    The Rimmed Shelf Sequence Stratigraphic Model 27
    The Escarpment Margin Sequence Stratigraphic Model 28
    Sequence Stratigraphic Model of Isolated Platforms 29
    High-Frequency Cyclicity on Carbonate Platforms: Carbonate Parasequences 32
THE CONSEQUENCES OF THE HIGH-CHEMICAL REACTIVITY OF CARBONATE
SEDIMENTS AND ROCKS DURING EXPOSURE AT SEQUENCE BOUNDARIES 33
    Carbonate minerals and their relative stability 33
    Controls over the mineralogy of carbonate sediments, today and in the past 34
    Mineralogy of ancient limestones: the concept of early progressive mineral stabilization and porosity evolution 35
  SUMMARY 36
CHAPTER 3. THE CLASSIFICATION OF CARBONATE POROSITY 37
  INTRODUCTION 37
  THE NATURE AND CLASSIFICATION OF CARBONATE POROSITY 37
    Choquette and Pray porosity classification 39

  The Lucia rock fabric/petrophysical carbonate porosity classification 43
THE NATURE OF PRIMARY POROSITY IN MODERN SEDIMENTS 48
  Intergrain porosity 48
  Intragrain porosity 49
  Depositional porosity of mud-bearing sediments 49
  Framework and fenestral porosity 50
SECONDARY POROSITY 52
  Introduction 52
  Secondary porosity formation by dissolution 52
  Secondary porosity associated with dolomitization 55
  Secondary porosity associated with breccias 58
  Secondary porosity associated with fractures 59
SUMMARY 59

CHAPTER 4. DIAGENETIC ENVIRONMENTS OF POROSITY MODIFICATION AND TOOLS FOR THEIR RECOGNITION IN THE GEOLOGIC RECORD 61

INTRODUCTION 61
  Marine environment 62
  Meteoric environment 63
  Subsurface environment 63
  Petrography-cement morphology 65
  Petrography-cement distribution patterns 68
  Petrography-grain-cement relationships relative to compaction 70
  Trace element geochemistry of calcite cements and dolomites 71
  Stable isotopes 78
  Strontium isotopes 84
  Fluid Inclusions 88

CHAPTER 5. NORMAL MARINE DIAGENETIC ENVIRONMENTS 93

INTRODUCTION 93
SHALLOW WATER, NORMAL MARINE DIAGENETIC ENVIRONMENTS 95
  Abiotic shallow marine carbonate cementation 95
  Recognition of ancient shallow marine abiotic cements 98
  Biologically mediated marine carbonate cementation and diagenesis 101
  Diagenetic setting in the intertidal zone 102
  Modern shallow water submarine hardgrounds 104
  Recognition and significance of ancient hardgrounds 106
  Diagenetic setting in the modern reef environment 106
  Recognition of reef-related marine diagenesis in the ancient record 110
  Early marine lithification of the Permian Capitan reef complex New Mexico, USA 113

Porosity evolution of the Golden Lane of Mexico and the Stuart City of Texas 115
Porosity evolution of Devonian reefs: Western Canadian Sedimentary Basin 123
SLOPE TO DEEP MARINE DIAGENETIC ENVIRONMENTS 128
Introduction to diagenesis in the slope to deep marine environment 128
Carbonate diagenesis associated with ramp to slope mud mounds 128
Carbonate diagenesis of mounds near hydrothermal and hydrocarbon vents. 133
Carbonate diagenesis associated with escarpment shelf margins: Enewetak Atoll .. 136
Carbonate diagenesis of steep escarpment shelf margins: Bahama Platform 139
SUMMARY 143

CHAPTER 6. EVAPORATIVE MARINE DIAGENETIC ENVIRONMENTS 145
INTRODUCTION 145
Introduction to diagenesis in evaporative marine environments 145
THE MARGINAL MARINE SABKHA DIAGENETIC ENVIRONMENT 148
Modern marginal marine sabkhas 148
Diagenetic patterns associated with ancient marginal marine sabkhas 151
Ordovician Red River marginal marine sabkha reservoirs, Williston Basin, USA ... 153
Mississippian Mission Canyon marine sabkha reservoirs, Williston Basin, USA 154
Ordovician Ellenburger marine sabkha-related dolomite reservoirs, west Texas, USA 158
Criteria for the recognition of ancient marginal marine sabkha dolomites 161
MARGINAL MARINE EVAPORATIVE LAGOONS REFLUX DOLOMITIZATION 162
The marginal marine evaporative lagoon as a diagenetic environment 162
Permian Guadalupian west Texas, USA: an ancient evaporative lagoon complex ... 164
Permian reservoirs of the South Cowden Field, Texas: reflux dolomitization 166
Jurassic Smackover dolomitization, Texas, USA: a reflux dolomitization event 170
Criteria for the recognition of ancient reflux dolomites 174
Regional evaporite basins, coastal salinas, their setting and diagenetic environment 175
The MacLeod salt basin 175
The Elk Point Basin of Canada 177
Michigan Basin, USA 178
Jurassic Hith Evaporite and the Arab Formation, Middle East 179
SUMMARY 182

CHAPTER 7. DIAGENESIS IN THE METEORIC ENVIRONMENT 185
INTRODUCTION 185
GEOCHEMICAL AND MINERALOGICAL CONSIDERATIONS 185
Geochemistry of meteoric pore fluids and precipitates 185
Isotopic composition of meteoric waters and meteoric carbonate precipitates 188
Mineralogic drive of diagenesis within the meteoric environment 190

    Implications of kinetics of the $CaCO_3$-H_2O-CO_2 system in meteoric environments 190
    Climatic effects 192
    Hydrologic setting of the meteoric diagenetic environment 194
THE VADOSE DIAGENETIC ENVIRONMENT 197
    Introduction 197
    Upper vadose soil or caliche zone 197
    Lower vadose zone 199
    Porosity development in the vadose diagenetic environment 199
THE METEORIC PHREATIC DIAGENETIC ENVIRONMENT 200
    Introduction 200
    Immature hydrologic phase: island-based and land-tied floating meteoric water lenses developed in mineralogically immature sediments 200
    Bermuda, a case study of an island with a permanent floating fresh water lens 202
    Meteoric diagenesis in Quintana Roo, Mexico strandplains 205
    The Oaks Field, meteoric diagenesis in a Jurassic shoreline, Louisiana, USA 206
    The mature hydrologic phase: regional meteoric water systems in sea level lowstands. 209
    Mature hydrologic system: regional system in a sea level lowstand: Bahama Bank 209
    Mature hydrologic system: regional gravity-driven system developed during relative sea level lowstand on a land-tied carbonate shelf 213
    Porosity development and predictability in regional meteoric aquifer environments 215
    Geochemical trends characteristic of a regional meteoric aquifer system 217
    Mississippian grainstones of southwestern New Mexico, USA: a case history of porosity destruction in a regional meteoric aquifer system 219
KARST PROCESSES, PRODUCTS AND RELATED POROSITY 222
    Introduction 222
    Solution, cementation, and porosity evolution in a mature karst system 225
    Yates Field, Texas, USA: case history of an unconformity-related karsted reservoir 225
DOLOMITIZATION ASSOCIATED WITH METEORIC AND MIXED METEORIC-MARINE WATERS 232
    Introduction 232
    Meteoric-marine mixing, or Dorag model of dolomitization 232
    Concerns about the validity of the mixing model of dolomitization 233
    Modern mixing zone dolomitization 234
    Pleistocene to Miocene mixing zone dolomitization 237
    Mississippian North Bridgeport Field, Illinois Basin, USA: mixed water dolomite reservoirs 238
    Where do we stand on mixing zone dolomitization today? 242
SUMMARY 243

CHAPTER 8. SUMMARY OF EARLY DIAGENESIS AND POROSITY MODIFICATION OF CARBONATE RESERVOIRS IN A SEQUENCE STRATIGRAPHIC AND CLIMATIC FRAMEWORK 245

INTRODUCTION 245

RESERVOIR DIAGENESIS AND POROSITY EVOLUTION DURING 3RD ORDER SEA LEVEL LOWSTANDS (LST) 246

Introduction 246
Diagenesis/porosity model of a carbonate ramp during sea level lowstand (LST) ... 246
Diagenesis/porosity model of a steep-rimmed carbonate shelf during sea level lowstand (LST) 249
Diagenesis/porosity model of a rimmed isolated carbonate platform during sea level lowstand (LST) 251

RESERVOIR DIAGENESIS AND POROSITY EVOLUTION DURING 3RD ORDER SEA LEVEL RISES (TST) 253

Introduction 253
Diagenesis/porosity model for a ramp in a rising sea level (TST) 253
Diagenesis/porosity model for a rimmed shelf during a rising sea level (TST) 255
Diagenesis/porosity model of a rimmed isolated carbonate platform during sea level rise (TST) 257

RESERVOIR DIAGENESIS AND POROSITY EVOLUTION DURING 3RD ORDER SEA LEVEL HIGHSTANDS (HST) 259

Introduction 259
Diagenesis/porosity model for a ramp during a sea level highstand (HST) 259
Diagenesis/porosity model for a rimmed shelf during a sea level highstand (HST) . 263
Diagenesis/porosity model for an isolated carbonate platform during a sea level highstand (HST) 265

DIAGENESIS AND POROSITY AT THE PARASEQUENCE SCALE 265

Introduction 265
Cumulative diagenesis associated with parasequence stacking patterns in a 3rd order sequence 266

DIAGENESIS AND POROSITY AT THE SUPERSEQUENCE (2ND ORDER) SCALE: SEQUENCE STACKING PATTERNS 268

Introduction 268
Progradational sequence set and diagenesis 268
Aggradational sequence set diagenesis 269
Retrogradational sequence set and diagenesis 269

DIAGENESIS AND POROSITY AT THE 1ST ORDER SCALE: ICEHOUSE VERSUS GREENHOUSE 270

Introduction 270

Long-term temporal changes in carbonate mineralogy: impact on diagenesis and porosity evolution 271
The architecture of sedimentary sequences and their diagenesis/porosity evolution as a function of long-term climatic cycles (icehouse versus greenhouse) 272
CASE HISTORIES 274
    Introduction 274
    Controls over porosity evolution in Upper Paleozoic shelf limestones, Southwest Andrews area, Central Basin Platform of the Permian Basin, USA: porosity development as a function of subaerial exposure during an icehouse time 275
    Stratigraphic controls over porosity development in an Albian ramp sequence, offshore Angola, Africa: Porosity evolution under greenhouse conditions in a mixed siliciclastic/carbonate/evaporite setting 281
SUMMARY 287

CHAPTER 9. BURIAL DIAGENETIC ENVIRONMENT 291
INTRODUCTION 291
THE BURIAL SETTING 291
    Introduction 291
    Pressure 292
    Temperature 293
    Deep burial pore fluids 295
    Tectonics and basin hydrology 295
PASSIVE MARGIN BURIAL REGIMEN 297
    Introduction 297
    Mechanical compaction and de-watering 297
    Chemical compaction 300
    Factors affecting the efficiency of chemical compaction 305
    Subsurface cements in a passive margin setting 307
    Petrography of burial cements 308
    Geochemistry of burial cements 310
    Impact of late subsurface cementation on reservoir porosity 312
    Subsurface dissolution in a passive margin setting 312
    The North Sea Ekofisk Field: a case history of porosity preservation in chalks 313
THE ACTIVE OR COLLISION MARGIN BURIAL REGIMEN 318
    Introduction 318
    Lower Ordovician Upper Knox Dolomite of the southern Appalachians, USA 318
    Dolomitization in Devonian carbonates, Western Canada Sedimentary Basin, Alberta, Canada 324
THE POST TECTONIC DIAGENETIC REGIMEN 329
    Introduction 329

Mississippian Madison Aquifer of the mid-continent, USA 329
Post orogenic aquifer system in Paleozoic carbonates of the Greater Permian Basin 332
PREDICTING CHANGES IN POROSITY WITH DEPTH 335
SUMMARY 339

CHAPTER 10. POROSITY EVOLUTION FROM SEDIMENT TO RESERVOIR: CASE HISTORIES 341

INTRODUCTION 341
THE MISSISSIPPIAN MADISON AT MADDEN FIELD WIND RIVER BASIN, WYOMING, USA 341
    Introduction 341
    General setting 342
    Stratigraphic and depositional setting 343
    Correlation of surface exposures into the deep subsurface at Madden 346
    Burial history of the Madison 350
    Porosity/permeability of Madison reservoirs and outcrop dolomites 350
    Diagenetic history of the Madison, Wind River Basin 351
    Conclusions and lessons learned from the Madison at Madden Field, Wyoming 354
THE UPPER JURASSIC SMACKOVER AND RELATED FORMATIONS, CENTRAL U.S. GULF COAST: A MATURE PETROLEUM FAIRWAY 355
    Introduction 355
    General setting 356
    Sequence stratigraphic and depositional framework of the Upper Jurassic 358
    Early diagenesis and porosity modification related to depositional sequences and sea level 365
    Burial diagenesis and subsurface porosity evolution, Oxfordian reservoirs, central Gulf of Mexico 367
    Upper Jurassic exploration and production constraints and strategies, central Gulf of Mexico 368
    Conclusions and lessons learned from the Smackover of the central Gulf of Mexico 371
THE TERTIARY MALAMPAYA AND CAMAGO BUILDUPS, OFFSHORE PALAWAN, PHILIPPINES: 3-D RESERVOIR MODELING 372
    Introduction 372
    Geologic setting 373
    Lithofacies and depositional model 375
    Diagenetic history 376
    Reservoir rock types as model input parameters 377
    The model 378
    Conclusions and lessons learned from the 3-D modeling of the Tertiary Malampaya

and Camago buildups 379
EPILOGUE 381
   INTRODUCTION 381
   THE LEGACY 381
   THE STATE OF THE ART 382
   WHERE DO WE GO FROM HERE? 384

REFERENCES 387

INDEX 425

Chapter 1

THE NATURE OF THE CARBONATE DEPOSITIONAL SYSTEM

INTRODUCTION

While this book is concerned primarily with porosity evolution and diagenesis in carbonate reservoirs, the reader and the author must ultimately share a common understanding of the fundamental characteristics of the carbonate realm. Therefore, these first introductory chapters are designed to highlight general concepts unique to the carbonate regimen, such as: 1) the biological origin enjoyed by most carbonate sediments; 2) the complexity of carbonate rock classification; 3) the bathymetric framework of carbonate depositional facies and environments; 4) the response of carbonate depositional systems to changes in relative sea level, i.e. concepts of sequence stratigraphy as applied to carbonates; 5) the diagenetic consequences of the high chemical reactivity of carbonates.

Readers desiring an in-depth review of carbonate depositional environments should read the review of carbonate facies by James, (1979) the extensive compilation edited by Scholle et al. (1983) and the most recent review presented in *Carbonate Sedimentology* by Tucker and Wright (1990). The recent book entitled *Sequence Stratigraphy,* edited by Emery and Myers (1996), will give the reader a reasonable overview of the basic concepts of sequence stratigraphy as well as its application to carbonate rock sequences.

THE BASIC NATURE OF CARBONATE SEDIMENTS AND SEDIMENTATION

Introduction

The basic characteristics of carbonate sediments can generally be traced to the overwhelming biological origin of carbonate sediments and the influences that this origin exerts on sediment textures, fabrics, and depositional processes such as the ability of certain organisms to build a rigid carbonate framework. The following section outlines these broad biological influences on carbonate sediments and sedimentation.

Origin of carbonate sediments

Well over 90% of the carbonate sediments found in modern environments are thought to be biological in origin and form under marine conditions (Milliman, 1974; Wilson, 1975; Sellwood, 1978; Tucker and Wright, 1990). Distribution of most carbonate sediments is di-

THE NATURE OF THE CARBONATE DEPOSITIONAL SYSTEM

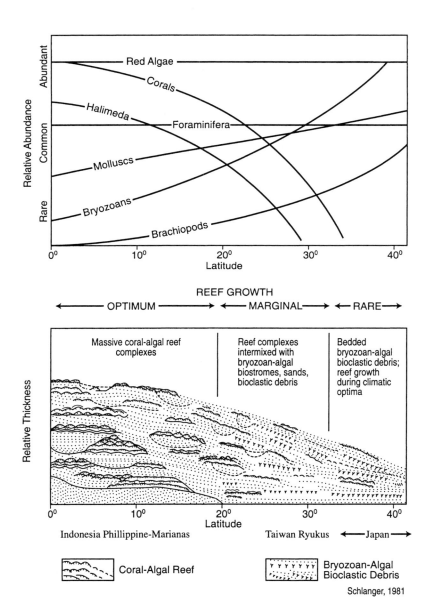

Figure 1.1. Latitudinal change from tropical to temperate carbonate facies in the northern Pacific. Used with permission of SEPM.

rectly controlled by environmental parameters favorable for the growth of the calcium carbonate secreting organisms. These parameters include temperature, salinity, substrate, and the presence/absence of siliciclastics (Lees, 1975). Schlanger (1981) beautifully illustrated the latitudinal control over the distribution of carbonate secreting organisms and its impact on total carbonate production. Carbonate production in tropical waters is much faster than in temperate waters because high carbonate producers such as hermatypic corals thrive in the warm, clear waters of the tropics (Fig. 1.1).

Since many hermatypic corals and other reef organisms are depth and light sensitive, maximum carbonate production is generally attained in the first 10m water depth in the marine environment (Fig. 1.2) (Schlager,

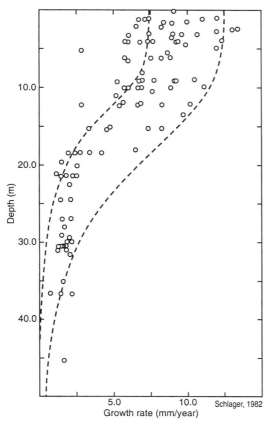

Figure 1.2. Predicted and observed rates of growth of the Caribbean reef coral Montastrea annularis. The predicted rates (dashed lines) are derived from a light-growth equation and compare well with the observed values (circles), indicating that light exerts a strong control on coral carbonate production. Used with permission of AAPG.

1992). Most carbonate sediments, therefore, were initially deposited or formed in shallow marine waters. The origin of larger carbonate grains is relatively easy to determine. The grain ultra-structure and overall shape allows the worker to easily determine the grain origin such as a fragment of an oyster, a whole foraminifer, or, perhaps, a chemically precipitated ooid. The difficulty arises when one considers the origin of the carbonate mud fraction. Does carbonate mud consist of fine particles derived by erosion of larger bioclastic material or is it a chemical precipitate (such as the whitings common to tropical carbonate environments) (Shinn et al., 1989; Macintyre and Reid, 1992)? Another possibility is that of aragonite needles released from green algae on death (Perkins et al., 1971; Neumann and Land, 1975). Finally, do bacteria play a role in the formation of carbonate muds (Drew,1914; Berner,1971; Chafetz, 1986; Folk, 1993)? Today most carbonate workers would concede that carbonate muds probably originate from all of the above mechanisms but are uncertain about the relative importance of each. We will re-visit these questions later when we consider mineralogical composition of carbonate sediments and in particular the role bacteria may play in marine diagenesis. Finally, it should be noted that most carbonate sediments are generally deposited near the site of their origin. This is in sharp contrast to siliciclastics, which are generally formed outside the basin of deposition, and are transported to the basin, where physical processes control their distribution. For siliciclastics, climate is no constraint, for they are found worldwide and are abundant at all depths, in fresh water as well as marine environments.

The reef: a unique depositional environment

The ability of certain carbonate-secreting organisms to dramatically modify their environment by encrusting, framebuilding and binding leads to the depositional environment unique to the carbonate realm, the reef (Fig. 1.3). In this discussion, the term reef will be used in its genetic sense, i.e. a solid organic framework that resists waves (James, 1983). In a modern reef, there is an organism-sediment mosaic that sets the pattern for organic

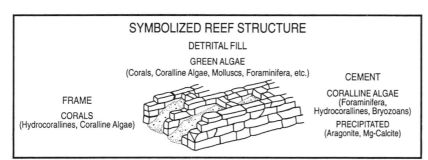

Figure 1.3. Diagram showing symbolized reef structure emphasizing interaction between frame, detrital fill, and cement. Used with permission of the Journal of Geology.

framework reef sequences. There are four elements: 1) the framework organisms, including encrusting, attached, massive, and branching metazoa; 2) internal sediment, filling primary growth as well as bioeroded cavities; 3) the bioeroders, which break down reef elements by boring, rasping, or grazing, thereby contributing sediment to peri-reef as well as internal reef deposits; and 4) cement, which actively lithifies and may even contribute to internal sediment (Fig. 1.3). While the reef rock scenario is complex, it is consistent. Chapter 5 presents a comprehensive treatment of marine cementation associated with reef depositional environments.

Today, corals and red algae construct the reef frame. Ancillary organisms such as green algae contribute sediment to the reef system. Reef organisms have undergone a progressive evolution through geologic time, so that the reef-formers of the Lower and Middle Paleozoic (i.e. stromatoporoids) are certainly different from those of the Mesozoic (rudistids and corals) and from those that are observed today (James, 1983). Indeed, there were periods, such as the Upper Cambrian, Mississippian, and Pennsylvanian, when the reef-forming organisms were diminished, or not present, and major reef development did not occur (James, 1983).

Framework reefs, then, while certainly influenced by physical processes are dominated by a variety of complex, unique, biological and diagenetic processes that have no siliciclastic counterpart.

Unique biological control over the texture and fabric of carbonate sediments

The biological origin of most carbonate

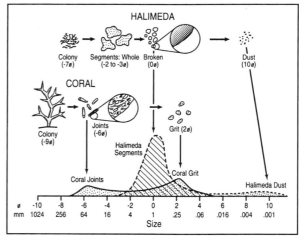

Figure 1.4. Bioclast ultrastructural control of grain size in beach sediments from Alacran Reef, Mexico. Halimeda breaks down into two main grain size modes, coarse sand composed of broken segments and micrite sized dust, composed of the ultimate ultrastructrural unit of the organism, single needles of aragonite. Corals also break down into two modes, large gravel sized joints, and sand sized particles which represent the main ultrastructural unit of the coral (Fig. 1.5). Used with permission of the Journal of Geology.

sedimentary particles places severe constraints on the utility of textural and fabric analysis of carbonate sediments and rocks. Size and sorting in siliciclastics are generally indicators of the amount and type of physical energy (such as wind, waves, directed currents and their intensity) influencing sediment texture at the site of deposition (Folk, 1968). Size and sorting in carbonate sediments, however, may be more influenced by the population dynamics of the organism from which the particles were derived, as well as the peculiarities of the organism's ultrastructure. Folk and Robles (1964) documented the influence of the ultrastructure of coral and Halimeda on the grainsize distribution of beach sands derived from these organisms at Alacran Reef in Mexico (Fig. 1.4).

In certain restricted environments, such as on a tidal flat, it is not uncommon to find carbonate grains composed entirely of a single species of gastropod (Shinn et al., 1969; Shinn, 1983). The mean size and sorting of the resulting sediment is controlled by natural size distribution of the gastropod population and tells us little about the physical conditions at the site of deposition. For example, one commonly encounters large conchs living in and adjacent to mud-dominated carbonate lagoons in the tropics. Upon death, these conchs become incorporated as large clasts in a muddy sediment. This striking textural inversion does not necessitate, as it might in siliciclastics, some unusual transport mechanism such as ice rafting, but simply means that the conch lived and died in a depositional environment dominated by carbonate mud.

Other textural and fabric parameters such as roundness also reflect biological control. Roundness in siliciclastic grains is generally thought to indicate distance of transport, and/

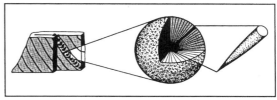

Figure 1.5. Ultrastructure of a coral septum composed of spherical bodies stacked one on top of the other in inclined rows. Each sphere is about 2 phi sized (grit) and consists of aragonite fibers arranged about a point center. Used with permission of Brill Publishing.

or the intensity of physical processes at the site of deposition (Blatt et al., 1980; Blatt, 1982). Roundness in carbonate grains, however, may well be controlled by the initial shape of the organism from which the grain is derived (for example, most foraminifera are round). In addition, an organism's ultrastructure, such as the spherical fiber fascioles characteristic of the coelenterates, may also control the shape of the grains derived from the coral colony (Fig. 1.5).

Finally, some grains such as oncoids, rhodoliths and ooids are round because they originate in an agitated environment where sequential layers are acquired during the grain's travels over the bottom, with the final product assuming a distinctly rounded shape (Bathurst, 1975; Tucker and Wright, 1990; Sumner and Grotzinger, 1993). Great care must be used when interpreting the textures and fabrics of carbonate sediments and rocks as a function of physical conditions at the site of deposition.

Carbonate grain composition

Skeletal remains of organisms furnish most of the coarse-grained sediments deposited in carbonate environments. It follows that the grain composition of carbonate sediments

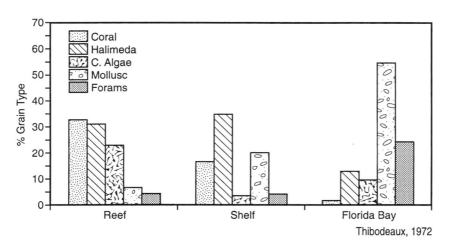

Figure 1.6. Graph showing percentage of major grain types in the dominant depositional environments of south Florida. Used with permission of the author.

and rocks often directly reflects their environment of deposition because of the general lack of transport in carbonate regimens and the direct tie to the biological components of the environment. A number of workers have documented the close correlation between biological communities, depositional environment and subsequent grain composition in modern carbonate depositional systems (Ginsburg, 1956; Swinchatt, 1965; Thibodaux, 1972) (Fig. 1.6). Our ability to determine the identity of the organism from which a grain originates by its distinctive and unique ultrastructure is the key element to the usefulness of grain composition for environmental reconstruction in ancient carbonate rock sequences (Bathurst, 1975). Carozzi (1967) and Wilson (1975) based their detailed microfacies studies on the thin section identification of grain composition, including detailed identification of the biological affinities of bioclasts.

In contrast, grain composition in siliciclastics is related to ultimate provenance of the sediment, climate, and stage of tectonic development of the source, rather than to conditions at the site of deposition (Krynine, 1941; Folk, 1954; Pettijohn, 1957; Blatt, 1982).

Carbonate rock classification

Carbonate rock classification parallels, in some respects, the classifications commonly used to characterize siliciclastics (Folk, 1968). Siliciclastics are normally classified on the basis of composition or texture, or both. Compositional classifications of sandstones generally are based on three end-members, Quartz+Chert, Feldspars and Unstable Rock Fragments (Blatt, 1982). As noted above, sandstone compositional classes generally reflect the tectonic setting, provenance, and climate of the extrabasinal source of the sandstones. In his textural classification, Folk (1968) uses three end-members, sand, mud, or gravel, or sand, silt, and clay, in the absence of gravel. The sediment and rock types recognized by such a texturally based classification are thought to reflect the general level of energy present at the site of deposition (Folk, 1968). This concept of interdependency of sediment texture and energy at the site of deposition has been incorporated into the two most widely accepted carbonate classifications. The first was published by Folk (1959) and the second by Dunham (1962).

Folk's classification is more detailed, en-

CARBONATE ROCK CLASSIFICATION

VOLUMETRIC ALLOCHEM COMPOSITION			>10% Allochems ALLOCHEMICAL ROCKS (I AND II)		<10% Allochems MICROCRYSTALLINE ROCKS (III)		UNDISTURBED BIOHERM ROCKS (IV)
			Sparry Calcite Cement > Microcrystalline Ooze Matrix	Microcrystalline Ooze Matrix > Sparry Calcite Cement	1-10% Allochems	<1% Allochems	
			SPARRY ALLOCHEMICAL ROCKS (1)	MICROCRYSTALLINE ALLOCHEMICAL ROCKS (2)			
	>25% Intraclasts (i)		Intrasparrudite (Ii:Lr) / Intrasparite (Ii:La)	Intramicrudite (IIi:Lr) / Intramicrite (IIi:La)	Intraclasts: Intraclast-bearing Micrite (IIIi:Lr or La)	Micrite (IIIm:L); if disturbed, Dismicrite (IIImX:L); if primary dolomite, Dolomicrite (IIIm:D)	Biolithite (IV:L)
	<25% Intraclasts	>25% Oolites (O)	Oosparrudite (Io:Lr) / Oosparite (Io:La)	Oomicrudite (IIo:Lr) / Oomicrite (IIo:La)	Oolites: Oolite-bearing Micrite (IIIo:Lr or La)		
		<25% Oolites / Volume Ratio of Fossils to Pellets	>3:1 (b)	Biosparrudite (Ib:Lr) / Biosparite (Ib:La)	Biomicrudite (IIb:Lr) / Biomicrite (IIb:La)	Fossils: Fossiliferous Micrite (IIIb:Lr, La, or L1)	Most Abundant Allochem
			3:1 - 1:3 (bp)	Biopelsparite (Ibp:La)	Biopelmicrite (IIbp:La)	Pellets: Pelletiferous Micrite (IIIp:La)	
			<1:3 (p)	Pelsparite (Ip:La)	Pelmicrite (IIp:La)		

Folk, 1959

Figure 1.7. Carbonate rock classification of Folk (1959). This classification is compositional as well as textural. Used with permission of AAPG.

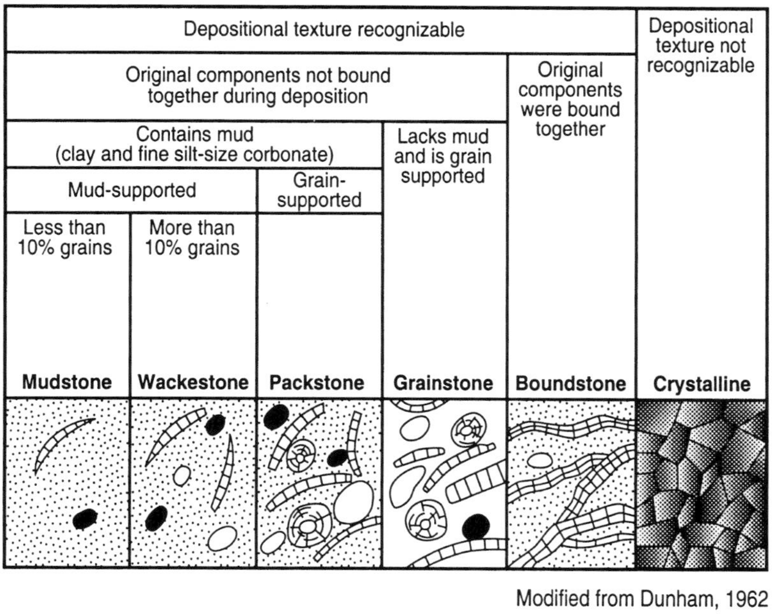

Depositional texture recognizable					Depositional texture not recognizable
Original components not bound together during deposition				Original components were bound together	
Contains mud (clay and fine silt-size corbonate)			Lacks mud and is grain supported		
Mud-supported		Grain-supported			
Less than 10% grains	More than 10% grains				
Mudstone	**Wackestone**	**Packstone**	**Grainstone**	**Boundstone**	**Crystalline**

Modified from Dunham, 1962

Figure 1.8. Dunham's carbonate classification (1962). This classification is primarily textural and depends on the presence of recognizable primary textural elements. Used with permission of Blackwell Science.

compassing a textural scale which incorporates grainsize, roundness, sorting, and packing, as well as grain composition (Fig. 1.7). The complexity of the Folk classification makes it more applicable for use with a petrographic microscope in a research setting. Dunham's classification, on the other hand, is primarily textural in nature, is simple, and is easily used in the field and by well-site geologists. The Dunham classification will be used exclusively in this book.

Dunham's major rock classes are based on the presence or absence of organic binding, presence or absence of carbonate mud, and the concept of grain versus matrix support (Fig. 1.8). The four major rock classes, mudstone, wackestone, packstone, and grainstone, represent an energy continuum (Dunham, 1962). The choice of these names by Dunham emphasizes the close affinity of his rock types to commonly used siliciclastic terminology. The term boundstone emphasizes the major role of organic binding and framework formation in carbonates, and is a rock type unique to the carbonate realm. As with Folk's classification, these rock names are modified by the most abundant constituents (such as ooid, pellet grainstone), further defining the biologic and

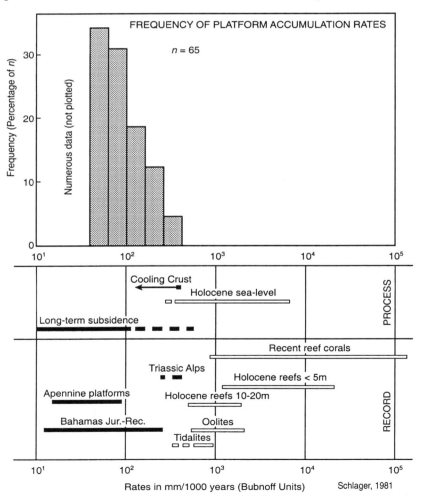

Figure 1.9. Growth rates of carbonate systems compared with relative sea level rise. Solid bars represent the geological record, open bars represent Holocene production rates and sea level rise. Used with permission of GSA.

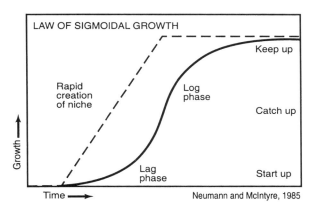

Figure 1.10. Population growth versus time, indicating 'start-up', 'catch-up' and 'keep-up' phases of Neumann and Macintyre (1985). Used with permission of Blackwell Science.

physical conditions at the site of deposition.

Efficiency of the carbonate factory and its impact on patterns of carbonate sedimentation

Today, under conditions of a relatively high sea level stand, in areas optimal for carbonate facies development, such as the tropical Bahama platform in the Atlantic, the rates of biologic production of carbonate sediments seem to far outstrip the rate of formation of new sediment accommodation space (Fig. 1.9).

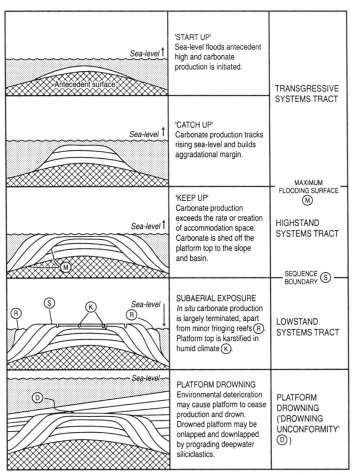

Figure 1.11. Schematic model for an isolated carbonate platform, showing idealized systems tract geometries and platform drowning. Used with permission of Blackwell Science.

The production of excess sediments leads to the transport of significant quantities of carbonate sediments into adjacent deep water (Neumann and Land, 1975; Hine et al., 1981; Mullins and Cook, 1986).

However, Neumann and McIntyre (1985) noted that carbonate sediment production was not a straight-line function during the flooding and the subsequent development of a high sea level stand over a platform, but followed what they termed the "law of sigmoidal growth." During the initial platform-flooding event, there is a lag during which carbonate production is relatively slow, and water depth increases, as new populations of carbonate secreting organisms are re-established on the platform. This is termed the "start-up phase" (Fig. 1.10). Once the carbonate factory is well developed on the platform, sediment production is maximized and available accommodation space is progressively filled. This is called the "catch-up phase." When available accommodation space is filled, it remains filled regardless of the rate at which new space is being generated. This phase is termed the "keep-up phase." When the rate of generation of new accommodation space slows, as sea level reaches its highstand, the volume of excess carbonate sediment increases, and the rate of transport of shallow-water carbonates into deeper water accelerates. These relationships lead to the following fundamental sedimentologic characteristics of carbonate sequences.

High-frequency carbonate cycles are normally asymmetrical (deepest water at the base) because of the initial lag phase during transgression. These same cycles often build up to exposure at their terminus because of the keep-up phase developed during the highstands of even short-duration eustatic cycles.

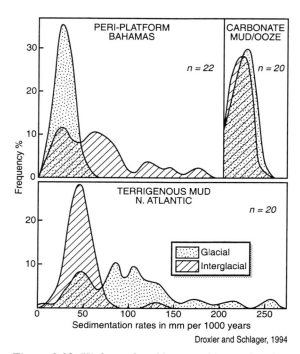

Figure 1.12. Highstand and lowstand basinal sedimentation rates for carbonate and siliciclastic systems. Used with permission of Geology.

During longer-term transgressions, sedimentation tracks sea level rise in the catch-up phase, resulting in aggradational platform margins (Fig. 1.11).

During sea level highstands, sedimentation out-strips accommodation in the keep-up phase, platform margins prograde, and significant highstand sediment shedding into deep water takes place (Fig. 1.11). This is in stark contrast to the lowstand shedding so characteristic of siliciclastic regimens. Droxler and Schlager (1994) documented contrasting highstand and lowstand carbonate and siliciclastic sedimentation rates in the Atlantic (Fig. 1.12).

One of the major consequences of the keep-up phase of the Neumann-McIntyre scheme, is the difficulty in drowning, and covering a carbonate platform with deeper marine shales once that carbonate platform has been

EFFICIENCY OF THE CARBONATE FACTORY

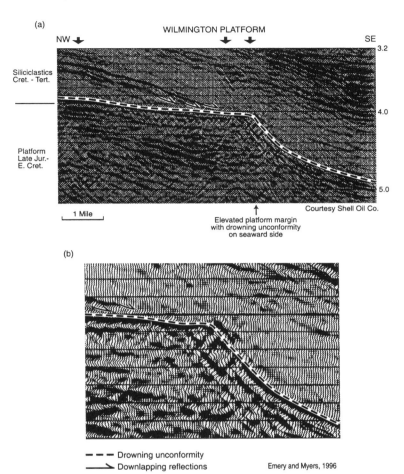

Figure 1.13. Drowning unconformity on the Wilmington Platform, offshore USA. Used with permission of AAPG.

Table 1.1

Nutrient supply
Low-nutrient environments are most favourable for organic carbonate growth, particularly reefs. In high-nutrient settings it is apparent that carbonate-secreting, framework-building corals are replaced by fleshy algae, sponges or soft corals. Modern examples include the East Java Sea, Indonesia, where corals are scarce below 15 m and are replaced by green algae. The lack of reefs is interpreted to be the result of upwelling nutrient-rich waters, which stimulate the growth of coral competitors (Roberts and Phipps, 1988).

Siliciclastic input
Clay particles in suspension in the water column will inhibit light penetration, significantly reducing or preventing reef growth. In addition, coral polyps are, and many predecessor framework building forms were, unable to cope with clay particles during feeding. Circumstantial evidence of carbonate demise by clastic input in the geological record is demonstrated by seaward stepping Miocene carbonate systems from south to north across the Luconia Province, offshore Sarawak, Malaysia, away from northward prograding delta systems (Epting, 1989; Figure 10.9). Note that changes in water salinity or nutrient content associated with a siliciclastic depositional system such as a delta, may be the prime cause of carbonate production demise.

Salinity variations
Changes of salinity may exert a dramatic effect on carbonate productivity, particularly that of reefs. In the Holocene, broad, shallow lagoons on flat-topped carbonate platforms acquired highly variable salinities. The seaward ebb flow of this 'inimical bank water' killed the reefs, which also became overwhelmed by seaward transport of carbonate fines from the lagoons (Neumann and Macintyre, 1985).

Oxygenation
When oxygen is depleted, reefs will die. However, the best evidence for massive carbonate demise resulting from oxygen deficiency is circumstantial. Schlager and Philip (1990) have used evidence for Cretaceous oceanic anoxia and compared this with the frequency of Cretaceous carbonate platform demise to indicate a possible causal link.

Predation
The importance of predation on carbonate communities is well-documented from modern day examples. The 'Crown of Thorns' starfish blight, which threatened to wipe out large tracts of the Great Barrier Reef is well-known. However, it is difficult, if not impossible, to demonstrate clearly the effect of predation in the fossil record, and is currently unknown on the scale of an entire carbonate platform.

Table 1.1. Causes of carbonate platform drowning. From Emery and Myers, 1996. Used with permission of Blackwell Science.

established. Figure 1.9 illustrates growth rates of modern carbonate producers compared to documented rates of relative sea level rise and accumulation rates of various ancient carbonate platforms (Schlager, 1992). Clearly, carbonate sediment production rates far outstrip even the fastest rate of relative sea level rise suggesting that major changes in environment, climate, and other factors must take place before a drowning event, or a drowning unconformity is initiated. Emery and Myers (1996) detailed the factors that might initiate a drowning event leading to a drowning unconformity (Table 1.1). These factors include nutrient supply, siliciclastic input, salinity variation, oxygenation, and predation.

Figure 1.13 illustrates a classic drowning unconformity on the Wilmington Platform, offshore USA where a Jurassic-Early Cretaceous carbonate platform is swamped by Cretaceous-Tertiary prograding siliciclastics (Schlager, 1992). As was noted by Emery and Myers (1996), while a rapid sea level rise might temporarily drown a carbonate platform, a more plausible explanation is a deterioration of the carbonate environment caused by one of the factors listed in Table 1.1.

Carbonate platform types and facies models

Over the past twenty years there has been much discussion concerning the terminology of large-scale carbonate depositional environments: Wilson, 1975; Read, 1985; Tucker and Wright, 1990; Emery and Myers, 1996. In this book we will generally follow the lead of Tucker and Wright (1990) using the term carbonate platform in a general sense for a thick sequence of shallow-water carbonates. Carbonate platforms develop in a variety of tectonic settings such as passive margins, failed rift systems, and foreland basins. Several broad categories of

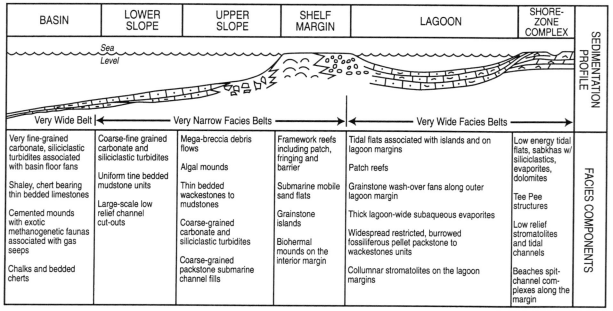

Figure 1.14. Standard facies belts on a rimmed shelf. Used with permission of Springer Verlag.

platforms are recognized: rimmed shelf, ramp, isolated platform, and eperic platform. There are a number of variations recognized by various workers within each broad category.

The rimmed carbonate shelf is characterized by a pronounced break in slope into deeper water at the shelf margin. Slope angles may vary from one degree to nearly vertical. The shelf may vary from 10's to 100's of kilometers in width. The shelf margin is a shallow-water high-energy zone marked by reefs and/or high-energy grain shoals. This shelf margin separates a shallow landward shelf lagoon from the deeper waters of the adjacent basin. Figure 1.14 illustrates a general facies model of a rimmed carbonate shelf.

The shelf margin facies tract is narrow, marked by shallow water, high-energy conditions with framework reefs, including barrier, fringing and patch reefs. Mobile grainstone shoals and storm-related islands are common. Biohermal mud mounds occur on the lagoon

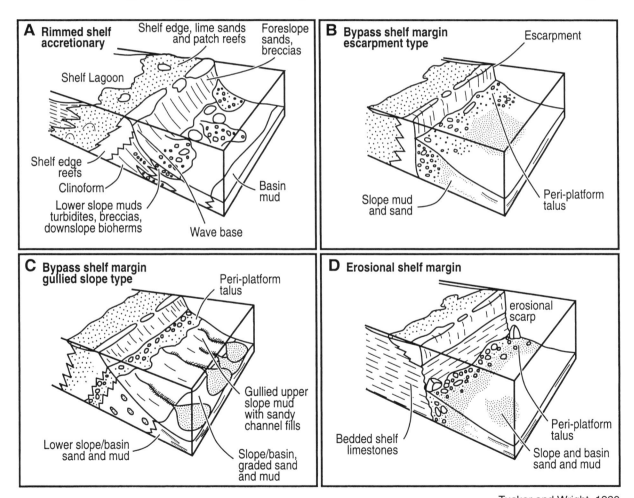

Tucker and Wright, 1990

Figure 1.15. Rimmed carbonate shelf types. A) Accretionary type. B) Bypass, escarpment type. C) Bypass, gullied-slope type. D) Erosional rimmed shelf margin. Used with permission of Blackwell Science.

margins. Slope to basin facies tracts are broader than the shelf margin facies tract but are still relatively narrow. The upper slope may have megabreccia debris flows, algal mud mounds, thin-bedded wackestones and carbonate and/or siliciclastic coarse to fine-grained turbidite packages. The lower slope is mud dominated with thin uniform bedding with carbonate or siliciclastic turbidite packages. The lower slope may exhibit large-scale low-relief channel cutouts. The basin is mud-dominated with thin-bedded, fine-grained basin floor fans. Shales, chalks, and cherts are common.

The shelf lagoon facies tract behind the shelf margin is broad. It generally exhibits deeper, quieter water than encountered at the margin. The lagoon is dominated by muddy sediments except for high-energy grainstones around localized bathymetric highs. Patch reefs are common. Under certain climatic conditions, the lagoon may turn hypersaline, and massive subaqueous evaporites may be deposited. The inner margins of the shelf lagoon are generally low-energy shoreline complexes marked by tidal flats, sabkhas, and occasionally beaches. If a source is available, siliciclastic shoreline complexes may replace carbonates as the dominant inner shelf lagoon facies. A number of different rimmed carbonate shelf types have been recognized (Fig. 1.15) based on geometry of the shelf margin and the geological processes operating along the shelf margin.

Modern examples of rimmed carbonate shelves include the Belize Platform in the Caribbean, the Bahama Platform in the Atlantic, and the Great Barrier Reef in the Pacific. The Belize rimmed shelf is marked by a well developed barrier reef along its margin, offshore isolated platforms developed on horst blocks, and a deep lagoon with numerous patch reefs

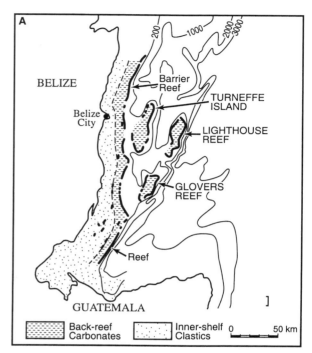

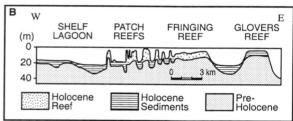

Figure 1.16. The Belize carbonate rimmed shelf. A) Map showing shelf-margin barrier reefs and reefs around isolated platforms (Lighthouse and Glovers Reefs and Turneffe Island). B) Sketch cross section from a seismic profile showing development of modern reefs upon pre-existing topographic highs of Pleistocene limestone (James and Ginsburg, 1979). Used with permission of Blackwell Science.

and inner shelf lagoon siliciclastics shed from the Mayan mountains opposite the southern reaches of the lagoon (Fig. 1.16).

The Lower Cretaceous carbonates occurring around the margin of the Gulf Coast of Mexico and the USA are excellent examples of ancient rimmed carbonate shelves. The high-energy shelf margin is marked by rudist, coral, and stromatoporoid reefs which restrict an enormous shelf lagoon (hundreds of miles wide in Texas) characterized by low-energy muddy carbonates (Bebout, 1974; Bebout, 1983; Moore, 1989). In the Albian/Aptian the lagoon goes hypersaline and a regional evaporite unit, the Ferry Lake Anhydrite, is deposited (Loucks and Longman, 1982). In Mexico, an isolated platform (the Tuxpan Platform) is developed on a graben structure offshore of the main shelf margin reminiscent of the modern isolated platforms in Belize (Enos et al., 1983). The Tuxpan Platform supports the historic "Golden Lane" oil fields of northeastern Mexico (Chapter 5).

A carbonate ramp, as originally conceived, is a gently sloping surface, generally less than one degree, on which high-energy shore line environments pass gradually into deeper water with no detectable change in slope (Ahr, 1973; Wilson, 1975; Tucker and Wright, 1990; Emery and Myers, 1996). Wilson (1975) suggested that carbonate ramps are generally an early evolutionary phase in the development of a rimmed carbonate shelf. As the high-energy shoreline complex of a ramp progrades into deeper water, it will ultimately take on the characteristics of a rimmed shelf, gaining an abrupt change in slope at the terminus of the prograding high-energy complex. See also the discussions in Tucker and Wright (1990) and Emery and Myers (1996). Once the geometry changes from ramp to shelf, it is exceedingly difficult to switch back to a ramp geometry, particularly since geometry of carbonate sedimentary sequences are generally frozen by early diagenetic processes such as cementation

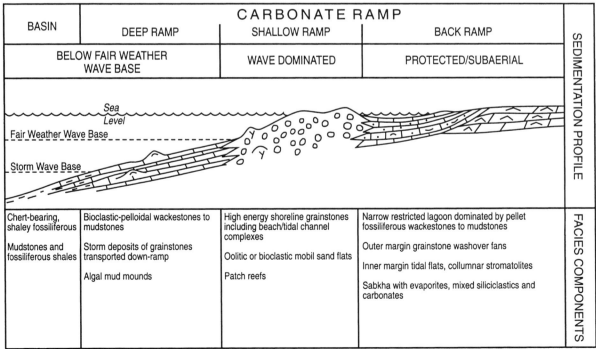

Figure 1.17. The carbonate ramp depositional model. Used with permission of Blackwell Science.

during perturbations of sea level. The best example of a modern carbonate ramp is the Trucial Coast of the Arabian Gulf.

Figure 1.17 illustrates the ramp depositional model. The inner shallow ramp is characterized by high-energy shoreline grainstones because of unimpeded wave travel across the shallower portions of the ramp. Barrier beach/tidal channel complexes reminiscent of siliciclastic shorelines are generally well developed. Through time, the inner ramp zone may expand by progradation giving rise to a broad high-energy facies tract. Grainstones may consist of ooids, bioclasts or peloids. Reefs or reef facies are not common but may locally occur as a result of coral/algal growth on hard substrates resulting from periodic exposure and cementation on and around bathymetric highs. The innermost ramp, behind the high-energy shoreline, is a protected environment characterized by shallow, narrow lagoons and subaerially exposed low-energy tidal flats and sabkhas. If there is a source of siliciclastics, they will tend to be trapped in this protected environment. Bioclastic and peloidal wackestones, packstones and mudstones characterize the deeper ramp, in front of the high-energy zone, and below fair weather wave base. Storm processes may transport ooids and bioclasts from the upper shoreface and siliciclastics from inner ramp lagoons onto the deeper shelf. Algal mud-mounds may develop on the deep ramp. See Tucker (1990) and Emery and Myers (1996) for further discussions on ramp depositional models. The Lower Cretaceous Aptian Cow Creek Limestone may be a good example of an ancient ramp sequence (Inden and Moore, 1983).

Isolated carbonate platforms are areas of shallow water carbonate deposition surrounded by deeper water. The Bahama Platform of the Atlantic is a well-known modern example of an isolated platform. These platforms may be of any size from the 700 km long Great Bahama Bank to the 35 km diameter Enewetak Atoll of the western Pacific. Larger isolated platforms generally occur in association with rifted margins such as the Lower Cretaceous Tuxpan Platform in Mexico (Enos et al., 1983), the Triassic and Jurassic platforms of Western Europe (Wilson, 1975), and the Miocene carbonate platforms of Sarawak and the Philippines (Emery and Myers, 1996). The smaller, atoll-like platforms are generally associated with oceanic volcanism. Isolated platforms are generally steep-sided rimmed platforms with reefing along windward margins of the larger platforms; or, in the case of the smaller platforms, they may be totally rimmed with reefs. Mature, well-developed isolated carbonate platforms are generally siliciclastic free. Their interior platform facies are generally low-energy mudstones and wackestones with possible patch reefs. Under arid conditions, the central lagoon, or platform interior may become hypersaline and develop evaporite deposition.

SUMMARY

The biological influence over the origin and nature of carbonate sediments is reflected in distinct climatic and environmental controls over their distribution, their general autochthonous nature, and the shape and skeletal architectural control over sediment texture. The commonly used Dunham carbonate rock classification is texturally based and generally parallels siliciclastic classifications. Reef rocks, however, are unique in origin and texture.

SUMMARY

Carbonate grain composition generally reflects the biological community present at the site of deposition.

The high efficiency of carbonate producing processes leads to massive off platform transport of sediments into surrounding deep basinal environments during sea level highstands as well as difficulty in drowning established carbonate platforms. Carbonate platform drowning events are generally ascribed to degradation of the marine environment.

The rimmed shelf and the carbonate ramp are the two common carbonate facies models used today. While grossly similar, the lack of a discrete change of slope into deeper water on the ramp necessitates a gradation of facies into deeper water, high-energy facies concentrated at the shoreline, no barrier reefing, and poorly developed lagoonal facies. Both ramp and rimmed platform may have siliciclastics at the shoreline. Isolated carbonate platforms generally are siliciclastic free.

Carbonate platform types are controlled and modified by marine cementation and reef framework growth along the platform margins, leading to escarpment margins and rimmed platforms.

In general, carbonate ramps ultimately evolve into rimmed platforms and escarpment platform margins.

Chapter 2

CONCEPTS OF SEQUENCE STRATIGRAPHY AS APPLIED TO CARBONATE DEPOSITIONAL SYSTEMS

INTRODUCTION

Stratigraphy is the science of rock strata. It includes lithostratigraphy, the organization of rock units on the basis of their lithologic character, and chronostratigraphy, the organization of rock units on the basis of their age or time of origin. Sequence stratigraphy is a relatively new, but wildly popular, stratigraphy subdiscipline which deals with the subdivision of sedimentary basin fill into genetic packages bounded by unconformities and their correlative conformities. These genetic packages are termed sequences. Sequences are composed of a series of genetically linked depositional systems (termed systems tracts) thought to have been formed between eustatic sea level fall inflection points (Posamentier, 1988). The linkage of the development of the unconformity-bounded sedimentary sequence and global eustatic events forms the basis for the chronostratigraphic framework of sequence stratigraphy, i.e. inter-basinal and inter-regional correlation of specific sequences deposited during a specific interval of time. Figure 2.1 illustrates the basic difference between lithostratigraphy and the chronostratigraphic significance of sequence stratigraphy.

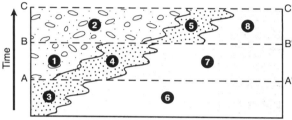

Figure 2.1. The difference between sequence stratigraphy, which has a geological time significance and lithostratigraphy, which correlates rocks of similar type. A lithostratigraphic correlation would correlate conglomerate units 1 and 2, sandstone units 3, 4 and 5 and mudstone units 6, 7 and 8. A sequence stratigraphic correlation would correlate time lines A-A', B-B' and C-C'. Used with permission of Blackwell Science.

SEQUENCE STRATIGRAPHY

Eustasy, tectonics and sedimentation: the basic accommodation model

Table 2.1 lists the major variables that are important in sequence stratigraphy. Tectonic subsidence controls basin geometry and the amount of new space available on the shelf for deposition of sediments, termed accommodation space by most authors. Eustatic sea level controls stratal patterns, such as on-lap, down-lap and top-lap as well as

Table 2.1

Variable	Controls
Tectonic subsidence	New space for sediments on the shelf and basin geometry
Eustatic sea level	Stratal patterns and lithofacies distribution
Sediment supply	Sediment fill and paleowater depth
Climate	Eustatic cycles and biological and chemical environment

Table 2.1. Major variables in sequence stratigraphy. Personal communication, John Sangre, 1994.

lithofacies distribution. Sediment supply controls sediment fill and paleowater depth. Finally, climate controls eustatic cycles, giving us glacial versus interglacial-driven sea level. Most importantly for carbonates, climate impacts biological and chemical environments.

Figure 2.2A outlines the difference between eustatic sea level, relative sea level, and water depth. Eustasy is measured between the sea surface and a fixed datum such as the center of the earth. Eustatic sea level can vary by changing ocean basin volume (e.g. by changing oceanic ridge volume) or by changing oceanic water volume (e.g. by releasing or capturing water during interglacial and glacial episodes). Any changes in eustatic sea level result in changes in alluvial base level on a global scale. A schematic eustatic sea level cycle is shown in Figure 2.2B. The two important points on the eustatic sea level curve are the falling inflection point, where eustatic sea level is falling at the fastest rate, and the rising inflection point, where eustatic sea level is rising at the fastest rate. These inflection points divide the curve into a falling and a rising limb. Relative sea level is measured between the sea surface and a local datum such as a

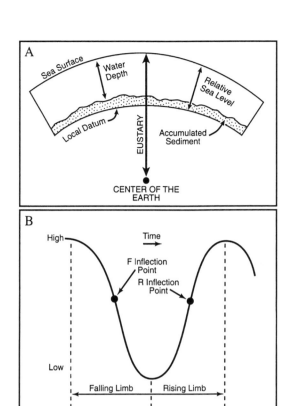

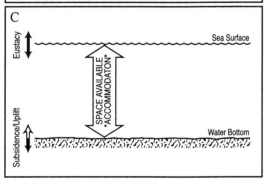

Posamentier et al., 1988

Figure 2.2. A. The difference between eustasy, relative sea level and water depth. Note that relative sea level is measured to an arbitrary datum within the crust and hence is tied to tectonic subsidence-uplift. B. A conceptual model of a sea level cycle through time. It is divided into a falling limb and rising limb separated by the low stand. The two inflection points represent points of the maximum rate of sea level movement. C. This cartoon represents the concept of the space available for sediment deposition hereafter called accommodation space. Used with permission of SEPM.

surface within the sediment pile (Fig. 2.2A). This local datum will move in response to tectonic subsidence, sediment compaction, or uplift. Relative sea level will rise during subsidence and fall in response to tectonic uplift. Relative sea level is obviously affected by changes in eustatic sea level as well. Changes in relative sea level can be thought of as changes in sediment accommodation space (Fig. 2.2C). As relative sea level rises, the volume of new accommodation space also rises. Conversely as relative sea level falls the volume of new accommodation space decreases.

Figure 2.3 illustrates the conceptual response of rate of addition of new accommodation space (i.e. relative sea level) to a eustatic sea level curve. In this model, the rate of change of eustasy varies through time while we assume that the rate of subsidence remains constant through the same period of time. This is a viable assumption because the time interval of the model is relatively short, 3 million years, and we are dealing with a single site on the shelf. If we look at the rate of eustatic fall in each of the three eustatic sea level cycles represented, we see that the rate of fall of the first two cycles is less than the rate of subsidence at the site. In this case, even though we are still adding new accommodation space,

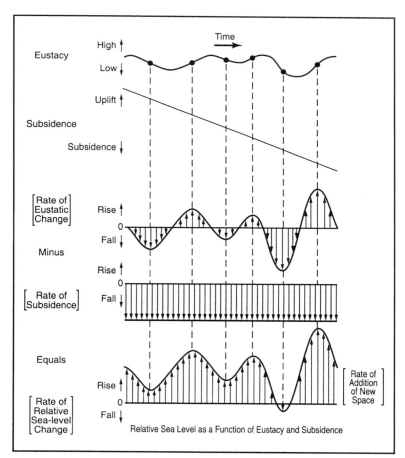

Figure 2.3. Basic accommodation model featuring varying eustatic cycles versus a constant subsidence rate over a time span of some 3 million years. Rate of relative sea level change is a measure of rate of addition of new accommodation space. Used with permission of SEPM.

Posamentier et al., 1988

the actual rate of addition is diminished. If this new accommodation space is filled with sediments, one should see a facies shift toward the sea marking each of the first two sequence boundaries rather than an unconformity as implied by the sequence definition given earlier. The sequence boundary is placed close to the eustatic falling inflection point where rate of addition of new accommodation space is at a minimum. In the third eustatic cycle represented in Figure 2.3, the rate of eustatic fall is greater than the rate of subsidence at the site and no new accommodation space is formed. In this situation an unconformity (erosion or non-deposition) marks the sequence boundary rather than the seaward facies shift

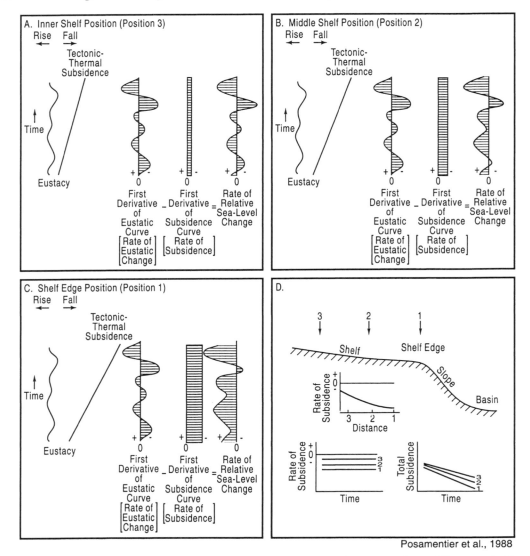

Posamentier et al., 1988

Figure 2.4. Basic accommodation model featuring varying eustatic cycles versus three different subsidence rates representing the inner shelf (A), the middle shelf (B) and inner shelf (C). The eustatic cycles and time span are the same as seen in Figure 2.3. Used with permission of SEPM.

seen at the first two sequence boundaries.

In Figure 2.3, it is quite clear that subsidence controls accommodation space while eustasy controls the stratal patterns that lead to the differentiation of the sediment pile into genetic units or sedimentary sequences.

The rate of subsidence will vary across most sedimentary shelves, with the highest rate of subsidence at the margin and the lowest subsidence rate in the shelf interior (Fig. 2.4). Using the same eustatic sea level curve used in Figure 2.3, the rate of development of new accommodation space (relative sea level) at three positions on the shelf illustrates the effects of variable subsidence rates on accommodation and stratal patterns (Fig. 2.4). Position 1 at the shelf margin exhibits the highest rate of subsidence. The resulting relative sea level curve (rate of change of accommodation) is the same as shown in Figure 2.3 (same rate of subsidence). Only the third cycle shows no accommodation space generated at the lowstand with a potential unconformity marking the sequence boundary. Position 3 is in the shelf interior and exhibits the lowest rate of subsidence. The rate of eustatic sea level fall is greater than the rate of subsidence in each of the three eustatic cycles, so no new accommodation space is generated at each lowstand and the sequence boundaries may all be represented by unconformities (erosive, or non-deposition). Position 2 in the center of the shelf has an intermediate subsidence rate so that the rate of eustatic sea level fall is greater than subsidence in cycles 1 and 3. Hence these sequences may be bounded by unconformities. This is exactly what one sees in the geologic record; unconformities on sedimentary shelves normally die out toward the basin.

During the early days of seismic stratigraphy, it was recognized that the seismic record was not uniform. Discrete packets of basin fill could be isolated on the basis of unique seismic reflector terminations, such as on-lap, down-lap and top-lap. These packets were predictably linked together into a basin fill package, which seemed to represent a discrete, unique sea level cycle. The large

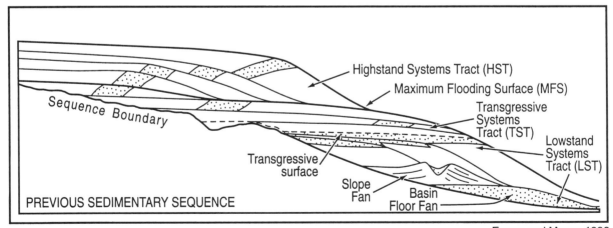

Emery and Myers, 1996

Figure 2.5. Stratal geometries in a type 1 sequence on a shelf-break margin. Five separate sedimentary packages are shown, traditionally assigned to three systems tracts; lowstand, transgressive and highstand. Used with permission of Blackwell Science.

package was termed a sedimentary sequence and the component packets were called systems tracts. An idealized sedimentary sequence is illustrated in Figure 2.5. The three basic systems tracts are lowstand system tract (LST), transgressive system tract (TST), and highstand systems tract (HST). Changing rates of sediment accommodation during a major relative sea level cycle forms them. The basic sequence stratigraphic model was initially applied almost exclusively to siliciclastic regimens. However, as will be seen in the following discussion, the model may be applied to carbonates with some modifications related to some basic differences between the sedimentology of siliciclastics versus the sedimentology of carbonates.

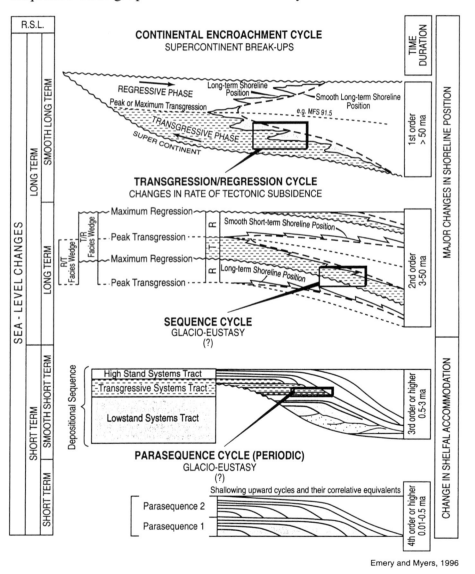

Figure 2.6. Hierarchy of stratigraphic cycles (after Duval, 1992). Used with permission of Blackwell Science.

Hierarchy of stratigraphic cycles

Figure 2.6 illustrates a useful hierarchy of stratigraphic cycles. The long term cycles, termed 1st and 2nd order cycles are driven by long-term tectonically controlled relative sea level changes such as super-continent break-ups (>50 ma duration) and changes in rate of subsidence associated with the breakup (3-50 ma duration) (Duval, 1992). The short-term 3rd order cycles (0.5-3 ma duration), presumably driven by glacio-eustasy, are the fundamental units of sequence stratigraphy. While we have no problems with glacio-eustatic cycles in those periods when ice caps were known to exist, there has long been a problem with glacio-eustatic forcing in presumably ice-free greenhouse episodes such as the Mesozoic. See discussion by Emery and Myers (1996) and later in Chapter 8, this book. The shorter-term, high frequency 4th order and higher cycles (0.01-0.5 ma duration) may be driven by climate-related glacio-eustasy or sediment-related auto cyclicity (James, 1984).

Introduction to carbonate sequence stratigraphic models

The following sections outline the general application of sequence stratigraphic concepts to carbonates, emphasizing those factors that specifically impact the behavior of carbonate sedimentation in response to relative sea level changes. These factors include climate, in-situ sediment production, rigid reef framework at shelf margins, and shelf margin marine cementation. In addition, the role of siliciclastics and evaporites will be discussed. Four main models will be considered: the ramp, rimmed shelf, escarpment margin, and isolated platform. Each model will be taken through a complete relative sea level cycle. While this discussion is necessarily brief, the interested reader may gain a more in-depth treatment by considering the works of Sarg (1988), Loucks and Handford (1993) and Emery and Myers (1996) and SEPM Memoir #63 edited by Harris, Saller and Simo (1999).

The ramp sequence stratigraphic model

During transgression, shallow water ramp facies, such as the marginal shoal, will initially onlap onto the bounding unconformity of the previous sequence. As the water deepens in the deeper ramp facies, however, carbonate production will decrease and shaly, organic-rich facies may be deposited (Fig. 2.7A). Under arid conditions, hypersalinity may cause density stratification resulting in anaerobic bottom environments with a high potential for organic material preservation and formation of hydrocarbon source rock. The Upper Jurassic lower Smackover of the US Gulf Coast is a homoclinal ramp and represents the transgressive systems tract of the Smackover/Norphlet sequence. The varved limestones of the lower Smackover are the source rock for Jurassic and Cretaceous reservoirs across the Gulf Coast. (Sassen et al., 1987; Moore and Heydari, 1993).

As the transgressive systems tract reaches its maximum flooding surface, coastal onlap reaches its inland-most position; and the inner ramp facies begin to prograde seaward over and downlap onto deeper ramp facies, initiating highstand sedimentation. An offlapping coastal zone followed by a narrow shallow lagoon/sabkha complex (Fig. 2.7B) characterizes the highstand. As these linked facies prograde into

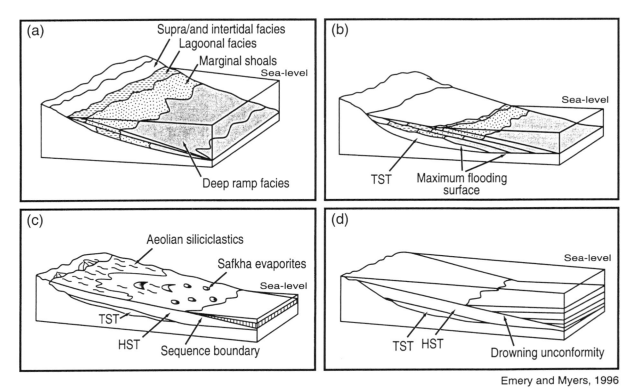

Figure 2.7. Sequence stratigraphic models for ramp systems. A) Transgressive systems tract B) High stand systems tract C) Low stand systems tract D) Drowning unconformity. Used with permission of Blackwell Science.

deeper water, the ramp becomes distally steepened with a distinct slope break into deeper water. During the transgression of the next sequence, this change in slope becomes the position of an evolving shelf margin and the ramp is transformed into a rimmed shelf (Wilson, 1975). If siliciclastics are in the system, they tend to be trapped in the back coastal lagoon. Under arid conditions siliciclastics are often associated with sabkha environments as coastal dune complexes.

The entire expanse of the ramp is generally exposed during the lowstand. If siliciclastics are present and the climate is humid, they may be transported across the ramp by alluvial processes with some incisement depending on magnitude of sea level fall (Fig. 2.7). If the climate is arid, siliciclastics will move across the exposed ramp as mobile dune fields or wadies (Fig. 2.7C). Lowstand prograding complexes are common features of the ramp during low sea level stands because of low slope angles. These lowstand prograding complexes generally have siliciclastics at the base and high-energy carbonate grainstones at the top and exhibit well defined clinoforms in seismic section. The lowstand systems tract will generally be dominated by siliciclastic slope and basin floor fans if a siliciclastic source is present. Under arid conditions, the shallow basin in front of the ramp may go hypersaline during sea level lowstand; and subaqueous evaporites may be deposited and onlap onto basin floor fans as well as onto the lowstand

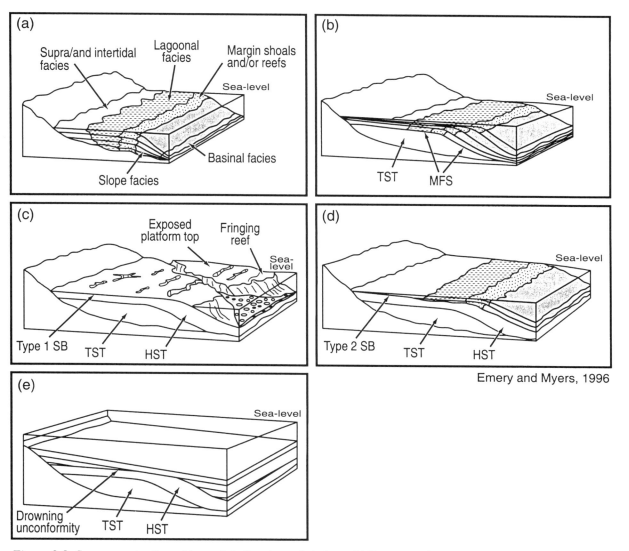

Figure 2.8. Sequence stratigraphic models for rimmed shelves. A) Transgressive systems tract B) Highstand systems tract C) Lowstand systems tract D) Type 2 unconformity and development of a shelf margin wedge E) Drowning unconformity. Used with permission of Blackwell Science.

prograding complexes developed on the outer ramp slope.

The rimmed shelf sequence stratigraphic model

As previously discussed, a rimmed carbonate shelf may evolve from a distally steepened ramp, or it may form over relict rimmed shelf topography inherited from an earlier sedimentary episode. During transgression, the high-energy shelf margin may well aggrade vertically in a keep-up mode, forming a relatively deep lagoon landward of the margin as is seen in Belize today (Purdy et al., 1975; Ginsburg and Choi, 1984) (Fig.

2.8A). Under arid conditions, this lagoon may be filled with evaporites during the transgression (Moore, 1989). Transgressive keep-up shelf margins are commonly reefal in nature.

During the sea level highstand, the aggradational shelf margin may shift to progradation, as the rate of formation of new accommodation space slows. Top sets progressively thin during this period and the lagoon may be filled with prograding tidal flats and inner shelf sand shoals (Fig. 2.8B). The amount of shelf margin progradation actually seen will be determined by the depth of the adjoining basin and the steepness of the slope into the basin. Carbonate turbidities and carbonate basin floor fans will be the rule in adjacent basinal environments.

At relative sea level lowstands, the entire shelf-top will be exposed, the carbonate factory shuts down, and meteoric diagenetic processes will impact the exposed shelf (discussed later in this chapter) (Fig. 2.8C). If siliciclastics are available, the climate during lowstand will determine how the clastics move across the exposed shelf and ultimately into the deeper waters of the adjacent basin. Under arid conditions, aeolian processes such as mobile dune complexes will be the preferred mode of transport. Under humid conditions, alluvial processes may transport the clastics to the shelf margin and onto the marine slope where gravity processes will furnish transport into deeper water with deposition on slope fans and basin floor fans. Lowstand prograding complexes such as were described in the preceding ramp section may form if the depth to the adjacent basin floor is small enough, and if relief on the shelf margin proper is not too steep. Generally, if the shelf margin is dominated by reefs, shelf margin relief will be too great and there will be no lowstand prograding complex. Under these conditions, the adjacent basin will be carbonate starved.

The escarpment margin sequence stratigraphic model

As noted in the previous section, a rimmed carbonate platform may evolve into a by-pass escarpment margin, particularly if framework reefs develop during aggradation. If oversteepning and erosion take place during transgression, the by-pass margin may evolve into an erosive margin, exhibit some backstepping, and produce a significant talus apron on the basin margin slope (Fig. 2.9A).

During highstand, the escarpment slopes are steep and little margin progradation can take place. However, most of the highstand carbonate production from the shelf will be transported over, or through the margin, to onlap the upper slope talus apron and be deposited as carbonate slope and basin floor fans (Fig. 2.9B). During the highstand, the top set beds of the platform will be progressively thinned as the rate of formation of accommodation space slows, approaching the late highstand and sequence boundary.

At the lowstand, the platform top will be exposed to subaerial conditions. As with the rimmed platform, the transport of any siliciclastics in the system will be determined by climate considerations: arid climate, aeolian processes; humid climate alluvial processes with potential incisement at the margin (Fig. 2.9C). The adjacent basin will generally be carbonate starved since there will be little chance for any carbonate production along the marginal escarpment. Siliciclastics will occur

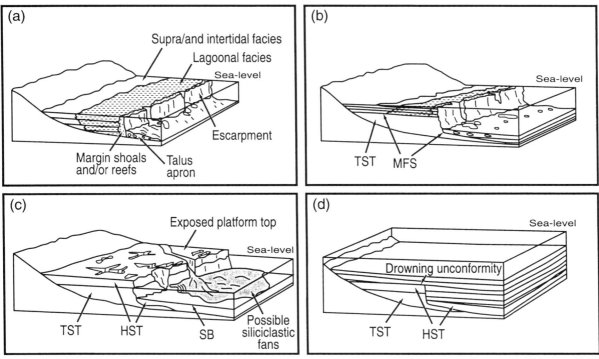

Emery and Myers, 1996

Figure 2.9. Sequence stratigraphic models for rimmed shelves. A) Transgressive systems tract B) High stand systems tract C) Low stand systems tract D) Drowning unconformity. Used with permission of Blackwell Science.

on the slope and in the basin as slope fans and basin floor fans. The removal of the buoyancy effect of seawater along the escarpment during lowstand may accelerate margin collapse. Lowstand collapse talus aprons mixed with siliciclastics are common on the upper slope of the escarpment margin. Because of the deep water and vertical slopes at the marginal escarpment, there will seldom be a lowstand prograding complex associated with an escarpment margin.

Sequence stratigraphic model of isolated platforms

Many isolated platforms are associated with structural elements such as horst blocks formed during rifting (see earlier discussion of isolated platform facies). Isolated carbonate platforms may have the characteristics of rimmed or escarpment margins, or be a combination of both types. One of the most singular features of isolated platforms is that they are generally siliciclastic free so that the strong reciprocal sedimentation patterns noted for continent-attached platforms is missing. Isolated platforms developed offshore from tectonically active areas, however, may ultimately be swamped by fine-grained siliciclastics prograding seaward from the adjacent growing land area. The Miocene platforms off the northwestern coast of Sarawak in Southeast Asia show such a pattern of siliciclastic encroachment (Fig. 2.10).

Isolated platforms tend to have steep margins. While ramps can be developed during

30   CONCEPTS OF SEQUENCE STRATIGRAPHY AS APPLIED TO CARBONATES

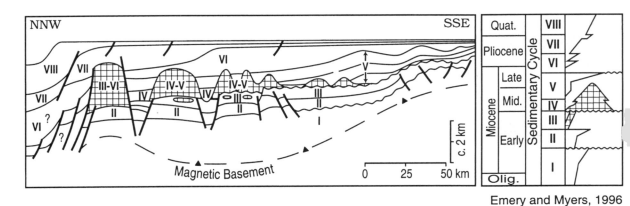

Figure 2.10. Tertiary carbonates of the Lucomia Province, offshore Sarawak. The carbonates young towards the north, as they back-step away from northward directed siliciclastic input (from Epting, 1989). Used with permission of Blackwell Science.

Figure 2.11. A 'pinnacle' type carbonate platform from the Miocene of offshore Philippines. It is impossible to break out different systems tracts from seismic data alone in this example, which shows continuous gradual backstepping and confused internal seismic character. The top of the platform is very rugose which represent karsting or the break-up of the platform into several smll pinnaceles prior to drowning. Used with permission of Blackwell Science.

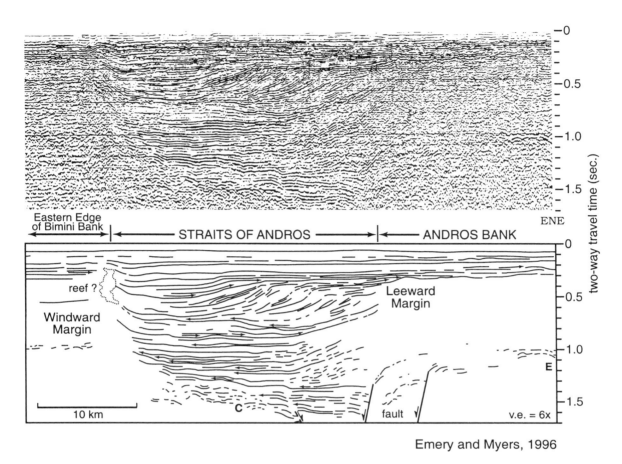

Figure 2.12. Windward (left) and leeward (right) platforms margins of the Bahamas (from Eberli and Ginsburg, 1987). Used with permission of Blackwell Science.

the early evolution of the platform, such as during initial transgression, the structural underpinnings of most isolated platforms drive them ultimately toward rimmed and escarpment margin geometries. In general, small symmetrical oceanic platforms show strong aggradation during transgression and slower aggradation during highstand, with minimal margin progradation. Backstepping during each transgressive episode, caused by environmental degradation, will ultimately lead to the pinnacle geometry so characteristic of some of the Tertiary isolated platforms of Southeast Asia (Fig. 2.11).

Larger isolated oceanic platforms that support major carbonate production on the platform during highstands are often asymmetrical, with windward reef bound margins that evolve toward escarpments and leeward carbonate sand rimmed margins that show strong margin progradation into deeper water. The present Great Bahama Platform is such an asymmetrical composite platform (Fig. 2.12). The Bahama Platform originally consisted of two isolated structurally controlled platforms, the Bimini Bank on the west, and the Andros Bank to the east separated by a deep oceanic channel termed the Straits of Andros

(Eberli and Ginsburg, 1987). Each platform developed reefs, and ultimately escarpment margins, on the windward side and sand shoal rimmed margins on the leeward side. The Straits of Andros was subsequently filled by westward prograding highstand sediments derived from the leeward margin of the Andros Bank leading to the coalescence of the two isolated platforms into the composite Great Bahama Platform (Fig. 2.12).

High-frequency cyclicity on carbonate platforms: carbonate parasequences

The preceding sections outlined the sequence stratigraphic models for 3rd order seismic scale cycles. Each of these 3rd order sequences is composed of a number of higher frequency cycles, termed parasequences, that are generally well developed on the tops of the platforms and form the basic building blocks of the larger, longer-term sequences (Fig. 2.6). The origin of these high-frequency cycles has been hotly debated for a number of years (Emery and Myers, 1996). Two main mechanisms are thought to be responsible for parasequences. Orbital perturbations force climate changes (Milankovitch cyclicity) and hence the advance and retreat of ice caps leading to high-frequency eustatic cycles. On the other hand, normal sedimentation processes in conjunction with subsidence may lead to high-frequency cycles termed autocycles. Progradation of tidal flat shorelines may cover the sediment source for tidal flat sedimentation, sedimentation ceases, subsidence causes transgression, and the cycle is repeated (James, 1984). Eustatic-forced cycles can be widely correlated, while autocycles will tend to be local in distribution.

High frequency eustatic cycles in carbonates respond predictably to changes in overall accommodation space generated during a 3rd order cycle by progressive changes in water depth and cycle thickness. During transgression, the rate of formation of new accommodation spaces increases. Sediments deposited at the terminus of the cycle are deposited in progressively deeper water, and in carbonates the cycles generally become

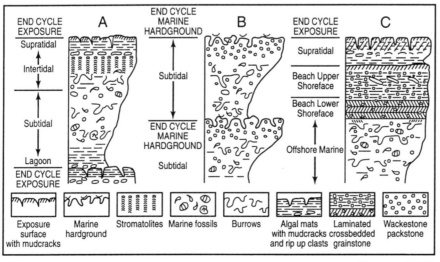

Figure 2.13. Conceptual models of common shoaling upward sequences. Models A and C have a terminal subaerial exposure phase, while model B type cycles are punctuated by marine hardgrounds developed on grainstones or packstones. Used by permission of Elsevier.

Moore, 1989

thicker (Fig. 2.7A). The resulting parasequence stacking pattern is termed retrogradational (following the usage of Van Wagoner, 1991). As the 3rd order cycle reaches highstand, the rate of formation of new accommodation space decreases, the facies of the terminus of the high-frequency cycle becomes progressively shallower water, and cycle thickness decreases (Fig. 2.7B). The resulting parasequence stacking pattern is termed progradational. Parasequence stacking patterns are commonly used to help unravel higher order cyclicity on carbonate platform tops, in outcrops, in cores, and on well logs (Montañez and Osleger, 1993; Saller et al., 1994).

Carbonate parasequences range in thickness from a meter to tens of meters. In form they are similar to siliciclastic parasequences starting with an abrupt marine flooding surface overlain by subtidal facies, then shoaling upward rapidly into shallower water facies (James, 1984). Most carbonate parasequences are asymmetrical because of the lag in establishment of the carbonate factory after initial flooding (law of sigmoidal growth, Chapter 1) resulting in the deepest water facies being deposited just above the marine flooding surface. In the catch-up phase most cycles will rapidly shallow up to near sea level, while in the keep-up phase the cycle may attain an intertidal to supratidal position depending on whether the parasequence is being developed in the transgressive (TST) or highstand (HST) phase. Figure 2.13 illustrates three common parasequence styles based on characteristics of the cycle terminus: low-energy tidal flat with exposure, stacked marine shoaling-upward parasequences without cycle terminus exposure but with marine cementation, and high-energy shoreface with terminal exposure. The geometry and the nature of the cycle terminal facies play a major role in the gross heterogeneity found in carbonate reservoirs as they do in siliciclastics. The diagenesis associated with cycle tops in carbonates, however, fixes the heterogeneity of reservoir porosity/permeability very early in its geologic history (Hovorka, 1993). This topic will be fully developed in Chapter 8 after we discuss the nature of diagenetic processes and their effect on porosity and permeability.

THE CONSEQUENCES OF THE HIGH-CHEMICAL REACTIVITY OF CARBONATE SEDIMENTS AND ROCKS DURING EXPOSURE AT SEQUENCE BOUNDARIES

Carbonate minerals and their relative stability

Calcite, crystallizing in the rhombohedral crystal system is the thermodynamically stable phase of calcium carbonate. Aragonite, calcite's orthorhombic polymorph, is about 1.5 times more soluble than calcite. Magnesian calcite is an important sub-group of carbonates that occurs as skeletal components and marine cements. While magnesian calcites have the gross crystallographic characteristics of calcite, magnesium substitution affects its lattice geometry and solubility. A magnesium calcite containing about 11 mole % $MgCO_3$ has approximately the same solubility as aragonite (Morse and Mackenzie, 1990).

Dolomite, while one of the most common carbonate minerals, is not well understood because it cannot be easily synthesized in the laboratory. Ideally, dolomite consists of equal molar ratios of $CaCO_3$ and $MgCO_3$. When dolomite exists in its ideal composition, it is stable over a relatively wide range of temperatures and pressures and is less soluble than calcite. When dolomite's composition is enriched in CaCO$_3$, however, its lattice geometry is affected and its solubility decreases (Morse and Mackenzie, 1990). These non-ideal dolomites are termed protodolomites and are characteristic of the rapid precipitation found today on tidal flats and supratidal flats (Land, 1985). At earth surface temperatures and pressures, therefore, common carbonate mineral species exhibit a hierarchy of solubility, with magnesian calcite the most soluble. Aragonite, calcite and ideal dolomite become progressively less soluble, or more stable. Protodolomite may actually be less soluble than calcite. As we will see in Chapters 5 and 7 of this book, this contrast in mineral solubility, particularly during exposure at sequence boundaries, is one of the major drivers in carbonate diagenesis and porosity/permeability evolution.

Controls over the mineralogy of carbonate sediments, today and in the past

Modern shallow marine carbonate sediments consist primarily of the minerals aragonite and magnesian calcite. Since most shallow marine carbonate sediments originate from organisms that live in the environment it is fair to state that carbonate mineralogy of these sediments is inherited directly from the mineralogy of the precursor organisms. What then, can we expect original carbonate mineralogy to have been for ancient sediments, since organisms have progressively evolved during the Phanerozoic? Figure 2.14 illustrates the mineralogical evolution of the major groups of carbonate producing organisms. These data suggest that Paleozoic carbonate sediments tended to be more calcite-rich while Mesozoic and Tertiary carbonates become progressively more aragonite-rich (Wilkinson, 1979).

What about the mineralogy of marine chemical precipitates such as muds (whitings), ooids, and marine cements? Modern tropical ocean surface waters are supersaturated with respect to all common carbonate species including calcite, dolomite, magnesian calcite and aragonite (Moore, 1989; Morse and Mackenzie, 1990). Modern shallow, tropical marine chemical precipitates such as ooids, marine cements and muds, however, seem to

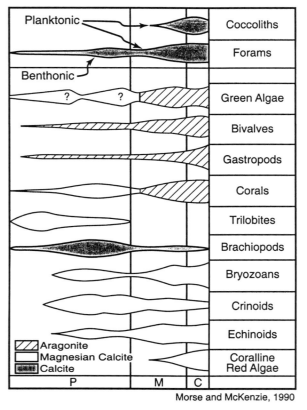

Figure 2.14 Mineralogical evolution of benthic and planktonic organisms during the Phanerozoic. Used with permission of Elsevier.

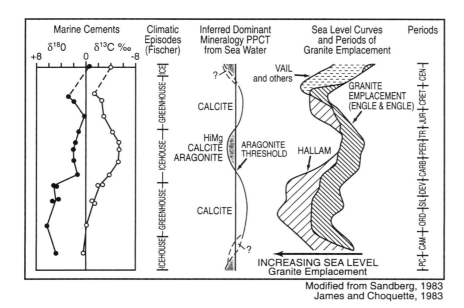

Figure 2.15. Temporal variation of stable isotopes and abiotic carbonate mineralogy as a function of sea level. Used with permission of Elsevier.

be dominated by aragonite, with subsidiary magnesium calcite cements in reef environments (Moore, 1989). Over the last 15-20 years a body of research has developed which strongly suggests that there has been a significant temporal variation of chemically precipitated carbonates (Mackenzie and Pigott, 1981; James and Choquette, 1983; Sandberg, 1983; Wilkinson and Given, 1986; Bates, 1991). Figure 2.15 outlines the temporal variation of stable isotopes and mineralogy of marine cements as a function of sea level.

Considerable discussion has developed concerning the specific drivers responsible for this mineralogical variation, but it seems certain that changes in atmospheric carbon dioxide and continental freeboard are involved (Wilkinson et al., 1985). As we will see in Chapter 8, these temporal changes in abiotic and biotic carbonate mineralogy have had significant impact on diagenetic overprint and early porosity/permeability evolution.

Mineralogy of ancient limestones: the concept of early progressive mineral stabilization and porosity evolution

While modern and most ancient carbonate sediments consist of relatively metastable aragonite and magnesian calcite, most ancient limestones are composed of calcite and dolomite, the most stable carbonate mineral species. It seems, therefore, that carbonate sediments undergo a progressive mineral stabilization during their transformation from sediment to rock (Moore, 1989). The very nature of carbonate cyclicity, as sketched in previous sections, assures that originally unstable sediments will come under the influence of meteoric, or modified sea water not once, but perhaps a number of times during the development of a single 3rd order sedimentary cycle. When meteoric water, undersaturated with respect to most carbonate mineral species, comes into contact with aragonite and magnesian calcite sediments, the aragonite and magnesian calcite are dissolved.

The dissolution of these mineral species increases the saturation of the water with respect to calcite (or dolomite) to the point where calcite (or dolomite) ultimately precipitates (Matthews, 1984; Moore, 1989; Morse and Mackenzie, 1990). If enough time is available, the entire original metastable mineral suite will become stabilized to calcite and dolomite prior to significant burial into the subsurface. It is during this mineral stabilization phase, early in the history of the sequence, that the nature and distribution of the pore systems of carbonate rocks are strongly modified by dissolution and cementation. Mineral stabilization is accompanied by the generation of new secondary porosity and destruction of both primary as well as secondary porosity (Moore, 1989). It is obvious, therefore, that the relative sea level history of a carbonate rock sequence plays a major role in the diagenesis and ultimate porosity evolution of the sequence. This is in marked contrast to siliciclastic sequences whose relatively stable mineral suites preclude significant early diagensis, so that porosity distribution is controlled primarily by favorable facies distribution rather than diagenetic overprint.

The nature of the diagenetic processes activated during relative sea level perturbations and their effects on carbonate pore systems will be one of the major thrusts of the remainder of this book.

SUMMARY

The general concepts of sequence stratigraphy may be applied directly to carbonate sedimentary sequences. In detail, however, there are some major differences between the application of sequence stratigraphy to carbonates versus its application to siliciclastics. Perhaps the major difference relates to highstand sediment shedding in carbonates versus lowstand shedding in siliciclastics. Efficiency of the carbonate sediment factory in tropical marine waters results in carbonate shelves and platforms that are difficult to drown. Marine cementation and framework reefing may lead to steep shelf margins that cannot support lowstand prograding complexes but often have megabreccia debris flows associated with shelf margin collapse. Reciprocal sedimentation, with siliciclastics deposited during lowstands and carbonates deposited during highstands, is a common feature of carbonate platforms tied to continental landmasses.

High-chemical reactivity of carbonate sediments is associated with intensive diagenesis and porosity modification at high-frequency cycle tops and sequence boundaries. Lithification associated with sequence boundaries suggests that alluvial valley incisement during lowstands will not be a common feature of carbonate sequences.

Chapter 3

THE NATURE AND CLASSIFICATION OF CARBONATE POROSITY

INTRODUCTION

Pore systems in carbonates are much more complex than siliciclastics (Choquette and Pray, 1970; Lucia, 1995b) (Table 3.1). This complexity is a result of the overwhelming biological origin of carbonate sediments and their chemical reactivity. The biological overprint results in porosity within grains, fossil-related shelter porosity and growth framework porosity within reefs. Chemical reactivity results in the common development of secondary porosity due to pervasive diagenetic processes, such as solution and dolomitization. While these processes affect the more chemically reactive carbonates throughout their burial history, the most dramatic porosity modification occurs early, driven by exposure to meteoric waters at sequence boundaries. This tie between diagenetic processes and porosity is the focus of the next several chapters.

THE CLASSIFICATION OF CARBONATE POROSITY

Introduction

The carbonate geological exploration model is primarily observational in nature and is based on outcrop, core and well log data used in the interpretation of depositional environments, depositional sequences and diagenetic overprint. The classification of carbonate porosity most often used in geological models is the genetic porosity classification of Choquette and Pray (1970) which emphasizes fabric selectivity. Because of its emphasis on the relationship of primary rock fabric to porosity and timing of porosity development, this classification is particularly well suited to geological models that integrate the depositional system with early to late diagenetic processes in order to determine porosity evolution through time. This type of model is particularly important to the explorationist who must predict the distribution of favorable reservoir rock at the time of hydrocarbon migration.

Modern carbonate reservoir characterization and modeling at the production scale, however, demand the full integration of quantitative petrophysical parameters such as porosity, permeability and saturation with a geological model where these parameters can be displayed in three dimensional space (Lu-

Table 3.1

Aspect	Sandstone	Carbonate
Amount of primary porosity in sediments	Commonly 25-40%	Commonly 40-70%
Amount of ultimate porosity in rocks	Commonly half or more of initial porosity; 15-30% common	Commonly none or only small fraction of initial porosity; 5-15% common in reservoir facies
Type(s) of primary porosity	Almost exclusively interparticle	Interparticle commonly predominates, but intraparticle and other types are important
Type(s) of ultimate porosity	Almost exclusively primary interparticle	Widely varied because of postdepositional modifications
Sizes of pores	Diameter and throat sizes closely related to sedimentary particel size and sorting	Diameter and throat sizes commonly show little relation to sedimentary particle size or sorting
Shape of pores	Strong dependence on particle shape -- a "negative" of particles	Greatly varied, ranges from strongly dependent "positive" or "negative" of particles to form completely independent of shapes of depositional of diagenetic components
Uniformity of size, shape, and distribution	Commonly fairly uniform within homogeneous body	Variable, ranging from fairly uniform to extremely heterogeneous, even within body make up of single rock type
Influence of diagenesis	Minor; usually minor reduction of primary porosity by compaction and cementation	Major; can create, obliterate, or completely modify porosity; cementation and solution important
Influence of fracturing	Generally not of major importance in reservoir properties	Of major importance in reservoir properties if present
Visual evaluation of porosity and permeability	Semiquantitative visual estimates commonly relatively easy	Variable; semiquantitative visual estimates range from easy to virtually impossible; instrument measurements of porosity, permeability and capillary pressure commonly needed
Adequacy of core analysis for reservoir evaluation	Core plugs if 1-in. diameter commonly adequate for "matrix" porosity	Core plugs commonly inadequate; even whole cores (~3-in. diameter) may be inadequate for large pores
Permeability porosity interrelations	Relatively consistent; commonly dependent on particle size and sorting	Greatly varied; commonly independent of particle size and sorting

Choquette and Pray, 1970

Table 3.1. Comparison f carbonate and siliciclastic porosity. Used with permission of AAPG.

cia, 1995b). Archie (1952) made the first attempt at integrating engineering and geological information for early carbonate reservoir models by developing a porosity classification where rock fabrics were related to petrophysical properties such as porosity, permeability and capillarity. Archie (1952) separated visible porosity from matrix porosity (Fig. 3.1) and utilized surface texture of the broken rock to estimate matrix porosity, permeability and capillarity. He described visible porosity according to pore size, with classes ranging from pinpoint to cutting size. While useful in estimating petrophysical characteristics, his classification was

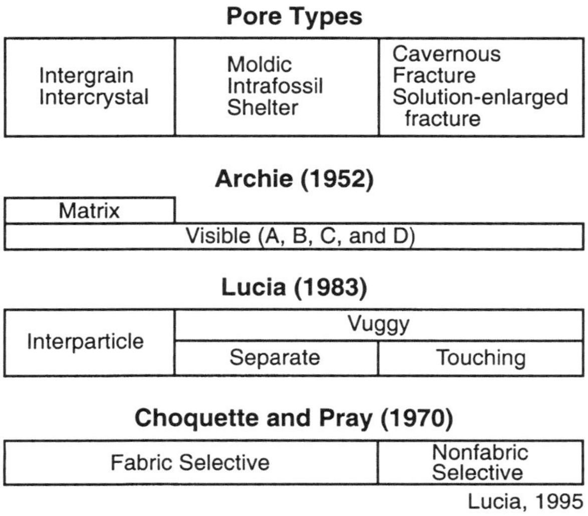

Pore Types

Intergrain Intercrystal	Moldic Intrafossil Shelter	Cavernous Fracture Solution-enlarged fracture

Archie (1952)

Matrix	
Visible (A, B, C, and D)	

Lucia (1983)

Interparticle	Vuggy	
	Separate	Touching

Choquette and Pray (1970)

Fabric Selective	Nonfabric Selective

Lucia, 1995

Fig. 3.1. Comparison of the classifications of carbonate pore space as developed by Archie (1952), Lucia (1983) and Choquette and Pray (1970). A= no visible pore space, and B, C, and D = increasing pore sizes. Used with permission of AAPG.

difficult to use in geological models because his classes cannot be defined in depositional or diagenetic terms.

Lucia (1983; 1995b; 1999) established a porosity classification, as an outgrowth of Archie's efforts, which incorporates both rock fabric, which can be related to depositional environments, and the petrophysical characteristics necessary for a viable engineering model (Fig. 3.1). While Lucia's porosity classification is useful for production reservoir characterization and modeling, it is difficult to use in geological exploration models where diagenetic history and porosity evolution is critical.

The following sections will first discuss the genetic, geologically based Choquette and Pray classification and then contrast it with Lucia's hybrid rock fabric/petrophysical porosity classification, emphasizing the strengths and weaknesses of each. The chapter will finish with discussions of the general nature of primary depositional environment-related porosity and diagenetically related secondary porosity.

Choquette and Pray porosity classification

The major features of the Choquette and Pray (1970) carbonate classification are illustrated in Figure 3.2. Fabric selectivity is the prime component of the classification. The solid depositional and diagenetic constituents of a sediment or rock are defined as its fabric. These solid constituents consist of various types of primary grains, such as ooids and bioclasts; later-formed diagenetic constituents, such as calcite, dolomite and sulfate cements; and recrystallization or replacement components, such as dolomite and sulfate crystals. If a dependent relationship can be determined between porosity and fabric elements, that porosity is referred to as fabric-selective. If no relationship between fabric and porosity can be established, the porosity is classed as not fabric-selective (Fig. 3.2).

It is important to assess fabric selectivity in order to better describe, interpret and classify carbonate porosity. Two factors determine fabric selectivity, the configuration of the pore boundary and the position of the pore relative to fabric (Choquette and Pray, 1970). In most primary porosity, pore boundary shape and location of the pore are determined completely by fabric elements. Primary intergranular pore space in unconsolidated sediments, therefore, is obviously fabric-selective because its configuration is determined solely by depositional particles. The same is true for primary intragranular porosity, which is controlled by the shape and location of cavities

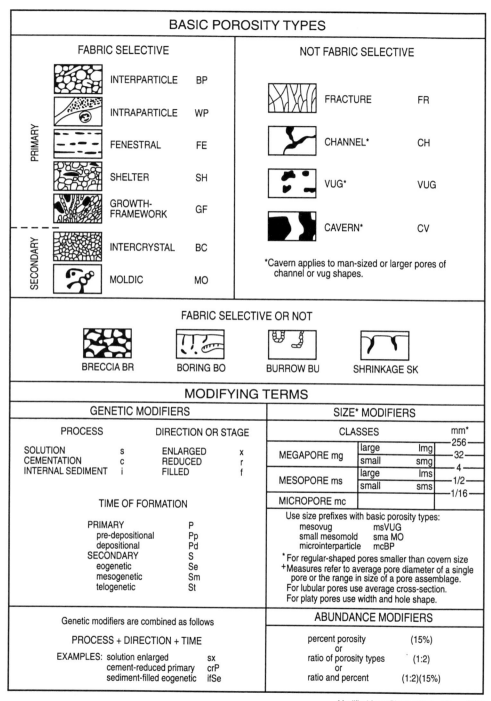

Fig. 3.2. *Classification of carbonate porosity. Classification consists of basic porosity types, such as moldic. Each type is represented by an abbreviation (MO). Modifying terms include genetic modifiers, size modifiers and abundance modifiers. Each modifier also has an abbreviation. Used with permission of AAPG.*

determined by the nature of growth of the organism giving rise to the particle.

In secondary pore systems, however, porosity may be either fabric-selective or not fabric selective, depending primarily on diagenetic history. As an example, moldic porosity is commonly fabric-selective because of the preferential removal of certain fabric elements from the rock, such as early removal of aragonitic ooids and bioclasts during mineral stabilization (Fig. 3.2). Another example is the later removal of anhydrite, gypsum, or even calcite from a dolomite matrix during the diagenetic history of a sequence. On the other hand, phreatic cavern development commonly cuts across most fabric elements, is not fabric-selective and is controlled primarily by joint systems.

Primary porosity is any porosity present in a sediment or rock at the termination of depositional processes. Primary porosity is formed in two basic stages, the predepositional stage and the depositional stage. The predepositional stage begins when individual sedimentary particles form and includes intragranular porosity such as is seen in forams, pellets, ooids, and other non-skeletal grains. This type of porosity can be very important in certain sediments.

The depositional stage is the time involved in final deposition, at the site of final burial of sediment or at the site of a growing organic framework. Porosity formed during this stage is termed depositional porosity and is important relative to the total volume of carbonate porosity observed in carbonate rocks and sediments (Choquette and Pray, 1970). The nature and quality of primary porosity will be covered more fully later in this chapter.

Secondary porosity is developed at any time after final deposition. The time involved in the generation of secondary porosity relative to primary porosity may be enormous (Choquette and Pray, 1970). This time interval may be divided into stages based on differences in the porosity-modifying processes occurring

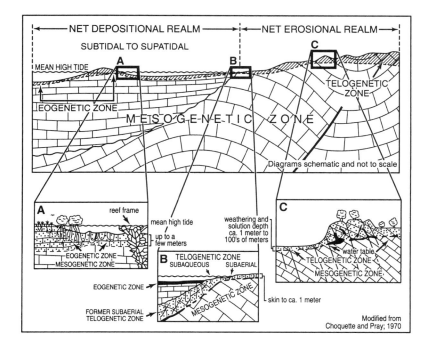

Fig. 3.3. Major environments of porosity evolution as developed by Choquette and Pray, 1970. Used with permission of AAPG.

in shallow surficial diagenetic environments versus those encountered during deep burial. Choquette and Pray (1970) recognized three stages: eogenetic, mesogenetic, and telogenetic (Fig. 3.3).

The eogenetic stage is the time interval between initial sediment deposition and sediment burial below the influence of surficial diagenetic processes. The upper limit of the eogenetic zone is generally a depositional interface, which can be either subaerial or subaqueous. In this book, the lower limit of the eogenetic zone is considered to be that point at which surface recharged meteoric waters, or normal (or evaporated) marine waters, cease to actively circulate by gravity or convection.

As noted in Chapter 2, the sediments and rocks of the eogenetic zone are generally mineralogically unstable or are in the process of stabilization. Porosity modification by dissolution, cementation, and dolomitization is quickly accomplished and volumetrically very important. Diagenetic environments that are active within the eogenetic zone include meteoric phreatic, meteoric vadose, shallow and deep marine, and evaporative marine. Each of these diagenetic environments and their importance to porosity development and quality will be discussed in Chapters 5-7.

The mesogenetic stage is the time interval during which the sediments are buried at depth below the major influence of surficial diagenetic processes. In general, diagenesis in the mesogenetic zone is marked by rather slow porosity modification and is dominated by compaction and compaction-related processes. While rates are slow, the time interval over which diagenetic processes are operating is enormous, and hence porosity modification (generally destruction) may well go to completion (Scholle and Halley, 1985; Heydari, 2000). The burial diagenetic environment coincides with the mesogenetic stage and will be discussed in detail in Chapter 9.

The telogenetic stage is the time interval during which carbonate sequences that have been in the mesogenetic zone are exhumed in association with unconformities to once again be influenced by surficial diagenetic processes. The term telogenetic is reserved specifically for the erosion of old rocks, rather than the erosion of newly deposited sediments during minor interruptions in depositional cycles. As such, sequences affected in the telogenetic zone are mineralogically stable limestones and dolomites that are less susceptible to surficial diagenetic processes. While most surficial diagenetic environments may be represented in the telogenetic zone, meteoric vadose, and meteoric phreatic diagenetic environments are most common. Telogenetic porosity modification will be discussed as a special case under the meteoric diagenetic environment in Chapter 7.

The Choquette-Pray (1970) porosity classification consists of four elements (Figs. 3.2 and 3.4): genetic modifiers, size modifiers, basic porosity types, and abundance modifiers.

Choquette and Pray (1970) recognize 15

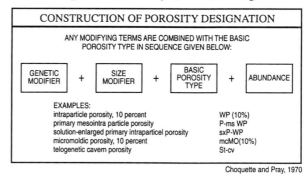

Fig. 3.4. Technique for construction a porosity designation. Used with permission of AAPG.

basic porosity types as shown in Figure 3.2 . Each type is a physically or genetically distinctive kind of pore or pore system that can be separated by characteristics such as pore size, shape, genesis, or relationship to other elements of the fabric. Most of these pore types are illustrated later in this chapter.

Genetic modifiers are used to provide additional information concerning the processes responsible for the porosity, or modification of the porosity (i.e. solution, and cementation). They also may be used to characterize the time of formation of the porosity (i.e. primary or secondary-eogenetic), and whether or not the porosity has been reduced or enlarged during its travels through the burial cycle.

Size modifiers are utilized to differentiate various size classes of pore systems, such as large pores (megapores) from small pores (micropores), while abundance modifiers are used to characterize percentages or ratios of pore types present in a carbonate rock sequence.

Porosity designations are constructed by linking these four elements in the sequence shown in Figure 3.4. The use of the shorthand designations shown on Figure 3.2, such as BP for interparticle and MO for moldic, allows rapid and precise porosity designation.

The Lucia rock fabric/petrophysical carbonate porosity classification

The Lucia classification (1983; 1995a; 1999) emphasizes the petrophysical aspects of carbonate pore space as does the earlier Archie (1952) classification. Lucia (1983), however, suggested that the most useful division of carbonate pore spaces for petrophysical purposes was the pore space between grains or crystals, which he termed interparticle porosity, and all other pore space, which he termed vuggy porosity. He further subdivided vuggy porosity into two groups: (1) separate vugs that are interconnected only through the interparticle pore network and (2) touching vugs that form an interconnected pore system (Fig. 3.1).

This is a major departure from the usage of Choquette and Pray (1970). They classified moldic and intraparticle pores as fabric selective porosity and grouped them with interparticle and intercrystal porosity (Figs. 3.1 and 3.2). While their classification is useful when trying to determine porosity genesis and evolution, Lucia's pore-type groupings more accurately reflects basic petrophysical differences. See Figure 3.1 for a comparison of the Archie, Choquette and Pray, and Lucia classifications. Table 3.2 contrasts the pore-type terminology of Lucia (1983) with that of Choquette and Pray (1970).

Table 3.2

Term	Abbreviations	
	Lucia (1983)	Choquette and Pray (1970)
Interparticle	IP	BP
Intergrain	IG	-
Intercrystal	IX	BC
Vug	VUG	VUG
Separate vug	SV	-
Moldic	MO	MO
Intraparticle	WP	WP
Intragrain	WG	-
Intracrystal	WX	-
Intrafossil	WF	-
Intragrain microporosity	μG	-
Shelter	SH	SH
Touching Vug	TV	-
Fracture	FR	FR
Solution-enlarged fracture	SF	CH*
Cavernous	CV	CV
Breccia	BR	BR
Fenestral	FE	FE

*Channel   Lucia, 1995

Table 3.2. Comparison of porosity terminology. Used with permission of AAPG.

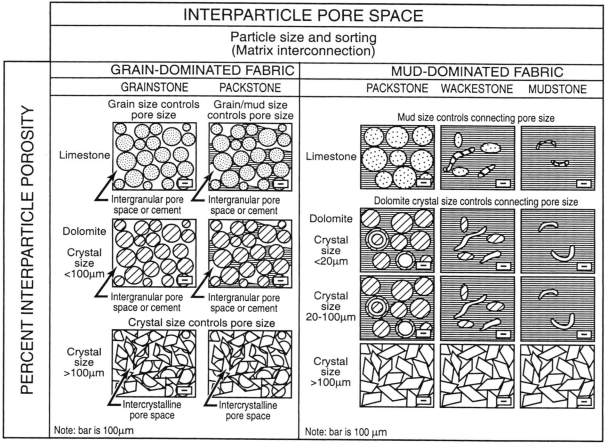

Fig. 3.5. Geological and petrophysical classification of carbonate interparticle pore space based on size and sorting of grains and crystals. Used with permission of AAPG.

The basic premise of the Lucia (1983) and Archie (1952) classifications is that pore-size distribution controls permeability and saturation and that pore-size distribution is related to rock fabric. One must remember, however, that the three pore-type groups of Lucia (interparticle, separate vug, or touching vug) have distinctly different petrophysical characteristics and that their separate volumes must be determined for each reservoir under study.

Figure 3.5 outlines Lucia's geological and petrophysical classification of carbonate interparticle pore space based on size and sorting of grains and crystals. Note that in general it follows the Dunham carbonate rock classification (Chapter 1) except that the packstone fabric group has been divided into grain-dominated and mud-dominated classes. This division more accurately portrays the petrophysical characteristics of mud-bearing grain supported rocks. Note also that dolomite is included in the classification, and that dolomite crystal size controls the pore size of the reservoir rock.

Figure 3.6 illustrates the relationship

LUCIA POROSITY CLASSIFICATION

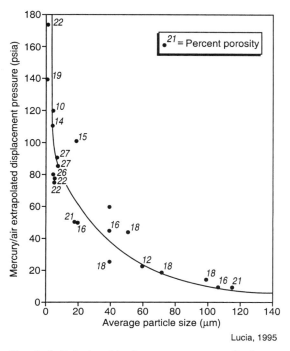

Fig. 3.6. Relationship between mercury displacement pressure and average particle size for nonvuggy carbonate rocks with greater than 0.1 md permeability. The displacement pressure is determined by extrapolating the data to a mercury saturation of zero. Used with permission of AAPG.

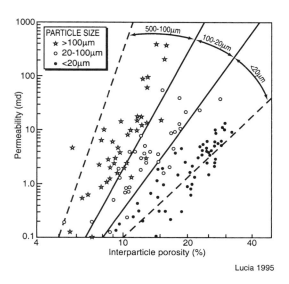

Fig. 3.7. Porosity-air permeability relationship for various particle-size groups in nonvuggy carbonate rocks. Used with permission of AAPG.

between mercury displacement pressure and average particle size for non-vuggy carbonate rocks. Note that this relationship is independent of porosity. The resulting hyperbolic curve suggests that there are important particle-size boundaries at 100 and 20 microns (Lucia,

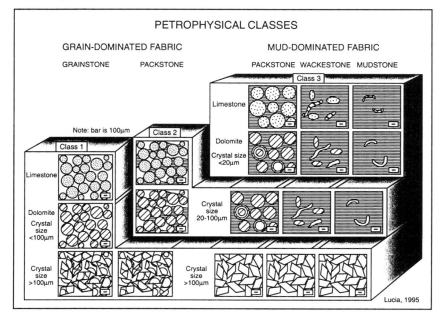

Fig. 3.8. Petrophysical and rock fabric classes of Lucia (1995) based on similar capillary properties and interparticle-porosity/permeability transforms. Used with permission of AAPG.

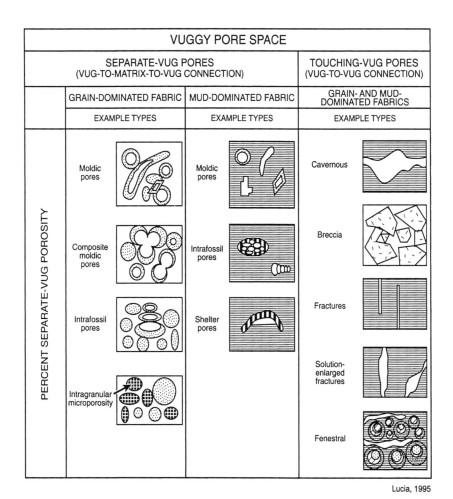

Fig. 3.9. Geological and petrophysical classification of vuggy pore space based on vug interconnection. Used with permission of AAPG.

1995b). Lucia (1983) determined that three porosity/permeability fields can be defined using these particle size boundaries (Fig. 3.7), a relationship that appears to be limited to particle sizes less than 500 microns. These three permeability fields form the basis for Lucia's (1983) petrophysical and rock-fabric classes (Fig. 3.8). These classes are termed class 1, with particle sizes from 500-100 microns, class 2, with particles from 100-20 microns, and class 3, with particles <20 microns.

Dolomitization can significantly change rock fabric. In fine-crystalline dolomite, original rock fabrics are easily determined. The coarser the dolomite crystals, however, the more difficult it becomes to determine original rock fabric. Petrophysically, however, dolomite crystal size, grain size, and sorting define the permeability field, and ultimately the petrophysical class (Fig. 3.8). In other words, in many coarse dolomites, the most important petrophysical parameters are dolomite crystal size and sorting, regardless of original rock fabric.

Figure 3.9 illustrates Lucia's (1983) geological/petrophysical classification of vuggy pore space based on vug interconnection. Vuggy pore space is divided into separate-vug pores and touching-vug pores.

Separate-vug pores are pores that are either within particles, or are significantly larger than the average particle size of the rock, and are interconnected only through the interparticle porosity. Separate-vugs (Fig. 3.9) are typically fabric selective in origin. Intrafossil porosity such as foraminiferal chambers, moldic pores developed by dissolution of ooids, fossils or pellets, microporosity within grains, molds caused by dissolution of dolomite or anhydrite crystals, and fossil shelter porosity are all examples of separate-vug pores.

The presence of separate-vug porosity in rocks containing interparticle porosity increases total porosity but does not significantly increase permeability (Lucia, 1983; Moore, 1989).

Lucia (1995b) classifies microporosity within fossils or grains as separate-vug porosity because it is connected by interparticle porosity. Separate-vug porosity is generally oil filled within the oil column because of relatively large pore size. Because of small pore size and high capillary forces, separate-vug microporosity may trap water and lead to high water saturation within a productive interval (Moore, 1989; Lucia, 1995b).

Touching-vug pores are defined as pores that are significantly larger than the average rock particle size and form a significant interconnected pore system. Touching-vugs are generally not-fabric selective, and include cavernous, breccia, fracture, and fenestral pore types (Fig. 3.9).

Touching-vug pores are generally thought

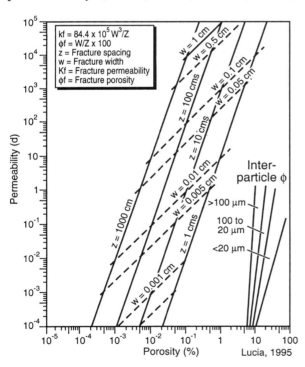

Fig. 3.10. Theoretical fracture air permeability/porosity relationship compared to the rock-fabric/petrophysical porosity, permeability fields. Used with permission of AAPG.

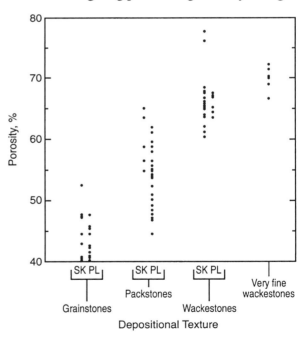

Fig. 3.11. Primary depositional porosity exhibited by various Holocene carbonate sediment textural types. Used with the permission of SEPM.

to be filled with oil in reservoirs and may increase permeability well above what may be expected from the interparticle pore system. Figure 3.10, constructed by Lucia (1995b) is a plot of fracture permeability versus fracture porosity as compared to the three interparticle permeability fields. These relationships show that permeability in touching-vug pore systems is related primarily to fracture width and is sensitive to very small changes in fracture porosity.

The following section describes the nature of porosity found in modern carbonate sediments.

THE NATURE OF PRIMARY POROSITY IN MODERN CARBONATE SEDIMENTS

Intergrain porosity

At the time of deposition, mud-free carbonate sediments, like their siliciclastic counterparts, are dominated by intergrain porosity (termed interparticle by Choquette and Pray (1970) and Lucia (1983) in their porosity classifications). These sediments exhibit porosities ranging from 40-50% (Fig. 3.11) (Enos and Sawatsky, 1981), which is near the upper limit of 48% expected in spherical particles with minimum packing (Graton, 1935). Excess porosity over the 27-30% expected in spherical particles showing close maximum packing and commonly observed in siliciclastic sediments (Graton, 1935) is the result of the wide variability of particle shape seen in carbonates. This shape variation seems to be a function of their biological origin (Dunham, 1962) and the common presence of intragrain porosity that may occupy a significant percentage of the bulk volume of the sediment (Enos and Sawatsky, 1981).

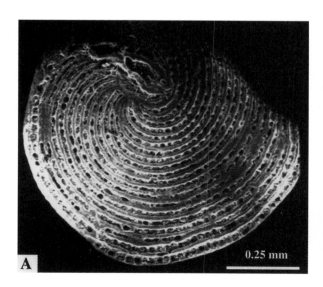
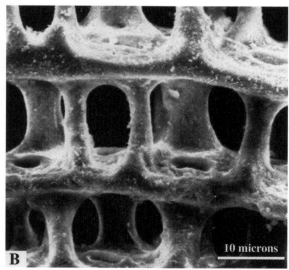

Fig. 3.12. Scanning electron photomicrograph of a large modern reef-related foraminifera showing large volume of intraparticle (WP) porosity. In an oil column, this microporosity would probably be filled with water (Lucia, 1999).

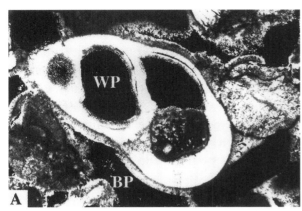 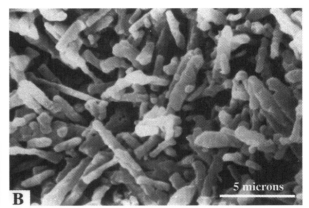

Fig. 3.13. Holocene sediments showing primary porosity. A. Living chamber of a Holocene gastropod, Florida showing intraparticle porosity (WP). The last chamber is filled with muddy internal sediment. Interparticle porosity (BP) is partially filled with marine aragonite cement. B. Ultrastructure of a Holocene Halimeda sp. showing intraparticle porosity (WP) between aragonite elements of the segment.

Intragrain porosity

Intragrain depositional porosity (termed intraparticle porosity by Choquette and Pray and separate-vug porosity by Lucia) is one of the fundamental differences between carbonate and siliciclastic porosity. Intragrain porosity may originate in a variety of ways. The living chambers of various organisms such as foraminifera, gastropods, rudists, and brachiopods often provide significant intragrain porosity (Fig. 3.12). The ultrastructure of the tests and skeletons of organisms—such as the open fabric of a green algal segment, consisting of a felted framework of aragonite needles, or the open structure of many coral polyps—can also provide intragrain porosity (Fig. 3.13). The ultrastructure of some abiotic grains, such as ooids and peloids consists of packed, needle-shaped crystals. This structure may also lead to significant intragrain porosity (Fig. 3.13) (Robinson, 1967; Loreau and Purser, 1973; Enos and Sawatsky, 1981). Finally, the activity of microboring algae and fungi may significantly increase the intragrain porosity of carbonate grains, before, during, and shortly after deposition (Perkins and Halsey, 1971).

Depositional porosity of mud-bearing sediments

Carbonate sediments containing mud range in porosity from 44 to over 75%. Grain-supported muddy sediments such as packstones

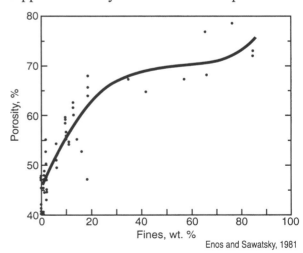

Fig. 3.14. Weight % of carbonate fines (silt plus clay) in Holocene sediments plotted against % porosity. Used with the permission of SEPM.

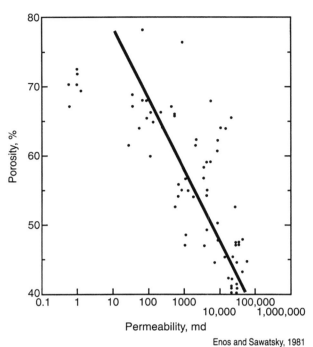

Fig. 3.15. Porosity-permeability plot of Holocene carbonate sediments. Used with the permission of SEPM.

show the lowest porosity range (44 - 68%), mud-supported sediments (wackestones) show porosities from 60-78% (Fig. 3.11) (Enos and Sawatsky, 1981), while deep marine oozes can have porosities of up to 80% (Schlanger and Douglas, 1974). The high porosities seen in the mud-supported shelf sediments are surely the effect of shape and fabric. For example, elongate needles that pack like jackstraws and result in high porosity dominate the mud fraction in modern sediments. In addition, oriented sheaths of water molecules responding to strongly electrically polarized aragonite crystals result in significantly higher porosities in aragonitic carbonate mud (Enos and Sawatsky, 1981). The exceptionally high porosities reported for deep marine oozes, however, are undoubtedly the result of the high intragrain porosity found in the dominant organic components of these sediments, such as the inflated chambers of pelagic foraminifera. Schlanger and Douglas (1974) reported 45% intragranular porosity and 35% intergranular porosity for pelagic oozes encountered during Deep Sea Drilling Program operations in the Pacific.

The high porosities reported for packstones by Enos and Sawatsky (1981), however, remain an enigma. Logically, one would assume that as the pores of grain-supported sediment are filled with mud, the porosity of the resulting mixture (packstone) should decrease. In reality, if the porosity of modern mud-bearing sediment is plotted against weight percent fines, one sees a steady increase in porosity with increasing weight percent fines, rather than the predicted porosity decrease in the grain-supported field (Fig. 3.14). Enos and Sawatsky (1981) believed that the sediments used in their study might have contained small, isolated domains of mud-supported sediment that could have masked the expected drop in porosity in the packstone field.

Permeability characteristics of modern grain-supported versus mud-supported sediments show an inverse relationship to porosity, with the lower porosity grain supported sediments exhibiting the highest permeability, and the higher porosity mud-supported sediments, the lowest permeability, as might be expected (Fig. 3.15) (Enos and Sawatsky, 1981).

Framework and fenestral porosity

Framework porosity, associated with the activity of reef-building organisms, can be an important depositional porosity type in the reef environment. Framebuilders, such as

 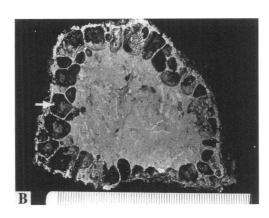

Fig. 3.16. A. Framework porosity developed within a modern coral reef, Grand Cayman, West Indies. Vertical dimension of the pore is 2 m. B. Boring sponge (Clionid) galleries (arrow) within modern coral framework, Jamaica, West Indies. Galleries are being filled with pelleted internal sediment and marine cement.

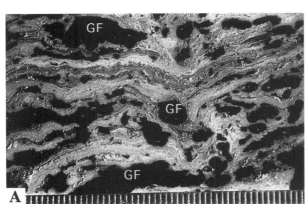 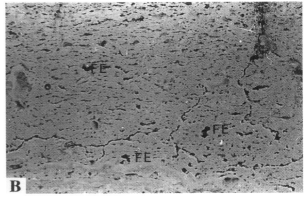

Fig. 3.17. A. Framework porosity (GF) developed in Holocene algal cup reef, Bermuda. Coralline algae show as white layers, porosity is black. Scale in cm, photo by R. N. Ginsburg. Reprinted by permission of AAPG. B. Fenestral porosity (FE) developed in semi-lithified muddy supratidal sequence, Sugarloaf Key, Florida. Photo by Gene Shinn. Used with permission of AAPG.

scleractinian corals, can construct an open reef framework, potentially enclosing enormous volumes of pore space during the development of the reef (Fig. 3.16). It is this depositional porosity potential that has long attracted the economic geologist to the study of reefs and reefing. Framework porosity potential, however, is dependent on the type of frame-building organism. While the scleractinian coral constructs an open framework reef, the coralline algae, stromatoporoids and sponges tend to erect a more closed framework structure with significantly less framework porosity because of the general encrusting mode of growth of these organisms (James, 1983; Tucker and Wright, 1990) (Fig. 3.17).

The borings of a number of organisms such as clionid sponges and pelecypods can generate a significant volume of porosity in the reef framework during the development of the reef complex (Fig. 3.16) (Moore et al., 1976; Land and Moore, 1980; Tucker and Wright, 1990). Reef related framework porosity tends to become filled quickly during early reef development in the depositional environment by coarse, as well as fine-grained, internal sediments leading to a complex depositional pore system(Land and Moore, 1980; Moore, 1989; Tucker and Wright, 1990).

Fenestral porosity, (Fig. 3.17B) commonly associated with supratidal, algal-related, mud-dominated sediments can be locally important (Shinn, 1968; Tucker and Wright, 1990). Enos and Sawatsky (1981) report depositional porosities from supratidal, algal-related wackestones of up to 65 % (Fig. 3.17). One would expect low permeabilities from such a mud-dominated system after lithification. Subsequent dolomitization, however, would tend to increase the permeability significantly, opening communication between the larger fenestral pores through the intercrystalline porosity developed in the matrix dolomite. Lucia (1995b), however, classifies fenestral porosity as a touching-vug pore system (see earlier discussion in this chapter) suggesting that the larger fenestral pores are actually interconnected without resorting to a matrix contribution.

SECONDARY POROSITY

Introduction

Primary depositional porosity is progressively lost during burial through a number of interrelated processes (Chapter 9). During this same evolutionary journey, porosity generating processes, which are dominantly diagenetic in nature, may also operate to increase the total pore volume or permeability of the evolving limestone. These processes are mainly of two types: 1) porosity generation by dissolution, and 2) porosity/permeability modification as a result of dolomitization. Fracturing, another important process, generally acts to increase permeability, rather than total pore volume (Lucia, 1995b).

Secondary porosity formation by dissolution

Dissolution of limestones and sediments may occur at any point in the burial history of the sequence (Chapter 9). The dissolution event, with its attendant increase in pore volume, generally occurs in response to a significant change in the chemistry of the pore fluid, such as a change in salinity, temperature, or partial pressure of CO_2 These changes are most likely to occur early in the history of burial (eogenetic stage), particularly at sequence boundaries in conjunction with the development of a meteoric water system (Saller et al., 1994). Later, after significant burial (mesogenetic stage), hydrocarbon maturation or shale dewatering may provide aggressive fluids for dissolution (Moore, 1989). Chapter 9 discusses the meteoric recharge of subsurface aquifers during post-orogenic uplift can result in significant late burial dissolution and cementation. Finally, dissolution may occur anytime during burial history whenever limestones have been exhumed in association with an unconformity (telogenetic stage) and placed into contact with meteoric waters (Loucks and Handford, 1992; Saller et al., 1994).

If original marine pore fluids are replaced by meteoric waters early in the burial history of a carbonate sequence, before mineral

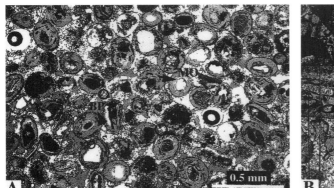
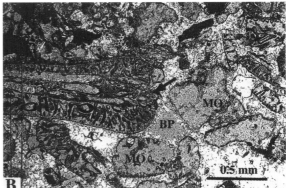

Fig. 3.18. A. Photomicrograph of moldic porosity (MO) developed in Pleistocene ooid grainstones, Shark Bay, western Australia. Plain light. Interparticle porosity (BP) shows as gray areas between ooids. Moldic porosity (MO) shows as gray areas within ooids. White areas within ooids are quartz grains. B. Photomicrograph of fossil moldic porosity developed in Pleistocene reef-related limestones from a core at Enewetak Atoll, western Pacific. Plain light. MO designates molds of coral fragments. The arrow points to a delicately dissolved Halimeda sp. with internal structure preserved. Calcite (C) cement reduces primary interparticle (BP) porosity.

stabilization, the process of dissolution leading to porosity enhancement can be distinctly fabric-selective and controlled by the mineralogy of individual grains (Moore, 1989). As was mentioned previously, modern shallow-marine carbonate sediments are composed of a metastable mineral suite consisting of aragonite and magnesian calcite. Both are unstable under the influence of meteoric waters but generally take contrasting paths to the stable phases, calcite and dolomite (Moore, 1989) (Chapter 7).

Both magnesian calcite and aragonite stability is achieved by the dissolution of the unstable phase and precipitation of the stable phase. Magnesian calcite generally dissolves incongruently, giving up its magnesium without the transport of significant $CaCO_3$, and

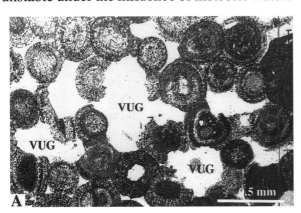

Fig. 3.19. Photomicrograph of late subsurface vuggy (VUG) porosity developed in Jurassic Smackover ooid grainstones in the Arco Bodcaw #1 well, 10, 894', southern Arkansas, USA. Plain light. These vugs may be solution enlarged interparticle porosity (sxBP). B. Core slab at 1400m from a deep core, Enewetak Atoll, western Pacific. Arrows indicate vuggy porosity (VUG) developed in medium crystalline dolomite.

generally does not exhibit an empty moldic phase (Moore, 1989; Morse and Mackenzie, 1990; Budd and Hiatt, 1993). On the other hand, aragonite dissolves, resulting in significant transport of $CaCO_3$ away from the site of dissolution and creating moldic porosity (Fig. 3.18) (Matthews, 1974; Moore, 1989). Moldic porosity is termed separate-vug porosity by Lucia (1995b) (see earlier discussion this chapter). It is this early moldic porosity formed during the rapid stabilization of aragonitic limestones in meteoric water systems, that may provide much of the early calcite cement that often occludes the intergranular depositional porosity of shallow marine carbonate sequences (Moore, 1989) (Chapters 7 and 8).

Dissolution occurring later in the burial history of a carbonate sequence, after mineral stabilization, will generally be characterized by not-fabric selective dissolution. In this case, the resulting pores cut across all fabric elements such as grains, cement, and matrix (Fig. 3.19). Choquette and Pray (1970) term these pore types, vugs, channels, and caverns, depending on size. Lucia (1995b), however, has a much wider definition of vugs. He refers to moldic pores and shelter voids as separate-vugs, as long as they are interconnected solely by interparticle porosity. In addition, Lucia (1995b) separates a second category of vugs which he terms touching-vug pores if the vugs are interconnected. This book, because of the emphasis on porosity evolution, generally follows the Choquette and Pray (1970) usage of vug pore systems.

This vuggy porosity development in exhumed limestones that are associated with unconformities (telogenetic) is the direct result of exposure to meteoric vadose and phreatic conditions where high partial pressures of CO_2 are common, and waters are undersaturated relative to most carbonate phases, including dolomite (James and Choquette, 1984). Vuggy or cavernous late secondary porosity associated with unconformities can be extensive, as well as economically important (Chapter 6) (Kerans, 1989; Moore, 1989; Loucks and Handford, 1992; Saller et al., 1994).

Late subsurface secondary porosity (mesogenetic) is a bit more difficult to explain because most subsurface fluids are believed to be supersaturated with respect to most carbonate phases due to long-term, extensive rock-water interaction (Moore, 1989; Morse and Mackenzie, 1990). However, high pressures, temperatures, hydrocarbon maturation, thermal degradation, post-orogenic meteoric recharge have been linked to the development of aggressive subsurface fluids that have the capability of developing significant subsurface secondary pore space in both carbonates and siliciclastics (Chapter 9). In some cases, the porosity development seems to have been substantial, such as is commonly seen in the Upper Jurassic Smackover Formation of southern Arkansas (Fig. 3.19) (Moore and Druckman, 1981; Moore and Heydari, 1993) (Chapter 6). This type of late porosity development is often described as a simple expansion of early secondary moldic, or primary intergranular porosity, to vuggy or cavernous porosity (Fig. 3.19) (Moore and Druckman, 1981; Choquette and James, 1987; Moore and Heydari, 1993).

In both subsurface and unconformity-related secondary porosity, the ultimate distribution of porosity is controlled either by preexisting porosity established earlier by original depositional environments or later by diagenesis, fractures, and perhaps even by distribution of stylolites.

Secondary porosity associated with dolomitization

Intercrystalline porosity (interparticle porosity of Lucia) associated with dolomites forms an important reservoir type in a number of settings ranging from supratidal/sabkha to normal marine sequences (Murray, 1960; Roehl and Choquette, 1985; Lucia, 1999). There has long been discussion concerning the role of dolomitization in porosity development and destruction (Fairbridge, 1957; Moore, 1989; Lucia, 1999). Murray (1960) documented the close relationship between percentage of dolomite and porosity in the Midale beds of the Charles Formation in Saskatchewan (Fig. 3.20). In this example, porosity initially dropped as dolomite percentage increased until 50% dolomite was reached, at which point, porosity increased with increasing dolomite percentage (Fig. 3.20). Murray (1960) explained the dolomitization control of porosity by concurrent dissolution of calcite to provide the carbonate for dolomitization. In Murray's example, the Midale beds were originally lime muds. In those samples with less than 50% dolomite, the undolomitized mud compacted during burial, and the floating dolomite rhombs occupied porosity, causing porosity to decrease with increasing dolomitization. At 50% dolomite, the rhombs act as a framework, thus preventing compaction; and as dolomite percent increases, porosity also increases.

The mode by which interrhomb calcite is lost is critical to the understanding of dolomite-related porosity. If the interrhomb calcite is lost by dissolution unrelated to dolomitization, the

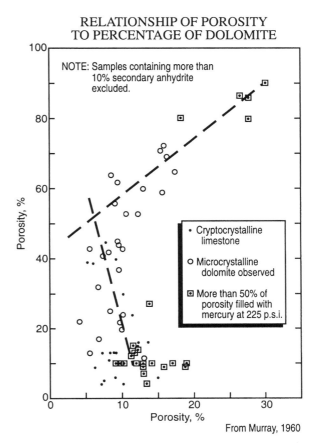

Fig. 3.20. Relationship of porosity to % dolomite in the Midale beds of the Charles Formation, Midale field, Saskatchewan, Canada. Used with the permission of SEPM.

increase in porosity is related to a specific diagenetic event that will affect only those dolomite sequences with appropriate geologic and burial histories, such as exposure along an unconformity with related meteoric phreatic dissolution. On the other hand, if the loss of calcite is associated with the dolomitization process itself (calcite providing the CO_3 for dolomitization), then most such dolomites should exhibit significant porosity (Fig. 3.21).

Weyl (1960) built a compelling case for local production of CO_3 by concurrent dissolution of calcite during dolomitization as

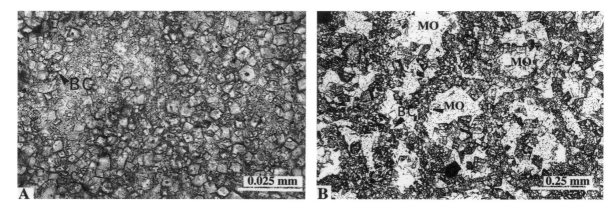

Fig. 3.21.A. Photomicrograph of fine crystalline tidal flat dolomite, Lower Cretaceous, central Texas. Intercrystalline porosity (BC) approximately 35%. Plain light. B. Photomicrograph of a Jurassic Smackover dolomite oil reservoir at Bryans Mill field, east Texas. Plain light. Moldic porosity (MO) developed by dissolution of ooids. Moldic pores are interconnected through intercrystalline porosity (BC). Porosity 25%, permeability 20 md.

the dominant source for dolomite-related porosity based on conservation of mass requirements. In addition, he noted that if dolomitization were truly a mole-for-mole replacement with local sourcing of carbonate, then the greater specific gravity of dolomite would lead to an increase of 13% porosity in the conversion from calcite to dolomite. If the dolomitization occurs early, prior to compaction, such as on a tidal flat, original porosity of the mud could exceed 60% (Enos and Sawatsky, 1981). These high initial porosities combined with the above solid volume decrease during dolomitization could easily lead to the 30-40% porosity exhibited by many sucrosic dolomite reservoirs (Murray, 1960; Choquette and Pray, 1970; Roehl and Choquette, 1985).

Lucia and Major (1994), however, in a paper on Bonaire dolomites assert: "The mole-for-mole theory of dolomitization does not apply to the Plio-Pleistocene carbonates of Bonaire. Indeed, there is no published geological evidence that the porosity through dolomitization theory applies to any sequence of carbonate rocks. Dolomite porosity values will always be equal to or less than the precursor values, suggesting that the characteristics of the precursor limestone are perhaps the single most important factor in the evolution of porosity during dolomitization."

In the same volume, Purser and others (1994) took a more balanced view. They suggested that while the characteristics of precursor limestones are indeed important to the ultimate porosity exhibited by dolomites, there are diagenetic conditions under which CO_3 seems to be locally sourced, and that indeed, under these conditions, porosity may be enhanced during dolomitization. They (Purser et al., 1994) refer to the Bahamas work of Beach (1982) in which Beach documented significant increases in porosity after dolomitization. However, both Sun, 1994, and Saller, 1998, present evidence of porosity loss during dolomitization.

The single most important diagenetic scenario where local sourcing of CO_3 may be most important, perhaps leading to increase in both porosity and permeability, is in dolomitization by post-gypsum brine reflux. In this situation, both Ca and CO_3 are severely depleted in the brines (because of precipitation of magnesian calcite, aragonite and gypsum) that are slowly refluxing through the underlying limestones. The only source of CO_3, and indeed Ca for dolomitization is the dissolution of the existing limestones (Moore, 1989; Moore and Heydari, 1993). Under these conditions, there should be little actual dolomite cementation. It is ironic, therefore, that brine reflux is the dolomitization model Lucia and Major used for their Bonaire dolomites. Based on their data, dolomite cementation has probably taken place; and Ca and CO_3 have no doubt been imported, which suggests that an alternative model of dolomitization should be invoked for the Bonaire dolomites (for example, Sibley, 1982; Vahrenkamp and Swart, 1994).

Porosity loss in association with dolomitization is perhaps most common in the subsurface. Remobilization of dolomite by dissolution, its transport by basinal fluids and subsequent porosity occluding precipitation as a saddle dolomite cement is a common late diagenetic phenomenon (Moore, 1989; Moore and Heydari, 1993)(Chapter 9).

While porosity enhancement by local sourcing of CO_3 can be debated, the potential for enhancement of porosity by dissolution of relict calcite or aragonite during fresh water flushing is great in most cyclic shallow marine-transitional dolomitization environments (Chapter 7).

An interesting variation on dolomite-related porosity is provided by Purser (1985; 1994), who describes relatively early dolomite dissolution by fresh meteoric water associated with the landward margins of a prograding coastal plain developed in middle Jurassic sequences of the Paris Basin. The dolomites were formed on tidal flats and were subsequently flushed by an unconformity-related meteoric water system that dissolved the dolomite rhombs (Fig. 3.22). Reservoirs at Coulommes field in the central Paris Basin, 35 km east of Paris, exhibit an average of 15% porosity developed in rhombic molds and interparticle porosity. A similar occurrence was described by Sellwood and others (1987) in the Jurassic Great Oolite reservoir facies in the subsurface of southern England (Fig. 3.23).

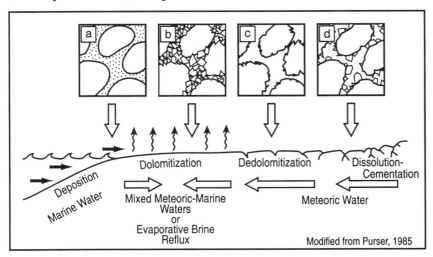

Fig. 3.22. Geologic setting for the dolomitization-dedolomitization of Jurassic ooid packstones from the Paris Basin. Final porosity consists of cement-reduced dolomite crystal-moldic porosity. Used with permission of Springer-Verlag, New York.

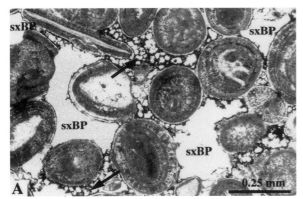 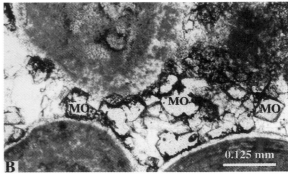

Fig. 3.23. A. Photomicrograph of solution enhanced intercrystalline porosity (sxBP) developed in Jurassic Bathonian ooid grainstones-packstones, Bath, England. Plain light. Arrows indicate concentration of dolomite molds between ooids. Photomicrograph of A showing close up of moldic porosity (MO) developed by dissolution of individual dolomite rhombs. Molds are outlined by bitumin. Plain light.

Secondary porosity associated with breccias

Brecciation of carbonate rock sequences can occur in a number of situations, including evaporite solution collapse, limestone solution collapse (Fig. 3.24), faulting, and soil formation (Blount and Moore, 1969). Limestone breccias, particularly those associated with evaporite orlimestone solution collapse, often result in enhanced porosity that may form either a reservoir for hydrocarbons or a host for mineralization. Loucks and Anderson (1985)(Chapter 7) describe porous zones developed in evaporite solution collapse sequences in the Ordovician Ellenburger at Puckett field in west Texas. These breccia zones developed during periodic subaerial exposure of the landward portions of the lower Ellenburger shoreline. The influence of meteoric waters, then, contributed significantly

Fig. 3.24. A. Breccia porosity (BR) indicated by arrows, associated with evaporite solution collapse breccia, Lower Cretaceous, central Texas, USA. B. Photomicrograph of stylolite-related slash fractures (see arrows) developed in Jurassic Smackover ooid grainstones in the Getty #1 Reddock well, at 13,692', Mississippi, U.S.A. Plain light. Calcite and anhydrite have filled fracture porosity (crFR).

to the productivity of Puckett field.

Porous breccia sequences and cavernous porosity developed during karstic limestone collapse are often associated with major unconformities and can produce reservoirs of prodigious size (Kerans, 1993). Giant karst-related reservoirs include the Mississippian Northwest Lisbon field, Utah (Miller, 1985) and the Permian Yates field (Craig, 1988; Tinker and Mruk, 1995; Lucia, 1999)(Chapter 7). Lucia (1995b) classifies both breccia and cavernous porosity as touching-vug pores with vug-to-vug interconnection.

Secondary porosity associated with fractures

Intense fracturing is present and affects the reservoir characteristics of some of the world's largest oil fields (Roehl and Choquette, 1985). While it is not always clear how much actual porosity is gained during the fracturing of carbonate reservoir rocks, because of the difficulty in measuring this type of porosity, there can be little doubt concerning the benefits that fractures can bring to ultimate reservoir production. Consider, for example, the case of Gaschsaran oil field in Iran, where a single well can produce up to 80,000 barrels of oil per day from fractured Oligocene Asmari Limestone with a matrix porosity of just 9% (McQuillan, 1985). Another example is the Monterrey Shale, at West Cat Canyon field, California, where matrix porosity is not measurable, but effective fracture porosity averages 12% and the field has 563 million barrels of oil in place (Roehl and Weinbrandt, 1985).

Fracturing is particularly effective and common in carbonate reservoirs because of the brittle nature of carbonates relative to the more ductile fine-grained siliciclastics with which they are often interbedded (Longman, 1985). Fracturing can take place at practically any time during the burial history of a carbonate sequence starting with shallow burial because of common early lithification. Fracturing can be associated with faulting, folding, differential compaction, solution collapse, salt dome movement, and hydraulic fracturing within overpressured zones (Roehl and Weinbrandt, 1985; Longman, 1985; McQuillan, 1985; Lucia, 1999).

Fractures in carbonates are commonly filled with a variety of mineral species including calcite, dolomite, anhydrite, galena, sphalerite, celestite, strontianite, and fluorite (Fig. 3.24). These fractures are, however, generally dominated by carbonate phases. Fracture fills are precipitated as the fracture is being used as a fluid conduit. CO_2 degassing during pressure release associated with faulting and fracturing in the subsurface can result in extensive, almost instantaneous calcite and dolomite precipitation in the fracture system (Roehl and Weinbrandt, 1985; Woronick and Land, 1985). These late carbonate fracture fills commonly have associated hydrocarbons as stains, fluid inclusions, or solid bitumen (Moore and Druckman, 1981; Roehl and Weinbrandt, 1985).

SUMMARY

Carbonate sediments and rocks generally have a much more complex pore system than do siliciclastics because of the wide variety of grain shapes common in carbonates, the presence of intragranular, framework, fenestral carbonate porosity, and the potential for the development of moldic and highly irregular dissolution-related carbonate porosity.

Two major carbonate porosity classifications have been developed to characterize these complex pore systems. The first, the genetic, geologically based classification of Choquette and Pray (1970) is particularly well suited for porosity evolution studies important for exploration efforts in carbonate terrains. The two major attributes of this classification are fabric selectivity and the genetic components of process, direction, and time of formation of the pores.

The porosity classification of Lucia (1983; 1995; 1999) based on the earlier classification of Archie (1952), combines rock fabric elements such as interparticle and vuggy porosity with their engineering attributes such as porosity, permeability, and saturation. While difficult to use in exploration, this classification is useful in carbonate reservoir characterization.

Primary porosity in carbonate sedimentary sequences is often higher than similar siliciclastic sequences because of the common occurrence of intragranular, fenestral, and framework porosity in carbonate sediments.

Dissolution processes that may affect the sequence at any time during burial dominate secondary porosity in carbonates. Early (eogenetic) fabric-selective, secondary porosity is common with its configuration controlled by individual grain mineralogy. While late (telogenetic or mesogenetic) porosity is generally not-fabric selective, its distribution is generally controlled by porosity existing at the time of the dissolution event.

Porosity evolution associated with dolomitization is a complex issue. Some workers suggest that no porosity is created during the dolomitization event because of net import of CO_3, and that dolomites are always less porous than their limestone precursors because of dolomite cementation (Lucia and Major, 1994). Others, however, outline diagenetic scenarios where CO_3 is locally sourced resulting in the net production of porosity during the dolomitization event (Moore, 1989; Purser et al., 1994).

Dissolution of excess calcite and aragonite in dolomitized sequences by fresh meteoric water during cycle-top exposure is perhaps the single most important porosity enhancing process in environment-related dolomitization.

Not-fabric-selective breccia and fracture porosity are exceedingly important porosity types in the subsurface. While brecciation may significantly enhance the total porosity of a sequence, fractures generally enhance permeability rather than total porosity.

Chapter 4

DIAGENETIC ENVIRONMENTS OF POROSITY MODIFICATION AND TOOLS FOR THEIR RECOGNITION IN THE GEOLOGIC RECORD

INTRODUCTION

As discussed in Chapter 3, Choquette and Pray (1970) sketched a broad temporal and spatial framework within which they considered the nomenclature and classification of carbonate porosity. They differentiated between primary porosity, developed before and during deposition, and secondary porosity, developed after deposition. It is post-depositional porosity evolution, and the ultimate predictability of this porosity, that will concern us for much of the remainder of this book.

Choquette and Pray (1970) recognized three zones (eogenetic, mesogenetic and telogenetic) in which post-depositional porosity modification and evolution occur (Fig. 3.3). While their scheme is useful in the discussion of the nomenclature and classification of carbonate porosity, it is really an inadequate framework within which to consider the genetic processes and products responsible for the porosity development and evolution in carbonate rock sequences after deposition because it ignores the basic fuel of diagenesis—water. The chemical characteristics of carbonate pore fluids, the rate of flux through the pore system, and the temperature and pressure regimen under which the resulting rock-water interactions are effected control the diagenetic processes that affect and modify carbonate porosity. These processes are primarily the dissolution of existing carbonate phases and/or the precipitation of new phases. Therefore, the environmental framework within which porosity development and evolution will be discussed is based on three broad families of waters and their distribution either on the surface or in the subsurface, as adapted from Folk (1974).

THE DIAGENETIC ENVIRONMENTS OF POROSITY MODIFICATION

Introduction

There are three major diagenetic environments in which carbonate porosity is formed or modified: meteoric, marine, and subsurface (Fig. 4.1). Two surface or near surface environments, the meteoric and marine (present in

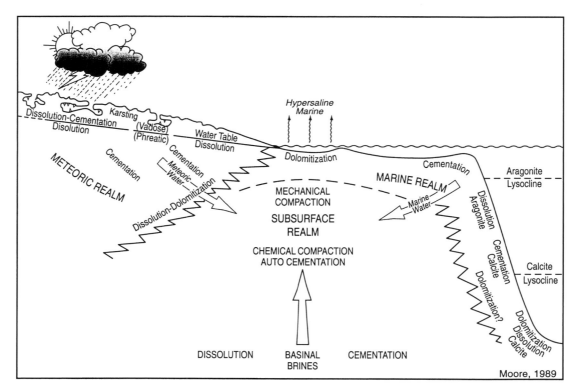

Fig. 4.1. Schematic diagram illustrating the common diagenetic environments within which post-depositional porosity modification and evolution occurs. Active diagenetic processes are identified in each of the three main diagenetic realms. Used with permission of Elsevier.

both the eogenetic and telogenetic zones of Choquette and Pray, 1970), are marked by the presence of pore fluids that are distinctly different from each other. The third environment, the subsurface (mesogenetic zone of Choquette and Pray, 1970), is characterized by mixtures of marine-meteoric waters or complex basinal-derived brines.

Marine environment

The marine environment, in which most carbonate sediments originate today, is characterized by normal or modified marine pore fluids generally supersaturated with respect to most carbonate mineral species (Moore, 1989; Morse and Mackenzie, 1990; Tucker and Wright, 1990)(Chapter 5). The marine environment, therefore, is potentially the site of extensive porosity destruction by marine cements (Land and Moore, 1980; James and Choquette, 1983).

As was mentioned in Chapter 2 (Fig. 2.15), however, the carbonate chemistry of marine waters may well have varied through time. During icehouse climatic periods, oceanic waters were characterized as aragonitic seas and presumably these waters were supersaturated with all carbonate mineral species as noted above. During greenhouse climatic periods, however, marine waters could well have been undersaturated with respect to aragonite but supersaturated with respect to calcite (Fig. 2.15). Under these conditions, aragonite grains may have dissolved, leading to the development of secondary porosity and the precipitation of calcite cements as is normally found in meteoric environments as well as the deep sea (Fig. 4.1)(Palmer et al., 1988).

The distribution of marine cementation is generally controlled by the rate of fluid movement through the sediment pore system and hence is dramatically affected by conditions at the site of deposition, such as energy levels, sediment porosity and permeability, and rate of sedimentation. Cementation, therefore, is not ubiquitous in the marine environment but occurs only in certain favorable sub-environments, such as within the shelf margin reef and the intertidal zone (Chapter 5). Secondary porosity development through dolomitization is common in association with evaporative marine waters (Chapter 6) and may also occur where normal marine waters are flushed through limestone sequences by thermal convection or marine water movement associated with Kohout convection (Vahrenkamp and Swart, 1994) (Chapter 5).

Meteoric environment

The meteoric diagenetic environment is characterized by exposure to subaerial conditions and the presence of relatively dilute waters that generally exhibit a wide range of carbonate mineral saturation states, from the strongly undersaturated to the supersaturated. Meteoric waters moving through the vadose zone (that zone above the water table where both air and water are present) are strongly undersaturated with respect to most carbonate mineral species. This undersaturation is fueled by the ready availability of soil zone carbon dioxide [Moore, 1989] (Chapter 6). If aragonite is present in conjunction with these undersaturated meteoric waters, the difference in solubility between aragonite and calcite will drive the mineral system toward the stable, less soluble calcite phase by dissolution of aragonite and precipitation of calcite cements (Matthews, 1974; Moore, 1989). The porosity modification potential of the meteoric environment is often enhanced by relatively rapid rates of fluid flow through the phreatic zone of the system (Hanshaw and others 1971; Back and others, 1979).

Subsurface environment

The subsurface environment is characterized by pore fluids that may either be a mixture of meteoric and marine waters (Folk, 1974), or a chemically complex brine resulting from long-term, rock-water interaction under elevated temperatures and pressures (Stoessell and Moore, 1983). Because of this extensive rock-water interaction, these fluids are generally thought to be saturated with respect to most stable carbonate species such as calcite and dolomite (Choquette and James, 1987). However, under the high pressure and temperature regimens of the subsurface, pressure solution is an important porosity destruction process that is often aided by cement precipitation in adjacent pore spaces due to the general supersaturation of the pore fluids. Finally, local areas of undersaturation related to thermal degradation of hydrocarbons may result in secondary porosity generation by dissolution.

Most diagenetic processes operate slowly in the subsurface because of the relatively slow movement of fluids under conditions of deep burial (Choquette and James, 1987) (Chapter

9). These processes, however, have enormous spans of geologic time within which to accomplish their work. Those subsurface sequences under the influence of post-orogenic meteoric recharge can suffer dramatic changes in formation fluid composition by mixing. These mixed fluids can cause dissolution of dolomite and precipitation of calcite if anhydrite is present and is dissolved (Chapter 9).

TOOLS FOR THE RECOGNITION OF DIAGENETIC ENVIRONMENTS OF POROSITY MODIFICATION IN THE GEOLOGIC RECORD

Introduction

Each of the diagenetic environments briefly described above is a unique system in which porosity is created or destroyed in response to diagenetic processes driven by the chemical and hydrological characteristics of pore fluids and the mineralogical stability of the sediments and rocks upon which these fluids act. The uniqueness of these environments is the basis for the predictive conceptual models that have been developed over the years to aid the economic geologist in anticipating the nature, quality, and distribution of carbonate porosity on a regional and a reservoir level.

In order to fully utilize the predictive value of these models, the geologist must be able to place the dominant pore modification event, whether dissolution or cementation, into this environmental framework. If, for example, one determines that a pore occluding cement was precipitated early in a regional meteoric phreatic environment, then one might predict the following: (1) the meteoric water system was probably associated with a major sea level lowstand, or stillstand; (2) it might also be reasonable to expect, based on the knowledge of this environment, to find more favorable porosity development in an up-dip, or up-flow direction (Longman, 1980; Moore and Druckman, 1981).

Today, the geologist in search of answers to those questions concerning the conditions under which porosity modification events take place has a broad spectrum of technology available for the hunt. Fundamental petrographic relationships between various cements, and among grains, cements, and matrix, as well as the relationship of carbonate phases to other introduced phases, such as sulfates, allow the geologist to construct relative paragenetic sequences, and to time porosity modification events relative to other diagenetic phases. Cement morphology, cement distributional patterns, and cement size can be related to certain aspects of the precipitational environment, such as chemistry of the pore fluids, rate of cement precipitation, and the relative saturation of the pore system with water (vadose zone versus phreatic zone).

The isotopic and trace element chemistry of pore-filling cements, as well as the chemistry of porosity-modifying dolomites can be related to the trace element and isotopic chemistry of the precipitating and dolomitizing fluids. Therefore, carbonate geochemistry can be used in conjunction with basic petrography to help identify the diagenetic environment, such as meteoric-phreatic, in which porosity modification was accomplished. Oxygen isotopic composition in conjunction with two-phase fluid inclusions can be used to estimate the temperature of formation of diagenetic phases, such as cements and dolomites, and therefore further refine paragenetic sequences as well as place more realistic time constraints on porosity modification events.

The remainder of this chapter is devoted to a general discussion of some of the most useful tools available to the geologist seeking to determine the environment in which important porosity modification events have occurred. This discussion is not meant to be definitive, because, obviously, entire books could be and have been written on each subject mentioned above. Instead, it is hoped that the reader will gain an introduction to each technique and observation, and, most importantly, an outline of the major problems that can seriously constrain their use. The observational petrographic techniques are stressed in this discussion because they are generally available to most working geologists in both industry and academia. The more sophisticated instrument-oriented techniques, such as isotopes and trace element geochemistry, are included in a most general way, so that the working geologist can become familiar with the manner in which the specialist currently uses them. Familiarity with the technique and its constraints will allow subsequent evaluation of the appropriateness of the technique's applicability to specific problems of porosity evolution.

Petrography-cement morphology

Numerous authors have commented on the diverse morphology of calcite and aragonite pore-fill cements and have attempted to relate crystal morphology to the chemical environment of precipitation (Lippmann, 1973; Folk, 1974; Lahann, 1978; Lahann and Siebert, 1982; Given and Wilkinson, 1985; González

et al., 1992; Dickson et al., 1993). The conventional wisdom of the '60s and '70s was that most fresh water calcite cements tend toward equant shapes, while marine calcite and aragonite cements tend toward elongate fibrous shapes. Folk (1974) developed a model directly relating the morphology of calcite cements to the Mg/Ca ratio of the precipitational fluid. Folk's model was based on the concept of the sidewise poisoning of the growing calcite crystal by the substitution of the Mg ion for Ca. The smaller ionic radius of the Mg ion,

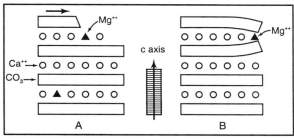

Fig. 4.2. Morphology of calcite crystals as controlled by selective Mg-poisoning (Folk, 1974). If as is shown in A, a Mg ion is added to the end of a growing crystal it can easily be overstepped by the next succeeding CO_3 layer without harm to the crystal growth. However, if as in B, the small Mg ion is added to the side of the crystal, the adjacent CO_3 sheets are distorted to accommodate it in the lattice, hampering further sideward growth, and resulting in the growth of small, fibrous crystals. Used with permission of SEPM.

relative to the Ca ion, causes lattice distortion at the edge of the growing crystal, stopping growth at the edge, and ultimately leading to the elongation of the crystal in the C-axis direction (Fig. 4.2). According to Folk's scheme, therefore, meteoric waters, generally characterized by low Mg/Ca ratios, would tend toward equant crystal shapes, while marine waters marked by high Mg/Ca ratios would produce the elongate calcite crystals noted in

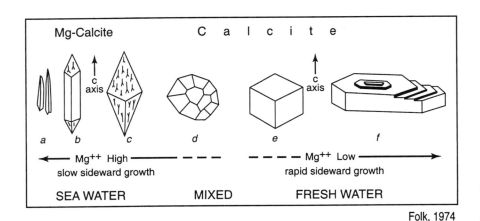

Fig. 4.3. Calcite crystal growth habit as a function of Mg/Ca ratio. Used with permission of SEPM.

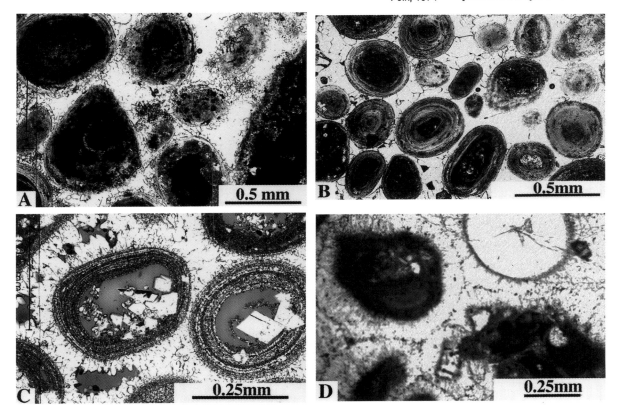

Fig. 4.4. Common calcite cement morphologies found in nature. (A) Small equant to rhombic calcite crystals forming a crust around grains. Meteoric phreatic zone, Joulters Cay, Bahamas. Plain light. (B) Equant calcite mosaic showing irregular distribution pattern. Sample taken at water table at Joulters Cay, Bahamas. Plain light. (C) Bladed calcite circumgranular crust cement in ooid grainstone that has seen meteoric water conditions. Upper Jurassic Smackover Formation, Murphy Giffco #1, Bowie Co. Texas, 7784' (2373m). Plain light. (D) Fibrous magnesian calcite marine cement in a Holocene grainstone collected from the Jamaica island slope at 260m below the sea surface by the Research Submersible Nekton. Note the polygonal suture patterns developed between adjacent cement crusts. Plain light.

marine beach rocks and reef sequences (Fig. 4.3). Figure 4.4 illustrates common calcite cement morphologies found in nature.

Lahann (1978), while agreeing with Folk that Mg could poison the growth of calcite due to lattice distortion, argued that the Mg poisoning effect could not be used to explain morphological differences in calcite crystals because the Mg poisoning should affect all growing surfaces equally, including the C-axis direction. Instead, Lahann called upon the differences in surface potential that develop on the different calcite crystal faces due to calcite crystallography to explain the range of morphologies observed in naturally formed calcites. For example, a crystal edge parallel to the C-axis will expose both Ca and CO_3 ions to the precipitating fluid, while the growing face normal to the C-axis will expose either Ca or CO_3 ions to the precipitating fluids, but not both at the same time (Fig. 4.2). In marine water, where there is a great excess of surface-active cations compared to surface-active anions, the C-axis face will normally be saturated with cations. This situation creates a strong positive potential on the C-axis face that will be greater than that found on the edge parallel to the C-axis, because both CO_3 as well as Ca will be exposed to the fluid. This positive potential will then ensure a greater concentration of CO_3 ions at the C-axis face, and will result in elongation of the calcite crystal parallel to the C-axis under normal marine conditions.

The Lahann model seems to satisfactorily

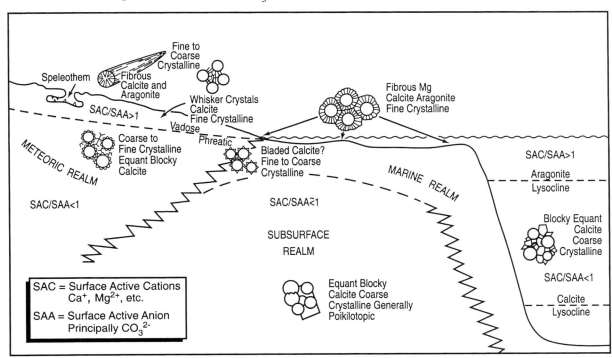

Fig. 4.5. Schematic diagram showing anticipated growth habits of pore-fill calcite cement in the principal diagenetic environments as controlled by the ratio of surface active cations (SAC) to surface active anions (SM). Used with permission of Elsevier.

explain the common occurrence of equant calcite crystals in most meteoric water situations. Under typical meteoric water conditions, both CO_3 and Ca ion concentrations are low. In this situation, saturation with respect to most carbonate phases would be at equilibrium or undersaturated. The differences in surface potential between faces would be minimal, resulting in equant calcite morphologies. The major exception is in the meteoric vadose environment where rapid CO_2 outgassing would lead to elevated saturation conditions. In this environment, elongate calcite crystals, such as are found in speleothems and the whisker crystal cements of the soil zone would be precipitated (Given and Wilkinson, 1985; González et al., 1992). Most subsurface calcite cements are also equant; and while Ca ion concentrations might be high, low CO_3 availability may result in slow equant crystal growth. A parallel situation exists in the deep marine environment, close to the calcite lysocline, where Ca ion concentrations are close to those found in surface waters. However, CO_3 availability is limited, and therefore these calcites exhibit an equant morphology (Schlager and James, 1978; Given and Wilkinson, 1985).

In summary, calcite crystal morphology seems to be related, by way of the CO_3 ion, to the rate of precipitation as controlled by the state of carbonate saturation within the environment. Each of the major diagenetic environments exhibits a wide range of saturation relative to calcite and can thus support a range of calcite morphologies. The mean water chemistry of each environment is distinctive enough so that meteoric phreatic waters generally precipitate equant calcite, while surface marine waters generally precipitate fibrous-to-bladed calcite, and subsurface waters almost always precipitate equant-to-complex polyhedral calcite cements.

Figure 4.5 summarizes the general calcite cement morphology that can be expected within our major diagenetic environments and subenvironments. This is obviously a simplistic view of a complex problem. Gonzalez et al (1992) stressed the importance of fluid flow, mineral saturation and number of nuclei in determining inorganic calcite morphology in speleothems. Dickson, however, stressed the importance of evolutionary crystal growth patterns in controlling ultimate calcite cement morphology (Dickson, 1993; Dickson et al., 1993). Kitano and Hood (1965) considered the effect of organics on the growing surfaces of crystals, as well as the effects of other components such as silica.

These problems notwithstanding, calcite cement morphology can be a useful index to the conditions of precipitation, and a valuable clue to the diagenetic history and porosity evolution of a rock. It must be emphasized here, however, that cement morphology must be used with appropriate caution within a solid geological framework and in conjunction with other tools such as geochemistry.

Petrography-cement distribution patterns

Under most normal conditions of calcite cement precipitation, the sediment or rock pore is totally saturated with water, and cement is precipitated equally around the periphery of the pore (Fig. 4.6C and 4.7C). In the vadose zone, above the water table, however, precipitation takes place in the presence of two phases, air and water. Under these conditions, water is held at the grain contacts, water movement is by capillarity, and cement distributional patterns are

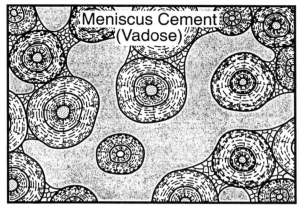
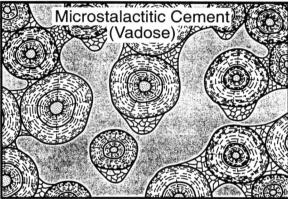
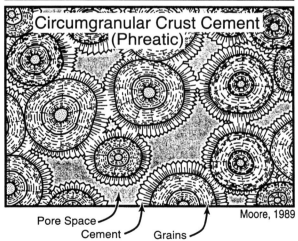

Fig. 4.6. Cement distributional patterns as a function of diagenetic environment. A. In the vadose zone, cements are concentrated at grain contacts and the resulting pores have a distinct rounded appearance due to the meniscus effect. The resulting cement pattern is termed meniscus cement. B. in the vadose zone immediately (Fig.

4.6 cont.) above the water table there is often an excess of water that accumulates at the base of grains as droplets. These cements are termed microstalactitic cements. C. In the phreatic zone, where pores are saturated with water, cements are precipitated as circumgranular crusts. Used with permission of Elsevier.

controlled by the distribution of the liquid phase (Dunham, 1971; Longman, 1980; James and Choquette, 1984; Moore, 1989; Tucker and Wright, 1990). Well within the zone of capillarity, calcite cement will tend to be concentrated at the grain contacts and will exhibit a curved surface that reflects the meniscus surface of the water at the time of precipitation caused by surface tension (Fig. 4.6A and 4.7A and B). This curved pattern of cementation is called meniscus cement (Dunham, 1971). In the more saturated areas immediately below the zone of capillarity and above the water table, more fluid is available, and excess water accumulates as droplets beneath grains. Calcites precipitated in this zone often show this gravity orientation and are termed microstalictitic cements (Longman, 1980)(Fig. 4.6B and 4.7D). The vadose cementation patterns discussed above may occur in both the meteoric environment and the marine environment (beach rock) (Moore et al., 1972; Moore, 1989) (Chapters 5 and 7). While these cementation patterns are quite useful in differentiating the vadose and phreatic subenvironments in Holocene and Pleistocene sequences (Halley and Harris, 1979), they are more difficult to utilize in ancient rock sequences. In ancient sequences, subsequent generations of cement often overgrow the original vadose cements, muting and ultimately destroying the distinctive vadose features. In addition, later partial dissolution of fine crystalline phreatic calcite cements can mimic the rounded pores associated with vadose cements (Moore and Brock, 1982)(Fig.

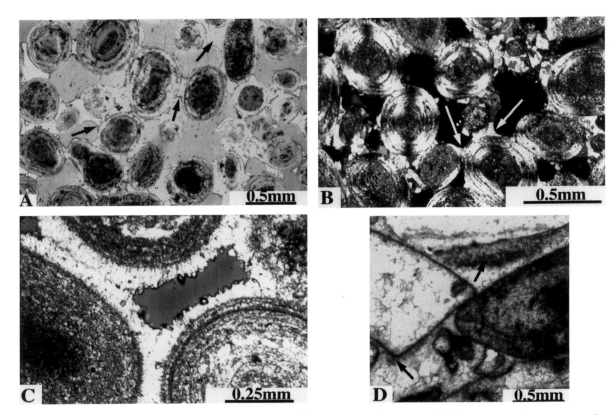

Fig. 4.7. Cement distribution patterns as a function of diagenetic environment. (A) Meniscus calcite cement characteristic of the vadose zone, Joulters Cay, Bahamas. Plain light. (B) Same as (A), crossed polars. (C) Circumgranular calcite crust cement characteristic of phreatic precipitation. Murphy Griffco#1, 7784' (2373m), Miller Co., Arkansas, USA. Plain light. (D) Microstalactitic calcite cement (Arrows) in Lower Cretaceous Edwards Formation, north central Texas. Cement precipitated as aragonite in the marine vadose zone (beach rock) associated with a prograding beach sequence. Plain light.

3.19A).

Petrography-grain-cement relationships relative to compaction

The timing of compaction events relative to cementation is both important and relatively easy to determine. If, for example, a cemented grainstone shows no evidence of compaction, such as high grain density, or grain interpenetration, the cementation event is most certainly early and before burial (Fig. 4.7C)(Coogan and Manus, 1975; Bhattacharyya and Friedman, 1984; Railsback, 1993; Goldhammer, 1997. Cements that are broken and involved in compaction are at least contemporary with the compaction event (Fig. 4.8A and 4.9A-C) and are probably relatively early (Moore and Druckman, 1981; Moore, 1985).

Cements that encase both compacted grains and earlier deformed cements, and themselves are not involved in compaction, surely represent later post-compaction subsurface cements (Fig. 4.8B and 4.9B and C) (Moore and Druckman, 1981; Moore, 1985). Finally, cross-

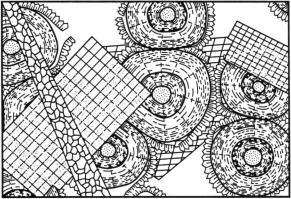

Moore, 1989

Fig. 4.8. Timing of diagenetic events by petrographic relationships. (A) Involvement of isopachous cement in compaction indicates that cement was early. (B) Cement that encases compacted grains and spalled isopachous cement is relatively late. (C) Cross cutting relationships of features such as fractures and mineral replacements are useful. Calcite filling fracture is latest event, replacement anhydrite came after the late poikilotopic cement, (Fig. 4.8 cont.) but before the fracture, while the isopachous cement was the earliest diagenetic event. Used with permission of Elsevier.

cutting relationships associated with fractures and mineral replacements can clearly establish the relative timing of each cementation and/or replacement event (Moore and Druckman, 1981) (Fig. 4.8C and 4.9D). The petrographic relationships outlined above can be used to determine the timing of porosity-modifying events relative to the general paragenesis of a sequence. The actual time, or the depth of burial under which the event took place, however, cannot be determined. For example, it can be seen that the rim cement in Figure 4.9C was precipitated before the rupture of ooids in this specimen. It is not known, however, how much overburden is necessary to accomplish the failure of the ooids or to cause the distortion and breakage of the cement itself (Moore, 1985). Some constraints, however, can be placed on the burial depth, temperature, and timing of porosity-related diagenetic events by utilizing trace element and isotope geochemistry and two-phase fluid inclusions.

Trace element geochemistry of calcite cements and dolomites

Directly precipitated carbonate cements incorporate various trace and minor elements proportionally to the particular element's concentration in the precipitating fluid. It is assumed that the element's incorporation is dominantly by substitution for Ca^{+2}, rather than interstitially between lattice planes, at site defects, as adsorbed cations, or within inclusions (Veizer, 1983b) (Dickson, 1990; Banner, 1995). Trace element incorporation by substitution of the trace for Ca^{+2} is controlled by the distribu-

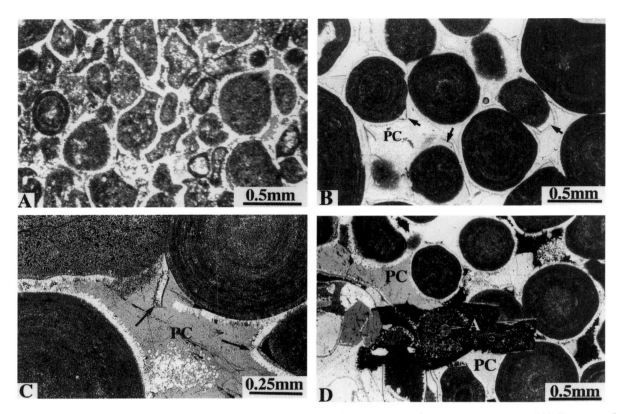

Fig. 4.9. Grain-cement relationships useful in determining relative timing of cementation events. (A) Compacted Jurassic Smackover pellet grainstone. Circumgranular crust cements obviously affected by compaction (arrows) so cement is early, before compaction. Walker Creek field, 9842' (3000 m), Columbia Co., Arkansas, USA. Plain light. (B) Distorted, and spalled, circumgranular crust calcite cements in a Jurassic Smackover ooid grainstone. These cements are early, pre-compaction. Poikilotopic calcite cements (pc) that are obviously later, subsurface cements have engulfed areas of spalled cement crust (arrows). Nancy field, 13,550' (4130m), Mississippi, USA. Plain light. (C) Close view of spalled calcite crust cement (arrow) of (B) engulfed by coarse poikilotopic calcite cement (pc). Poikilotopic cement post-compaction. (Crossed polars). (D) Anhydrite (arrow) replacing both ooids and post-compaction poikilotopic calcite cement, suggesting that anhydrite is a late diagenetic event. Note inclusions of calcite that outline ghosts of grains and cements. Jurassic Smackover, 13,613' (4149m), Nancy field, Mississippi, USA. (Crossed polars).

tion coefficient D, as outlined by Kinsman (1969), where mTRc and mCa represent molar concentrations of the trace or minor element of interest and molar concentrations of Ca in the fluid and solid phases (Fig. 4.10). This illustrates the calculation of trace element composition of a calcite precipitated from waters of known composition. The distribution coefficients used in Figure 4.10 were experimentally determined.

The distribution coefficient of an element's incorporation into the solid phase is generally determined experimentally by precipitating the solid phase from a solution of known concentration and by measuring the concentration of the element in the solid phase (Kinsman and Holland, 1969). A popular modification of this technique, designed to minimize the effects of

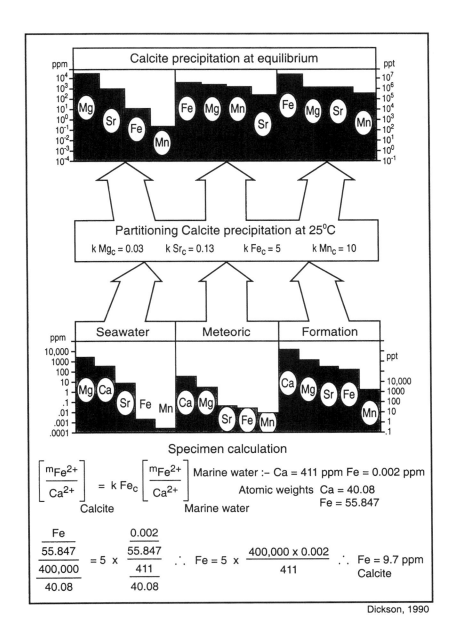

Figure 4.10. Diagram illustrating the calculation of trace element abundance in calcite precipitated from waters of known composition using partition coefficients. The cation concentrations of sea water are well known, while the meteoric water approximates the mean cation concentrations of river water and formation water approximates cation concentration of an oil field brine. The partition coefficients used in this figure are arbitrary. Used with permission of Blackwell Scientific Publishing.

precipitation rate, involves the transformation of aragonite or dolomite into calcite under controlled conditions of temperature, pressure, and trace element composition (Dickson, 1990; Morse and Mackenzie, 1990; Banner, 1995). Obviously, this process presents a problem in the case of dolomites, which are difficult to synthesize under earth-surface temperature and pressure conditions (Veizer, 1983). McKenzie et al., (1995) have, however, synthesized dolomite at low temperatures by utilizing microbial mediation. Behrens and Land (1972) reasoned that the D for Sr and Na in dolomite should be half that for calcite because dolomite contains half as many Ca sites as an equivalent amount of calcite; this reasoning has

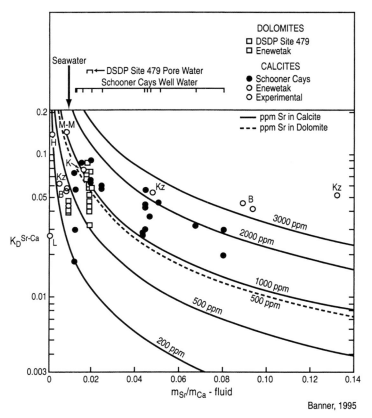

Fig. 4.11. K_D^{Sr-Ca} vs. molar Sr/Ca ratio in fluid. Contours are for calcite and dolomite Sr concentrations in ppm. Given any two of the three values (K_D^{Sr-Ca}, Sr/Ca ratio in fluid or carbonate mineral Sr concentration), the third value can be calculated and estimated by inspection from this diagram. Reported K_D^{Sr-Ca} and Sr/Ca pairs are plotted for studies of experimental and natural systems. K_D^{Sr-Ca} values for natural systems determined from analysis of (1) dolomites and pore waters from cores at DSDP site 479 (Baker and Burns, 1985), (2) calcites and well waters from Schooner Cays, Bahamas (Budd, 1984) and (3) dolomites and calcites from Enewetak Atoll and seawater (Quimby-Hunt and Turekian, 1983; Saller, 1984). Natural fluid compositions (seawater, DSDP pore waters and Schooner Cays well water) are plotted relative to abscissa only. Experimental values (O) are determined by the following studies: B=Baker et al. (1982) two determinations each at 60° and 80°C; H=Holland et al. (1964) and L=Lorens (1981): 25°C value for lowest precipitation rate, equilibrium fluid compositions not reported; K=Kitano et al. (1971) average K_D^{Sr-Ca} value at 20°C; Kz=Katz et al. (1972): average K_D^{Sr-Ca} value for selected experiments at high and low fluid Sr/Ca ratios all at 40C°; M-M=Mucci & Morse (Mucci and Morse, 1983): average K_D^{Sr-Ca} value using synthetic fluid with seawater composition. Used with permission of Sedimentology.

become common practice (Fig. 4.11) (Veizer, 1983b; Banner, 1995).

The size of the crystal lattice obviously plays a major role in determining the magnitude of D for the carbonates. The large unit cell of orthorhombic aragonite crystals can accommodate cations larger than Ca (such as Sr, Na, Ba, and U), while the smaller unit cell of the rhombohedral calcite preferentially incorporates smaller cations (such as Mg, Fe, and Mn). Hence, the distribution coefficient for Sr in aragonite is much larger than it is for Sr in calcite (Table 4.1).

Initially, the use of trace elements in determining the diagenetic environment of processes such as cementation held great promise (Kinsman, 1969). Sr and Mg are of particular interest because aragonite (Sr rich) and magnesian calcite (Mg rich) dominate metastable carbonate mineral suites from modern shallow marine environments. Stabilization of these shallow marine carbonates to calcite and dolomite involve a major reapportionment of these elements between the new diagenetic carbon-

Table 4.1

Trace element	Reported D	Recommended D	Sources of data
Calcite			
Sr	0.027-0.4 (1.2)	0.13 D.P. 0.05 A→dLMC HMC→dLMC 0.03 LMC→dLMC	Holland et al. (1964), Holland (1966), Kinsman and Holland (1969), Katz et al. (1972), Ichikuni (1973), Lorens (1981), Kitano et al. (1971), Jacobson and Usdowski (1976), Baker et al. (1982)
Na	0.00002-0.0003		Müller et al. (1976), White (1978)
Mg	0.013-0.06 (0.0008-0.12)		Winland (1969), Benson and Matthews (1971), Alexandersson (1972), Richter and Füchtbauer (1978), Kitano et al. (1979b), Katz (1973), Baker et al. (1982)
Fe	$1 \leq x \leq 20$		Veizer (1974), Richter and Füchtbauer (1978)
Mn	5.4-30 (1700)	6 D.P. 15 A→dLMC HMC→dLMC 30 LMC→dLMC	Bodine et al. (1965), Crocket and Winchester (1966), Michard (1968), Ichikuni (1973), Lorens (1981)
Aragonite			
Sr	0.9-1.2 (1.6)		Kitano et al. (1968, 1973), Holland et al. (1964), Kinsman and Holland (1969)
Na	~0.00014 (3-4 x $\underline{D}^{Na}_{calcite}$)		Calculated from figure 1 of White (1977), Kitano et al. (1975)
Mg	~0.0006-0.005		Brand and Veizer (1983)
Mn	0.86 (~1/2-1/3 $\underline{D}^{Mn}_{calcite}$)		Raiswell and Brimblecombe (1977), Brand and Veizer (1983)
Dolomite			
Sr	0.025-0.060		Katz and Matthews (1977), Jacobson and Usdowski (1976)
Na	(as $\underline{D}^{Na}_{calcite}$) (?)		White (1978)

D.P. = Direct precipitation. dLMC = diagenetic low-Mg calcite   Modified from Veizer, 1983

Table 4.1. Commonly accepted distribution coefficients for the incorporation of trace elements in calcite, aragonite, and dolomite. The full references to the data sources may be found in Veizer, 1983. Used with permission of SEPM.

ates and the diagenetic fluids. Na is of interest because it is a relatively major cation in seawater and brines and a minor constituent in dilute or meteoric ground waters. Fe and Mn are multivalent, are both sensitive to Eh and pH controls, and occur in very low concentrations in sea water, but are present in higher concentrations in ground water and oil field brines (Veizer, 1983). Waters from the major diagenetic environments then, are so different in their trace and minor element compositions (Table 4.2) that the cements precipitated from these diverse waters should easily be recognized by their trace element signatures. As interest in the utilization of trace elements in carbonate studies increased, widespread concern surfaced over their validity and the controls over the magnitude of the most commonly used distribution coefficients (Bathurst, 1975; Land, 1980; Moore, 1985; Banner, 1995).

Most carbonate geologists realize that distribution coefficients are, for the most part, relatively crude estimates and can be affected dramatically by temperature, major element

Table 4.2

Composition g/l	Subsurface Smackover Formation[a]		North American[b] River Systems		Seawater[c]
	Range	Mean	Range	Mean	
Cl	31.0-223.0	171.7	0.0008 -1.5	0.04	19.0
Na	5.0- 87.4	67.0	0.002 -0.9	0.04	10.5
Ca	6.8- 55.3	34.5	0.0004 -0.6	0.04	0.4
Mg	0.1- 8.6	3.5	0.0008 -0.2	0.009	1.35
Sr	0.1- 4.7	1.9	0.00001-0.009[d]	0.00067[d]	0.008
Mn		0.03		0.00008[g]	0.00002[g]
Fe	—	0.04		0.00040[g]	0.00020[g]
SO$_4$	0.0- 4.0	0.4	0.0008-1.0	0.091	2.7
TDS[e]	51.0-366.0	279.00	0.05 -1.1	0.3	35.0
	Equivalent Ratios				
Mg/Ca	0.05 -0.52	0.17	0.082 - 0.89	0.37	5.3
Na/Cl	0.45 -0.77	0.60	0.72 -61.6	11.65	0.86
Sr/Ca	0.0025-0.085	0.025	0.0009- 0.0091[d]	0.0035[d]	0.0086[f]
Na/Ca	1.06 -2.41	1.69	0.09 - 5.23	0.70	23.09

[a]Data calculated from Collins, 1974
[b]Data calculated from Livingstone, 1963
[c]Data from Riley ane Skirrow, 1965
[d]Data from Skougstad and Horr, 1963
[e]Total dissolved solids
[f]Value from Kinsman, 1969
[g]Data from Veizer, 1983

Modified from Moore and Druckman, 1981

Table 4.2. The mean chemical composition of Jurassic formation waters from the Gulf Coast, U.S.A., north American river waters, and seawater. This table illustrates the wide range of compositional differences seen between meteoric, subsurface, and marine waters. Used with permission of AAPG.

composition of the solid phase and the rate of precipitation of the cement (Mucci and Morse, 1983; Given and Wilkinson, 1985; Banner, 1995). Table 4.1 outlines some of the reported range of distribution coefficients for the most commonly used trace and minor elements in the three major carbonate minerals as compiled by Veizer in 1983. Figure 4.11 shows the compilation of experimental distribution coefficients for the incorporation of Sr in calcite and dolomite furnished by Banner in 1995. Figure 4.11 also includes distribution coefficients for Sr incorporation in calcite and dolomite determined from naturally occurring marine cements and dolomites. Table 4.1 and Figure 4.11 illustrate the basic problem facing the geologist who wishes to utilize trace elements in carbonate diagenetic studies; that is, reported distribution coefficients exhibit a huge range of values. In short, the natural system is so complex that we generally do not understand, and cannot quantify, the basic factors that control the magnitude of these distribution coefficients in nature (Banner, 1995) (Humphrey and Howell, 1999).

Moore (1985) presents a well-documented case history of Jurassic subsurface calcite cement. Based on stable isotopes, two-phase fluid inclusions, and radiogenic Sr isotopic studies, Moore concluded that these cements precipitated from present oil field brines at elevated temperatures. Table 4.3 presents a comparison of measured trace element composition of the cement to the cement trace element composition that should be expected from the present subsurface fluids. Distribution coefficients that span the range of D's found in Ta-

Table 4.3

Element		Smackover Brine[1]	Calculated Trace Element Composition of a Calcite Precipitated from Smackover Brine		Measured Composition of Post-Compaction Calcite (ppm)
Strontium		Equivalent Ratio Sr/Ca	Calculated[2] Sr (ppm) in calcite ppct. from this water with indicated distribution coefficients at 25°C		
			.14 (Kinsman, 1969)	.054 (Katz et al., 1972)	
	Range	.0025-.085	307-10,426	118-4022	
	Mean	.025	3000	1350	<100
Magnesium		Equivalent Ratio Mg/Ca	Calculated[2] Mg (ppm) in calcite ppct. from this water using Katz's (1973) distribution coefficients at 25°C and 70°C		
			(0.573) 25°C	(.0973) 70°C	
	Range	0.05-0.52	696-7,244	1183-12,200	
	Mean	0.17	2368	4021	3160 ± 498
Manganese		Equivalent Ratio Mn/Ca	Calculated[2] Mn (ppm) in calcite ppct. from this water using Michard's (1968) distribution coefficient 5.4 at 25°C		
	Range	2.15×10^{-5} - 1.07×10^{-3}	64-3174		
	Mean	3.2×10^{-4}	949		498 ± 85
Iron		Equivalent Ratio Fe/Ca	Calculated[2] Fe (ppm) in calcite ppct. from this water using Richter and Fuchtbauer (1978) distribution coefficient of 1 at 25°C		
	Range	2.1×10^{-5} to 4.39×10^{-3}	92-2450		
	Mean	1.42×10^{-3}	793		2014 ± 326

Modified from Moore, 1985

[1]Equivalent ratios from Table 4.1
[2]Calculated composition of calcite from expression m Sr/Ca Calcite = Distribution Coefficient x m Sr/Ca of Precipitation Fluid (Kinsman, 1969)
[3]Mean composition in ppm determined by electron microprobe analysis is at the L.S.U. Department of Geology, Microanalysis Laboratory by C. H. Moore. Number of Analyses Mg, Fe, Sr. 100; Mn 55.
[4]Equivalent ratios from Meyers, 1974

Table 4.3. Comparison of the measured trace element composition of Jurassic Smackover post-compaction calcite cements to the calculated trace element composition of a calcite precipitaated from an average Jurassic formation brine using several commonl used distribution coefficients. Used with permission of SEPM.

ble 4.1 for each of the element's incorporation into calcite were used in construction of the table. It is clear that the trace and minor element composition of these late cements, particularly Sr, does not adequately identify present brines as the precipitational fluid, regardless of the distribution coefficient used.

This case history illuminates clearly the problems presently facing the geologist using and trying to interpret trace element compositions of carbonates with the idea of reconstructing original diagenetic environments. Are trace element studies, then, an exercise in futility? They certainly are, at the present stage of understanding, if specific knowledge is sought concerning the composition of the diagenetic fluid responsible for a diagenetic event, such as a pore-fill cement. On the other hand, if used intelligently within a solid geologic and petrologic framework, trace elements can provide the geologist with some useful information concerning gross characteristics of the diagenetic environments responsible for cementation and dolomitization events.

For example, Banner (1988; 1995) suggests using covariation of trace elements and isotopes to construct quantitative models of diagenesis that give pathways on bivariate plots of changing trace element abundance and isotopic compositions in response to fluid-rock interaction. Trace elements used in concert with isotopes can be independent of distribution coefficients over a wide range, and seem to be diagnostic of the general type of fluid involved in the diagenetic system.

Trace elements may be useful in determining the original mineralogy of rock components, such as ooids, bioclasts, and cements. Sandberg (1973; 1983; 1985) used high Sr concentrations in concert with petrography to infer an original aragonite composition for a variety of materials throughout the Phanerozoic. Moore and others (1986) used petrography and the trace elements Sr and Mg to suggest that Upper Jurassic ooids in southern Arkansas were originally both aragonite and magnesian calcite. The inferred aragonite ooids exhibited high Sr concentrations, while inferred magnesian calcite ooids yielded low strontium but relatively high magnesium concentrations.

As will be discussed in Chapter 7, detailed variations in the Sr and Mg, compositions of meteoric cements may reflect the mineralogical composition of the sediments and rocks through which the diagenetic fluids have passed. As diagenetic fluids pass through, and interact with, metastable mineral suites, such as aragonite and magnesian calcite, the Sr and Mg composition of the fluids increases. Cements precipitated from these fluids down hydrologic flow will reflect these dissolution events (Dickson, 1990).

The pH-Eh sensitivity of Fe and Mn has been used extensively in recent years to infer the environmental conditions under which cements have been precipitated (Chapter 7). The oxidizing conditions of the vadose environment favor the Fe^{+3} and Mn^{+4} states and preclude the incorporation of divalent Fe and Mn in the calcite lattice. The general reducing conditions found in the phreatic zone, however, support the incorporation of both in calcite. The influence of Mn and Fe over the cathodoluminescence characteristics of carbonates has made cathodoluminescence a useful tool in determining changes or gradients in oxidation states in ancient hydrologic systems (Machel, 1985; Hemming et al., 1989) (Chapter 7).

Finally, trace elements may be used to determine chemical gradients that give useful information concerning direction of paleo-fluid flow in diagenetic terrains (Moore and Druckman, 1981), as well as estimates of rock: water ratios (open versus closed diagenetic systems) during diagenesis (Brand and Veizer, 1980; Veizer, 1983a).

Stable isotopes

There are about 300 stable isotopes and

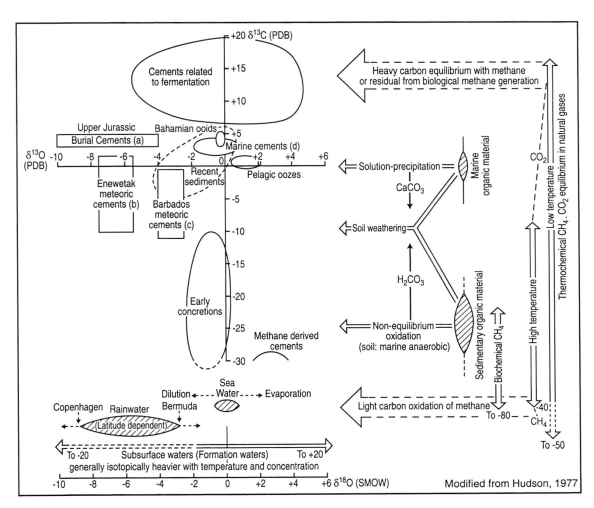

Fig. 4.12. Distribution of carbon and oxygen isotopic compositions of some carbonate sediments, cements, and limestones with some of the factors that control them. Data sources as outlined by Hudson (1977) except as follows: a. Moore, 1985. b. Saller, 1984. c. Matthews, 1974. d. James and Choquette, 1983. Used with permission of the Geological Society of London.

1200 radioisotopes (unstable isotopes) of the 92 naturally occurring elements. The nucleus of each of these elements contains a certain number of protons, but the isotopes of that element have different numbers of neutrons. This results in the element and its isotope having slight differences in mass and energy giving rise to differences in physical and chemical properties. These differences are greatest in elements of low atomic number and may cause significant differences in distribution of the isotopes during natural processes, such as evaporation, condensation, photosynthesis, and phase transformations. These distributional differences, or fractionation, form the basis for the utilization of light stable isotopes in diagenetic studies (Anderson and Arthur, 1983; Dickson, 1990).

For carbonates, the two most naturally abundant isotopes of oxygen, ^{18}O and ^{16}O, and carbon, ^{13}C and ^{12}C, are commonly used, while sulfur isotopes are used in sequences containing evaporites and sulfide mineralization. The discussion here will be limited to the stable isotopes of oxygen and carbon. Because mass spectrometric analysis can determine isotope ratios more precisely than individual isotope abundance, variations in $^{18}O/^{16}O$ and $^{13}C/^{12}C$ ratios between samples are examined. Variations in these ratios in most natural systems are small but can be precisely measured. The ∂ notation is useful for expressing these small differences relative to a standard value. ∂-values are reported in parts per thousand, which is equivalent to per mil (Hudson, 1977). For oxygen:

$$\partial^{18}O = (^{18}O/^{16}O_{Sample}) (^{18}O/^{16}O_{Standard})/ (^{18}O/^{16}O_{Standard}) \times 1000$$

The standard commonly used for both oxygen and carbon in carbonates is referred to as PDB. The standard is derived from a belemnite collected from the Cretaceous Pee Dee Formation in South Carolina. Using the formulation above, the PDB standard would have $\partial^{18}O$ and $\partial^{13}C = 0$. Most marine carbonates are close to this value and hence are small numbers. Oxygen isotopic compositions can also be referred to SMOW (Standard Mean Ocean Water). The conversion of $\partial^{18}O$ values from the PDB to the SMOW scale is approximately (Anderson and Arthur, 1983):

$$\partial^{18}O \text{ SMOW} = 1.03\partial^{18}O \text{ PDB} + 30.86$$

Oxygen and carbon isotopic compositions for the same samples are commonly plotted against each other, making relationships and trends between samples relatively easy to determine. A $\partial^{18}O$ vs. $\partial^{13}C$ cross-plot with the extension of the axes through their intersection at 0,0 (Fig. 4.12) facilitates the comparison of ∂-values as either negative or positive. It should be noted that today many isotope geochemists do not use "depleted" or "enriched" to describe stable isotope values because a ^{13}C enrichment may actually be a ^{12}C depletion. Instead, they advocate "high" or "low" $\partial^{18}O$, $\partial^{13}C$ values. Figure 4.12, taken from (Hudson, 1977), is such a cross-plot, incorporating along the base of the diagram ranges of the $\partial^{18}O$ (SMOW) composition of natural waters from our three diagenetic realms: meteoric, marine, and subsurface. Isotopic values of marine waters vary several per mil about a mean of 0, depending on temperature, evaporation, and dilution from sea ice (Anderson and Arthur, 1983). Under conditions of heavy evaporation, such as is evident in salt ponds and on sabkhas, marine waters can exhibit relatively high $\partial^{18}O$ values. Meteoric waters are relatively low, but show a wide range of oxygen isotopic compositions due to latitude and altitude effects (Hudson, 1977; Dickson, 1990). Subsurface fluids also have a wide range of compositions from $\partial^{18}O = -20$ to $+12$ per mil (SMOW) (Land and Prezbindowski, 1981), with most subsurface oil-related brines generally being on the high side of the range (Hudson, 1977). The carbonates precipitated from these waters as cements will generally reflect the oxygen isotopic composition and temperature of the precipitating fluid, as controlled by the temperature-dependent isotopic fractionation between the liquid and mineral phase. Meteoric calcite cements, therefore, will tend to exhibit relatively low oxygen isotopic ratios (Fig. 4.12, Chapter 7), while shallow-water marine cements will have relatively high oxy-

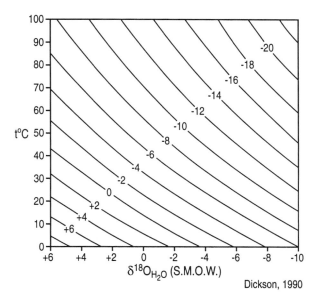

Figure 4.13 Equilibrium relationship between $\partial^{18}O$ of calcite, temperature and $\partial^{18}O$ of water: x-axis represents temperature between 0 and 100°C. The curved lines represent constant $\partial^{18}O$ values (PDB) for calcite calculated from expression in Craig (1965). Taken from Dickson, 1990. Used with permission of Blackwell Science.

gen isotopic ratios (Chapters 5 and 6). The oxygen isotopic fractionation between water and calcite is highly temperature-dependent (Friedman and O'Niel, 1977) with the $\partial^{18}O$ of the calcite becoming increasingly lower with increasing temperature (Fig. 4.13). Calcites precipitated in the subsurface, even though precipitated from pore fluids with heavy oxygen isotopic composition, can have relatively low $\partial^{18}O$ values that may generally overlap those of meteoric waters (Moore, 1985). Therefore, in any interpretation of the oxygen isotopic composition of a calcite cement, the geologist must have independent evidence of the temperature of formation (i.e. petrography, two-phase fluid inclusions) in order to estimate the original isotopic composition of the water of precipitation and hence the diagenetic environment. Conversely, if the water's isotopic composition is known, or can be assumed, a temperature of precipitation for the calcite can be estimated using the isotopic composition of both the solid and liquid phases and entering Figure 4.13 and solving for temperature.

Dolomite presents a special problem because the fractionation factor is poorly known and has been, for a number of years, a matter of debate (Land, 1980; Anderson and Arthur, 1983; Dickson, 1990). While extrapolation of

Table 4.4

Reservoir	Mass	$\delta^{13}C‰$
1) Atmosphere (pre-1850 CO_2) (~290 ppm)	610 (range: 560-692)	−6.0-7.0
2) Oceans — TDC	35,000	0
DOC	1000	−20
POC	3	−22
3) Land Biota (biomass)	(range: 592-976)	−25
	(~10% may be C_4? plants)	−12
4) Soil Humus	(range: 1050-3000)	−25
5) Sediments (Total)	(range: 500,000-10,000,000)	—
inorganic C	423,670	+1
organic C	86,833	−23
6) Fossil Fuels	>5000	−23

From Anderson and Arthur, 1983

Table 4.4. Major carbon reservoirs, the mass of carbon found in each, and their mean carbon isotopic composition. Mass units, 10^{15} grams carbon. $\partial^{13}C$ permil PDB. Used with permission of SEPM.

high temperature experimental data would give a concentration of ^{18}O in the neighborhood of 6 per mil higher than contemporaneous calcite, Land (1980) suggests that a 3-4 per mil increase of ^{18}O in dolomite relative to syngenetic calcite is more reasonable. In addition, as Land (1980; 1985) points out, dolomite commonly originates as a metastable phase (protodolomite) that seems to be susceptible to recrystallization and hence isotopic re-equilibration during burial and diagenesis. Reeder (1981; 1983) has shown the presence of complex mineralogical domains in dolomites that surely affect not only the ultimate stability of dolomites, but their isotopic composition as well. These factors seriously impact the interpretation of the isotopic composition of dolomite relative to the nature of the dolomitizing fluids and the diagenetic environment under which dolomitization was accomplished. Owing to the original metastability of many dolomites, the isotopic composition of a dolomite might well reflect the nature of recrystallizing fluids, rather than the composition of the original dolomitiz-

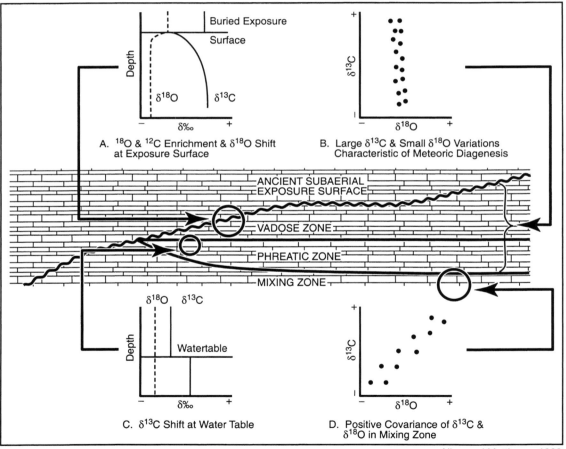

Fig. 4.14. Schematic diagram showing the anticipated carbon and oxygen isotopic shifts across meteoric diagenetic interfaces such as exposure surfaces, water tables, and meteoric-marine water mixing zones, as compared to the trend expected in the meteoric phreatic zone. Used with permission of the International Association of Sedimentologists.

ing solution.

Organic processes often govern the geochemical behavior of carbon, with photosynthesis playing a major role. The range of $\partial^{13}C$ values found in nature is outlined on the right side of Figure 4.12. Organic carbon exhibits low values of $\partial^{13}C$ (-24 per mil PDB) relative to the oxidized forms of carbon found as CO_2 (-7 per mil) and marine carbonates (0 to +4 per mil). Extremes in $\partial^{13}C$ values generally involve methane generation either by biochemical fermentation in the near surface, or by thermochemical degradation of organic matter in the subsurface at temperatures greater than 100°C (Hudson, 1977; Anderson and Arthur, 1983). Methane generated from fermentation produces very low $\partial^{13}C$ values (-80 per mil), but the residual organic matter shows high $\partial^{13}C$ values. Oxidation of the methane with subsequent cementation will result in cements with low $\partial^{13}C$ values (Roberts and Whalen, 1977), while oxidation of the residual organic material and subsequent cementation incorporating carbon from this source will lead to cements with high $\partial^{13}C$ (Fig. 4.12) (Curtis et al., 1972).

Thermochemically derived methane can indirectly lead to the precipitation of subsurface calcite cements with very low $\partial^{13}C$ values (Anderson and Arthur, 1983; Sassen et al., 1987; Heydari and Moore, 1993; Moore and Heydari, 1993).

Soil weathering and carbonate mineral stabilization involving dissolution of marine limestones and sediments and subsequent precipitation of calcite cements in the vadose and the shallow phreatic zones will generally result in cements and limestones with moderately -$\partial^{13}C$ compositions (Fig. 4.12) (Allen, 1982; James and Choquette, 1984). Table 4.4 outlines the estimated mass of carbon in the various natural reservoirs found in and adjacent to the earth's crust. It is obvious that most carbon is tied up in sediments and limestones of the lithosphere and that the majority of this carbon is marine in origin and carries a relatively high $\partial^{13}C$ signature (Hudson, 1977; Anderson and Arthur, 1983). Therefore, in any diagenetic process that involves rock-water interaction, such as solution-reprecipitation under burial conditions, the relatively heavy rock carbon tends to dominate and to buffer low $\partial^{13}C$ contributions from biological processes. The lack of appreciable temperature fractionation associated with carbon isotopes supports this buffering tendency (Anderson and Arthur, 1983).

Cross-plots of $\partial^{18}O$ versus $\partial^{13}C$, as seen in Figure 4.12 are commonly used to distinguish the diagenetic environment responsible for specific cements. Similar cross plots detailing the isotopic composition of individual limestone components, such as fossils or ooids collected on a regional basis, are useful in determining the pathways of diagenetic change suffered by a limestone, as well as determining the original isotopic composition of the marine components that will give an estimate of the isotopic composition of the paleo-oceanic waters (Lohmann, 1988).

The two most common stable isotope trends observed on these types of cross-plots are: 1) the J-shaped curve of Lohmann (Lohmann, 1988), showing the covariant trend $\partial^{18}O$ as well as $\partial^{13}C$ toward lower values, during progressive diagenesis under meteoric water conditions; and 2) the dominant trend toward lighter $\partial^{18}O$ values, during burial caused by increasing pore-water temperature and rock buffering of the $\partial^{13}C$ composition of the cement (Moore, 1985; Choquette and James, 1987).

Tightly controlled vertical stable isotopic sampling of cores and outcrops has been used to detect geochemical gradients that reflect vadose zones and water table positions (Fig. 4.14). A rapid change in $\partial^{13}C$ composition toward lighter values seems to be characteristic of the vadose zone (Allan and Matthews, 1982). This technique can be useful if applied carefully and adequately constrained by geologic setting and petrography.

Finally, the $\partial^{18}O$ values of marine limestones and cherts have tended to decrease significantly with increasing age (Fig. 4.15). The root cause of this temporal variation is a matter of some controversy. Has the temperature of the oceans been cooling, driving the isotopic composition of these limestones and cherts toward heavier values? The alternative is that cherts and carbonates have continually re-equilibrated during diagenesis and that the $\partial^{18}O$ composition of these cherts and limestones represent diagenetic conditions rather than the temperature and isotopic composition of the seawater from which they originated (Anderson and Arthur, 1983).

The uncertainties surrounding the causes of the temporal variation of oxygen isotopic composition of limestones certainly makes the interpretation of stable isotopic data more difficult. It is obvious from our discussions above, and with a glance at Figure 4.12, that a calcite cement or a limestone might obtain the same stable isotopic signature under strikingly different diagenetic environments, depending on the temperature of the pore fluid and the rock: water ratio at the time of precipitation. These and other inherent complexities of stable isotopic geochemistry in natural systems call for careful use of the technique in carbonate diagenetic studies. The problem must be constrained within a solid geologic and petrographic framework. In addition, complimentary supporting technology, such as trace elements, radiogenic isotopes, and fluid inclusions must be used (Moore, 1985). Finally, it should be pointed out, that when dealing with limestones or dolomites that have undergone complex diagenetic histories, each fabric component of the rock (grains, and each diagenetic phase) should be analyzed for stable isotopes separately to avoid whole rock average values that can be meaningless, or misleading.

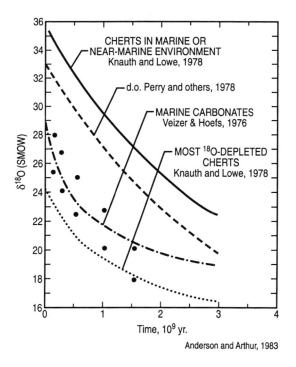

Fig. 4.15. Oxygen isotopic composition of marine carbonates and cherts plotted against time. While the marine carbonates plot near the most depleted $\partial^{18}O$ chert curve, the shapes of the curves and the range of change are similar. Used with permission of SEPM.

Strontium isotopes

The '80s saw the development of consid-

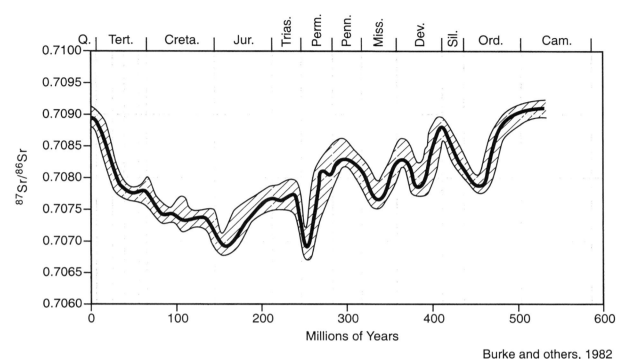

Fig. 4.16. $^{87}Sr/^{86}Sr$ ratios of marine carbonates plotted against time. The ruled area represents the data envelope. Originally published in Geology, v.10, pp 516-519. Used with permission.

erable interest in the utilization of strontium isotopes in problems of carbonate diagenesis (Stueber et al., 1984; Moore, 1985; Banner et al., 1988). There are four stable isotopes of strontium: ^{88}Sr, ^{87}Sr, ^{86}Sr, and ^{84}Sr. ^{87}Sr is the daughter product of the radioactive decay of ^{87}Rb, and its abundance has therefore increased throughout geologic time (Burke et al., 1982; Stueber et al., 1984). Since carbonate minerals generally contain negligible rubidium, once ^{87}Sr is incorporated into a carbonate, the relative abundance of ^{87}Sr in the carbonate will remain constant. It should be noted, however, that non-carbonate phases such as clays can have high Rb contents, and therefore small amounts of such contaminants in an ancient carbonate can significantly affect measured ^{87}Sr contents. It is therefore critical for Sr isotopic studies of carbonates to obtain pure samples or to evaluate the effects of impurities that may be present (Banner, 1995). The relative abundance of ^{87}Sr is generally expressed as the ratio $^{87}Sr/^{86}Sr$. Finally, isotopic fractionation during natural processes, such as encountered in carbon and oxygen isotope systems in diagenetic studies, is negligible for strontium isotopes (Banner, 1995).

Sr content in the oceans is constant (8ppm) and the concentration of ^{87}Rb is very low. Since the residence time of Sr is much shorter than the half-life of ^{87}Rb the amount of ^{87}Sr produced in the oceans is negligible (Dickson, 1990). However, it is well known that the $^{87}Sr/^{86}Sr$ of seawater has varied through time (Fig. 4.16)(Burke et al., 1982). The $^{87}Sr/^{86}Sr$ signature in seawater is derived from the mixing of two end member sources, which have distinctly different $^{87}Sr/^{86}Sr$. These two sources are the

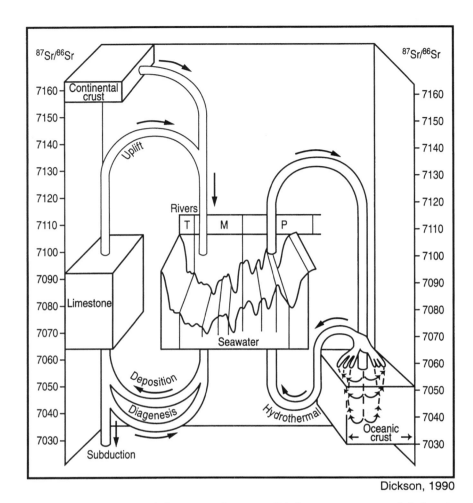

Fig. 4.17. The Phanerozoic marine strontium cycle. Sea water curve from Burke et al. (1982). See Fig. 4.16, this book for an enlarged view of the sea water curve. P=Paleozoic, M=Mesozoic and T=Tertiary. "Pipes" filled with aqueous fluid except subduction limb which removes carbonates from the immediate cycle. Used with permission of Blackwell Science.

continental crust and the oceanic crust. Rb has been preferentially sequestered in the continental crust leading to very high $^{87}Sr/^{86}Sr$ relative to the oceanic crust (Fig. 4.17). Continental weathering and oceanic hydrothermal processes deliver $^{87}Sr/^{86}Sr$ to seawater. Global tectonic activity controls the relative proportions of the contribution from each source, hence the gross variation of $^{87}Sr/^{86}Sr$ in seawater through time. Seawater $^{87}Sr/^{86}Sr$ is buffered by contributions from ancient limestones which track mean seawater isotopic compositions (Fig. 4.17)(Dickson, 1990).

The strontium present in marine limestones and dolomites, then, should have an isotopic composition that depends on the age of the rocks, reflecting the temporal variations in oceanic $^{87}Sr/^{86}Sr$, rather than representing the in situ decay of ^{87}Rb. The latest charts of $^{87}Sr/^{86}Sr$ versus time (for example, Fig .4.16) are detailed enough to allow the use of strontium isotopes in high-resolution stratigraphy (Saller and Koepnick, 1990). Additionally, strontium isotopes may be employed as an important tool in solving a variety of diagenetic problems.

If the $^{87}Sr/^{86}Sr$ of a marine carbonate sample differ from the value estimated for seawater at the time of deposition of the sample, then this difference can be attributed to diagenetic alteration by younger seawater or to diagenesis by non-marine fluids. $^{87}Sr/^{86}Sr$ can be used to date diagenetic events such as dolomitization, if the dolomitization event involved marine waters and no subsequent recrystallization took place. Saller (1984), using $^{87}Sr/^{86}Sr$, determined that marine dolomites in Eocene sequences beneath Enewetak Atoll started forming in the Miocene, and are probably still forming today. Petrography, trace elements, and the stable isotopes of oxygen and carbon were used to determine the marine origin of the dolomites.

During early meteoric diagenesis, the $^{87}Sr/^{86}Sr$ of the calcites precipitated as the result of the stabilization of aragonite and magnesian calcite will generally be inherited from the dissolving carbonate phases. This rock-buffering is the result of the enormous difference between the high Sr content of marine carbonates relative to the low levels of Sr normally seen in fresh meteoric waters (Moore, 1985; Banner et al., 1989). Sr derived from marine carbonates, then, generally swamps any contribution of Sr from meteoric waters that might exhibit high $^{87}Sr/^{86}Sr$ gained by interaction with terrestrial clays and feldspars. The resulting calcite cements or dolomites will normally exhibit a marine $^{87}Sr/^{86}Sr$ signature compatible with the geologic age of the marine carbonates involved in the diagenetic event. Any enrichment of $^{87}Sr/^{86}Sr$ above the marine value in such a meteoric system would indicate that an enormous volume of meteoric water had to have interacted with the rock to overcome the Sr-buffering of the marine limestones (Moore et al., 1988). This type of diagenetic modeling, using $^{87}Sr/^{86}Sr$ with the support of other stable isotopes and trace elements has been successfully applied to a variety of diagenetic problems (Banner et al., 1989; Kaufman et al., 1991; Cander, 1995).

Oil field brines and other deep subsurface waters are often highly enriched with Sr relative to that of sea water, presumably because of extensive interaction with the rock pile through geologic time (Collins, 1975). The Sr in these waters is generally enriched with ^{87}Sr, indicating interaction with siliciclastics sometime during their evolutionary journey (Stueber et al., 1984). Because of these considerations, the $^{87}Sr/^{86}Sr$ of calcite cements and dolomites formed in the subsurface can give some insight into the timing of the cementation or dolomitization event. If post-compaction cements or dolomites associated with brines with elevated $^{87}Sr/^{86}Sr$ have ratios near the seawater values expected for the enclosing limestones, it can be inferred that the cements or dolomites received most of their Sr from the enclosing limestones, rather than the brines. This indicates that the diagenetic event took place prior to arrival of the waters containing Sr enriched in ^{87}Sr. Phases enriched in ^{87}Sr would presumably have been transformed, or precipitated, in the presence of the present radiogenic brine, or a similar pore fluid in the past, and obviously would represent a later event.

Figure 4.18 illustrates the strontium isotopic composition of rock components, diagenetic phases, and oil field brines from the Lower Cretaceous Edwards Formation, U.S. Gulf Coast. Lower Cretaceous seawater had a $^{87}Sr/^{86}Sr$ of 0.7073. Edwards oil field brines have a strontium isotopic composition of 0.7092, presumably a result of interaction with associated rubidium-bearing shales and sandstones resulting in elevated isotopic ratios. Matrix limestone from the Edwards has a $^{87}Sr/^{86}Sr$ near that of

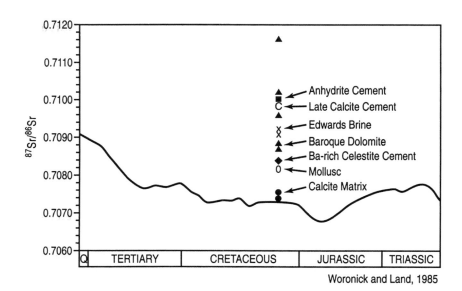

Fig. 4.18. Strontium isotopic composition of limestone constituents and subsurface brines from Lower Cretaceous subsurface sequences in south Texas plotted against the Burke et al. (1982) curve. All components except the calcite matrix plots above the Lower Cretaceous marine curve suggesting involvement of basinal waters that had seen siliciclastics before arriving and interacting with these limestones. Used with permission of SEPM.

Lower Cretaceous seawater. Post-compaction calcite cements, baroque dolomites, and other diagenetic phases, however, have elevated ratios indicating that the composition of these phases was influenced by present Edwards brines (Woronick and Land, 1985).

In summary, strontium isotopic studies, while not definitive when used alone, can provide important constraints on the origin and timing of porosity-modifying diagenetic events as well as the evolution of associated waters. Rapid advances in thermal ionization mass spectrometry during the '90s have, however, allowed enhanced Sr isotope stratigraphy. The low concentration levels of Sr found in most ancient calcites and dolomites, combined with the necessity of separating diagenetic phases for analysis gives further impetus for the continued advance of this important technology (Banner, 1995).

Fluid Inclusions

Carbonate cements often entrap inclusions of the precipitating fluid at lattice defects or irregularities. These are termed primary fluid inclusions. If the inclusions were trapped as a single phase (e.g., one liquid phase) at relatively low temperatures (< 50°C), the inclusions would remain as single-phase fluid inclusions. If, however, the inclusions were trapped as a single phase under elevated temperatures (> 50°C), they would separate into two phases, a liquid and vapor as the inclusion cools, because the trapped liquid contracts more than the enclosing crystal and a vapor bubble nucleates to fill the void. This is a consequence of the different coefficient of thermal expansion for liquids versus solids (Fig. 4.19) (Roedder, 1979; Klosterman, 1981). Secondary fluid inclusions can also form as a result of micro-fractures or crystal dislocations affecting the crystal after it has finished growing (Roedder, 1979; Goldstein and Reynolds, 1994).

The minimum temperature of formation of the crystal containing a two-phase fluid inclusion can be estimated by heating up the sample until the vapor bubble disappears. This is called the homogenization temperature. The following basic constraints must be met before

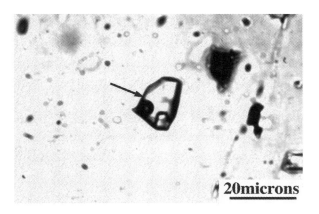

Fig.4.19.Two-phase fluid inclusion (arrow) in poikilotopic post-compaction calcite cement, Jurassic Smackover at 10,170' (3100 m). Walker Creek Field. Columbia Co. Arkansas. USA. Plain light.

two-phase fluid inclusion homogenization temperatures can be properly used to estimate temperature of formation of the enclosing crystal (Roedder and Bodnar, 1980; Goldstein and Reynolds, 1994): 1) the inclusion must have been trapped initially as a single phase; 2) only primary inclusions can be utilized; 3) the volume of the cavity cannot change significantly by dissolution or precipitation during burial; 4) the inclusion cannot stretch or leak during burial history, and hence change the volume of the cavity; 5) the fluid in the inclusion must consist of water, or be a water-salt system; and 6) the burial pressure on the inclusion at the time of entrapment must be known in order to allow pressure correction of the homogenization temperature.

Other information pertinent to the nature of the fluids responsible for cement precipitation can be developed through the freezing behavior of the fluids within the inclusion. Dissolved salts depress the freezing point of fluids. The salinity of the fluid within an inclusion, and, by extension, the salinity of the precipitating fluid, can be estimated by its freezing temperature (Klosterman, 1981; Goldstein and Reynolds, 1994). The gross chemical composition of the fluid within an inclusion can also be estimated by freezing behavior, as well as by a number of non-destructive analytical techniques such as Laser Raman Microprobe, Nuclear Magnetic Resonance, Proton Microprobe and others (Goldstein and Reynolds, 1994).

At first blush, the utilization of fluid inclusions in diagenetic studies seemed to hold great promise, because the inclusion may be a direct window into the characteristics of the diagenetic fluid responsible for the enclosing crystal. During the early '80s a number of diagenetic studies utilized inclusion data to help constrain conditions of formation of subsurface calcite cements and dolomites (Klosterman, 1981; Moore and Druckman, 1981; Moore, 1985; O'Hearn, 1985). During that period, however, there arose a number of troubling questions concerning the validity of the results of inclusion studies of carbonate minerals such as calcite and dolomite. For instance, Moore and Druckman (1981) reported seemingly primary two-phase fluid inclusions present in early, near-surface, meteoric cements from the Jurassic Smackover. In addition, many reported homogenization temperatures seemed consistently too high relative to the geological setting of the enclosing rocks (Klosterman, 1981; O'Hearn, 1984, Moore, 1985). In 1987, Prezbindowski and Larese (1987) presented compelling experimental evidence that fluid inclusions in calcite may well stretch during progressive burial, changing the volume of the inclusion and hence significantly increasing the apparent temperature of formation of the inclusion. These results seem to explain the consistently high homogenization temperatures observed by earlier workers, as well as the presence of two-phase fluid inclusions in cements precipitated at the surface under low temperature conditions, but buried later and exposed to higher burial temperatures (Moore and Druckman, 1981).

In the mid 90's Goldstein and Reynolds (1994) presented an awesome compendium on the systematics of fluid inclusions in diagenetic minerals including their utility in diagenetic studies of carbonates. In this primer, they outlined, clearly, all major constraints on the system; they presented an exhaustive and useful petrographic treatment for the recognition of primary inclusions and detailed a number of case histories.

Two-phase fluid inclusion data continues to be generated and used today in carbonate diagenesis studies (Balog et al., 1999). It seems to this writer, however, that many of us are using this tool in a rather casual way without truly coming to grips with the complexities of the system as outlined so well by Goldstein and Reynolds (1994). Generated carefully, by a well-trained analyst, two-phase fluid inclusion data can give the carbonate geologist valuable information concerning the temperature regime and nature of diagenetic fluids. Two phase fluid inclusion data generated by poorly trained individuals unaware of the rigorous constraints on the system, however, is worse than useless, since the data can be misleading and its generation an enormous time sink.

SUMMARY

There are three major diagenetic environments in which porosity modification events such as solution, compaction, and cementation are active: the marine, meteoric, and subsurface environments.

The meteoric environment with its dilute waters, easy access to CO_2, and wide range of saturation states relative to stable carbonate phases, has a high potential for porosity modification, including both destruction by cementation as well as the generation of secondary porosity by dissolution.

Modern shallow marine environments are particularly susceptible to porosity destruction by cementation because of the typically high levels of supersaturation of marine waters relative to metastable carbonate mineral phases. In the geologic past, however, shallow marine waters might well have been undersaturated with respect to aragonite, and dissolution of aragonite grains in conjunction with calcite precipitation, might well have taken place. Changes of carbonate saturation with depth may lead to the precipitation of stable carbonates as cements, as well as the development of secondary porosity by dissolution of aragonite in oceanic basins and along their margins.

The subsurface environment is generally marked by the destruction of porosity by compaction and related cementation. Thermal degradation of hydrocarbons and the slow flux of basinal fluids during progressive burial drive later porosity modification by dissolution and cementation.

The key to the development of viable porosity modification models is the ability to recognize the diagenetic environment in which the porosity modification event occurred. There are a number of tools available to the carbonate geologist today. If used intelligently, they can provide the information necessary to place a diagenetic event in the appropriate environment.

Petrography and geologic setting provide a cheap, readily available, relative framework for porosity modification. Trace element and stable isotopic geochemistry of cements and dolomites provide insight into the types of waters involved in the diagenetic event, and hence the event's link to a specific diagenetic environment. Two-phase fluid inclusions can be used to assess the temperature of cement or dolomite formation and the composition of the precipitating or dolomitizing fluid, and hence may provide a tie to a specific diagenetic environment, such as the subsurface.

Geochemically based techniques, however, are relatively expensive, time-consuming, and constrained by significant problems, such as uncertainties in distribution coefficients, temperature fractionation effects, and low concentration values. Two-phase fluid inclusion studies are complex, extremely time-consuming and exhibit significant problems, such as stretching of inclusions during burial, the recognition of primary inclusions, and the necessity for pressure corrections.

None of these tools should be utilized in isolation. Rather, they should be used as complimentary techniques that can provide a set of constraints that allows the carbonate geologist to assess the environment in which a porosity-modifying diagenetic event took place, providing they are used within a valid petrographic and geologic framework.

In the following chapters we shall use many of these tools in our consideration of porosity evolution in each major diagenetic environment.

Chapter 5

NORMAL MARINE DIAGENETIC ENVIRONMENTS

INTRODUCTION

The marine diagenetic realm is an environment of variable diagenetic potential even though the gross chemical composition of marine waters is relatively constant at any one instant in geologic time. This variability is fueled by four factors: 1) Variations in kinetic energy of the surface waters commonly occur at the sites of carbonate deposition. These variations in energy influence both the flux of diagenetic fluids through the sediments and the rate of CO_2 outgassing. Both processes are major controls over abiotic carbonate precipitation. 2) Water temperature and pressure change as one moves from relatively warm surface waters down into cold, deep oceanic waters along shelf margins. These changes impact the saturation state of common carbonate minerals and hence their diagenetic potential. 3) The intervention of marine plants and animals in the CO_2 cycle through respiration and photosynthesis dramatically impacts carbonate diagenetic potential both in terms of precipitation and dissolution of carbonate mineral species. 4) Microbial life processes such as sulfate reduction and chemosynthesis fueled by methane may have a huge impact on diagenetic processes in shallow, as well as deep, marine environments.

Today, normal marine surface waters between 30° north and south latitude are supersaturated with respect to aragonite, magnesian calcite, calcite and dolomite (James and Choquette, 1990; Morse, 1990). Abiotic carbonate precipitation, therefore, is thought to be one of the major operational porosity modification processes across tropical, equatorial, shallow marine shelves. Under these conditions, variations in rate of water flux through the sediments control the gross distribution of major cementation events. This rather clean distributional pattern is complicated by the potential of significant microbial mediated carbonate precipitation.

The saturation state of surface marine waters relative to $CaCO_3$ changes dramatically with increasing depth as a function of changes in partial pressure of CO_2, progressive temperature decrease, and progressive pressure increase with depth(James and Choquette, 1990; Morse, 1990). While pressure is basically a straight-line increase with depth in the marine water column, temperature decreases rapidly in relatively shallow waters across the thermocline, immediately below a well-oxygenated zone of relatively constant temperature termed the mixed zone (Fig. 5.1A). Aragonite becomes undersaturated in the ther-

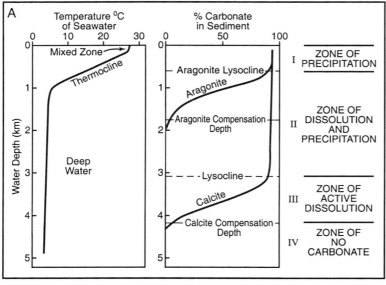

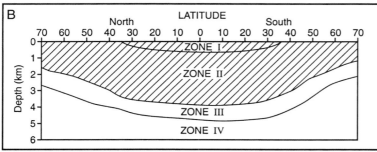

Fig. 5.1A. General relationships of temperature and carbonate mineral saturation as a function of depth in oceanic water. B. Variations in the different zones of sea-floor diagenesisis in the modern ocean. Used with permission of the Geological Association of Canada.

mocline, while the waters remain saturated with respect to calcite and dolomite. Within this zone (zone II, Fig. 5.1) on the sea floor, or as is most likely along a platform margin, aragonite sediment and rock components dissolve and calcite precipitates as void fill cement (Saller, 1984b). Below approximately 3 km (depth varies between ocean basins), the water becomes undersaturated with respect to calcite but remains supersaturated with respect to dolomite (zone III, Fig. 5.1). In this zone, calcite rock and sediment components dissolve, often to be replaced by dolomite (Saller, 1984b). Little carbonate, other than dolomite, remains below the calcite compensation depth (zone IV, Fig. 5.1).

In temperate regions north and south of latitude 30°, surface water temperatures cool, and carbonate sediments are dominated by the Foramoral assemblage of temperate and polar seas. Under these conditions, carbonate production decreases dramatically (James and Choquette, 1990; James and Clark, 1997). In these temperate-polar seas, carbonate saturation states change in response to the lower water temperatures. As a result, zone II, the zone of solution of aragonite and precipitation of calcite (Fig. 5.1A and 5.1B), will generally bathe shallow shelves resulting in lower diagenetic potential for the calcite-dominated sediments which are accumulating in this system.

The following discussion will cover the detailed diagenetic potential of two normal marine sub-environments: the shallow surface waters of the equatorial and temperate regions, and the deep marine environment of the shelf margin. This discussion will: 1) emphasize the nature and distribution of porosity modification processes active in the environment and assess their effectiveness; 2) develop criteria for the recognition of similar diagenetic environments in ancient carbonate rock sequences; and 3) present several case histories of marine diagenesis where cementation has played a major role in the porosity evolution of economically important ancient carbonate rock sequences.

SHALLOW WATER, NORMAL MARINE DIAGENETIC ENVIRONMENTS

Abiotic shallow marine carbonate cementation

Abiotic marine cementation has long been thought to be the major diagenetic process actively modifying porosity under shallow, normal marine conditions (Moore, 1989; James and Choquette, 1990; Tucker and Wright, 1990). The cements precipitated are the metastable phases, magnesian calcite and aragonite. Magnesian calcite cements are generally most prevalent in reef-related environments (Land

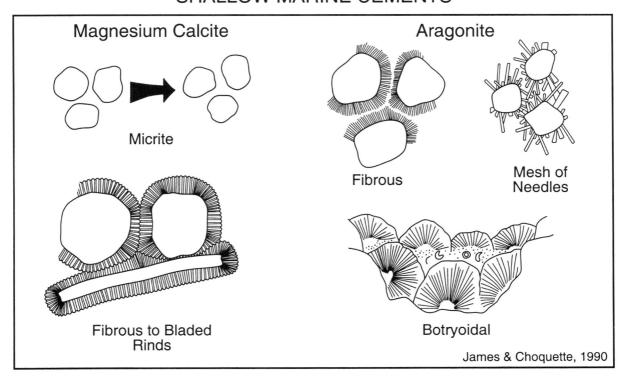

Fig. 5.2. *Common morphology of modern marine carbonate cements. Used with permission of the Geological Association of Canada.*

and Moore, 1980; Schroeder and Purser, 1986), while aragonite cements are most common in intertidal beach rock and hardground environments (Moore, 1977).

Fig. 5.3. Characteristic marine cements.(A) Aragonite needle cement, modern beach rock St. Croix, U.S. Virgin Islands. Crossed polars. (B) SEM view of (A). (C) Botroyoidal aragonite cement deep fore reef, Belize. Courtesy N.P. James. (D) Bladed magnesian calcite cement (12 mole% $MgCO_3$), modern beach rock St.Croix, U.S. Virgin Islands. Plain light. (E) SEM view of D. (F) Fibrous marine magnesian calcite cement, reef-derived debris flow dredged from 900 m (2952') on the island slope of Grand Cayman, British West Indies. Note polygonal suture patterns in cement filled pores (arrows). Plain light.

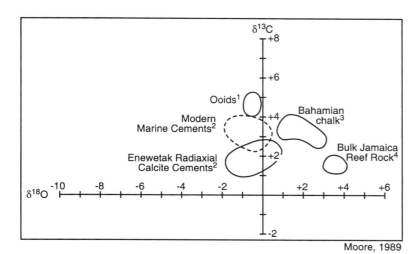

Fig. 5.4. Oxygen, carbon isotope composition of various Holocene abiotic marine cements and carbonates. Data sources as follows: 1 and 3, James and Choquette, 1990; 2. Saller, 1984; 4. Land and Goreau, 1970. Used with permission of Elsevier.

In general terms, most modern shallow marine abiotic cements precipitate as fibrous-to-bladed circumgranular crusts of either aragonite or magnesian calcite mineralogy (Fig. 5.2, discussion, Chapter 4). Magnesian calcite can occur as either pelleted micrite crusts, or as fibrous-to-bladed rinds around bioclasts or coating the surfaces of pores or cavities (Fig. 5.2). Aragonite cements can occur as fibrous crusts, a mesh of rather coarse needle crystals or large massive botryoids occluding larger pores and voids in reef-related sequences (Fig. 5.2 and 5.3). If the pore space is completely occluded by fibrous or bladed marine carbonate cement, a polygonal suture pattern that seems to be characteristic of shallow marine cements is developed (Shinn, 1969)(Fig. 5.3F). Figure 5.3 shows light and SEM photomicrographs of characteristic abiotic marine cements.

The stable isotopic composition of modern marine aragonite and magnesian calcite cements as compared to modern ooids, chalks, and bulk reef rock, is shown on Figure 5.4. Modern marine cements generally show high values of $\partial^{18}O$, which are in the same compositional range as modern ooids and radiaxial calcite cements from Enewetak, but have considerably lower values than bulk Jamaica reef rock. The $\partial^{13}C$ values of modern marine cements are in the same general positive range as other marine precipitates and bulk reef rock from Jamaica. These marine cements are enriched with respect to Mg, Sr, and Na, and depleted relative to Fe and Mn, reflecting the gross stable isotopic and trace element composition of marine waters (Table 4.2). Modern, shallow marine abiotic carbonate cements are generally nonluminescent because marine cementation environments commonly exhibit high-energy, oxidizing conditions that preclude the incorporation of Fe and Mn in the carbonate lattice (Chapter 4).

Because most surface waters are supersaturated with respect to $CaCO_3$, abiotic cementation events are linked directly to the flux of these supersaturated fluids through the sediments. It would require 10,000 pore volumes of saturated marine water to completely fill a pore with carbonate cement, assuming that all available $CaCO_3$ was precipitated from the fluid. With a 10% precipitation efficiency, not unrealistic for a natural system, over 100,000 pore volumes would be required (Scholle and Halley, 1985). Cementation, then, will be fa-

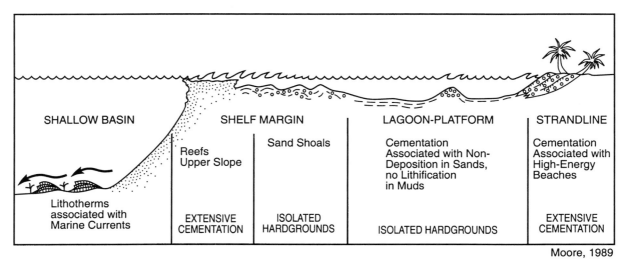

Fig. 5.5. Distribution of abiotic marine cementation associated with shallow marine depositional environments. Used with permission of Elsevier.

vored under the following conditions: 1) in a high-energy setting where sediments are both porous and permeable; 2) where sediment water flux is maximized and CO_2 degassing can operate; 3) in areas of low or restricted sedimentation rate, or restricted sediment movement, so that the sediment-water interface is exposed for an appropriate length of time to assure adequate pore-water exchange; 4) in locales of high organic activity where framework organisms can stabilize substrates and erect large pore systems, and organic activity can impact the partial pressure of CO_2; 5) and finally, in regions where surface waters exhibit higher than normal saturation values for $CaCO_3$.

The kinetic factors necessary for abiotic marine cementation obviously limit the volume and distribution of cementation under shallow normal marine conditions. Figure 5.5 outlines the distribution of abiotic cementation that has been observed in modern marine environments. Significant cementation that seriously impacts porosity is concentrated at the shelf margin in conjunction with reef development and the upper-most shelf slope. Grammer et al. (1999) report rapid aragonite cementation at water depths to 60m along the eastern shelf margin of Lee Stocking Island. While relatively deep, these sites are still above the aragonite lysocline.

The beach high-energy intertidal zone and hardgrounds associated with areas of nondeposition in ooid shoals are other, less important environments of shallow marine abiotic cementation. Each of these cementation environments will be discussed in subsequent sections of this chapter.

Recognition of ancient shallow marine abiotic cements

The major problem in recognizing ancient marine cements rests in the fact that all modern marine cements are metastable phases that will ultimately stabilize to calcite and dolomite through some process of dissolution and re-precipitation. How much of the unique tex-

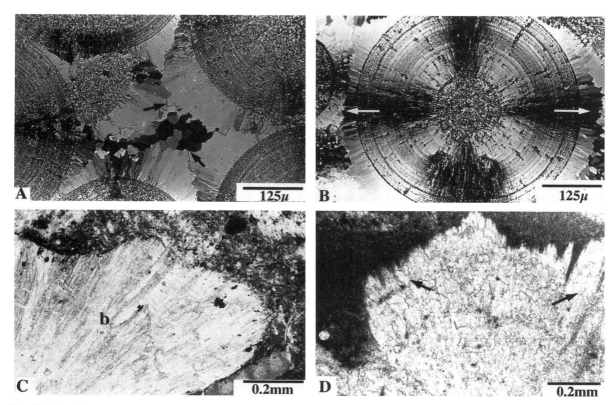

Fig. 5.6. Ancient marine cements. (A) Jurassic fibrous marine calcite cement inferred to be originally magnesian calcite (Moore et al., 1986). Note the polygonal suture pattern where these cements come together in the center of the pore (arrows). Jurassic Smackover Formation, subsurface, south Arkansas, U.S.A. Crossed polars. (B) Jurassic fibrous marine calcite cement precipitated in optical continuity with a radial ooid. It is thought that both the cement and ooid were originally magnesian calcite. Same sample as (A). Crossed polars. Photos courtesy Tony Dickson. (C) Preserved Pennsylvanian botryoidal aragonite (b), Sacramento Mountains, New Mexico. Plain light. (D) Lower Permian (Labrocita Formation) botryoidal aragonite (?) that has been recrystallized to calcite. Note the fine crystalline calcite mosaic (arrows) through which the individual botryoidal rays are seen as ghosts. Compare to preserved aragonite in (C). Plain light.

Table 5.1[1]

1. *Still aragonite.* Encountered in quite old rocks.
2. *Mosaic of generally irregular calcite spar containing oriented aragonite relics.* Crystal boundaries commonly cross-cut original structure defined by organic or other inclusions. Replacement crystals normally $10-10^3$ times larger than replaced aragonite crystals.
3. *Calcite mosaic as for Criterion 2, but without aragonite relics.* Original aragonite mineralogy supported by elevated Sr^{2+} content, relative to levels reasonably expected in dLMC resulting from alteration of originally calcitic constituents.
4. *Calcite mosaic as for Criterion 3, but Sr^{2+} not elevated, or not measured.* Differing Sr^{2+} content in the resulting calcites can reflect differing degrees of openness of the diagenetic system.
5. *Mold or subsequently calcite-filled mold.* Common for aragonitic constituents, but equivocal because observed for non-aragonitic constituents in undolomitized limestone.

[1]Arranged in order of decreasing reliability, from Sandberg (1983)

Table 5.1. Criteria for the recognition of ancient abiotic aragonites. (Used with permission from Nature, Vol. 305, pp. 19-22. Copyright (C) 1983, Macmillan Journals Limited). Arranged in order of decreasing reliability.

tural and geochemical information carried by the original marine cement is lost during this transformation to its more stable phase? Magnesian calcite generally transforms to low magnesian calcite with little loss of textural detail (Moore, 1989). Ancient shallow inferred marine magnesian calcite cements, then, should exhibit the characteristic bladed-to-fibrous textures enjoyed by most of their modern counterparts (Compare Figs. 5.3D to 5.6A and B). During this transformation, however, some geochemical information is invariably lost. If stabilization takes place in an open system in contact with meteoric waters, much of the original trace element and isotopic signal contained in the original marine cement may be lost. If, however, stabilization is accomplished in a closed system in contact with marine waters, much of the original marine geochemical information may be retained. Therefore, inferred magnesian calcite cements should be expected to exhibit relatively heavy stable isotopic compositions and to be enriched relative to Sr and Na (Moore, 1989). Lohmann and Meyers (1977) documented the correlation of the presence of microdolomite to original magnesian calcite mineralogy in Paleozoic crinoid plates and associated fibrous cements.

Aragonite cement, however, shows significantly more textural loss than does magnesian calcite during stabilization, and indeed in some cases where fresh waters are involved in stabilization, all textural information may be destroyed by dissolution before calcite is reprecipitated in the resultant void (Moore, 1989). In the case of total dissolution, the presence of precursor aragonite marine cement is impossible to determine. When stabilization takes place in marine pore fluids or in waters nearly saturated with respect to $CaCO_3$, however, ghosts of the original aragonite cement texture and some of the original geochemical information may be retained within the replacive calcite mosaic.

Sandberg (1983) developed a set of criteria useful in determining original aragonite mineralogy of grains and/or cements (Table 5.1). One of the cornerstones of this set of criteria is the common occurrence of original aragonite as relicts in enclosing coarse calcite mosaics. These calcite-after-aragonite marine cements exhibit elevated Sr composition and often show high $\partial^{18}O$ values relative to meteoric derived cements (James and (Moore et al., 1986; Heydari and Moore, 1994). Massive botryoids of mosaic calcite associated with relict aragonite and high Sr compositions are believed to be examples of seafloor cements that were once composed of aragonite (Fig. 5.6C and D) (Sandberg, 1985; Kirkland et al., 1998). Similar botryoids occur in Mississippian to Lower Jurassic rocks and have also been reported in the Lower Cambrian and the Middle Precambrian (Sandberg, 1985).

As noted in Chapter 2, the nature of abiotic marine precipitation, such as ooids and marine cements, has changed in a cyclical fashion through the Phanerozoic. These cycles apparently reflect oscillations in P_{CO_2} driven by global tectonics and result in relatively long periods of time during which abiotic precipitation is dominated by calcite, followed by periods such as are found in today's ocean, where aragonite is the dominate marine precipitate (Fig. 2.15). In the Lower Paleozoic and in the Mesozoic periods, which are believed to have been marked by calcite seas, marine cements should dominantly have been calcite and should exhibit well preserved textural and geochemical information. In the Upper Paleozoic and the

Tertiary, aragonite and, perhaps, magnesian calcite marine cements should dominate; and recognition of these metastable precursors may be more difficult.

Finally, as indicated in Chapter 4, there is also some evidence that the stable isotopic composition of marine waters has systematically changed through time, as reflected in the plot of $\partial^{18}O$ and $\partial^{13}C$ compositions of marine cements from lower Cambrian to Holocene (Fig. 2.15). These changes parallel the results of Veizer and Hoefs (Veizer and Hoefs, 1976)(Fig. 4.15) and must be taken into account when utilizing stable isotopic compositions of cements to help determine precipitational environment.

Biologically mediated marine carbonate cementation and diagenesis

One of the most compelling and controversial topics in carbonate diagenesis today is the active as well as passive role of microorganisms, such as algae, cyanobacteria, fungi, and bacteria in sea floor cementation. We have long been aware of the relationship of cyanobacteria and algae and the marine water lithification of algal mats and stromatolites in the intertidal and near subtidal environments (Dill et al., 1989; Monty, 1995).

In the 90's, there was a veritable explosion of interest in the formation of marine micrite and marine micrite cements in association with microbial biofilms (Riding, 1991; Lees and Miller, 1995; Monty, 1995). Microbial biofilms are microbial cells embedded in an organic matrix of microbial origin, essentially a mucilaginous microbial ooze (Wilderer and Characklis, 1989). Folk (1993) discovered tiny particles that he has described as nannobacteria in carbonate rocks and sediments. These observations, coupled with the observations, experiments, and discussions of Chafetz (1986; 1992) (Buczynski and Chafetz, 1993), have certainly increased the carbonate community's interest in the role that bacteria may play in the precipitation of carbonate cements in marine environments.

The actual mechanism of carbonate precipitation associated with microbial microfilms is still debated. Photosynthesis by cyanobacteria, nitrate reduction, sulfate reduction, ammonification, bacterial methane oxidation, and passive nucleation on bacterial cells are all possibilities (Monty, 1995). As we will see in the following discussions, there are certain instances, such as carbonate deposits associated with deep water methane seeps, where stable isotopic data can be more definitive in outlining the role microbes may play in precipitation and sea floor diagenesis. However, in many cases geochemical data gathered on cements and micrites only suggest precipitation under normal marine conditions.

Many authors have used the presence of bacteria-like structures entombed within peloids and pelleted micrites as direct evidence that bacteria induce carbonate precipitation (Chafetz and Buczynski, 1992; Folk, 1993; Monty, 1995; Kirkland et al., 1998). The presence of these structures alone does not directly tie bacterial living processes to carbonate precipitation. Land and Goreau's 1970 and Land and Moore's 1980 stable isotopic work on pelleted high magnesium calcite crusts within a Jamaica reef framework, however, suggests biologic mediation during sub sea lithification of the crusts. In addition, Land and Goreau (1970) concluded that the fatty acids present in these pelloidal crusts have identical spectra to fatty acids thought to come from bacteria.

 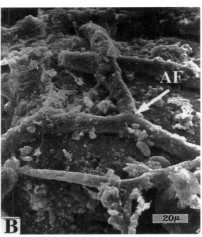

Fig. 5.7. SEM photomicrographs of algal cements in beach rock, St. Croix, U.S. Virgin Islands. (A) AF is an algal filament, G is a grain. (B) Close view of (A) showing algal filament AF.

Finally, Land and Goreau (1970) noted that these pelloidal crusts often defied gravity and were ubiquitously penetrated by boring organisms such as Clionid sponges strongly suggesting lithification under marine conditions (Fig. 5.11B and C).

These petrographic relationships have been described from a number of modern and ancient reefs (Land and Moore, 1980; Kirkland et al., 1998) (Fig. 5.11) and when combined with Land and Goreau's geochemical evidence certainly seem to implicate bacteria and/or biofilms in some marine cementation processes.

In the following sections, we will assess the importance of possible microbial involvement in each of the marine diagenetic environments discussed.

Diagenetic setting in the intertidal zone

The beach setting is ideal for the precipitation of abiotic marine cements. High-energy conditions, including wave and tidal activity, as well as the presence of relatively coarse, highly porous, and permeable sediments, ensure that adequate volumes of supersaturated marine water are able to move through the pore system to accomplish the cementation. While beach rock cementation often occurs in zones of mixing between meteoric and marine waters (Moore, 1973, 1977), it is believed that CO_2 degassing, rather than the mixing phenomenon itself, is the prime cause for precipitation (Hanor, 1978). Tidal pumping and wave activity provide the most logical mechanisms for the CO_2 outgassing in the beach shoreface zone.

Some beach rock cementation seems to be associated with biological activity in the beach shoreface. The resultant cements are pelleted micrite masses that coat the grains and sometimes almost fill the pore (Fig. 5.7). Petrographic and SEM observations reveal that the micrites are generally endolithic algal filaments binding the grains together (Fig. 5.7A and B). The actual cement is fine crystalline magnesian calcite coating each filament as well as filling the holes in the grains made by the filaments (Fig. 5.7B). The cementation seems to be the result of the living processes of the algae themselves although stable isotopic data for these samples is not available. The pelleted texture may be the result of the two dimensional cut of the thin section through the anastamos-

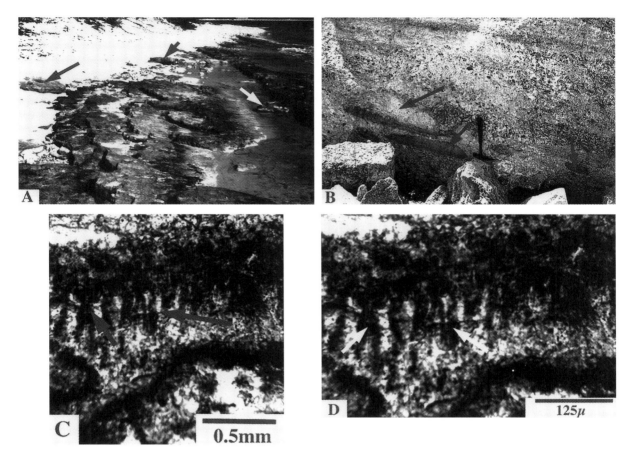

Fig. 5.8. (A) Beach rock, Grand Cayman, British West Indies. Note the blocks (arrows) that have been torn up and are being reincorporated into the beach. (B) Beach rock clasts incorporated in a Lower Cretaceous beach sequence (arrows), northcentral Texas, USA. (C) Microstalactitic cement (arrow) occurring in beach rock clasts from outcrop shown in (B). (D) Close up of (C) showing the fibrous ghosts (arrows) occurring in microstalactitic calcite mosaic suggesting that the beach rock cement was aragonite. Plain light.

ing mass of algal filaments or be related in some manner to a biofilm as described by Kirkland et al. (1998).

Active beach progradation will generally not allow beach rock formation because the sediment-water interface is not exposed long enough for significant cementation to occur. In carbonate environments, however, coarse beach shoreface sediments are often out of equilibrium with the day-to-day energy levels occurring at the site. Instead, the beach shoreface sediments often represent storm events. Under these conditions, the beaches do not prograde on a day-to-day basis; nor are the coarse sediments significantly disturbed by the daily activity of low energy waves. Thus, both abiotic and biologically mediated cementation can occur.

Cemented zones are typically thin (< 1 m in the Caribbean, but can be thicker in areas of high tidal range such as the Pacific), and occur in the upper shoreface, with an upper limit controlled by mean high tide (Fig. 5.8A). While most cementation in the beach shoreface takes place in the marine phreatic zone, in areas of relatively high tidal range, marine vadose cements often occur in beach rock (Fig. 5.8C and D). Cementation in the shoreface is often followed by the destruction of the beach rock bands during storms and results in the formation of boulder and breccia units associated with the beach environment (Fig. 5.8A and B).

Beach rock cements are generally fibrous-to-bladed crusts that may completely occlude available pore space (Fig. 5.3A). Today, directly precipitated beach rock cements are both aragonite and magnesian calcite, with aragonite generally dominant. Microbial/algal pelleted microcrystalline cements in beach rock are commonly magnesian calcite (Moore, 1973, 1977).

Beach rock cementation is common in ancient carbonate shoreface sequences (Inden and Moore, 1983) and is relatively easy to recognize in outcrop and core because of its association with suites of sedimentary structures characteristic of shoreface sedimentation. In addition, the presence of large clasts with fibrous-to-bladed cements similar to present shallow marine cements (Fig. 5.8B), and fibrous pendant cements (Fig. 5.8C and D) indicating early cementation in the intertidal zone, are characteristic of early marine shoreface cementation (Inden and Moore, 1983; James and Choquette, 1983). Porosity modification in the beach intertidal zone by cementation is generally limited to the upper part of the upper foreshore (Moore et al., 1972).

Beach rock cementation, then, has a tendency to vertically compartmentalize reservoirs. Perhaps the best example of this type of modification in a reservoir setting can be found in the Upper Jurassic Smackover Formation at Walker Creek Field in southern Arkansas (Brock and Moore, 1981). In this field, the reservoir rock is an ooid grainstone with preserved primary intergranular porosity. The field is developed over subtle salt structures that were active during deposition of the Smackover ooid sands. Well-developed fibrous-to-bladed marine cements are only present in association with early structure and presumably formed as beach rock during ephemeral island development over the crests of the salt structures. The cemented zones are thin (less than a meter), but usually the cement completely occludes the pore space (Fig. 5.6A and B) and helps compartmentalize the reservoir into multiple-producing horizons (Brock and Moore, 1981).

Modern shallow water submarine hardgrounds

Thin, discontinuous, submarine-cemented crusts are common in certain modern carbonate environments such as the Bahama Platform, Shark Bay in western Australia (Fig. 5.9), and along the margins of the Persian Gulf. In the Bahamas, submarine hardgrounds are most common along the margins of the platform, associated with relatively high-energy conditions such as ooid or grapestone sand shoals. The frequency of hardground development seems to decrease into the platform interior away from high-energy bank margins (Dravis, 1979). In the Persian Gulf, hardground development seems to be much more widespread and may even involve siliciclastics. In the Persian Gulf, the Bahamas, and Shark Bay, the dominant cement mineralogy and morphology is fibrous aragonite (Fig. 5.9B), with subordinate magnesian calcite. Modern hardgrounds invari-

 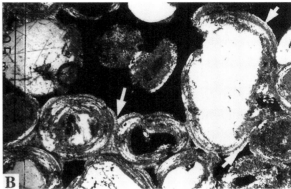

Fig. 5.9. Marine hardground, Shark Bay, western Australia. (A) Hardground slab on ship deck, note 55mm lens cap for scale. (B) Thin section photomicrograph of (A) showing thin crust aragonite cement. Note that some ooids have quartz as nucleus.

ably show cementation gradients decreasing away from the sediment-water interface, and with the exception of the microbial-related magnesian calcite, form directly by precipitation from supersaturated marine waters (Fig. 5.10). It would seem that sedimentation rate and/or substrate stabilization play the major role in hardground formation, with hardgrounds forming where sediments are either not being deposited, or where a substrate has momentarily been stabilized by subtidal algal or bacterial mats (Shinn, 1969; Dravis, 1979). Cessation of sedimentation or substrate stabilization need not involve a long period of time before the development of a hardground by cementation. Both Shinn (1969) and Dravis (1979) observed that hardgrounds could form in the space of months.

The regional distribution of hardgrounds is probably related to surface water saturation relative to $CaCO_3$. The margin of the Bahama Platform is an area of higher $CaCO_3$ saturation caused principally by CO_2 degassing driven by high wave and tidal energy, progressive heating of colder marine water, and photosynthetic activity of shallow-water algae. Hence, hard-

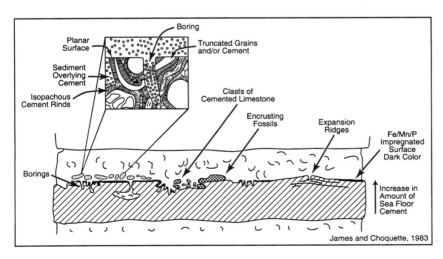

Fig. 5.10. Criteria for the recognition of marine hardgrounds. Used with permission of Elsevier.

ground development would be expected to die out toward the platform interior, as reported by Dravis (1979). Development of hardgrounds in the Persian Gulf is more widespread than in the Bahama Platform. This is apparently because of the very high state of supersaturation of Persian Gulf surface water with $CaCO_3$, resulting from the progressive development of hypersalinity along the margins of the Gulf.

Cool water carbonates, north and south of 35° latitude, generally show little early marine cementation or hardground development (Nicolaides and Wallace, 1997). Occasional hardgrounds occur in areas of very low sedimentation and exhibit minor calcite cementation since cool marine waters are generally undersaturated with respect to aragonite (Figure 5.1B).

Recognition and significance of ancient hardgrounds

Shallow marine hardgrounds are common in the geologic record (James and Choquette, 1990; Durlet and Loreau, 1996). They occur as an integral part of ooid-bearing ancient platform margin sequences (Halley et al., 1983; Durlet and Loreau, 1996), within coarse grainstones in normal marine cratonic sequences (Wilkinson et al., 1982), or as the terminal phase of shallowing-upward sequences (Chapter 2).

Hardgrounds may be recognized in core by the presence of an abraded, sometimes irregular, bored surface. This surface may exhibit the holdfasts of sessile organisms and an encrusting fauna on the surface with intraclasts derived from the hardground incorporated in the sedimentation unit above (Rassmann-McLaurin, 1983; James and Choquette, 1990). The surface may be stained with manganese; and glauconite is often found associated with the hardground, perhaps marking a decrease in sedimentation rate. Petrographically, hardgrounds commonly show typical marine cement textures that may be penetrated by marine endolithic algae and that may be associated with marine internal sediments. Compaction effects that involve adjacent sedimentation units are commonly absent within hardgrounds (Rassmann-McLaurin, 1983) (Fig.5.10).

While the actual volume of marine cement associated with hardgrounds is small and the effect on the total porosity of a sedimentary sequence is minimal, hardgrounds, like beach rock, can act as a reservoir seal, effectively compartmentalizing a reservoir and slowing or stopping vertical fluid migration.

Diagenetic setting in the modern reef environment

The locus of marine cementation in shallow marine carbonate environments is the shelf-margin reef. The framework reef provides a remarkably favorable environment for the precipitation of marine cement. The reef itself furnishes a porous stabilized substrate that facilitates the uninhibited movement of supersaturated marine water through its pores, allowing these waters to interact with a variety of carbonate material. Reefs are generally sited along shelf margins, in areas of high wave and tidal activity, ensuring the constant movement of vast quantities of marine waters through the reef framework. The shelf-margin site supplies the reef with unlimited volumes of fresh, cool marine water. This water is rapidly degassed relative to CO_2 by agitation, warming, and or-

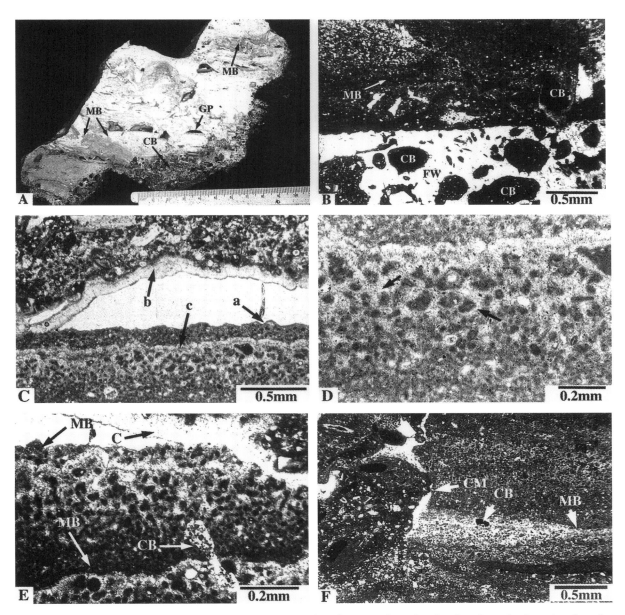

Fig. 5.11. Slab and thin section photomicrographs of modern framework reef rock taken from the Jamaica deep fore reef during submersible operations by the author and others in 1970. These samples were recovered from 212m of water. (A)The entire rock consists of cemented internal sediments. Several geopetals are formed by cemented internal sediments (GP). The darker areas are probably pelleted microbial biofilms (MB). The periphery of the sample is riddled with clionid borings (CB). (B) Photomicrograph of a portion of (A) showing coral framework (FW) riddled by clionid borings (CB) some of which are filled by cemented pelleted internal sediments. Microbial biofilm (MB) has been penetrated by a clionid boring. Plain light. (C) Reef cavity perhaps formed by a microbial crust, subsequently filled with pelleted internal sediment (c). Each pellet is coated with a halo of magnesian calcite cement. Pelleted microbial mat (a) shows gravity defying stromatolitic-like surface. Magnesium calcite cement

(Fig. 5.11, cont.) crust coats the top surface of the cavity (b). Plain light. (D) Close view of C showing dark micrite pellets encrusted with a halo of magnesian calcite cement (arrows). Plain light. (E) Close view of another portion of A showing possible microbial biofilms (MB). Coarse magnesian calcite crust on the upper surface of the cavity (C) and a clionid boring (CB) penetrating pellets and cements indicating marine lithification. Plain light. (F) Larger view of 212m sample showing a large cavity (CM) perhaps cut by a clionid filled with laminated cemented pelleted internal sediment cut by a clionid boring (CB). Several possible microbial biofilms (MB) interrupt the internal sediment. Plain light.

ganic activity, leading to further saturation with respect to $CaCO_3$ and ultimately to massive cementation within the reef frame. This process has been well documented by numerous workers in modern reefs at sites around the world including: Jamaica (Land and Moore, 1980); Bermuda (Ginsburg and Schroeder, 1973); Red Sea (Dullo, 1986); Belize (James and Ginsburg, 1979); Panama (Macintyre, 1977); Great Barrier Reef (Marshall, 1986); Mururoa Atoll (Aïssaoui et al., 1986); south Florida (Lighty, 1985); St. Croix (Zankl, 1993) and many other sites and authors.

While both aragonite and magnesian calcite cements are associated with modern marine reef lithification, magnesian calcite is today the dominant marine cement mineralogy (Purser and Schroeder, 1986). Fibrous aragonite commonly occurs as epitaxial overgrowths on aragonite substrates such as coral septa (James and Ginsburg, 1979)(Fig. 5.2 and 5.3). Aragonite may grow into larger cavities in a spherulitic habit that can develop into spectacular large botryoidal masses, such as were described by James (1979) from the deeper fore reef (>65m) of Belize (Fig. 5.2 and 5.3).

Modern magnesian calcite reef-related cements occur as two distinct types: peloidal aggregates of micrite-sized crystals bearing a radiating halo of coarser scalenohedral crystals (Fig. 5.11D), and thick fibrous-to-bladed magnesian calcite crusts lining larger voids (Fig. 5.11C and E). As noted earlier in this chapter, the peloidal magnesian calcite often occurs as gravity defying crusts on and within the reef framework (Fig. 5.11A, C, and E). The pelleted nature of this most common feature of submarine lithification of reefs is generally thought to be the result of microbial activity within the reef framework (Kirkland et al., 1998).

The stable isotopic composition for selected magnesian calcite submarine, reef-related cements shows higher $\partial^{18}O$ and $\partial^{13}C$ values than abiotic marine cements, supporting some organic intervention in their precipitation (Land and Goreau, 1970; Land and Moore, 1980) (Fig. 5.4, this book). Aragonite cements, including the massive botryoidal cements of Belize, have an average Sr composition of 8300 ppm, and an average Na composition of 2000 ppm, both within the range expected from aragonites precipitated from normal marine waters (James and Ginsburg, 1979).

Cement distribution in modern reefs is strongly skewed toward the seaward margins of the reef, and indeed the leeward edges are frequently completely uncemented (Marshall, 1986). This distribution seems to be in response to higher-energy conditions and cooler oceanic water temperatures along the seaward edges of the reef. The large volume of water moved through the reef framework by waves and tides, and CO_2 degassing by warming and agitating the cold marine waters, would result in the intense cementation observed along the seaward margins of modern reefs. This distribution pattern also clearly shows that reef-related marine cementation is a complex process that cannot be explained solely by microbial living processes, but surely involves physical-chemical processes as well.

In detail, reef cementation often seems to be concentrated close to the reef-water interface, and may die out toward the interior of the reef, sheltering zones of high porosity, particularly if framework accretion is rapid (Lighty, 1985). In areas with slower reef accretion rates, such as in Jamaica, or in the deeper fore-reef zones of Belize, the entire seaward margin of the reef can be solidly lithified, destroying most porosity (Moore, 1989). During drilling operations on the Belize barrier reef at Carrie Bow Key, Shinn and others (1982) encountered pervasively cemented reef framework on the seaward margin of the reef and uncemented framework less than 600m landward of the reef crest.

The actual reef lithification process, and by extension, early porosity evolution, is much more complex than biologically mediated passive marine cementation. Three distinct processes, operating simultaneously, affect early lithification and porosity modification in the reef environment: bioerosion, internal sedimentation, and marine cementation. During the initial development of reef framework, porosity potential is practically unlimited because of the following factors: 1) common formation of large shelter voids during the development of the framework of the reef (some of these voids are so large that a person can actually swim through them, as shown in Figure 3.16A); 2) the presence of copious intragranular porosity within the framework organisms, such as corals (Chapter 3, Figs. 3.12 and 3.13); 3) the presence of significant intergranular porosity within associated sand-dominated sediments; 4) and finally, the development of extensive secondary porosity within the reef organic framework, as a result of the activity of boring organisms such as pelecypods, sponges, algae, and fungi (Chapter 3, Fig. 3.14).

Reef-associated marine waters contain suspended sediments derived from the physical and biological breakdown of the reef framework in high-energy environments as well as calcareous phytoplankton and zooplankton (Moore and Shedd, 1977). This suspended material is swept into, and is ultimately deposited within, the framework of shelf margin reefs as internal sediments. These sediments are carried by the flow of immense volumes of marine water through porous reef framework driven by tides and wave activity (Ginsburg et al., 1971). Internal sediments consist of unidentifiable silt and clay-sized aragonite and magnesian calcite, as well as coccoliths, planktonic foraminifera, planktonic mollusks, and silt-sized chips derived from the activity of clionid sponges (Moore et al., 1976).

These internal sediments are cemented, dominantly by finely crystalline magnesian calcite, very shortly after they are deposited within the voids developed in the framework (Fig. 5.11). The extent of microbial involvement in the cementation of these internal sediments is unknown at the present time. Deposition seems to be episodic, and cements become much coarser at times of nondeposition along the sediment water interface in larger cavities (Fig. 5.11C-E). The final fabric is one of fine, intergranular cements interrupted periodically by horizontal, geopetal crusts of coarser magnesian calcite cement that can be traced up the walls and across the roof of the remaining cavity (Fig. 5.11) (Land and Moore, 1980; James and Choquette, 1983). Regardless of origin, the internal sediment fills porespace, is cemented, and is an important component of early reef lithification and porosity modification.

Reef framework, cemented internal sediments, and marine cements are penetrated and destroyed by the galleries and borings of various marine organisms, including lithophagus pelycepods and clionid sponges. These galleries and borings can displace more than 50% of the lithified framework of the reef with relatively large organism-filled voids at any one time (Moore and Shedd, 1977) (Fig. 5.11). Upon death of the organism, an empty void is produced that immediately begins to be filled with internal sediment that is quickly cemented (Fig. 5.11). The bacteria involved in the degradation of the dead rock boring organisms that made the void are no doubt involved in this rapid cementation. In reefs where framework accretion is not exceptionally fast, the process is repeated innumerable times until an original porous, recognizable reef framework is reduced to a relatively nonporous pelleted lime wackestone within which little reef framework is preserved (Fig. 5.11) (Land and Moore, 1980).

The repetitive application of this tri part set of concurrent destructive-constructive processes is not only the cause for massive early porosity reduction in reef sequences, but also the difficulty geologists frequently encounter when trying to recognize framework reefing in cored subsurface material. Because the shelf margin reef is the one environment where these three processes are active over relatively long periods of time, the recognition of the activity of concurrent internal sedimentation, marine cementation, and bioerosion can be useful criteria for the recognition of the reef environment in cored subsurface sequences (Land and Moore, 1980).

Recognition of reef-related marine diagenesis in the ancient record

Detailed studies of ancient reef sequences indicate that many of the same textural and porosity modification patterns seen in modern reefs are mirrored in their ancient counterparts. These studies suggest that concurrent internal sedimentation, abiotic and microbial marine cementation, and bioerosion have long been the most important factors controlling the nature of early marine lithification within the reef environment (Kirkland et al., 1998). Clionid borings have been observed as far back as the Devonian (Moore et al., 1976), and have played a major role in textural modification during early marine lithification in Jurassic coralgal reefs across the Gulf of Mexico (Crevello et al., 1985), Cretaceous rudist reefs of central Texas (Petta, 1977), and Upper Cretaceous of Tunisia (M'Rabet et al., 1986). Thus the recognition of clionid borings and their concurrent penetration of framework, internal sediments, and cements may be a prime criterion for the recognition of early marine lithification of ancient reef sequences through most of the Phanerozoic.

Pervasive early cements, believed to have been precipitated abiotically directly from marine waters, have been described from reef sequences throughout the Phanerozoic (Schroeder and Purser, 1986). In ancient rock sequences these inferred marine cements associated with reef environments exhibit two basic fabrics: 1) isopachous radiaxial fibrous calcite with perfectly preserved, but sometimes

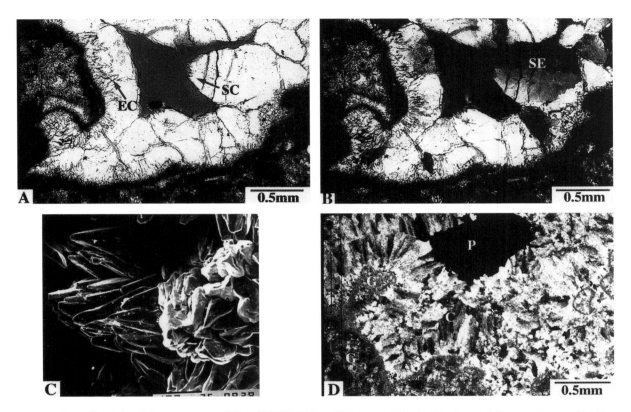

Fig. 5.12. Radiaxial calcite cement, at 640m (2100') F-1 well Enewetak Atoll. Rocks are Miocene in age. (A) Note inclusion rich zone (EC) and serrated termination (SC). Plain light. (B) Same as A. Note irregular extinction patterns (SE). Characteristic of radiaxial calcite under crossed polars. (C) SEM photomicrograph of A showing that the serrated termination actually is the reflection of a compound crystal consisting of a bundle of individual elongate crystals. Bar is 100 μ. (D) Photomicrograph illustrating the complex extinction patterns of radiaxial calcites under crossed polars. Photos courtesy of Art Saller.

complex fibrous crystal structure usually with undulose extinction (Fig. 5.12); 2) spherulitic fibrous isopachous fringes and botryoidal masses, recrystallized into a fine to course calcite mosaic with the original fibrous crystal morphology preserved as ghosts outlined by fluid and solid inclusions (Fig. 5.6 C and D) (Sandberg, 1985).

Radiaxial fibrous calcites show a range of textures, as schematically presented in Figure 5.13. Radiaxial fibrous calcite proper (as originally defined by Bathurst in 1959) is identified by a pattern of subcrystals within each crystal that diverge away from the substrate in an opposing pattern of distally-convergent optic axes, giving a corresponding curvature to cleavage and twin lamellae (Kendall, 1985). Fascicular-optic calcite (Kendall, 1977) is a less common type distinguished by the presence of a divergent pattern of optic axes coinciding with that of the subcrystals. Radial fibrous calcites (Mazzullo, 1980) do not have undulose extinction but are formed of turbid crystals. A variety of original mineralogies have been proposed for radiaxial fibrous calcite cements. They were originally believed to represent the

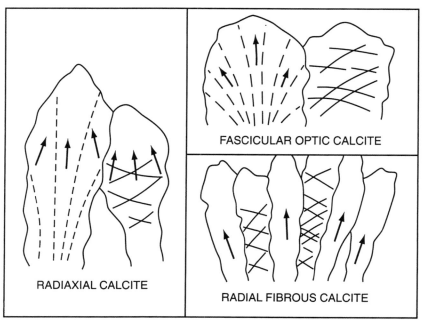

Fig. 5.13. Schematic diagrams illustrating the differences between radiaxial, fascicular optic, and radialfibrous calcites. For each diagram arrows record fast vibration directions, narrow continuous lines represent twin-planes, and discontinuous lines represent subcrystal boundaries. Used with permission of Elsevier.

recrystallization fabric of an acicular marine aragonite precursor (Bathurst, 1982). Basing their interpretations on petrographic considerations, Sandberg (1985) and Kendall (1985) suggest that these cements were originally precipitated as a calcite cement that may have had an original high magnesian calcite mineralogy because of the common occurrence of microdolomite inclusions. Saller (1986) describes radiaxial fibrous calcite cements of low magnesian calcite mineralogy that are precipitated into aragonite dissolution voids in Miocene limestones from cores taken along the margin of Enewetak Atoll (Figs. 5.12 and 5.37). Saller develops compelling isotopic (Fig. 5.40) and petrographic evidence (Fig. 5.12) to indicate that these cements were precipitated directly from deep, cool marine waters located beneath the aragonite compensation depth and above the calcite compensation depth.

The second common group of reef-related cement fabrics, the spherulitic fibrous fringes and botryoidal masses described by James and Klappa (1983), Kirkland (1998), and others, are generally thought to have originally been aragonite marine cement subsequently recrystallized to calcite (Fig. 5.6) (Sandberg, 1985). These inferred aragonite marine cements are only common during certain spans of geologic time: Middle Pre-Cambrian (Grotzinger and Read, 1983), Early Cambrian (James and Klappa, 1983), Late Mississippian (Mazzullo, 1980), to Early Jurassic (Burri et al., 1973), and mid-to-late Cenozoic (Sandberg, 1985) (Fig. 5.6).

The distinctly clumped temporal distribution pattern of aragonite cements strongly parallels the oscillatory temporal trends in abiotic carbonate mineralogy established by Mackenzie and Pigott (1981) and Sandberg (1983)(see introduction to this chapter; Fig. 2.15). These trends were based primarily on the interpretation of the original mineralogy of Phanerozoic ooids. The periods when inferred aragonitic ooids are common coincide with the documented occurrences of botryoidal aragonite. During these intervals of apparent aragonite inhibition, calcite seems to have been the dominant precipitational phase in marine waters, and hence radiaxial calcite cements dominated the shallow and deep reef environments. To date, modern radiaxial low-magnesian calcite cements have only been found within shelf-margin reef sequences bathed with cold marine waters below the aragonite compensation depth (Saller, 1986).

The third and perhaps the most important type of ancient reef-related cement fabrics are the pelleted, originally magnesian calcite, microbial crusts described by numerous authors from reefs and mud mounds from the Pre-Cambrian to the Recent (Monty, 1995) (Kirkland et al., 1998). These distinctive and volumetrically important cement fabrics occur in reefs with both the aragonite botryoids and calcite radiaxial fabrics throughout the Phanerozoic. They occur regardless of the interpreted sea water chemistry, strongly supporting a biologically mediated origin for the cement fabric.

In the following pages we will discuss three case histories of economically important ancient reef sequences. The first is an outcrop study of the Permian Capitan reef exposed on the flanks of the Delaware Basin in New Mexico and west Texas, USA. This study illustrates and emphasizes the importance of microbial-related cements to the lithification and porosity evolution of upper Paleozoic shelf margin reefs. The second two case studies emphasize the importance of early marine cementation to ultimate reservoir development in well-known reef sequences. They are the Lower-Middle Cretaceous reef-bound shelf margin of the Gulf of Mexico, anchored on the south by the Golden Lane of Mexico, and the middle Devonian reef trends of western Canada.

Early marine lithification of the Permian Capitan reef complex New Mexico and west Texas, USA

The Permian Capitan reef complex is exposed in the Guadalupe Mountains of west Texas and New Mexico, where outcrops form part of the western margin of the Delaware Basin (Fig. 5.14A). The Capitan Formation is, in general, the shelf margin reef facies equivalent of a restricted shelf interior represented by the Seven Rivers, Yates, and Tansill formations. The Capitan evolved from a dominantly progradational buildup on a ramp (lower Capitan) to a steep fronted shelf margin reef (middle Capitan) to patch reefs in the upper Capitan (Garber et al., 1989; Kirkland et al., 1998). The Bell Canyon Formation is the basinal equivalent of the Capitan (Fig. 5.14B). While absence of a lateral seal for the reef facies limits actual oil production from the Capitan, associated shelf and basinal units such as the Yates and Bell Canyon are major oil producers in the west Texas area. The economic importance of the units associated with the reef, as well as its excellent exposures along the Guadalupe Mountain front, form the basis for the long term, intense interest of the geological community in the Capitan (Garber et al., 1989).

The major sources for controversy relative to the depositional environment represented by the Capitan can be traced to uncertainties concerning the nature of, or lack of the reef framework and the apparent mud dominated fabrics characteristic of this unit (Kirkland and Moore, 1996). Many workers felt that the erect calcareous sponges typical of the Capitan Reef

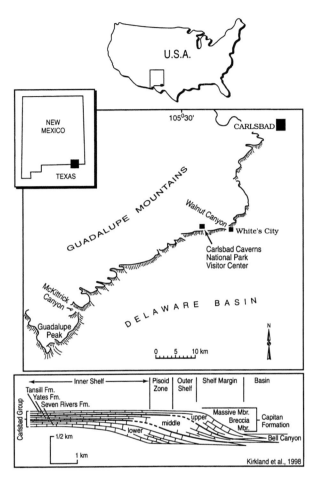

Figure 5.14. Geologic setting of the Permian Capitan reef complex. (A) Location map showing the Guadalupe Mountain front. (B) General relationship of the lithostratigraphic units of the Carlsbad Group and their paleogeographic setting. Used with permission of SEPM.

acted as bafflers, rather than true framebuilders (Achauer, 1969). The preponderance of sponges and mud-dominated fabrics led many workers to view the entire Capitan as a deepwater mud mound system, rather than a true framework reef. This interpretation led workers to reassess the geometry of the reef complex, leading to the deep mound, shelf crest model of Dunham (1972).

Subsequent detailed paleoecologic studies by Kirkland et al. (1998) and Wood et al. (1994; 1996) have confirmed that the Capitan was a wave-resistant framework reef formed under moderate to high-energy conditions in shallow to moderate water depths. The key observation is the recognition of small to large reef cavities supporting a diverse cryptic calcareous sponge fauna occurring as pendant cryptobionts attached to the lithified cavity walls and ceilings (Figs. 5.15 and 5.16) (Wood et al., 1994). Bryozoans, calcareous sponges and possible red algae form the primary framework and framework porosity of the reef. Cemented, pelleted microbial crusts form secondary encrustations that add to the strength of the reef (Figs. 5.15, 5.16 and 5.17).

The remaining framework porosity is filled with pelleted, gravity-deposited, marine-cemented internal sediment, cemented microbial mats, neomorphosed marine aragonite botryoidal cements and radiaxial calcite cements (Figs. 5.15, 5.16 and 5.17). Marine cementation and internal sedimentation during and shortly after deposition destroy almost all original framework porosity.

Kirkland et al. (1998) interpreted the microstratigraphy of the middle Capitan to be the result of deposition in waters ranging in depth from 10 to 140m. The middle Capitan supports large framework cavities, a diverse cryptobiont fauna and large volumes of botryoidal and microbial cements. The upper Capitan is thought to have been deposited under much shallower, high-energy conditions (<1 to 10m water depth). It exhibits much smaller framework cavities bound by thin red algal and clotted, microbial encrustations with vague dendritic structures (Fig. 5.18). The middle and upper Capitan framework porosity is largely occluded by early marine diagenetic processes

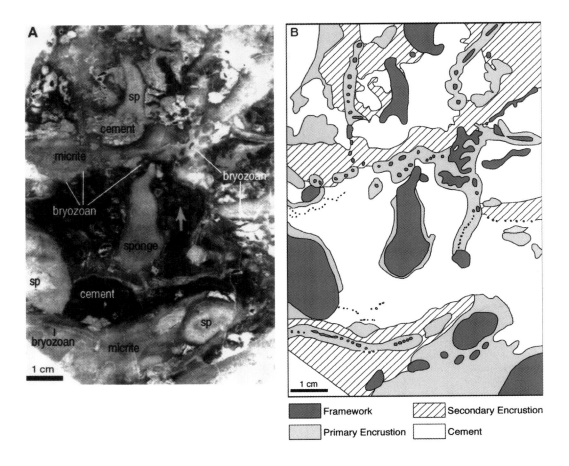

Fig.5.15. (A) Photograph of middle Capitan polished slab, McKittrick Canyon, Guadalupe National Park. Arrow indicates way up. A bryozoan near the center of the photograph is cut in both longitudinal section (white net-like shape) and cross section (small white dots). The sponge in the center of the photo was attached to that bryozoan and hung below it. Micrite accumulated above the same bryozoan. Thus this bryozoan forms the roof of one cavity and the base of another. Fine, light gray layers of <u>Archaeolithoporella</u> encrust the sponges and bryozoans. Botryoidal aragonite (cement), now neomorphosed, filled most of the remaining pore space. (B) Line drawing of the slab shown in A. Bryozoans and sponges form the framework. <u>Archaeolithoporella</u> is the most common primary encruster. Microbial micrite is the secondary encruster. Abundant botryoidal aragonite along with minor radiaxial fibrous calcite and meteoric spar filled the remaining pore space. Used with permission of SEPM.

that Kirkland et al. (1998) feel are dominated by microbial activity.

Porosity evolution of reef-related Lower-Middle Cretaceous shelf margins: the Golden Lane of Mexico and the Stuart City of south Texas

The Lower-Middle Cretaceous shelf margin is one of the most dominant geologic fea-

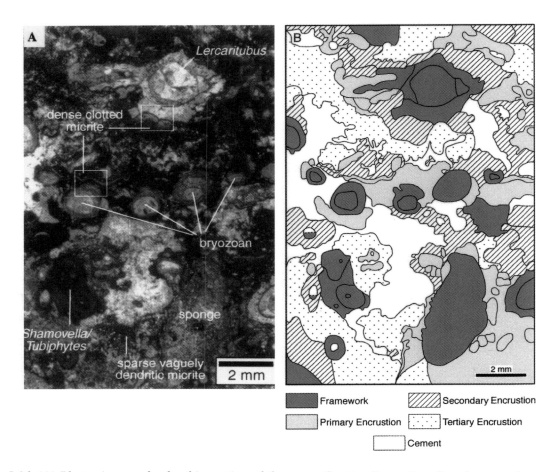

Fig. 5.16. (A) Photomicrograph of a thin section of the upper Capitan Formation. Sample was taken near the transition from reef to outer shelf, McKittrick Canyon, Guadalupe National Park. Bryozoans, sponges, <u>Shamovella/Tubiphytes</u>, and <u>Lercaritubus</u> form the framework. Another type of <u>Shamovella/Tubiphytes</u>, <u>Archaeolithophyllum</u> and an unknown organism form primary encrustations. Dense, clotted micrite contains possible cyanophyte fossils and forms secondary encrustations. Sparse, in some places vaguely dendritic, micrite forms tertiary encrustations. Cements (radiaxial fibrous calcite, and meteoric spar) fill remaining voids. (B) Line drawing of the thin section shown in A. Used with permission of SEPM.

tures of the Gulf of Mexico, forming a 2500-km-long structural and sedimentological boundary around the Gulf of Mexico basin (Buffler and Winker, 1988) (Fig. 5.19). The southwestern segment is a shallow marine carbonate shelf that can be traced in outcrop from southeastern Mexico to central Texas. The shelf margin is exposed at the surface in Mexico as a result of Laramide tectonism, but rapidly dips into the subsurface to the north and south.

The Lower Cretaceous shelf margin sequence of south Texas and northeastern Mexico consists of two distinct progradational to accretionary reef-dominated shelf margin sequences called the Stuart City (El Abra in Mexico) and Sligo (Guaxcama in Mexico), separated by a regional drowning event marked by the Pearsall Shale (Fig. 5.19). In Texas, the

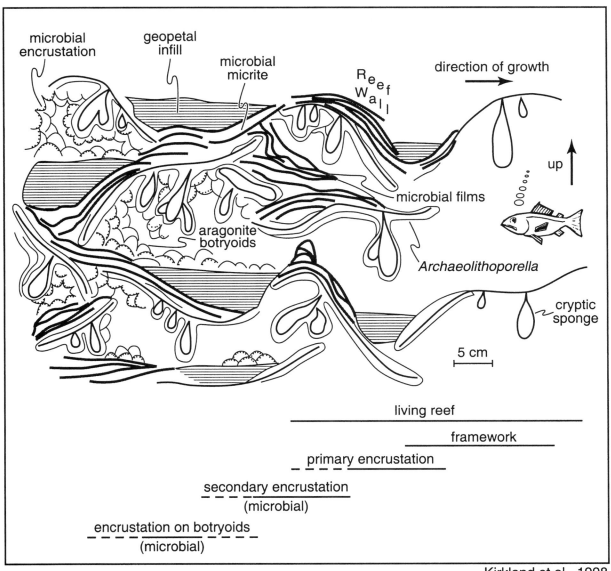

Fig. 5.17. Schematic diagram depicting cross section into vertical wall of living Capitan reef. Open water of basin is to the right. The reef wall is close to vertical. The outer rims of cavities in the reef wall were lined by <u>Archaeolithoporella</u> (thin lines). Darker sections of the cavities were encrusted by microbial biofilms (thicker lines), which were the site of accumulation and/or precipitation of peloidal microbial micrite. Horizontal lines represent geopetals composed of internal sediment often encrusted with aragonite botryoids. Initial layers of botryoids were commonly encrusted by microbial layers (dots). Used with permission of SEPM.

Stuart City is characterized by the development of persistent, thick (up to 800 m), rudist dominated reefs and reef-related facies buried some 5000 to 6000 m in the subsurface (Bebout and Loucks, 1974; Achauer, 1977). Seaward of the shelf margin, the sequence passes rapidly into

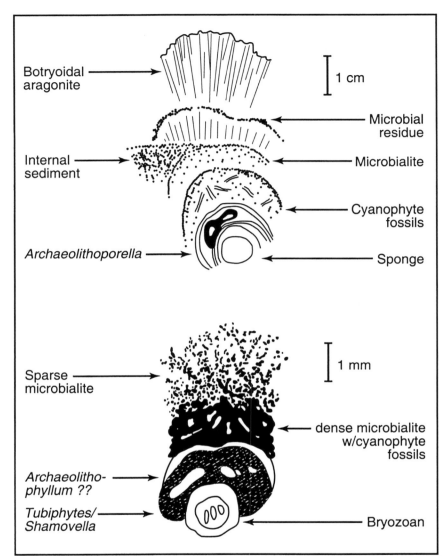

Fig. 5.18. Schematic diagram illustrating idealized microstratigraphic cycles for the middle Capitan (above) and the upper Capitan (below). Note difference in scale. The middle Capitan sample is interpreted as forming in the quiet conditions of the deepest part of the reef. In contrast, the upper Capitan sample is interpreted as forming in agitated conditions in the shallowest part of the reef. Used with permission of SEPM.

dark, pelagic, foraminiferal limestones, termed the Atascosa Formation or Group (Winter, 1962). Shelfward, the Stuart City changes to the relatively low-energy shelf sequences of the Glen Rose Formation and the Fredericksburg Group (Fig. 5.19B). These units ultimately crop out across a wide region of central and west Texas. The subsurface Stuart City extends west and south in the subsurface into Mexico where it is termed the El Abra Formation.

In southeastern Mexico, reef-dominated shelf margin development was continuous from the Lower through Middle Cretaceous and perhaps extended up into the Upper Cretaceous. These shelf margins were developed in the El Abra Formation on a series of large, structurally-controlled, epicontinental platforms called the Valles-San Luis Potosi and Tuxpan (Gold-

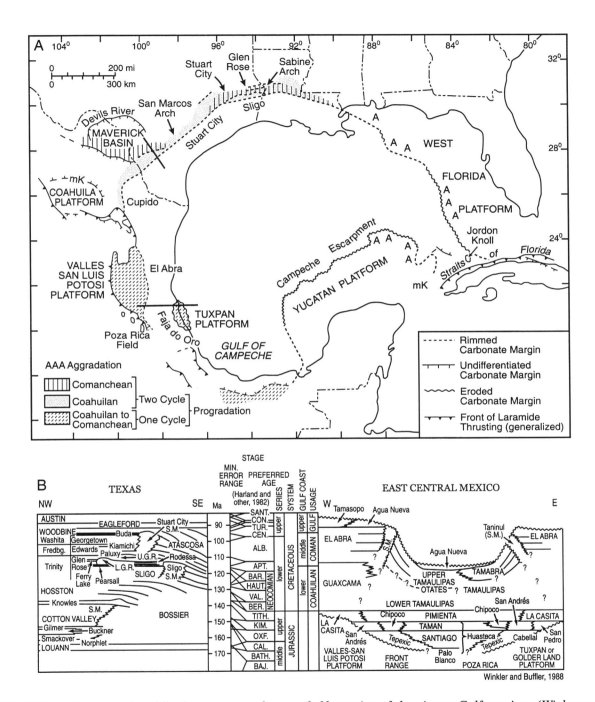

Fig. 5.19. A. Lower and middle Cretaceous carbonate shelf margins of the circum-Gulf province (Winker and Buffler, 1988). Location of two cross sections indicated by solid black lines. B. Stratigraphic relationships of the Mesozoic, south Texas showing the two-cycle shelf margin. (Winker and Buffler, 1988). (Lower right) Stratigraphic relationships of the Mesozoic in east central Mexico and their correlation with south Texas (Winker and Buffler, 1988). Used with permission of Elsevier.

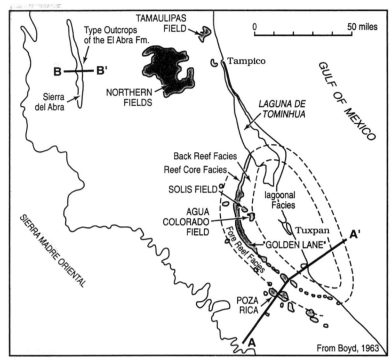

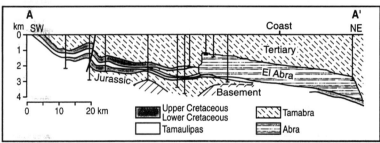

Fig. 5.20. (Top) General location map of the Golden Lane area. Major oilfields shown in stippled pattern. Line A-A' is location of cross section below. Line B-B' is location of cross section across the type outcrops of the El Abra Formation. (Bottom) Subsurface cross section across the Golden Lane to the Sierra Madre Oriental. (Top) Used with permission of Elsevier.

en Lane) platforms. The El Abra margin was exposed by Laramide tectonism along the Valles-San Luis Potosi Platform (Figs. 5.19 and 20) (Buffler and Winker, 1988).

The subsurface Stuart City trend in Texas was extensively penetrated during hydrocarbon exploration in the late '50s and early '60s in the search for a south Texas "Golden Lane." While the depositional setting and facies were believed to be similar to those encountered in the El Abra Formation of the Golden Lane in northeastern Mexico (Rose, 1963) (Fig. 5.21),

the south Texas Stuart City trend has been an economic disappointment. It has produced only small quantities of gas to date, in contrast to the billions of barrels of oil recovered from the El Abra (Bebout and Loucks, 1974).

The Stuart City is a marginal hydrocarbon trend because of a general lack of porosity and permeability in its potential reservoir facies. The Stuart City reef and reef-related sequences were formed, as similar sequences are today, with extensive primary growth porosity, as well as intergranular and intragranular porosity.

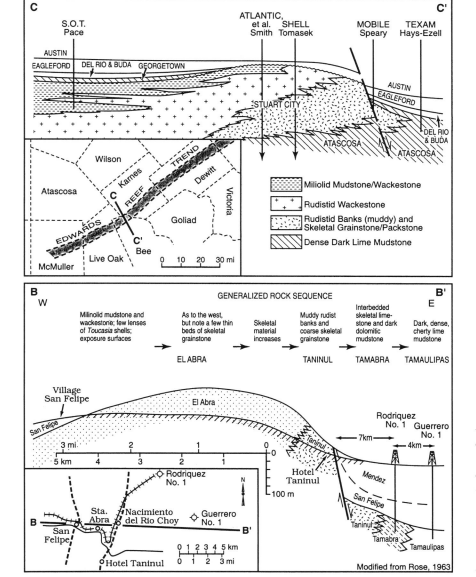

Fig.5.21. Comparative cross sections across the Middle Cretaceous shelf margins in south Texas (Top) and southeast Mexico at El Abra (Bottom). Locations shown in inset maps, lithologic patterns shown in top figure. Used with permission of Elsevier.

Petta, (1977), in a study of an outcrop Glen Rose rudist reef complex equivalent to the subsurface Stuart City, found that internal sedimentation, marine cementation, and clionid bioerosion developed fabrics identical to those encountered in modern framework reefs, and were the dominant processes causing early porosity loss in this Cretaceous reef sequence.

Prezbindowski (1985) documented massive early porosity loss in the Stuart City shelf-margin rudist reef sequences by cementation. Achauer (1977) noted that most of the porosity loss was the result of marine cementation by radiaxial calcite and indicated that these cements were primarily encountered along the shelf margin.

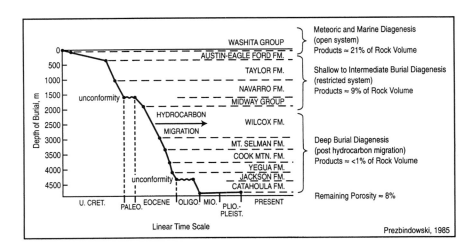

Fig.5.22) Approximate burial curve for the Lower Cretaceous Stuart City reef trend, south Texas. (Right) Diagenetic environments and porosity evolution with burial. Used with permission of Elsevier.

Prezbindowski (1985) estimated that some 21% of the Stuart City's primary porosity was lost by cementation (out of a total of 38%) before the sequence was buried 100 m (Fig. 5.22). While some of this early porosity loss seems to have been by cementation under meteoric phreatic conditions, the majority of the porosity loss is by marine cementation since there is little evidence for substantial subaerial exposure of the Stuart City (Prezbindowski, 1985). Only 9% of the remaining porosity was lost during the next 4000m of burial, leaving an average porosity encountered in the subsurface of some 8% (Fig. 5.22). This porosity is generally unconnected, occurring as intragranular porosity associated with either the living cavities of rudists or the tests of foraminifera. Due to the lack of pore connectivity, very low average permeability values are encountered in the south Texas Stuart City (Bebout and Loucks, 1974).

The Golden Lane is a trend of highly productive oil fields developed in what is generally considered to be reefal facies of the El Abra Formation along the margins of the Tuxpan Platform, located some 75 km east of the main Lower-Middle Cretaceous shelf margin (Boyd, 1963) (Fig.5.20). While this trend has been producing since 1908, little is actually known about the nature of the reservoir rock because most wells are completed in the uppermost El Abra. The correlation of environment and facies with the south Texas Stuart City is based on outcrop studies of exposures of El Abra platform and related facies north and west of the productive trend along the margins of the Valles-San Luis Potosi Platform (Enos, 1986)(Fig. 5.21). At this site the platform-to-basin facies tracts can be traversed, and shelf-margin rudist reefs change to fore-reef talus and ultimately to the dark, laminated, basinal limestones of the Tamalipas Formation. The facies tract is similar to the facies tract encountered in the shelf-margin wells penetrating the Stuart City in south Texas (Fig. 5.21).

The syndepositional diagenetic history of the El Abra shelf margin also seems to parallel that of the south Texas Stuart City with massive marine cementation, dominated by radiaxial calcite, which forms an early porosity destructive stage (Shinn et al., 1974). Shortly after deposition, however, the subsequent geologic history of these analogous sequences dramatically diverges. The Golden Lane platform only 70 km east of the Sierra Madre Oriental experienced intense deformation associated with Laramide tectonics shortly after deposition of the El Abra. The platform was uplifted and tilted toward the east, causing subaerial exposure and intense karsting along the western margin of the platform, in turn resulting in the development of extensive cavernous porosity in the El Abra shelf-margin facies (Coogan and others, 1972) (Fig. 5.20).

In contrast, the south Texas Stuart City shelf-margin trend is located some 500km from the locus of Laramide tectonism. The Stuart City was rapidly buried into the deep subsurface, with no opportunity for significant subaerial exposure or karsting. It is this exposure and karsting related to tectonics that is the key to the phenomenal hydrocarbon production seen in the Golden Lane since the early 1900's.

Porosity evolution of Middle Devonian reef complexes: Leduc, Rainbow, "Presquile" and Swan Hills reefs, Western Canadian Sedimentary Basin

The western Canadian Sedimentary Basin extends over 1900 km (1200 miles) north to south and is over 550 km (350 miles) wide. The Precambrian Shield forms the eastern margin, and the Cordilleran tectonic front outlines the western border (Klovan, 1974) (Fig. 5.23). The Middle and Upper Devonian sedimentary sequences deposited within and along the margins of this enormous intracratonic basin harbor 62% of Alberta's recoverable, conventional crude oil in reef-related limestones and dolomites (Davies, 1975). The Middle and Upper Devonian Givetian and Frasnian series represent a basic north-to-south transgression of the basin with progressive onlap of reef-bound shelves through time (Fig. 5.23). There are three major intervals of reef development: 1) The Givetian "Presquile" shelf-margin reef complex fronts the Hare Indian Shale Basin and has a broad shelf lagoon extending to the east and south containing high relief Rainbow reef complexes. This shelf lagoon was highly restricted, and became evaporitic. The Rainbow reefs were ultimately buried by massive Muskeg evaporites. 2) Frasnian age, low-relief Swan Hills reefs were developed on the Slave Lake carbonate shelf along the margins of the Peace River Arch. 3) The Frasnian age Leduc reefs were developed as moderate relief, linear trends, on the Cooking Lake platform. The Leduc reefs were ultimately drowned by Ireton shales and associated Duperow evaporites (Fig. 5. 23).

Devonian reefs of western Canada consist of a bank margin of stromatoporoid coral framework and fore-reef debris. The bank interior facies consist of low-energy lagoonal wackestones and mudstones, mobile sandflats, and some low-energy supratidal sequences (Walls and Burrowes, 1985). Subaerial exposure and submarine hardgrounds often interrupt buildup growth. Reef growth is generally initiated on submarine hardgrounds (Wendte, 1974).

The diagenetic history of Devonian reefs has been the subject of numerous studies emphasizing various aspects of the diagenetic history of isolated reef complexes, generally without regard to ultimate porosity evolution or any attempt to discover unifying principles that might be useful as predictive tools (Walls and Burrowes, 1985). The synthesis of Walls and Burrowes (1985) provides a comprehen-

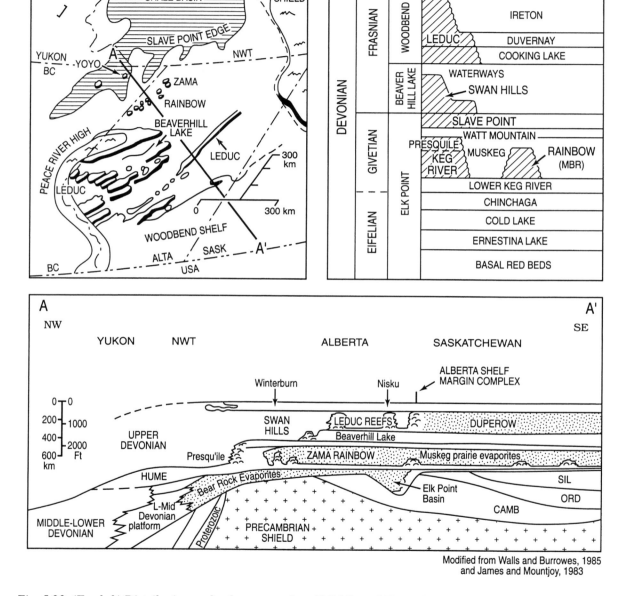

Fig. 5.23. (Top left) Distribution and paleogeography of Middle and Upper Devonian reef carbonates in western Canada. (Top right) Devonian stratigraphy in western Canada illustrating stratigraphic position of reef carbonates. (Bottom) Subsurface schematic cross section along line A-A' (shown in top left) through the western Canadian sedimentary basin illustrating the stepped onlap mode during Late Devonian time. Used with permission of Elsevier.

sive view of the diagenetic history of four of the most economically important western Canadian Middle Upper Devonian buildups. Their work reveals that the margin of each bank has quite a different diagenetic history, and hence porosity evolution, than the adjacent bank interior (Walls and Burrowes, 1985). These differences are illustrated in Figures 5.24 and 5.25, where porosity evolution is traced from deposition to the present and cementation environment is inferred. A direct comparison of reservoir quality and characteristics of four reef complexes is compiled in Table 5.2.

Walls and Burrowes (1985) conclude that marine cementation dominated by radiaxial calcite is the single most important diagenetic process affecting porosity in Middle and Upper Devonian reefs of western Canada. In the Leduc and Rainbow reef complexes, marine cementation almost completely occludes depositional porosity along the reef margins, while the volume of marine cement in the bank interiors is less than half that encountered along the margins. The lack of massive marine cementation and the presence of secondary porosity developed during intermittent subaerial exposure of the bank interiors give Leduc and Rainbow bank interiors better reservoir quality than their adjacent bank margins (Fig. 5.25, Table 5.2). However, the bank interiors of Leduc and Rainbow reefs generally do show more subsurface cements associated with pressure solution than do the bank-margin sequences. Pressure solution, cementation, and multiple exposure surfaces tend to horizontally compartmentalize the favorable porosity/permeability trends of these bank interiors (Table 5.2).

The "Presquile" and Swan Hills reef complexes contrast sharply with Leduc and Rainbow (Fig. 5.24 and 5.25, Table 5.2). The

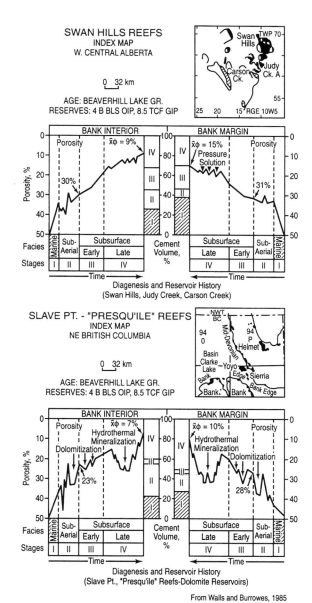

Fig. 5.24. Summary of diagenesis and reservoir history for the Swan Hills reefs (Top) and the Slave Pt. Presquile reefs (Bottom). Bank interiors and bank margins are compared. Porosity evolution through time, and cement volume in % are estimated. Used with permission of Elsevier.

volume of marine cements found in the margins of these two reef complexes is generally half that found at Leduc and Rainbow. As a result, subsurface cements and dolomitization are much more important porosity modification processes in the Presquile and Swan Hills sequences. Finally, because of preserved primary porosity and dolomitization, the best quality reservoir rock is found in "Presquile" and Swan Hills bank margins, rather than in their bank interiors, as seen in Leduc and Rainbow.

Why do the Leduc and Rainbow reefs exhibit so much more intense marine cementation than "Presquile" and Swan Hills? Two factors seem important. Both Leduc and Rainbow are closely associated with evaporitic conditions (Duperow and Muskeg evaporites), possibly implying that marine waters present during deposition and syngenetic diagenesis were tending toward hypersalinity, and hence could have been more saturated with respect to $CaCO_3$. Both Leduc and Rainbow show moderate-to-high relief relative to the adjacent seafloor, and both face broad open-water shelves that might result in higher-energy conditions, with high marine water flux, and a propensity to cementation.

"Presquile" and Swan Hills seem to be developed under lower-energy situations because of bank-margin geometry and relative relief above the seafloor; therefore, they are less likely to exhibit massive marine cementation. While these contentions have yet to be tested by a comprehensive biofacies assay, it is obvious that the ability to predict these types of diagenetic trends on a regional basis is extremely important to future exploitation and exploration strategy in the western Canadian sedimentary basin, as well as similar basins in the geologic record.

These examples from the Permian Basin, Gulf of Mexico and Canada illustrate clearly the importance of syngenetic marine diagenetic processes, including possible microbial activities, on porosity evolution in shallow marine reef, and reef-related sequences. The message

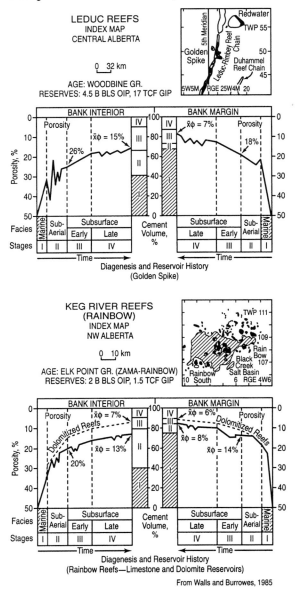

Figure 5.25. Summary of diagenesis and reservoir history for the Leduc reefs (Top) and the Keg River (Rainbow) reefs (Bottom). Bank interiors and bank margins are compared. Porosity evolution through time, and cement volume in % are estimated. Used with permission of Elsevier.

Table 5.2

Reef Complex		φ Mean (%)	K Mean (md)	Reservoir Character
Leduc (Golden Spike)	M	7	25	Poor Perm.; Vert. & Horiz. Perm. Barriers
	I	12	60	"Stratified"; Horiz. Perm. Barriers
Swan Hills	M	15	170	Good Perm.; Few Perm. Barriers
	I	9	25	"Stratified"; Horizontal Perm. Barriers
Rainbow Mbr	M	-8—LS 6-DOL.	35 250	Variable Perm. Lensoid Reservoirs
	I	13-LS 7-DOL.	90 310	Variable Perm. Limited Perm. Barriers
Slave Pt.— "Presqu'ile" (Dolomite)	M	10	75	Good Vertical Perm.; Lensoid Reservoirs
	I	7	50	Limited Perm.

M = Reef Margin   I = Reef Interior   From Wall and Burrowes, 1985

Table 5.2. Comparison of the reservoir quality and characteristics of various Devonian reef margins and interiors. Used with permission of Elsevier.

to the explorationist is clear. *While marine cementation often destroys the original depositional porosity of reef sequences, some reef trends have been found to be enormously productive. These productive reef trends invariably have seen massive diagenetic overprints that have rejuvenated original depositional porosity by dissolution, dolomitization, and fracturing.* Fortunately, the original aragonitic mineralogy of many reef-building organisms enhances the effectiveness of secondary porosity generation by meteoric-related dissolution. The original magnesian calcite mineralogy of other reef-related organisms may well enhance the dolomitization potential of reef sequences. The common location of reefs at the terminal phase of shallowing upward sequences improves the probability of meteoric diagenetic enhancement of porosity. Finally, the shelf margin position of reefs is favorable for the movement of later, basinal-derived fluids through and around the reefs. These fluids are often capable of dolomitization and dissolution, and are commonly the carriers of hydrocarbons formed in basinal shales. *Reefs can be attractive exploration targets but require, perhaps more than any other carbonate depositional setting, a total rock history approach incorporating diagenetic models before success is assured.*

The discussion will now move from warm surface marine water and focus on the cold, dark world of the transitional slope to deep marine environment and examine the porosity modification processes operating there.

SLOPE TO DEEP MARINE DIAGENETIC ENVIRONMENTS

Introduction to diagenesis in the slope to deep marine environment

In this section we will consider three diagenetic settings. The first consists of deep ramp to upper slope low-energy, mud-dominated environments in water depths generally exceeding 100m. While this setting is usually below the photic zone it may extend by accretion into well-oxygenated well-lighted relatively shallow waters. These environments support extensive, important Phanerozoic algal mud mounds including the Waulsortian mounds of Europe and North America.

The second setting is found on the deep slope to basin floor. At these sites, hydrothermal processes associated with tectonic activity or cold petroleum seeps act in concert with biological activity in the presence of methane to create a unique and important diagenetic environment. In both these settings microbial activity is an exceedingly important diagenetic process.

The final deep marine diagenetic settings that we will consider are the steep escarpment shelf margins associated with major reef-bound shelves and platforms. These steep escarpments face deep open marine basins. In this situation, coarse, porous shallow marine carbonate sequences composed of unstable mineralogies interact with progressively colder marine waters that may become undersaturated with respect to magnesian calcite, aragonite and even calcite with depth. While porosity may be dramatically reduced along the escarpment margins, the hydrologic setting is such that enormous volumes of marine waters may be driven through the platform interior or outer shelf by thermal and/or Kohout convection cells. These waters have high diagenetic potential, particularly relative to dolomitization.

Carbonate diagenesis associated with ramp to slope mud mounds

Pre-Cambrian to Mesozoic carbonate mud mounds (Fig. 5.26) have long been enigmatic features of deep ramp, upper slope environments. Bosence and Bridges (1995) suggest a tripartite division of carbonate buildups (Fig. 5. 27) that include: 1) framework reefs (Fig. 1.3), 2) microbial mud mounds, and 3) biodetrital mud mounds. In this section we will primarily discuss microbial mounds. Figure 5.28 illustrates the distribution, characteristics and biota associated with mud mounds through time. Microbial mud mounds have a number of features in common including clotted-to-pelleted mud, stromatactis, thrombolitic, stromatolitic, fenestrate fabrics, and large volumes of fibrous cement (Fig. 5.29).

Many of these fabrics detailed in Figure 5.29, such as stromatactis and fenestrae, in combination with slumps and fissures suggest some rigidity of the structure during deposition. The presence of large volumes of early marine fibrous cement clearly indicates that cementation plays a significant role in development of these mounds. Clotted and pelleted muds associated with stromatolitic fabrics in a deeper marine low-energy environment suggest that much of this cementation may be the result of microbial processes rather than the passive abiotic cementation characteristic of shallower framework reefs.

Lees and Miller detailed a conceptual diagenetic model for Waulsortian mounds (1995)

Figure 5.26. Slope to basin carbonate mounds. (A) Early Devonian Kess Kess mounds of the eastern High Atlas, Morocco along the Hamar Laghdad ridge (see Fig.5.32). These mounds are thought to be related to hydrothermal vents. Mound in foreground is 32m high (arrow). Photo from Belka, 1998, used with permission of SEPM. (B) Middle Devonian carbonate mound of the Bader Basin (Jebel el Otfal) in the eastern Anti Atlas, Morocco. This 40m mound is thought to be dominated by microbial activity and was developed in relatively deep water. Photo from Kaufmann, used with permission of SEPM. (C) 10m high Cretaceous Campanian carbonate mounds located near Pueblo, Colorado, USA. These mounds are thought to be related to cold basinal petroleum seeps.

that is probably a reasonable view of the early diagenesis suffered by most Phanerozoic mud mounds sited in relatively deep water. The model is in two phases; the first relates to the processes operative along the surface layer of the mound (Fig. 5.30) and the second, the processes operating within the mound during early burial (Fig. 5.31).

On the surface of the mound, active biofilms drive the mound growth and early lithification by production of in situ $CaCO_3$ mud and fine clotted cements (Fig. 5.30). Lithified and semi-lithified substrates provide colonization sites for encrusters such as bryozoans and epifauna such as crinoids, sponges, brachiopods, and other taxa that further strengthen the frame of the mound.

Just below the surface of the active biofilm, bacterial degradation leads to pH changes resulting in both silica and carbonate dissolution. Carbonate dissolution includes the enlargement of primary voids such as fenestrae as well as the development of secondary solution voids and the local collapse of primary

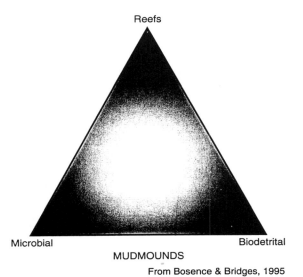

Fig. 5.27. Tripartite classification of carbonate mounds.

matrix (Fig. 5.31). Dissolved bicarbonate sourced from the active biofilm at the surface leads to the precipitation of large volumes of marine botryoidal and radiaxial cements in both primary and secondary voids. As the diagenetic site moves from oxic to suboxic conditions, bacterial sulfate reduction provides pyrite; and methogenic bacteria oxidize methane to provide more bicarbonate for void-fill marine cement. Primary voids within the mounds are no doubt interconnected and seem to have maintained some connection to the surface of the mound as shown by laminated pelleted geopetal void fills (Fig. 5.29).

Figure 5.29 is a schematic diagram

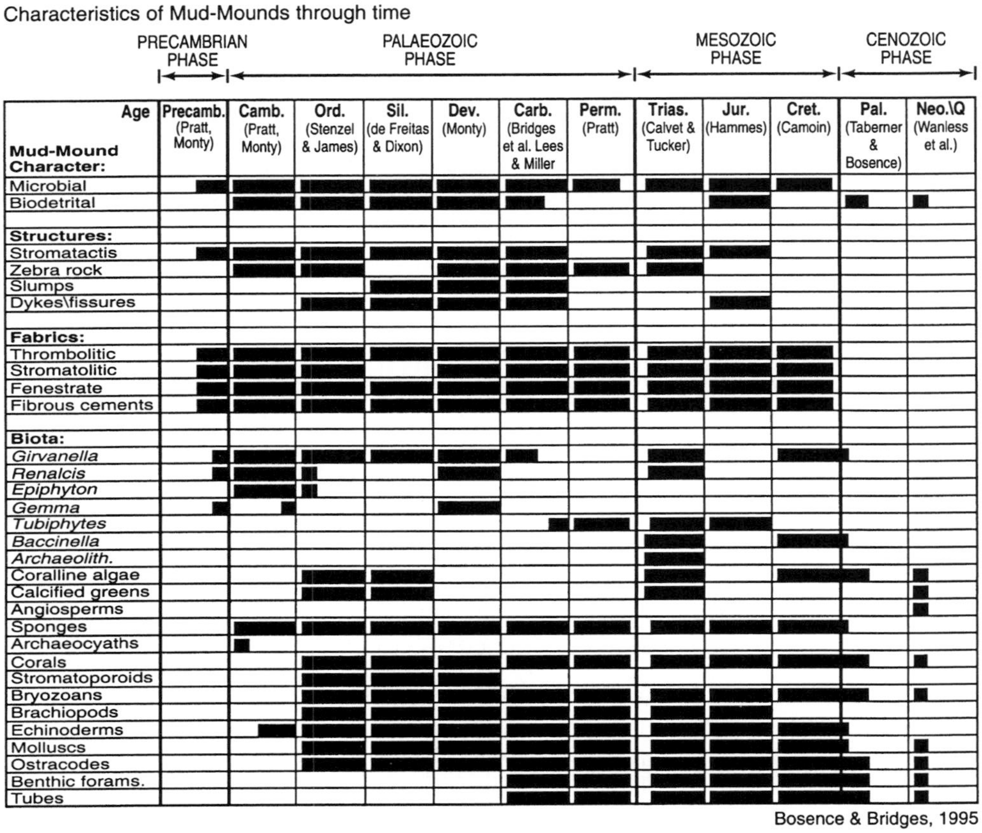

Fig. 5.28. Distribution, characteristics and biota associated with carbonate mounds through time. Used with permission of the IAS.

showing the four early marine diagenetic stages suffered by Waulsortian mounds, and by extension, many Phanerozoic mud mounds. While the major diagenetic event is precipitation of marine cements, dissolution, as a by-product of biologic degradation, can be a significant process.

The porosity of most mud mounds seems to have been lost by biologically mediated cementation on the sea floor. There are, however, a number of instances where carbonate mud mounds are hydrocarbon reservoirs with significant porosity and permeability. The Pennsylvanian phylloid algal mounds of the Paradox basin, for example, are among the reservoir rocks of the giant Aneth Field (Grammer et al., 1996). It is easy to visualize the development of meteoric secondary dissolution porosity in mounds that have accreted into shallow water, like some of the shallower water mounds of the Paradox, where they may be subaerially exposed during lowstands of sea level. It is more difficult to see how mud mounds deposited in 100's of meters of water could ever have been exposed to porosity-enhancing meteoric dissolution processes. Under

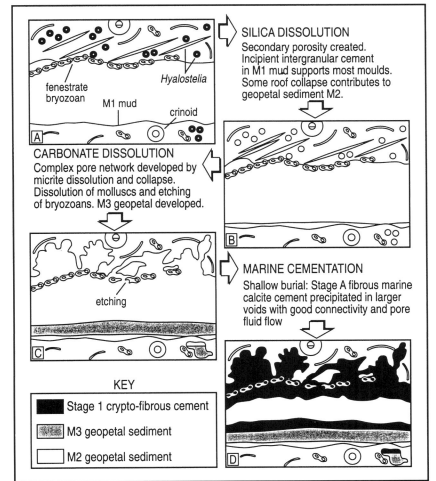

Fig. 5.29. Common diagenetic fabrics of carbonate mounds. Note that Stage 1 crypto-fibrous cement (see lower right panel) is what is typically termed stromatactis. The elongate pores shown in the lower right panel will be seen in slab and outcrop as fenestrae. Used with permission of the IAS.

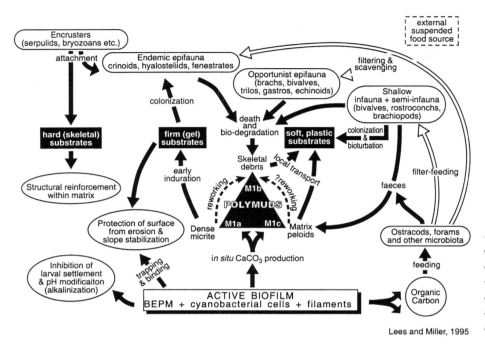

Fig. 5.30. Conceptual diagenetic model for Waulsortian mounds. Processes operative along the surface layer of the mound. Used with permission of the IAS.

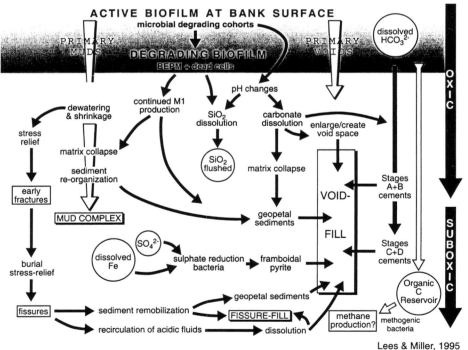

Fig. 5.31. Conceptual diagenetic model for Waulsortian mounds. Processes operative witin the mound. Used with permission of the IAS.

these conditions, it is conceivable that the microbial-related dissolution events described above could provide the porosity seen in these mud mound reservoirs, particularly if the carbonate saturation potential in the surrounding marine waters was lowered by the reduced temperatures found in deeper marine environments.

Carbonate diagenesis associated with mounds located near sea floor hydrothermal and cold hydrocarbon vents.

Carbonate mounds similar to the carbonate mud mounds described above form in relatively deep-water, shale-prone environments as a result of hydrothermal hot springs and petroleum, sulfur and methane cold seeps on the sea floor. The main diagenetic drivers in this extreme environment seem to be chemoautotrophic bacteria. These bacteria oxidize various reduced chemical species such as sulfides and convert CO_2 into organic carbon that other organisms such as tubeworms can utilize in their living processes. One of the most important processes relative to carbonate mound formation at these sites is the bacterial oxidation of methane to form CO_2 (Kaufman et al., 1996), ultimately leading to extensive submarine cementation, mound formation and preservation.

Spectacular conical Lower and Middle Devonian mud-mounds occur in Morocco and Algeria (Kaufmann, 1997; Belka, 1998). The Middle Devonian mounds in both Morocco and Algeria are typical Paleozoic microbial-dominated mud mounds with stromatactis similar to the slope mounds described in the previous section (Kaufmann, 1997). The Early Devonian "Kess Kess" mounds in Morocco, however, seem to be directly related to hydrothermal venting (Belka, 1998). The mounds occur in the northeast corner of the Anti-Atlas at Hamar Laghad (Fig. 5.32). They occur during a relatively short time period within the Early Devonian Emsian (Fig. 5.33) and are encased in coeval crinoidal mudstones to packstones,

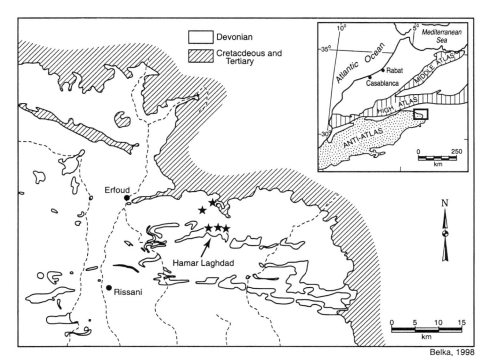

Fig. 5.32. Location map of northeastern Morroco showing the distribution of the Early Devonian Kess Kess carbonate mounds (see Fig. 5.26) in the Anti Atlas Mountains. Used with permission of SEPM.

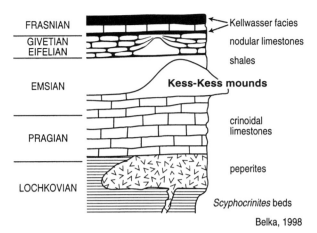

Fig. 5.33. Stratigraphic setting of the Early Devonian Kess Kess carbonate mounds. Used with permission of SEPM.

marls, and shales. While the mud mounds are dominated with stromatactis, they have intercalations of biogenic packstones that exhibit high species diversity but no stromatoporoids, algae or bryozoa (Belka, 1998).

The mounds developed on sea floor topographic highs which were formed by a submarine volcanic eruption in the Lochkovian (Fig. 5.34 and 5.36). The subsequent uplift of the area by a lacolith intrusion resulted in two sets of intersecting faults, a radial and a tangential set. The mounds developed over these fault intersections (Fig. 5.34). Stable isotope data (Fig. 5.35) suggest that hydrothermal fluids (strongly depleted $\partial^{18}O$ values) as well as normal Devonian seawater ($\partial^{18}O$ values near -5) were involved in formation and lithification of the mounds. Strongly depleted $\partial^{13}C$ values (Emsian seawater has a stable isotopic composition of +0.62 to +2.43 for $\partial^{13}C$), however, suggest that bacterial methane oxidation may also have played a major role in their formation and lithification. Figure 5.36 is a conceptual model of the "Kess Kess" mound formation.

Roberts and others (1989) describe modern carbonate mounds related to cold petroleum seeps from the continental slope of the northern Gulf of Mexico. These seep-related mounds support exotic chemosynthetic biota and exhibit extensive carbonate cementation in an otherwise deep-water shale environment. While oxygen isotopic values of carbonate mound components are consistent with cold marine bottom water, their carbon isotopic values are significantly depleted suggesting bacterial methane oxidation.

In the ancient record, mounds with similar geochemical and biological characteristics, and occurring in similar environmental settings, have been described from the Miocene of

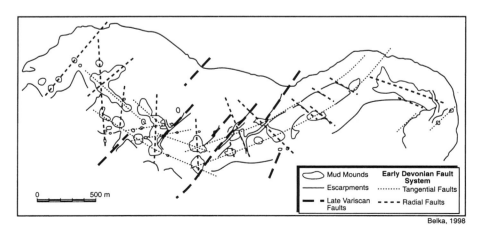

Fig. 5.34. Distribution of Kess Kess mounds relative to fault network at Hamar Laghdad. Used with permission of SEPM.

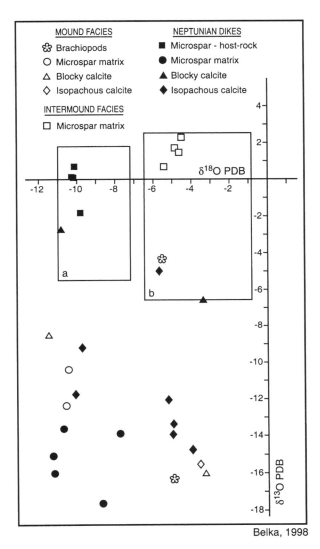

Fig. 5.35. Carbon and oxygen isotope values measured on calcite samples from the Hamar Laghdad area. Boxes indicate $\partial^{13}C$ and $\partial^{18}O$ ranges of finely crystalline carbonate matrix of the mounds (a) and of nonluminescent cements in crinoidal limestones underlying the mounds and in stromatactis cavities (b) as reported by Mounji et al., 1996). Used with permission of SEPM.

northern Italy, the Cretaceous Campanian of Colorado, U.S.A., the Cretaceous Albian of the Canadian Arctic and the Jurassic Oxfordian of southeastern France (Gaillard et al., 1992).

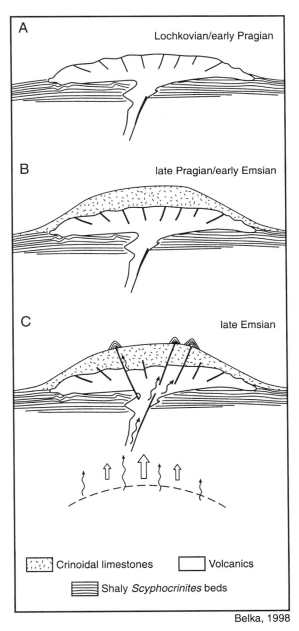

Fig. 5.36. Model for origin of the Kess-Kess mud mounds. (A) Submarine basaltic eruption. (B) Crinoid colonization and accumulation of thick-bedded crinoidal packstones. (C) Doming of the paleohigh due to reactivation of the magmatic source in the subsurface; ascent of hydrothermal fluids; formation of carbonate mounds at sites of venting. Not to scale. Used with permission of SEPM.

Carbonate diagenesis associated with steep escarpment shelf margins: Pacific Enewetak Atoll studies illustrating dissolution, cementation and dolomitization driven by thermal convection

Saller's work on Enewetak (1984a; 1984b; 1986) provided the first clear view of the effects of deep marine waters on the coarse, mineralogically metastable carbonate sequences exposed on the steep constructional escarpment margins of platforms and oceanic atolls. A series of observational drill holes penetrate the entire carbonate section of Enewetak Atoll (some 1400m) through the Upper Eocene, which lies directly on volcanic basement (Fig. 5.37). The facies encountered in cores from these drill holes mirror present Enewetak Atoll margin facies, including lagoon margin, backreef, reef crest, and fore-reef facies. This observation suggests that the atoll margin has retained similar facies and has been located near its present site since initiation of the atoll, atop a marine volcano in the Eocene. Lithologies range from boundstones to grainstones and packstones and were originally composed of magnesian calcite and aragonite (Saller, 1984b; 1986).

The upper 60-100m of the carbonate sequence on Enewetak representing the Holocene

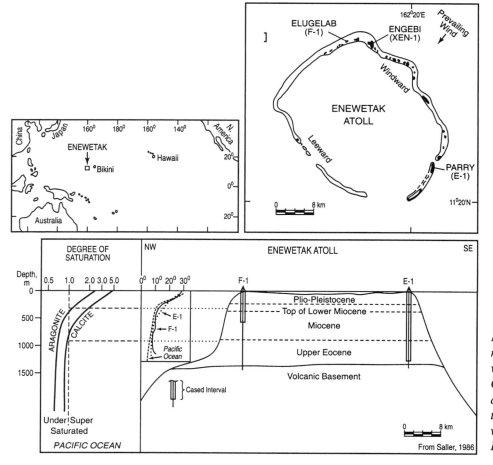

Fig. 5.37. Location map of Enewetak Atoll, western Pacific (top). Geologic (lower right) and oceanographic setting (lower left). Used with permission of Elsevier.

and Pleistocene show the effects of repeated exposure to subaerial conditions during sequential glacial-related sea level lowstands. There is some evidence of subaerial exposure during the Middle-to-Upper Miocene. There is, however, no evidence of major unconformities in the section below 200m. The carbonate section of the Lower Miocene and Upper Eocene, then, has been continuously under the influence of progressively deeper marine water of the western Pacific Basin since the Upper Eocene (Saller, 1984b; 1986).

One of the deep drill holes on Enewetak exhibits tidal fluctuations and amplitudes in phase with marine tides, even though it is located up to 3 km from the adjacent ocean and cased to below 600m. These observations indicate free marine circulation with adjacent ocean waters through deep carbonate units. In addition, the thermal profile of this well tracks that of adjacent ocean water, reinforcing the conclusion that deep carbonate sequences on the margin of the atoll are in open communication with adjacent cold marine waters below the aragonite lysocline (375m, as shown on Figure 5.37) (Saller, 1984b).

In cores from these wells taken at depths between the present aragonite lysocline depth (375m) and the present calcite lysocline (1000m), Saller (1984b; 1986) documents the pervasive dissolution of aragonite, and the precipitation of radiaxial and bladed calcite ce-

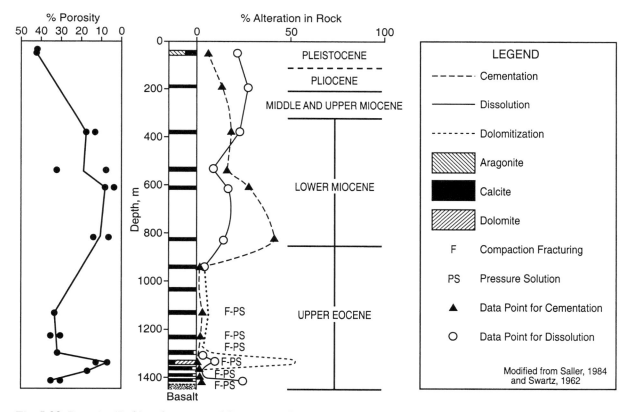

Fig. 5.38. Porosity (Left) and patterns of diagenetic alteration (Right) exhibited by the F-1 deep drill hole at Enewetak Atoll. Used with permission of Elsevier.

ments containing an average of 3.2 mole % $MgCO_3$ (Fig. 5.13). These cements are often precipitated into bioclast molds, formed by dissolution of *Halimeda* segments and other aragonite grains; they are isotopically in equilibrium with present cold, deep marine waters (Fig. 5.4). Saller (1986) also documents the progressive loss of magnesium from magnesian calcites through this zone, confirming the observations of Schlager and James (1978) concerning mineralogical changes associated with the lithification of peri-platform oozes below the aragonite lysocline depth along the margins of Tongue of the Ocean in the Bahamas.

Figure 5.38 illustrates Saller's (1984b) estimates for the percentage of diagenetic alteration (as cementation, dissolution, and dolomitization) seen in one of the two deep Enewetak core holes. In the upper section where subaerial exposure was common, dissolution was the major active diagenetic process. Between the aragonite and calcite lysocline depths, cementation was the dominant process. Below the calcite lysocline depth, dolomitization became dominant. Porosity data from this well (Swartz, 1962), as plotted on Fig. 5.38, seem to indicate that radiaxial calcite cementation in relatively deep water along platform margins is important and can have a significant impact on porosity development. As indicated earlier in this chapter, ancient radiaxial calcite cements often occur in sequences that were, or could have been, in contact with circulating marine waters during or shortly after deposition in settings similar to the Enewetak situation. These sequences include Paleozoic mud mounds (Lees and Miller, 1995), the Devonian in Alberta (Carpenter and Lohmann, 1989), the Devonian reef complex of western Australia (Hurley and Lohmann, 1989), and finally, the Lower Cretaceous shelf margin, south Texas and Mexico (Prezbindowski, 1985; Enos, 1986).

Below the calcite lysocline depth, cementation and dissolution seem to cease. Porosity, relative to the overlying radiaxial calcite cementation zone, increases dramatically, presumably because of lack of pore-fill cement rather than an increase in dissolution (Fig. 5.38). Calcite dissolution and pervasive dolo-

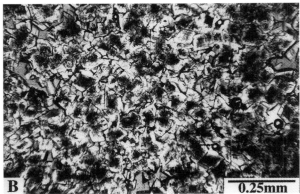

Fig. 5.39. (A) Slab photo showing pervasive dolomitization of Upper Eocene sequence in well F-1 at 1350 m (4429'). Note solution vugs (arrows). (B) Thin section photomicrograph of (A) showing cloudy centers and clear rims in each dolomite rhomb. Note that significant porosity (p) is still present. Photographs furnished by Art Saller.

mitization begin to affect the section in the Enewetak cores at about 1200m, fairly close to the base of the carbonate sequence (Fig. 5.39). Saller (1984a) explains these dolomites as the result of dolomitization by cold, deep marine waters, undersaturated with respect to calcite but still saturated with respect to dolomite, flushing through porous reef-related carbonate sequences. Stable isotope (Fig. 5.40) and trace element compositions of the dolomites are compatible with precipitation from normal marine waters. Strontium isotope composition of the dolomites contained in Upper Eocene strata indicates that dolomitization commenced in the Miocene and probably continues on to the present (Saller, 1984a).

Thus the atoll margins at Enewetak have undergone extensive dissolution of aragonite, precipitation of huge volumes of porosity plugging calcite, and at the base of the carbonate pile, dolomite replacement of calcite, all accomplished by normal marine waters. How does one move the enormous volumes of marine water through the margins of the atoll that are needed to accomplish this solution, precipitation, and dolomitization? The main zones of marine diagenesis are too deep to call upon wave, tidal and current activity as a hydrologic pump. Saller (1984a) used thermal convection to move the large volumes of marine fluids through the atoll that are needed to accomplish the marine diagenesis observed at Enewetak. Saller (1984a) based his model on the earlier work of Swartz (1958) who postulated the presence of thermal convection systems within Enewetak and Bikini atolls.

Carbonate diagenesis associated with steep escarpment shelf margins: marine water dolomitization of Bahama Platform by Kohout or topography-driven hydrologic flow

Simms (1984) applied a marine-driven Kohout (thermal) convection model to the Bahama Platform to explain the pervasive dolomitization of Tertiary-to-Pleistocene carbonates within the core of the platform (Figs. 5.41 and 5.42). The Kohout model is an open-cycle thermal convection cell which develops because of a strong horizontal density gradient between the cold ocean water around the platform and the warmer interstitial waters heated by the geothermal gradient within the platform (Fig. 5.43B) (Kohout, 1965). While Simms assumed that the dolomitizing fluids were marine, these platform dolomites were previously believed to have been the result of fresh marine water

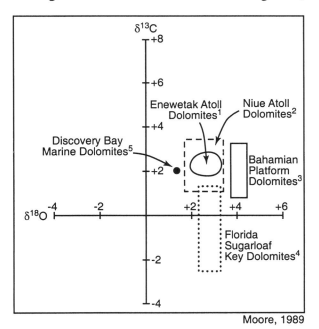

Fig. 5.40. Oxygen and carbon isotopic composition of Enewetak Atoll dolomites compared to other reported marine dolomites. Data from 1. Saller, 1984, 2. (Aharon, 1987), 3. (Vahrenkamp, 1994 #13329), 4. (Carballo, 1987), 5. (Mitchell, 1987).

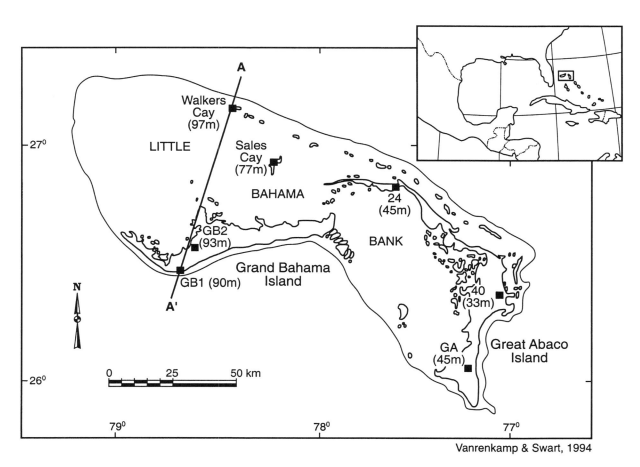

Fig. 5.41. Location and core lengths of cores on Little Bahama Bank that penetrate pre-Pleistocene sediments. Cores from Walkers Cay, Sale Cay, Grand Bahama Ialand and Great Abaco Island recovered dolomites. Used with permission of IAS.

mixing. Simms (1984) makes a persuasive case for the importance of Kohout convection-related dolomitization along the steep seaward margins and into the cores of major carbonate platforms and shelves.

In subsequent studies on Little Bahama Bank, Vahrenkamp and others (1988; 1990; 1991; 1994) detailed the distribution, petrography and geochemistry of massive platform dolomites within Little Bahama Bank based on a suite of 6 core holes across the platform (Figs. 5.41 and 5.42). Stable isotopic data clearly suggest that the dolomitizing fluid is normal marine water (Fig. 5.44). Their isotopic and trace element data suggest distinct diagenetic gradients from platform margin to platform interior Vahrenkamp and Swart, (1987). What sort of pump can one call upon to move the prodigious volumes of marine water through the platform to completely dolomitize over 80m of limestone? Three potential hydrologic models are shown in Figure 5.43. Vahrenkamp and Swart's geochemical studies seem to negate the evaporative reflux model. Vahrenkamp and Swart (1994) favor the mixing zone-induced seawater circulation model because of the geometry

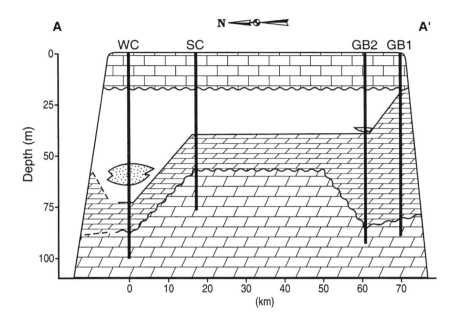

Fig. 5.42. North-south cross section of Little Bahama Bank along line (A-A') of Fig. 1. A Pleistocene limestone (Lucayan Formation) of uniform thickness covers pre-Pleistocene limestones and dolomites. The lower dolomite strata were dolomitized during the early Late Miocene (phase I), the upper dolomite during the Late Pliocene and latest Pliocene/early Pleistocene (phases II and III). The timing of replacement dolomitization suggests a Middle Miocene and a pre-Late Pliocene age for the respective stratigraphic units. The contact between lower and upper dolomite strata indicates significant depositional and/or erosional topography on Little Bahama Bank before deposition of the precursor sediments of the upper dolomite. Used with permission of IAS.

of the dolomite mass within the platform (Fig. 5.42).

This model only works during low sea level stands, and large floating fresh water lenses must be developed in the upper platform. Movement of fresh water toward the margins of the platform, driven by topography generated by the lowered sea level entrains subjacent marine water. Marine water flow from the platform margin into the interior is initiated in order to replace the water lost along the mixing zone.

Wheeler and Aharon (1993; 1997) suggested that the dolomites on Niue Atoll in the south Pacific were also the result of mixing zone induced seawater circulation. Aharon et al. (1987) had previously contended that the Niue dolomites were the result of thermal convection much like the situation on Enewetak (Saller, 1984a). It would seem premature, however, to abandon Simms' Kohout convection model for the Little Bahama Bank dolomites

A. REFLUX

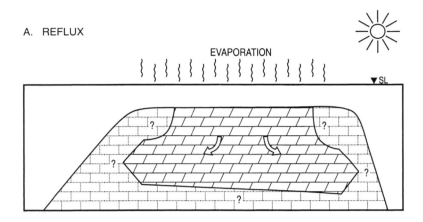

B. GEOTHERMAL CONVECTION

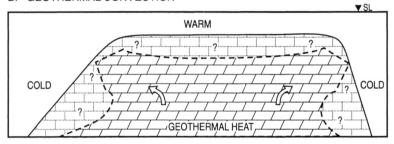

C. MIXING ZONE-INDUCED SEAWATER CIRCULATION

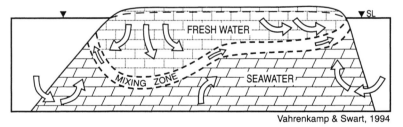

Vahrenkamp & Swart, 1994

Fig. 5.43. Three groundwater flow models with seawater circulation which have been associated with platform dolomitization. (A) Sea water refluxing and (B) Geothermal convection require (partial) platform submergence for shallow-burial near-surface dolomitization, whereas mixing zone-induced seawater circulation (C) can only develop during (partial) emergence. Asymmetry of flow regimes and hence dolomite bodies may be the result of density gradient, platform shape, topography, permeability differences, etc. Used with permission of IAS.

solely on the basis of dolomite geometry because its distribution under the platform is based on just 6 core holes.

By computer modeling, Kaufmann (1994) determined that either the Kohout thermal, or the mixing zone seawater circulation model, could pump enough seawater through the platform to dolomitize the platform interior. In the case of Kohout circulation, the Mg flux associated with the flux of marine waters is so slow that the steep platform margin must be in existence for millions of years in order to accomplish the dolomitization. This constraint is not a problem for the Bahamas or for the atolls in the Pacific. On the other hand, Kaufmann (1994) suggests that mixing zone marine cir-

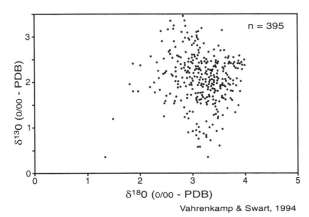

Fig. 5.44. Cross-plot of oxygen versus carbon isotopic of Little Bahama Bank dolomites. Used with permission of IAS.

culation, tied to eustatic pumping, can mobilize a large volume of Mg-bearing seawater over a long period of time depending on eustatic periodicity. In this case, the Mg flux would be relatively high and conducive to platform-wide dolomitization in a geologically reasonable time frame.

The Middle Cretaceous Tamabra Limestone at Poza Rica, was thought by Enos (1977) to represent deep-water, mass-flow deposits from the El Abra-Golden Lane platform margin to the east. These deep-water sequences are commonly dolomitized, and Enos (1988) favors dolomitization as the result of deep circulation of meteoric waters from the surface of the adjacent, exposed platform. As indicated by Enos (1988), however, the geologic setting, patterns of diagenetic modification, and stable isotopic composition of the dolomites are all compatible with dolomitization by marine waters driven through the platform margin by Kohout convection. The Poza Rica trend, then, could well represent an economically significant example of porosity enhancement by deep marine dolomitization driven by Kohout thermal convection.

SUMMARY

Porosity modification in the marine diagenetic environment is centered on cementation and dolomitization with some dissolution porosity enhancement beneath the aragonite and calcite lysoclines. Marine cementation may be either abiotic or mediated by microbial processes. There are two major subenvironments where these marine diagenetic processes are operative: the shallow water, normal marine diagenetic environment within the warm, well mixed, oceanic surface zone; and the deep slope to basin diagenetic environment, in the density-temperature stratified oceanic water column below the surface mixing zone.

Porosity evolution in the shallow water, normal marine environment is dominated by porosity loss through cementation, both abiotic and microbial. The two major sites of marine cementation are the high-energy intertidal zone and the shelf-margin framework reef, where high-energy, stable substrates, high organic activity, and high porosities/permeabilities ensure a high rate of fluid flux and the elevated saturation states necessary for significant cementation. While intertidal cementation (as well as cementation associated with hardgrounds) tends to be vertically and laterally restricted, these cemented zones can serve as reservoir seals and may vertically compartmentalize reservoirs. Marine abiotic and microbial cementation in reefs, combined with bioerosion and internal sedimentation, can, depending on reef type and accretion rate, totally destroy the high initial porosities enjoyed by most reef sequences.

The keys to the recognition of marine cementation in the ancient record are the occurrence of distinctive cement textures and fabrics, marine geochemistry, and geologic setting. Reef-related marine cementation is a major factor in the porosity evolution of a number of economically important ancient reef trends.

The deep marine environment from slope to basin can be especially important relative to porosity evolution. Today active, rapid aragonite cementation on the upper slope extends to depths of at least 60m. In the Paleozoic and Mesozoic, carbonate mud mounds developed on upper shelf slopes and in the deeper waters of distal ramps. Microbial processes centered on massive early cementation-dominated marine diagenesis in these mud-mounds. Porosity enhancement by early biochemical dissolution may allow some of these mud mounds to become important reservoir rocks.

Mud mounds formed in conjunction with hydrothermal and cold petroleum vents on the sea floor are sites of massive cementation and mineralization driven by chemosynthetic bacteria. Their economic significance is uncertain at the present time.

Steep margins of carbonate platforms fronting oceanic basins are sites where the thermocline and the carbonate compensation depth impinge on previously deposited carbonate sequences. Dissolution of aragonite, precipitation of radiaxial calcite cements, dissolution of calcite, and, finally, dolomitization can occur and modify porosity as the shelf margin encounters progressively deeper and colder waters toward the basin.

Kohout thermal convection and mixing-zone-induced sea water circulation driven by eustatically coupled topography are the most likely hydrologic pumps enabling the movement of large volumes of marine water through steep platform margins. These newly refined models of platform-scale marine diagenesis are important concepts in our continuing search for understanding massive platform dolomitization.

Chapter 6

EVAPORATIVE MARINE DIAGENETIC ENVIRONMENTS

INTRODUCTION

Marine and marine-related evaporites are precipitated and/or deposited in a number of marine and marginal marine environments (Schreiber et al., 1982; Kendall, 1984; Schreiber, 1988; Tucker and Wright, 1990). The following discussion, however, does not exhaustively survey the complex world of evaporites. Instead, the focus is on those general situations where the marine evaporative regime significantly impacts the porosity evolution of carbonate marine sequences, either by porosity modification related to dolomitization, or by porosity occlusion and reservoir sealing due to precipitation of evaporites.

There are two general evaporitic settings that will be considered in this chapter. The first is the marginal marine sabkha, where evaporites are generally formed within fine-grained carbonate tidal flat sequences. The second setting is the subaqueous hypersaline marginal marine environment, such as the lagoon or the salina, where evaporites are generally formed in the water column.

The following pages will be a review of the special diagenetic conditions found in conjunction with waters that have been modified by evaporation, followed by a discussion of each of the two major diagenetic settings in turn. Emphasis will be on those processes and products having the potential for significant impact on porosity development and evolution in carbonate reservoirs.

Introduction to diagenesis in evaporative marine environments

When marine waters are evaporated in a closed laboratory system, such as in the seminal experiments of Usiglio (1849), a predictable series of mineral species are incrementally precipitated, starting with $CaCO_3$ and advancing through the evaporative mineral suite to include $CaSO_4$ and $NaCl$. The series ends with the precipitation of the bittern salts (KCl and others). During the evaporative process, the remaining waters become progressively denser (Fig. 6.1). In the more open natural carbonate systems found bordering modern oceans today, the processes become more complex with the evolving evaporative brines often interacting with adjacent sedimentary carbonate phases and mixing with other waters including continental meteoric waters.

The single most important factor relative to carbonate porosity modification in the evap-

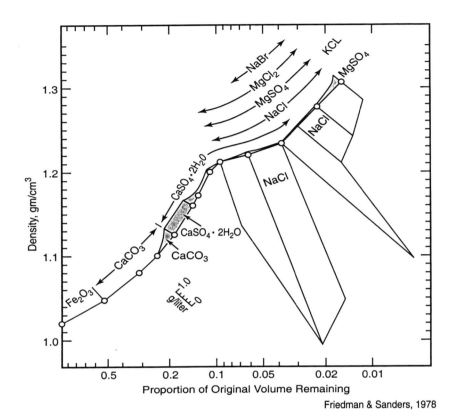

Fig. 6.1. Graph of density versus proportion of original volume remaining in closed-system evaporation of Mediterranean seawater by Usiglio (1849) showing physical conditions under which each mineral is precipitated. Amount of each mineral (in grams per liter) is shown by the lengths of lines drawn perpendicular to curve. Used with permission of John Wiley and Sons.

orative marine environment is the common association of dolomite with evaporite minerals. While most modern marine surface waters are supersaturated with respect to dolomite, formation appears to be particularly favored in evaporated marine waters (Hardie, 1987; Sun, 1994). Figure 6.2 illustrates the interaction of two of the main parameters that may influence the kinetics of the growth of dolomite in natural surface waters: the Mg/Ca of the precipitating fluid, and the salinity of the precipitating fluid (Folk and Land, 1975). Under conditions of elevated salinity such as may be found in marine environments, Mg/Ca ratios must be significantly higher in order to overcome the propensity of Mg^{2+} to form hydrates before dolomite can precipitate (Morrow, 1982).

The precipitation of $CaCO_3$ as aragonite and $CaSO_4$ as gypsum during the earlier stages of evaporation highstand 6.1) is, from the standpoint of porosity modification, the most important segment of the evaporative series. While the gypsum precipitate will ultimately form an aquaclude, and often acts as a reservoir seal, the progressive extraction of Ca^{2+} from the evaporating seawater to form aragonite and gypsum drives its Mg/Ca ever higher, favoring dolomite precipitation. Those evaporative marine waters associated with the precipitation of aragonite and gypsum, then, are not only chemically suited for the formation of dolomite, but also their higher density makes them a particularly effective Mg^{2+} delivery system for the dolomitization of adjacent carbonates.

The dolomites forming today under sur-

INTRODUCTION

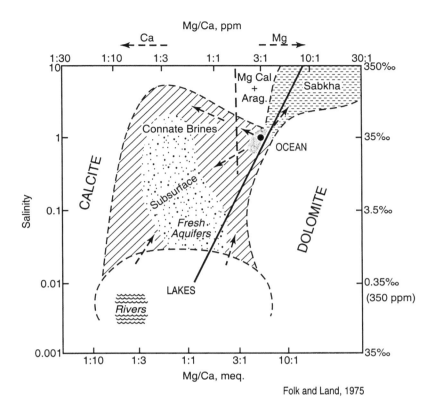

Fig. 6.2. Fields of occurrence of common natural waters plotted on a graph of salinity vs. Mg/Ca ratio. Fields of preferred occurrence of dolomite, aragonite, Mg-calcite and calcite are also shown. At low salinities with few competing ions and slow crystallization rates, dolomite can form at Mg/Ca ratios near 1:1. As salinity rises, it becomes more difficult for ordered dolomite structure to form, thus requiring progressively higher Mg/Ca ratios; in the sabkha, Mg/Ca ratios of 5 to 10:1 are necessary before dolomite can crystallize owing to abundance of competing ions and rapid crystallization. Used with permission of AAPG.

face evaporative marine conditions are generally poorly ordered, contain excess calcium, and have been termed protodolomites, or calcian dolomites (Gaines, 1977; Reeder, 1983). Land (1980) and Hardie (1987) make the point that ancient dolomites are generally more ordered and less soluble than their modern surface-formed disordered counterparts. They both strongly emphasize that surface formed dolomites tend to reorganize structurally and compositionally with time, "giving surface-formed dolomites a recrystallization potential that is increased with increasing temperature and burial" (Hardie, 1987). Subsequent workers (Moore et al., 1988; Gao et al., 1992; Kupecz et al., 1992; Kupecz and Land, 1994; Vahrenkamp and Swart, 1994; Gill et al., 1995; Malone et al., 1996) have documented extensive later re-

crystallization of early-formed dolomites from a wide variety of geologic ages and settings.

During the 70's and 80's the utilization of stable isotopes and trace elements to determine the characteristics of ancient dolomitizing fluids became increasingly more common and popular. Evaporative surface marine waters seem to leave a distinct geochemical fingerprint on dolomites formed under their influence. The stable isotopic composition of several suites of samples from a number of Holocene evaporative environments is presented in Figure 6.3. Modern surface-formed dolomites are generally characterized by enrichment with respect to ^{18}O and ^{13}C (PDB) (Land, 1985). In addition, they tend to be enriched with Na and Sr (Land, 1985; Allan and Wiggins, 1993).

The geochemical fingerprint approach

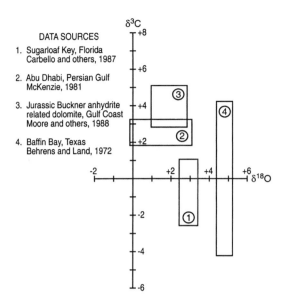

Fig. 6.3. Oxygen and carbon isotopic composition of evaporative-related marine dolomites. Data sources as indicated. Used with permission of AAPG.

recrystallization of surface-formed dolomites compounds the problem. The isotopic and trace element composition of ancient evaporite-related dolomite sequences, then, may actually only reflect a gross representation of the chemistry of recrystallizing fluids, rather than the composition of original precipitational fluids.

The following paragraphs discuss modern as well as ancient evaporative marine diagenetic environments and attempt to develop integrated stratigraphic-petrographic-geochemical criteria that can be used to recognize the major evaporative marine diagenetic environments and their associated dolomites.

THE MARGINAL MARINE SABKHA DIAGENETIC ENVIRONMENT

Modern marginal marine sabkhas

The marginal marine sabkha is perhaps one of our most important and best-characterized depositional environments (Tucker and Wright, 1990). Because of its transitional position between marine and continental condi-

described above must be used with extreme caution when dealing with ancient dolomitized sequences. While there is a great deal of uncertainty concerning trace element distribution coefficients, and fractionation factors in dolomites (Chapter 4), the prospect of extensive

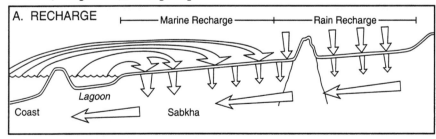

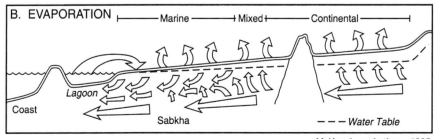

Fig. 6.4. A. Diagram of the sabkha illustrating water movement during recharge, mainly in winters or springs with shamal storms. B. Origin of water and circulation pattern during times of evaporation, which is the situation most of the year. Used with permission of Elsevier.

tions, the sabkha's highly evaporative pore fluids often have a complex origin. They may be derived from the sea by periodic surface flood recharge, from the land via gravity-driven continental aquifer systems, and finally, from the air via coastal rainfall (Patterson and Kinsman, 1977; McKenzie et al., 1980) (Fig. 6.4).

Modification of marine and continental waters by progressive evaporation drives sabkha pore fluids to gypsum saturation and leads to the ultimate precipitation of aragonite and gypsum within lagoon-derived aragonitic sabkha muds (Butler, 1969; McKenzie et al., 1980) (Fig. 6.5).

Gypsum and aragonite precipitation from sabkha pore fluids dramatically increase the molar Mg/Ca ratio of the pore fluids from marine values of 5 to 1 to over 35 to 1 as reported from modern sabkhas of the Trucial Coast (Fig. 6.5). These elevated Mg/Ca ratios are favorable for the formation of dolomite, either by the replacement of aragonitic sabkha muds (Patterson and Kinsman, 1977; McKenzie, 1981) or as a direct interstitial precipitate (Hardie, 1987).

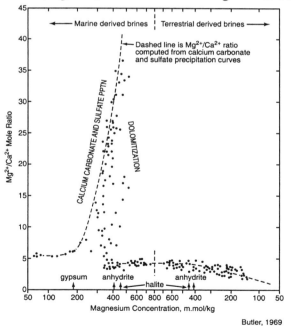

Fig. 6.5. Variation in magnesium-to-calcium ratio with magnesium concentration across the sabkha. Used with permission of SEPM.

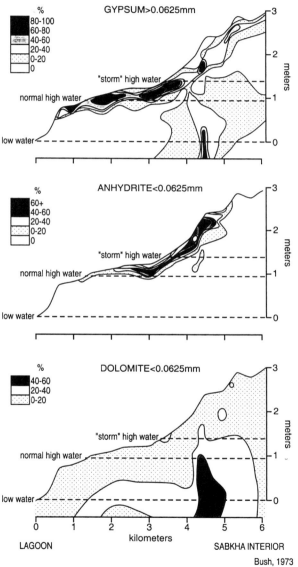

Fig. 6.6. Distribution of gypsum, anhydrite and dolomite across a sabkha traverse taken southeast of Abu Dhabi from point of low water (sea level) to 3 meters above low water, some 6 km from the lagoon shore. Used with permission of SEPM.

At levels of higher chlorinity, such as are found in the interior of the sabkha, waters are in equilibrium with anhydrite and may reach halite saturation. At the continental margins of the sabkha, retrograde diagenesis driven by continental brines may dominate, with sulfate equilibrium swinging back into the gypsum field (Butler, 1969; McKenzie et al., 1980).

As seen today in modern settings, such as Abu Dhabi along the Arabian Gulf, the sabkha sequence is relatively thin, averaging little more than 3 meters in thickness, with a lateral extent of some 12 kilometers. Syngenetic dolomites formed at Abu Dhabi are concentrated across a 3-5 km swath of the interior of the sabkha starting just landward from the lagoon, coincident with the highest reported pore fluid salinity values. Dolomitization in this area affects a 1 to 2m sequence of muddy lagoonal and intertidal sediments starting about a half-meter below the sabkha surface highstand 6.6). While dolomite percentages of up to 60% have been reported, 20 to 40% dolomitization is more common (Bush, 1973). Gypsum is generally concentrated in the upper meter of the sabkha sequence but can be found in quantity throughout the sequence in the interior of the sabkha in areas of highest chlorinity highstand 6.6). Anhydrite can occupy the entire upper half-meter of the sequence in the interior reaches of the sabkha highstand 6.6). Halite deposits, as crusts, are ephemeral and are generally destroyed by infrequent rains or seawater flooding (Butler, 1969; Bush, 1973). The interiors of marginal marine sabkhas are often marked by rapid intercalation with continental siliciclastics. This is a rather simplistic view of modern sabkha systems with

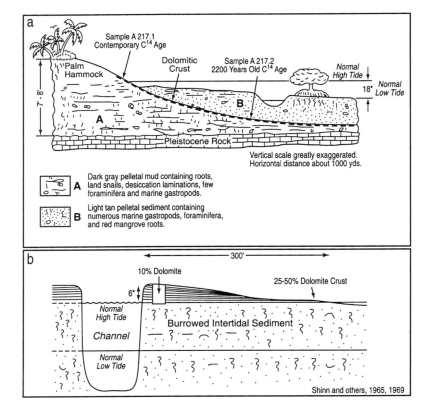

Fig. 6.7. a. Stratigraphic cross section across the flank of a palm hammock on Andros Island, the Bahamas. Note variation in age along the dolomitic crust. b. Schematic cross section across a tidal channel natural levee on the Andros Island tidal flats. Note that dolomite % is greater in the thin flanks of the levee than in the thicker channel margin part of the levee due to a dilution effect caused by the greater rate of sedimentation along the channel margins. Used with permission of SEPM.

emphasis on the diagenetic processes operating in the system. For an in-depth sedimentological analysis the reader is referred to Kendall and Warren (1988).

More humid conditions with greater meteoric water input will, of course, limit the production and preservation of evaporites in a supratidal environmental setting. It follows that reduced evaporite production may diminish the dolomitization potential within a supratidal environment. Modern humid supratidal sequences, such as have been described from south Florida and Andros Island in the Bahamas, exhibit sparse dolomite. This dolomite is generally concentrated in relatively thin surface crusts associated with the natural levees of tidal creeks, and adjacent to palm hammocks highstand 6.7) (Shinn et al., 1965; Shinn, 1983). There is some work that suggests that these limited dolomite crusts may be forming from normal marine waters aided by tidal pumping, rather than evaporative marine waters (Carballo et al., 1987). Mazzullo and others (1987; 1988; 1995) have described similar dolomite crusts with a similar origin in Belize.

Finally, Gebelein and others (1980), detected calcium-rich protodolomites beneath the tidal flats of Andros Island in the Bahamas. These dolomites occur in a spotty distribution in concentrations generally between a trace and 5%. These dolomites occur beneath all depositional environments but seemingly are more concentrated beneath palm hammocks. Initially thought to be associated with fresh-marine water mixing zones, these dolomites also seem to be normal marine in origin, with no apparent affinities to the surface dolomite crusts or evaporative fluids.

Diagenetic patterns associated with ancient marginal marine sabkhas

Ancient sabkha sequences are important oil and gas reservoirs, with dolomitized supratidal and subtidal sediments forming the reservoir, and associated evaporites often forming the seal. When one compares a variety of these economically important ancient sabkhas to their modern analogues from the Persian Gulf, one is invariably struck by the greater volume of dolomite present in the ancient sequences as compared to their modern counterparts. In the typical ancient shoaling upward sequence capped with sabkha deposits, the muddy subtidal and more open marine materials, as well as the overlying sabkha, are often

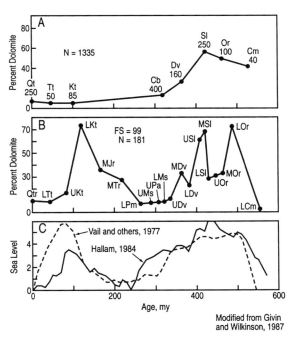

Fig. 6.8. A. Percentage dolomite as a function of age as reported by Chilingar (1956). B. Phanerozoic dolomite abundances recalculated in part by Given and Wilkinson (1987) from data in Chilingar (1956). C. Estimates of global sea level from Vail and others (1977)(1977) and Hallam (1984)(1984). Used with permission of SEPM.

totally dolomitized highstand 6.12, 6.15, 6.18) (Loucks and Anderson, 1985; Ruzyla and Friedman, 1985; Chuber and Pusey, 1985). As was mentioned above, dolomitization on modern Persian Gulf sabkhas is restricted laterally as well as vertically and seldom exceeds 60% of the sediment by volume. The lateral distribution of dolomitization may be extended by the sedimentologic progradation of the sabkha environment in a seaward direction. The vertical extension of dolomitization to previously deposited subtidal marine sediments below the sabkha, however, seems to necessitate the downward and lateral reflux of evaporite-related brines with elevated Mg/Ca through the porous, but muddy subtidal sediments. The modern sabkha hydrological system seems compatible with such a scenario (McKenzie and others, 1980; McKenzie, personal communication; Land, 1985). While modern muddy subtidal sediments display extremely high porosities that could support the displacement of marine pore fluid by heavy evaporative brines (Chapter 3), the system's operation has not been adequately documented.

Indeed, when one reflects on the reported skewed distribution of dolomites through geologic time (Fig. 6.8A), one must consider the possibility that the carbonate chemistry of ancient oceans was more favorable to dolomitization (Ingerson, 1962; Zenger, 1972; Mottl and Holland, 1975; Zenger and Dunham, 1980).

Given and Wilkinson (1987), however, recently re-evaluated much of the data on dolomite distribution versus age and have developed a compelling case for a cyclic distribution of dolomite through time (Fig. 6.8B), rather than the exponential increase with age reported by previous authors. Distinct modes of high dolomite percentage in the Mesozoic and Lower Paleozoic correlate well with extensive continental marine flooding expected during times of relatively high sea level. Given and Wilkinson (1987), assuming that most dolomitization is accomplished by marine waters, related these patterns of dolomitization to periods of higher P_{CO_2}, lower CO_3^{-2} concentrations, and hence

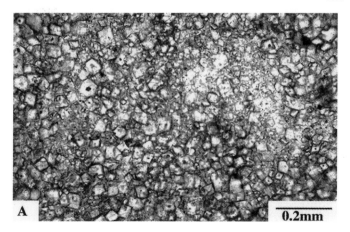

Fig. 6.9. (A) Sucrosic dolomite from a surface exposure of a Texas Lower Cretaceous supratidal sequence. Dolomite is thought to have been flushed by meteoric water during its early diagenetic history. Porosity is 34%, permeability is 18md. Plain light. (B) SEM photomicrograph of same sample showing loose arrangement of individual dolomite rhombs.

lower levels of calcite saturation in marine waters that would act to promote dolomitization (Machel and Mountjoy, 1986; Sun, 1994).

Porosity development and modification in sabkha-related reservoirs centers about dolomite and the dolomitization of the original mud-dominated sediments characteristic of the supratidal sabkha and adjacent subtidal marine-to-lagoonal environments. The common occurrence of moldic porosity in association with fine crystalline dolomite in subtidal and sabkha sequences (Moore et al., 1988; Allan and Wiggins, 1993) suggests preferential dolomitization of muddy matrix, as well as the introduction of fresh water into the environment, perhaps shortly after dolomitization. Indeed, fresh water flushing of partially dolomitized muddy sediment would tend to dissolve the remaining undolomitized aragonite, magnesian calcite or calcite, concentrating the initially floating dolomite rhombs into a crystal-supported fabric. This process will result in the porous sucrosic dolomite texture so common to many ancient sabkha sequences (Ruzyla and Friedman, 1985)(Fig. 6.9). In addition, fresh water would tend to dissolve evaporites associated with the dolomites, leading to a dramatic increase in porosity by the development of vugs after gypsum and anhydrite. In the upper parts of sabkha sequences, where gypsum and/or anhydrite are more concentrated, sulfate removal can lead to solution-collapse breccias and the development of significant additional porosity (Loucks and Anderson, 1985), (Fig. 6.19). Finally, the common introduction of fresh meteoric water shortly after or during sabkha progradation is the ideal vehicle to stabilize, through recrystallization, the metastable dolomites commonly associated with marginal marine sabkhas (Land, 1985).

Fresh water influence in a sabkha sequence is the normal result of the transitional nature of the sabkha, a dominantly subaerial depositional environment standing midway between continental and marine conditions. Roehl (1967) coined the term diagenetic terrain for the subaerially exposed landward margins of the sabkha perpetually under the influence of continental and meteoric waters. At the end of a high frequency cycle, or during the lowstand at the beginning of a third-order sequence, the diagenetic terrain would be expected to develop and to diagenetically modify the marginal marine sabkha. *The common occurrence of fresh water influx and the development of diagenetic terrains at the end of each sabkha cycle is perhaps the single most important factor controlling the development of economic porosity within sabkha-related sequences.*

Ordovician Red River marginal marine sabkha reservoirs, Williston Basin, USA

One the best-known, economically important ancient sabkha sequences is the Ordovician Red River Formation. It produces from a number of fields in the U.S. portion of the Williston Basin. The Red River is preferentially dolomitized around the margins of the Williston Basin and across paleohighs such as the Cedar Creek Anticline, coincident with thick cyclical evaporite-bearing sabkha sequences (Ruzyla and Friedman, 1985; Allan and Wiggins, 1993). The Cabin Creek Field, developed along the crest of the Cedar Creek Anticline in southeast Montana, (Fig. 6.10) is a typical Red River sabkha-related reservoir.

The Cabin Creek Field has produced over 75 million barrels of oil from dolomitized Low-

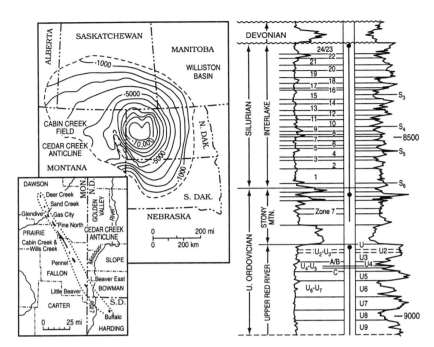

Fig. 6.10 (Left) Location map of the Cabin Creek Field on the Cedar Creek Anticline in the southwestern margins of the Williston Basin (Roehl, 1985). (Right) Type radioactivity log of Lower Paleozoic formations associated with the Cedar Creek Anticline. Major zones and markers are shown. Solid circles and lines in the center of the log indicate productive zones (from Roehl, 1985). Reprinted with permission from Carbonate Petroleum Reservoirs, Springer-Verlag, New York.

er Ordovician Red River and Silurian Interlake sequences with an average porosity of some 13%. The field is a combined structural-stratigraphic trap occurring at an average depth of 9000ft (2750m). The upper 50m of the Red River at Cabin Creek consists of 3 shoaling-upward cycles (4th order depositional sequences) capped by sabkha complexes with diagenetic terrains developed at the sequence-bounding unconformities (Fig. 6.11) (Ruzyla and Friedman, 1985). While all the supratidal sequences are dolomitized, subtidal carbonates exhibit both limestone as well as dolomite (Fig. 6.11). Although most of the effective porosity is concentrated within the dolomitized supratidal cap, good porosity occasionally occurs in dolomitized subtidal intervals toward the base of a sequence (Fig. 6.11). A schematic diagram of a typical Red River sabkha-capped sequence outlines the general distribution of diagenetic features, porosity, and pore types that might be expected in similar sequences (Fig. 6.12). Anhydrite is lost toward the base of the sequence, while the entire sequence might be dolomitized. Porosity quality and porosity type is variable and may be facies selective in response to fresh water leaching of evaporites, aragonitic grains, or undolomitized mud matrix. Table 6.1 summarizes the porosity, permeability, and pore type for each environment at Cabin Creek, while Figure 6.13 outlines the diagenetic processes responsible for these pore systems. The dominance of secondary solution-related porosity such as vugs and molds clearly illustrates the influence of freshwater flushing in the development of porosity at Cabin Creek.

Mississippian Mission Canyon marginal marine sabkha reservoirs, Williston Basin, USA

The Mississippian Mission Canyon Formation is also a prolific oil producer in the Williston Basin. The Mission Canyon is a shal-

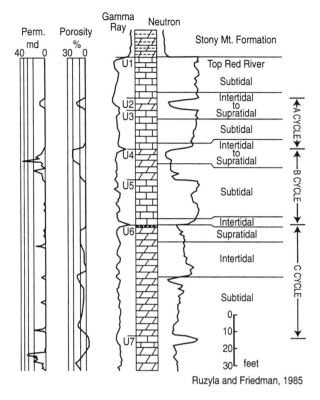

Fig.6.11. Typical columnar section of the upper Red River showing depositional environments, porosity, and permeability determined by core analysis and gamma ray-neutron curves. Reprinted with permission from Carbonate Petroleum Reservoirs. Springer-Verlag, New York.

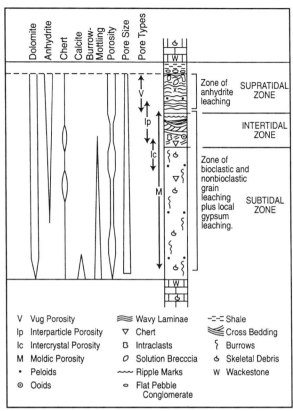

Fig.6.12. Depositional and diagenetic features of the upper Red River dolomites in the Cabin Creek field. Reprinted with permission from Carbonate Petroleum Reservoirs Springer-Verlag, New York.

Table 6.1

Zone & Environment	Porosity %	Perm (md)	N_c	Oil Sat. %	Water Sat. %	Total Sat. %	Pore Size (mm)	Relative Amounts of Each Pore Type[1]			
								V	M	Ip	Ic
U_4-U_5 supratidal	16.8	11.7	59	29.3	21.4	50.7	0.13	48%	13%	13%	26%
U_6-U_7 supratidal	7.4	0.6	110	16.3	47.4	63.7	0.04	95%	5%	—	—
U_4-U_5 supratidal	15.3	7.1	49	30.2	23.5	53.7	0.19	19%	81%	—	—
U_6-U_7 supratidal	7.1	3.6	156	17.2	39.4	56.6	0.13	31%	31%	26%	11%
U_4-U_5 supratidal	10.7	1.4	110	15.3	41.3	56.6	0.19	32%	64%	—	4%
U_6-U_7 supratidal	9.4	2.3	156	16.2	44.3	60.5	0.17	25%	44%	10%	21%

[1] V = vug, M = moldic, Ip = interparticle, Ic = intercrystal
N_c = number of core analyses
N_{ts} = number of thin-section analyses

Table 6.1. Porosity, permeability, saturation, and thin section data for dolomite reservoirs in the upper Red River Formation. Used with permission of Elsevier.

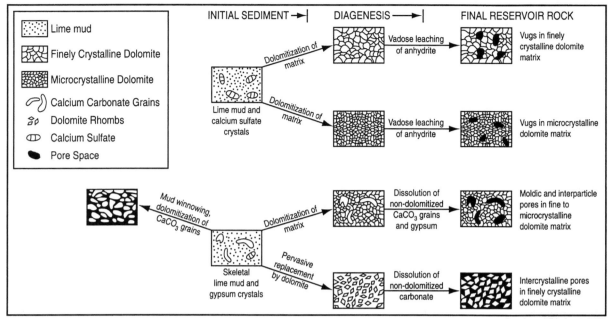

Fig. 6.13 Diagenetic processes involved in the development of upper Red River pore systems. Reprinted with permission from Carbonate Petroleum Reservoirs, Springer-Verlag, New York.

lowing-upward 3rd order depositional sequence with facies tracts ranging from basinal deep-water carbonates to evaporite-dominated coastal sabkhas and evaporative lagoons at the upper sequence boundary (Fig. 6.14) (Lindsay and Roth, 1982; Lindsay and Kendall, 1985). A stratigraphic cross section across the North Dakota portion of the Williston Basin illustrates the lateral facies change from evaporites on the northeast margin of the basin to carbonates toward the basin center (Fig. 6.14) (Malek-Aslani, 1977). The presence of thick continuous anhydrites with halite indicates that at least part of the northeast margin could have been a subaqueous marginal marine evaporative lagoon, separated from the marine shelf by barrier shoreline complexes. The Little Knife field, located in the center of the Williston Basin of western North Dakota (Fig. 6.14) is a typical Mission Canyon evaporite-related reservoir.

The Little Knife field has produced over 31 million barrels of oil from dolomitized Mission Canyon reservoirs, with porosity averaging 14% at an average depth of 9800ft (3000m). The field is a combined structural-stratigraphic trap developed on a northward-plunging anticlinal nose, with closure to the south by lateral facies change into non-porous lithologies (Lindsay and Kendall, 1985). Figure 6.15 depicts a typical log profile of the Mission Canyon in the center of the Little Knife field, and correlates major porosity-modifying diagenetic events with depositional environments. Pervasive dolomitization is present at the top of the cycle, closely associated with evaporites of the capping sabkha-lagoonal sequence, and decreases dramatically toward the base, or toward the more normal marine portions of the sequence. Anhydrite replacement of allochems is also common in the upper 3 zones (4th order

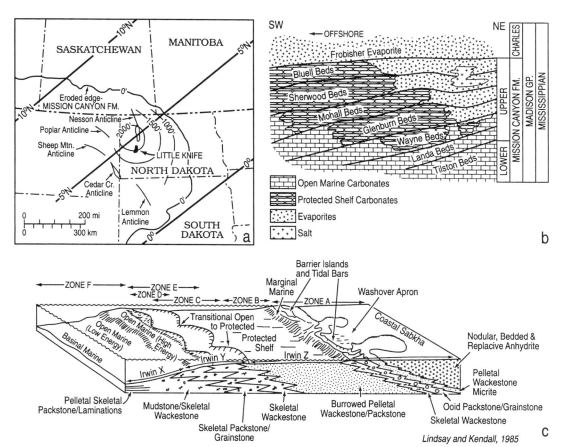

Fig.6.14. (Upper left) Index map of the Williston Basin showing: eroded edge of the Mission Canyon Formation; major surface and subsurface structural features; isopach thickness of the Madison Group; generalized Carboniferous paleolatidude lines; and the location of the Little Knife field. (Upper right) Regional cross section of Mission Canyon Formation across the North Dakota portion of the Williston Basin. (Lower) Idealized depositional setting of the Mission Canyon Formation at Little Knife Field. Informal log zonations A-F, and Irwin's (1965) epeiric sea energy zones (XYZ) illustrate respective positions occupied in the depositional system. (Lindsay and Kendall, 1985). Reprinted with permission from Carbonate Petroleum Reservoirs, Springer-Verlag, New York.

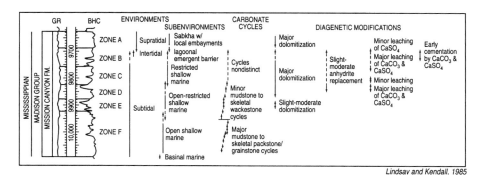

Fig.6.15. Type gamma-ray/sonic log of Mission Canyon Formation in the central portion of the Little Knife field. Depositional environments, cycles, and diagenetic modifications are shown to the right of the log. Reprinted with permission from Carbonate Petroleum Reservoirs, Springer-Verlag, New York.

sequences?) of the sequence. Dolomitization and anhydritization of the sequence is clearly the result of reflux of evaporative brines generated on adjacent marginal marine sabkhas, or within marginal marine evaporative lagoons. The lagoonal model would certainly provide a larger volume of dolomitizing brines for reflux through adjacent mud-dominated marine sequences. The distribution of dolomite seems to have been controlled by porosity/permeability trends developed in the original sedimentary sequence by burrowing and winnowing (Lindsay and Kendall, 1985). Again, early fresh water influx, probably associated with subaerial exposure at the end of the main cycle, had a major impact on porosity development, dissolving remnant undolomitized matrix, as well as metastable and anhydritized allochems, and resulting in significantly enhanced porosity in the reservoir.

Ordovician Ellenburger marginal marine sabkha-related dolomite reservoirs, west Texas, USA

The Ordovician Ellenburger Dolomite is a major exploration target in southwest and west central Texas. The Ellenburger was deposited on a broad shallow marine shelf of some 800km width during a major Cambrian-Ordovician transgression over the Precambrian basement (Barnes et al., 1959) (Fig. 6.16). The lower Ellenburger passes conformably into the Cambrian Bliss Sandstone below. Figure 6.17A, and B portrays the general depositional model for the Bliss-Ellenburger as conceived by Loucks and Anderson (1985). The Bliss and Lower Ellenburger probably represent a 3rd

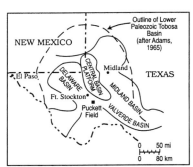
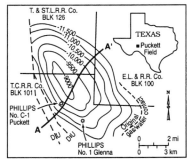
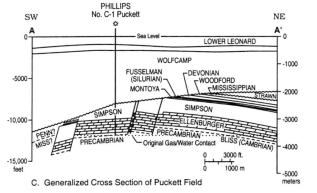

Fig. 6.16. Geologic setting of the Puckett Field, Permian Basin, U.S.A. Reprinted with permission from Carbonate Petroleum Reservoirs, Springer-Verlag, New York.

order depositional sequence. The Bliss may be the lowstand systems tract (LST) while the overlying marine-sabkha complex represents the transgressive (TST) and highstand systems tracts (HST) of the depositional sequence (Fig. 6.17). The overlying depositional sequence, the middle and upper Ellenburger, consists of a series of high-frequency cycles (4th order sequences?). These cycles contain shallow-marine shelf deposits building upward into marginal marine supratidal-tidal channel complexes. The cycles are generally capped with diagenetic terrains developed during terminal exposure of the cycle prior to the development of the subsequent cycle (Fig. 6.17). The Puckett field, located in the transition

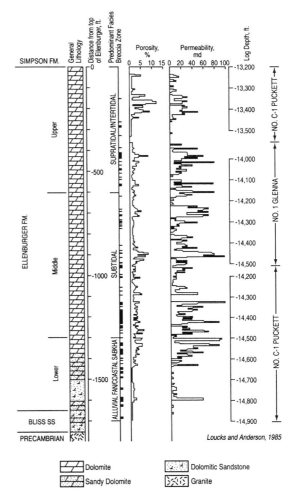

Fig. 6.18. Profiles of porosity and permeability vs. depth in the Phillips No. C-1 Puckett well. Predominant facies refers to the facies comprising the largest proportions of individual cycles. Reprinted with permission from Carbonate Petroleum Reservoirs, Springer-Verlag, New York.

between the Delaware and Val Verde basins in far-west Texas, is a representative Ellenburger field (Fig. 6.16).

The Puckett Field is a structural trap developed along a large faulted anticlinal feature with more than 760m of closure (Fig. 6.16). The field has produced more than 2.6 trillion cubic feet of gas from Ellenburger dolomites with an average porosity of 3.5% at a depth

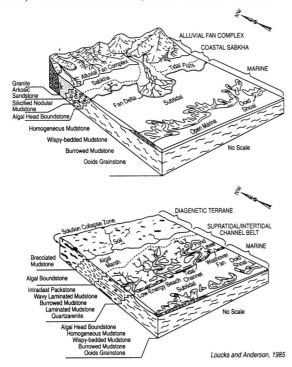

Fig. 6.17. (Top) Depositional model for the Bliss and lower Ellenburger at Puckett field. (Bottom) Depositional model for the middle and upper Ellenburger at Puckett Field. Reprinted with permission from Carbonate Petroleum Reservoirs, Springer-Verlag, New York.

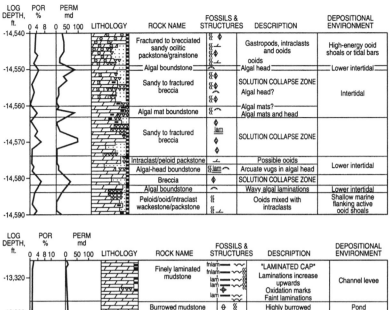

Fig. 6.19. (Top) Core description, environmental interpretation, and porosity-permeability profile of a section of the alluvial fan/sabkha sequence in the lower Ellenburger and upper Bliss. Phillips No. C-1 Puckett well, 14,797 to 14,846 feet (4510-4525m) (Bottom) the same for a section of the low-energy subtidal sequence in the middle Ellenburger. Phillips No. 1 Glenna well, 14,306 to 14,359 feet (4361-4377m). Reprinted with permission from Carbonate Petroleum Reservoirs, Springer-Verlag, New York.

averaging 3600m. The original gas column was over 490m thick (Loucks and Anderson, 1985). The entire Ellenburger at Puckett field is dolomitized. Subtidal facies dolomitization is, at least in part, related to reflux of evaporative brines from adjacent marginal marine sabkha and supratidal sequences. However, economic porosity and permeability occur in a cyclic fashion through the entire sequence (Fig. 6.18). A detailed look at typical cored sequences at Puckett reveals that major porosity zones in the lower Ellenburger generally coincide with solution-collapse breccias while porosity zones in the upper Ellenburger are developed below diagenetic terrains or within channel sequences (Fig. 6.19). The solution-collapse breccias in the Lower Ellenburger seem to be related to the removal of marginal marine sabkha evaporites during exposure at the sequence boundary. Porosity development in the upper Ellenburger is enhanced by mold and vug generation. These molds and vugs are associated with intercycle diagenetic terrains and may also be facies controlled. Another type of facies-controlled porosity is the interparticle porosity development seen in grainstone-packstone channel sequences (Fig. 6.18). Fractures in all reservoir facies significantly enhance perme-

Criteria for the recognition of ancient marginal marine sabkha dolomites

Table 6.2 outlines some of the major criteria that should be satisfied before an ancient dolomite is assigned to a marginal marine sabkha origin. The criteria are listed in decreasing order of reliability.

Perhaps the most important constraint on dolomite origin and usually the most difficult to determine, is the timing of the dolomitization event. Fabric and textural destruction often precludes the dolomite's precise assignment in a relative paragenetic sequence. The close relationship of dolomite fabrics to recognizable early diagenetic features has been used with varying success to suggest the penecontemporaniety of various dolomitization events. These fabrics include demonstrable vadose cements (Ward and Halley, 1985), the absence of two-phase fluid inclusions (Moore et al., 1988; Allan and Wiggins, 1993), dolomitization fabric selectivity (Beales, 1956; Kendall, 1977), and relationship to early compactional phenomena (Moore and others, 1988).

The easiest and most widely applied criterion is the incorporation of the dolomite into a sequence that exhibits the sedimentologic and mineralogic characteristics of the marginal marine sabkha depositional environment, such as algal laminations, mudcracks, ripup clasts, thin interbedded nodular anhydrite, or in their absence, solution-collapse breccias.

The patterns of dolomitization associated with a marginal marine sabkha should reflect the complex, but localized, sedimentologic and diagenetic environment normally found in this setting. For an example, ancient sabkha-related dolomitization should normally exhibit local, rapid shifts into undolomitized limestones, anhydrites and siliciclastics, both laterally as well as vertically. One should not expect thick, regional, pervasive dolomitized sequences to be associated with the sabkha environment.

Modern sabkha dolomites often exhibit small crystal sizes (less than 5 microns, with occasional aggregates ranging up to 20 microns). These smaller dolomite crystals are thought to reflect the numerous nucleation sites available in mud-dominated sabkha-related sequences (Land, 1980; Land, 1985). However, ancient sabkha-related dolomites commonly display much coarser crystal sizes, perhaps due to later recrystallization of the initial metastable calcian dolomite.

As has been discussed earlier, the trace

Table 6.2

1. Dolomite must be syngenetic

2. Dolomite must be associated with a sequence exhibiting sedimentologic and mineralogic characteristics of the marginal marine sabkha depositional environment.

3. Patterns of dolomitization should reflect the complex but localized sedimentologic, diagenetic, hydrologic setting of the sabkha. Dolomitization tends to be spotty and interbedded with evaporites, limestone, and siliciclastics.

4. Marginal marine sabkha dolomites crystal sizes tend to be small, but may gain significant size during later diagenesis.

5. Dolomite geochemistry may reflect original evaporative pore fluids if sequence buried in relatively closed system. Normally, however chemical and isotope composition seem to have been re-set by later recrystallization.

Table 6.2. Criteria for the recognition of ancient marine sabkha dolomites. Listed in decreasing order of reliability.

element and isotopic composition of ancient sabkha-related dolomites may not adequately reflect the evaporative nature of the original dolomitizing fluids due to the probability of recrystallization of the earlier metastable dolomite.

Land (1980) further suggests that most carbonate platforms have been flushed early in their history by meteoric waters due to carbonate platform sedimentation generally being punctuated by upward-shoaling cycles separated by subaerial exposure surfaces. This propensity for frequent platform subaerial exposure leaves sabkha terminal sequences susceptible to recrystallization in dilute meteoric waters. This recrystallization is generally reflected in the chemistry of the dolomites by low trace element compositions and light stable isotopic ratio sequences (Land, 1980; Land, 1985; Allan and Wiggins, 1993). These relationships have no doubt contributed to the tremendous rush toward the utilization of the Dorag, or schizohaline model of dolomitization (Hardie, 1987; Sun, 1994). After all, most dolomites found on carbonate platforms might well exhibit a meteoric water geochemical fingerprint, regardless of origin.

MARGINAL MARINE EVAPORATIVE LAGOONS AND SALINAS REFLUX DOLOMITIZATION

The marginal marine evaporative lagoon as a diagenetic environment

Structural or sedimentologic barrier or sill development in a marginal marine setting often modifies free hydrologic exchange between adjacent marine waters and back-barrier shelf-lagoonal environments and, in the case of total evaporative drawdown, desiccated playas. In the case of the shelf lagoonal environment, the stage is set for the development of hypersalinity, evaporite precipitation, and potentially strong density gradients. Figure 6.20 illustrates the basic characteristics of a barred basin or lagoon. Given the proper climatic conditions, in-flowing seawater is progressively evaporated as it moves from the sill toward the shoreline, setting up strong horizontal concentration (density) gradients. The heavier brines ultimately sink as they approach the landward margins of the lagoon, leading to a return flow of heavy brines toward the sill. This return flow establishes a seaward-sloping pycnocline separating the in-flowing marine waters from seaward-flowing dense brines (Adams and Rhodes, 1961; Logan, 1987; Kendall, 1988). Depending on the position of the sill relative to the pycnocline, the dense returning brine will either flow over the sill or be trapped behind it. If trapped, the brine will be refluxed laterally through the sill, and/or downward through

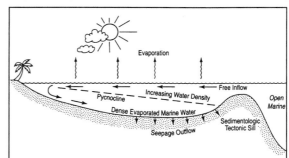

Fig.6.20. Schematic diagram of an evaporitic silled basin, or silled shelf lagoon model. As water is evaporated, increasing water density, a pycnocline is established, separating dense evaporative fluid below, from less dense marine waters above. The dense evaporative brines may reflux through the basin or lagoon floor or through the sill if porosity and permeability allows. The sill may be either sedimentologic such as a reef build-up or high-energy sand complex, or tectonic, such as a horst block. Used with permission of Elsevier.

THE EVAPORATIVE LAGOON AS A DIAGENETIC ENVIRONMENT

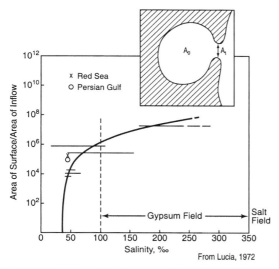

Fig. 6.21. Relationships between the ratios of the areas of an inlet and an evaporite basin and the maximum salinity developed in the basin. Reprinted with permission from Evaporites and Hydrocarbons, 1985, Columbia University Press, New York.

subjacent units forming the lagoon, or basin floor. Kendall (1988) questioned the efficacy of the barred basin model on the basis of the difficulty of maintaining the balance between inflow of fresh marine water and reflux of evaporative brines. Kendall's concerns are is based on the discussions of Lucia (1972) and Kinsman (1974) that suggest that slow rates of evaporation impose stringent limits on the size of the evaporite basin inlet. Lucia (1972) calculated that the ratio of surface area of a basin to the cross-sectional area of the inlet must exceed 10^6 before gypsum precipitation and at least 10^8 before halite precipitation (Fig. 6.21). This ratio is basically a measure of the balance between the rate of water loss and gain in the basin. This conclusion is based on an inflow of normal salinity marine water. If the inflow marine water is already evaporated to a salinity between carbonate and gypsum precipitation, the restrictions on ratio of the surface area of the basin and cross-sectional area of the inlet would be considerably reduced and the barred basin model becomes more viable. Cratonic basins (such as the Permian Delaware and Midland basins, and interior salt basins such as the Upper Jurassic East Texas and North Louisiana basins) are important examples of settings where the barred basin model is appropriate.

Adams and Rhodes (1961) used the fundamental elements of the barred basin model to develop their concept of dolomitization by seepage refluxion based on the Permian of west Texas. In their model (Fig.

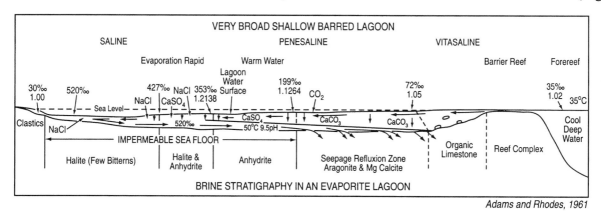

Fig. 6.22. Idealized section of brine stratigraphy in evaporite lagoon based on the Permian reef complex of New Mexico and Texas, U.S.A. Reprinted with permission of AAPG.

6.22) seawater is progressively evaporated as it moves from sill toward the shoreline across a relatively shallow, but wide, barred-shelf lagoon. Gypsum saturation is reached in the lagoon interior, and evaporites are deposited along the shoreward margins of the lagoon. At this point, brine density is high enough to set up density-driven brine return flow toward the sill, and horizontal density stratification is established with a seaward-dipping pycnocline. While the landward margin of the lagoon floor is sealed by the precipitation of evaporites, the dense brines are able to reflux down through the lagoon floor seaward of the zone of evaporite precipitation. The refluxing dense brines exhibit an elevated Mg/Ca ratio and proceed to dolomitize the lagoonal floor and subjacent porous limestone units as they displace lighter marine pore fluids.

If the sill is sedimentologic in nature, such as a reef tract, or ooid shoal complex, and an equilibrium is established between sedimentary accretion, relative sea level rise, and evaporite precipitation, thick sequences of evaporites can be precipitated along the landward margins of shelf lagoons. Enormous volumes of dolomitizing fluids are subsequently available to reflux through, and to pervasively dolomitize, adjacent porous units (Moore and others, 1988).

As a result of modeling studies, Kaufman (1994) suggested that the silled lagoon model (barred basin model) was hydrologically sound and was an efficient Mg transport system for widespread dolomitization of subjacent limestone sequences. Kaufman (1994) calculated that a silled lagoon with brines evaporated past gypsum precipitation could dolomitize 0.08% (or 8m) of a 1000m thick subjacent limestone sequence with a porosity of 10% by brine reflux in 100,000 years (Table 6.3). In 10,000 years 0.8m of limestone would have been dolomitized. If the porosity of the underlying limestones is greater than 10%, a significantly thicker dolomite sequence could be formed.

A difficulty with the silled lagoon-reflux model of dolomitization is the general lack of an adequate modern analogue. While dolomites associated with silled lagoons seem to be common in the geologic past, this setting is relatively rare today. Salt ponds such as Pekelmeer, on the island of Bonaire (Murray, 1969; Lucia and Major, 1994) and East Salina on West Caicos (Perkins et al., 1994) in the Caribbean have been used as settings for reflux dolomitization. These hypersaline ponds, however, are essentially desiccation features and more appropriately considered to be salinas and playas, being totally barred from the sea by a substantial barrier. We will discuss the salina/playa model later in this chapter after the following discussion of ancient silled lagoonal reflux dolomitization.

Table 6.3

Fluid Composition (molal NaCl)	Mg Mass Flux[1] (moles/m^{2}yr)	% Dolomite Formed in 100,000 yr[2]
0.7	0.05	0.04
2.5	0.9	0.8
5.0	5.0	4.1

[1]Vertical flux rate.
[2]Original limestone body is 1 km^3 with 10% porosity, and is composed entirely of calcite.

From Kaufman, 1994

Table 6.3. Vertical Mg mass flux rates during reflux and (%) converted to dolomite formed in 100,000 years. Used with permission of SEPM.

The Upper Permian Guadalupian of west Texas, USA: an ancient marginal marine evaporative lagoon complex

Perhaps the world's best-exposed carbonate shelf margin-to-evaporite transition occurs in the Upper Permian outcrops along the Guadalupe Mountains of New Mexico and Texas. Here, the massive biogenic limestones of the Capitan Formation mark the margin of the Delaware Basin, which extends across the southeastern corner of New Mexico and into far west Texas (Fig. 6.23). Landward from the steep, abrupt Capitan shelf margin, extensive Upper Permian shelfal carbonates and evaporites (termed the Artesia Group and composed of the Seven Rivers, Yates, and Tansill formations) were deposited.

Figure 6.23B illustrates the facies distribution across the Capitan shelf margin during deposition of the lower Seven Rivers Formation. This map view extends across a portion of the northwestern shelf at Walnut Canyon near Carlsbad New Mexico. The carbonates of the shelf margin give way to evaporites (gypsum) within 15 km in a shelfward direction. Most of the carbonates shelfward of the Capitan margin consist of replacive, fine-crystalline dolomite that preserves the fabrics and textures of the original packstones, grainstones, and wackestones that were deposited along the margins of an extensive evaporative shelf lagoon (Sarg, 1981).

Sarg (1976; 1977; 1981) in his studies of the carbonate-evaporite transition within the Seven Rivers along the Guadalupe Mountains came to the conclusion that the evaporites were dominantly precipitated under subaqueous conditions and that the evaporite facies of the Guadalupian shelf represented an enormous shelf lagoon. This interpretation supported the earlier environmental interpretations of Lloyd (1929) and Bates (1942). It does, however, conflict directly with the interpretations of Kendall (1969), Silver and Todd (1969), Hills (1972) and Meissner (1972), who proposed that the Seven Rivers evaporites and related dolomites were of sabkha origin.

Sarg (1981) based his interpretation on: the considerable thickness of the gypsum sequences relative to the thin evaporite sequenc-

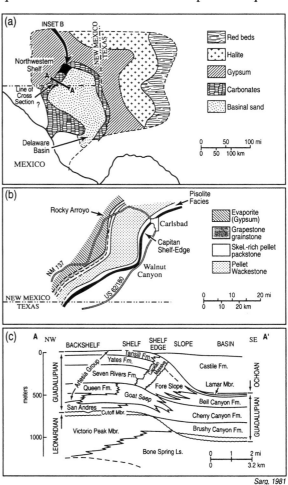

Fig. 6.23. Geologic setting of the Upper Permian Guadalupian of west Texas and New Mexico. (a) Major facies of the Guadalupian shelf surrounding the Delaware Basin. Note that the carbonate shelf is rather narrow, and gives way to evaporites toward the land, and siliciclastics into the basin. (b) Lithofacies map of the Guadalupian shelf margin in the vicinity of Carlsbad, New Mexico showing the transition from shelf limestones to evaporites. (c) The stratigraphy of the Upper Permian, shelf-to-basin along line of section A-A' (see (a) above for location). Used with permission of SEPM.

es seen in modern sabkhas; lateral facies relationships with mesohaline facies that show no evidence of subaerial exposure; lack of any evidence of subaerial exposure either in the evaporite or related carbonate facies; the maintenance of the gypsum carbonate transition at the same geographic location through most of Seven Rivers time (a relationship not compatible with a prograding sabkha sequence); and finally, the lack of any characteristic sabkha vertical sequences.

Consequently, Sarg (1981) calls upon seepage reflux of dense brines from a regional evaporative shelf lagoon, as originally envisioned by Adams and Rhodes (1960), to dolomitize the associated carbonate facies of the Seven Rivers as well as the overlying Queen Formation. Similar evaporative lagoons of slightly younger age produced the brines that dolomitized the younger (overlying) Yates and Tansill formations. In the nearby subsurface these and associated Permian units form some of the most prolific oil producing reservoirs in west Texas and are of enormous economic importance (Silver and Todd, 1969).

Dolomitized Permian reservoirs of the South Cowden Field west Texas: reflux dolomitization from a shelf lagoon

The Permian South Cowden Field is located on the eastern margin of the Central Basin Platform (Fig. 6.24). Its reservoirs are developed in Guadalupian Grayburg dolomites (Fig. 6.25). Figure 6.26 shows the location of the unit of interest (Moss Unit), a structure map on top of the Grayburg and locations of a seismic line (Fig. 6.27), and a stratigraphic cross section (Fig. 6.28) across the Moss unit. The Grayburg in Cowden Field was originally

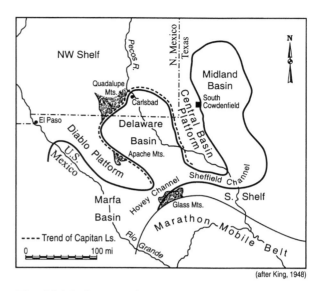

Fig. 6.24. Index map showing the location of the South Cowden Field on the eastern margin of the Central Basin Platform. Modified from King (1948). Used with permission from AAPG.

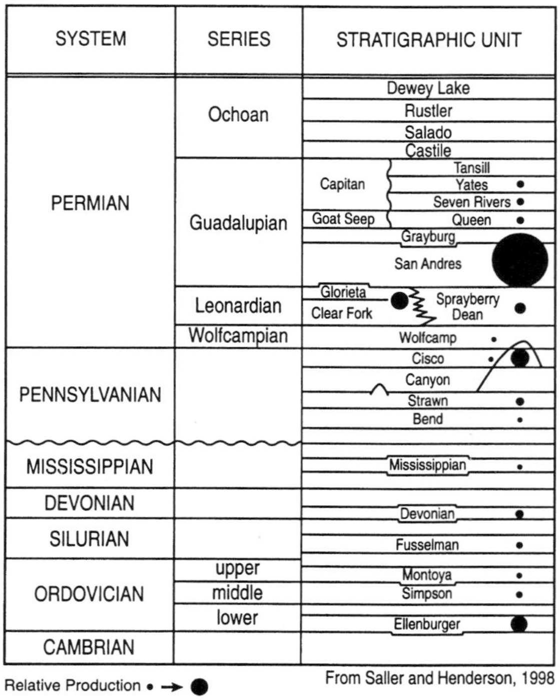

Fig. 6.25. Stratigraphic setting and relative hydrocarbon production from Paleozoic sequences west Texas, USA. Used with permission from AAPG.

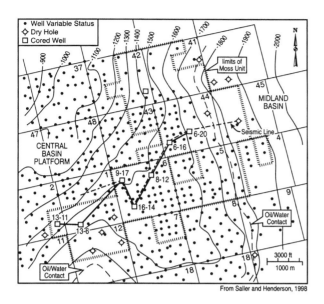

Fig. 6.26. Structure map on top of the Grayburg Formation at the Moss unit South Cowden Field. Location of seismic line (Figure 6.27) and cross section (Figure 6.28) are shown. Used with permission from AAPG.

deposited as limestone across the Central Basin Platform to Midland Basin transition. The well-defined Grayburg shelf margin is shown in the seismic section (Fig. 6.27). The west to east stratigraphic cross section across South Cowden Field shows that two sedimentary sequences can be recognized in the Grayburg in this field, the Grayburg A and Grayburg B, separated by well-developed exposure surfaces (Saller and Henderson, 1998). The shelf margin would seem to be located between well 6.16 and well 6.20 (Fig. 6.27). In the stratigraphic cross section (Fig. 6.28) dramatic changes in sequence thickness and onlap of the Grayburg A sequence onto Grayburg B sequence are located between these two wells. Obviously both sequences are strongly progradational into the Midland Basin to the east (Saller and Henderson, 1998).

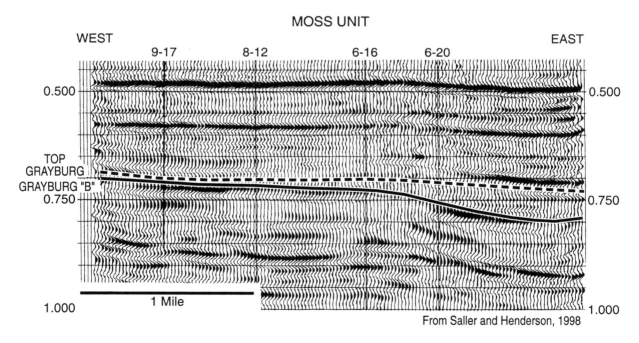

Fig. 6.27. Segments of a three dimensional seismic survey that approximate a west-to-east profile through most cored wells in the Moss unit, South Cowden field. See Figure 6.25 for location. Vertical scale is in seconds of two-way traveltime. Line segment provided by UNOCAL. Used with permission from AAPG.

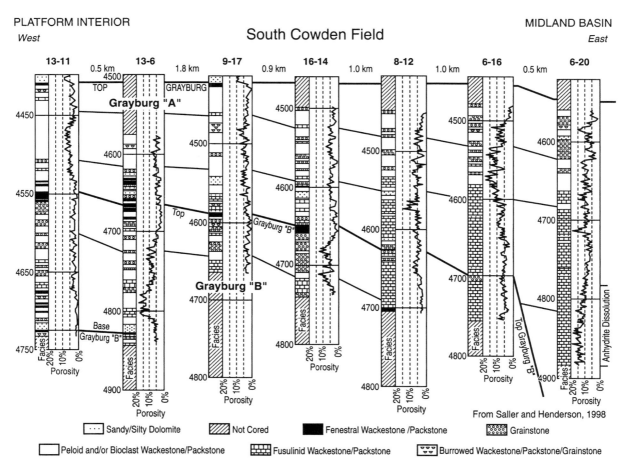

Fig.6.28. Stratigraphic cross-section of porosity in the Grayburg Formation at the Moss unit, South Cowden Field, showing depositional facies and core-measured porosity. Porosity greater than 5% is shaded. Note how porosity generally increases basinward (to the right) in the Grayburg A. Correlation lines between wells are not necessarily cycle tops. See Figure 6.25 for location of wells on cross section. Datum is top of Grayburg except in well 6-20. Used with permission of AAPG.

While practically all Grayburg rocks in South Cowden are dolomitized, enough fabric details remain to allow reconstruction of original limestone fabrics and to construct a depositional model of the Grayburg at South Cowden. Figure 6.29 details the original carbonate facies present at South Cowden and the depositional environments in which they formed. The main features of the model are the grainstone/packstone complex forming a potential sill at the shelf margin and the muddy sediments of the broad shelf-lagoon and landward-tidal flats developed on the Central

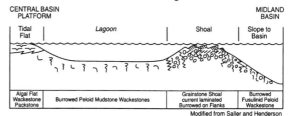

Fig.6.29. General depositional model for the Clear Fork and Grayburg formations. Modified with artistic license from Saller and Henderson (1998).

Basin Platform proper (Fig. 6.29).

The pervasive dolomitization exhibited by these rocks seems to have been early and associated with evaporative marine water refluxed basinward from the shelf lagoon shortly after or during deposition (Saller and Henderson, 1998). A heavy (lagoonward) to light (basinward) carbon and oxygen isotopic gradient is seen in the South Cowden dolomites (Fig. 6.30). These data seem to suggest the movement of heavy dolomitizing fluids from the lagoon into the grainstones of the shelf margin where they mix with isotopically lighter basin-derived waters.

Porosity and permeability in the dolomites of the Grayburg Formation at South Cowden are very heterogeneous, with porosity ranging from 0.2 to 20% and permeability ranging from 0.01 to 200md (Fig. 6.31). However, the comparison of porosity/permeability from platform interior to margin shows a striking

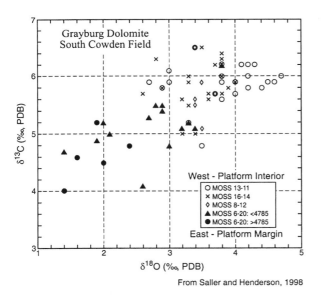

Fig. 6.30. Plot of stable isotopes for the Grayburg Formation in the Moss Unit. Generally lower isotopic values occur in more basinward wells. See Figure 6.26 for location of wells. Used with permission of AAPG.

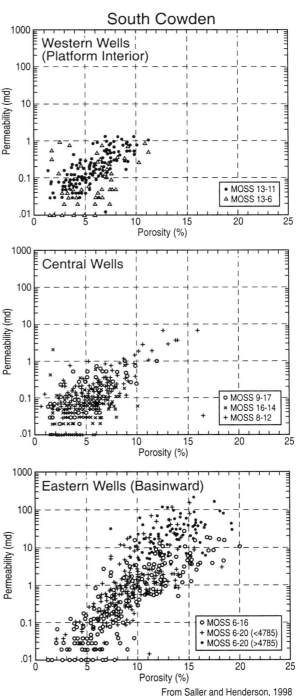

Fig. 6.31. Plots of permeability vs. porosity for the Grayburg A in cored wells in the Moss unit. Used with permission of AAPG.

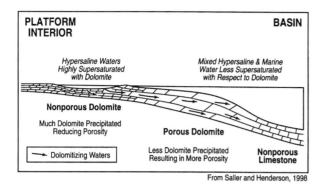

Fig. 6.32. Saller and Henderson's 1998 model for dolomitization and porosity development along the margins of the Central Basin Platform dolomitized by refluxing brines generated by evaporation of seawater in restricted lagoons. Used with permission of AAPG.

increase in reservoir quality toward the shelf edge (Fig. 6.31). Saller and Henderson (1998) suggest that this increase in porosity toward the margin is the result of massive dolomite cementation of the shelfward rocks during initial dolomitization. As the residual brines reflux seaward dolomite supersaturation decreases resulting in a decrease in dolomite cementation with an attendant increase in porosity and permeability (Fig. 6.32).

This follows closely the Bonaire model of Lucia and Major (1994) discussed at length in Chapter 3. Since there are significant age-equivalent evaporites across the Central Basin Platform (E. Stoudt, personal communication, 2000), however, it seems reasonable to speculate that the decrease in reservoir quality seen along the shelf margin at South Cowden is the result of a facies change from muddy lagoonal sediments to porous shelf margin grainstones rather than porosity loss by dolomite cementation in the lagoon. Similar facies-related porosity gradients have been seen by other workers along the margins of the Central Basin Platform in both the San Andres and Glorietta formations (E. Stoudt, personal communication, 2000).

The following section will detail evaporative reflux dolomitization sourced from an Upper Jurassic Oxfordian silled lagoon where the refluxing brines provided Mg^{+2} while the limestones below provided the Ca^{+2} and CO^{-3} necessary for dolomitization, and porosity/permeability was enhanced.

Upper Jurassic Smackover platform dolomitization, east Texas, USA: a reflux dolomitization event

The Oxfordian Upper Jurassic Smackover Formation is an important oil and gas trend across the northern Gulf of Mexico, from Mexico to Florida. Reservoir rocks are developed in ooid grainstones accumulated on high-energy platforms developed along the landward margins of a series of interior salt basins (Fig. 6.33) (Moore, 1984). Upper Smackover ooid grainstones have suffered pervasive platform dolomitization over a large area of east Texas, while the upper Smackover in adjacent areas of Louisiana and Mississippi is still limestone (Fig. 6.33).

The stratigraphic setting of the Smackover is shown in Figure 6.34. The upper Smackover ooid grainstones represent the terminal late highstand systems tract of Jurassic depositional sequence II (Moore and Heydari, 1993). A regional unconformity exposed the upper Smackover to subaerial conditions during the lowstand of Jurassic depositional sequence III. During the subsequent transgression of Jurassic sequence III, an ooid-rimmed evaporative shelf lagoon was developed. The evaporites deposited in this lagoon in east Texas are termed the Buckner Formation. An isopach of the

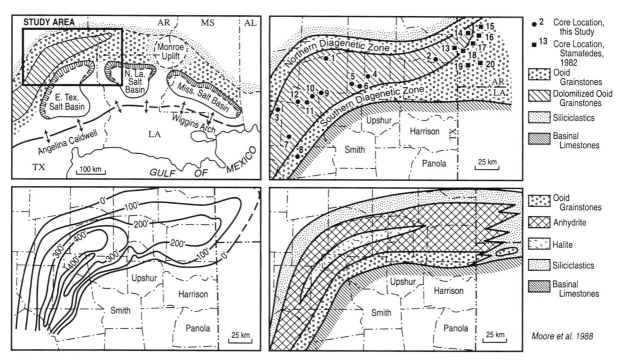

Fig. 6.33. Geologic setting of the Upper Jurassic, Gulf Coast, U.S.A. (Upper left) Structural setting at the time of deposition of the Upper Jurassic. High-energy ooid grainstone shelves developed landward of interior salt basins. These grainstones dolomitized to the east in Alabama, and to the west in Texas. Location of study area in east Texas indicated. (Upper right) Upper Smackover shelf lithofacies, emphasizing distribution of dolomitized grainstones in east Texas. Sample control indicated. (Lower left) Isopach map of the Buckner evaporite (gypsum plus halite). (Lower right) Lithofacies developed in the Buckner Formation. Used with permission of SEPM.

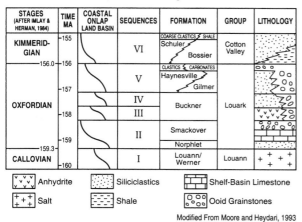

Fig. 6.34. Stratigraphy of the Jurassic, central Gulf Coast, USA. Used with permission of AAPG.

Buckner and its lateral facies relationships, from siliciclastics in the shelf interior to an ooid grainstone barrier facies along the basin margin, is shown in Figure 6.33. Buckner evaporites consist dominantly of anhydrite, with over 30m of halite present in the Buckner depocenter (Moore et al., 1988).

The close correspondence of upper Smackover platform dolomitization and the distribution of evaporites in the overlying Buckner suggest that reflux dolomitization is the most appropriate model that can be applied to the Smackover in east Texas. Petrographic relationships and the lack of two-phase fluid inclusions in the dolomite indicate a relatively early timing for the dolomitization event. Finally, dolomite distribution patterns in the upper Smackover, showing a dolomitization

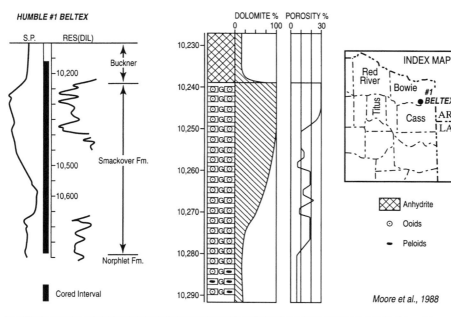

Fig. 6.35. Log and core description of the Humble #1 Beltex well, Bowie County, Texas, U.S.A., showing distribution of dolomite relative to the overlying evaporites of the Buckner Formation. Used with permission of SEPM.

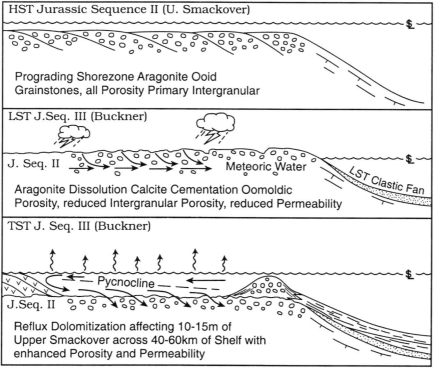

Fig. 6.36. Conceptual diagenetic model across the Jurassic sequence II-III boundary. Note that dolomitization and evaporite deposition took place during depositional sequence III transgressive systems tract (TST). Reflux of brines from lagoon above enhanced by the high porosities and permeabilities developed in the highstand systems tract (HST) of depositional sequence II by exposure to meteoric waters during the low stand systems tract of depositional sequence III.

gradient increasing upward toward overlying Buckner evaporites, are compatible with a reflux model (Fig. 6.35).

The reflux model proposed by Moore and others (1988) and Moore and Heydari (1993) for the upper Smackover platform dolomites is shown in Figure 6.36. A high energy Buckner barrier was developed during the TST of Jurassic sequence III forming an evaporative shelf lagoon. Anhydrites were precipitated in the shelf interior, density stratification developed in the lagoon, and Mg^{+2}-rich, Ca^{+2} CO_3^{-2}-poor brines refluxed down through underlying porous upper Smackover ooid grainstones, ultimately dolomitizing the central portion of the upper Smackover platform across most of east Texas. It seems that the dolomitizing fluids used the $CaCO_3$ available in the calcite cements surrounding oomolds since the dolomite mimics the rimming calcite cement and no dolomite occurs within the oomolds. In this situation, dolomitization seems to be porosity-enhancing mole-for-mole replacement rather than a porosity-destructive cementation event as described by Lucia and Major (1995) and Saller and Henderson (1998). When sea level reached the HST of Jurassic sequence III, sedimentation along the lagoon barrier shut off the free flow of water into the lagoon, evaporative drawdown occurred, the halite facies was deposited, the floor of the lagoon is sealed, and reflux ceased.

However, the geochemistry of these Smackover dolomites does not reflect an evaporative origin, but may be more characteristic of meteoric waters (Moore and others, 1988). Smackover dolomites display low Sr and Na compositions and depleted $\partial^{18}O$ values relative to dolomites encased in Buckner anhydrites above (Fig. 6.37). Geochemical gradients, particularly in Fe, Mn, and $^{87}Sr/^{86}Sr$, established by Moore and others (1988) in the Smackover platform dolomites, indicate the influence of a regional meteoric water system acting from platform interior to platform margin (Fig. 6.37).

These geochemical relationships strongly suggest a major recrystallization event of an original evaporative reflux platform dolomite by a major meteoric water system. The meteoric aquifer probably developed during exposure of the landward, siliciclastic facies of the

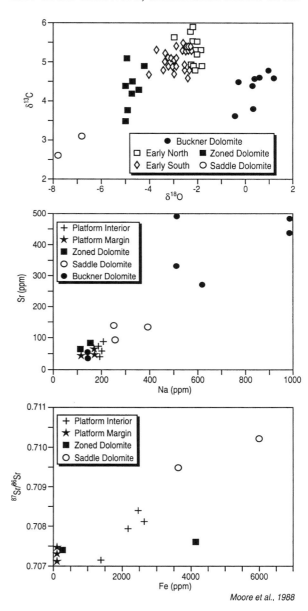

Fig.6.37. Isotopic and trace element composition of Smackover and related dolomites, east Texas. Used with permission of the SEPM.

upper Smackover carbonate sequence during a major unconformity developed during the Cretaceous (Murray, 1961). The inconsistencies in the petrography, geography, and geochemistry of the Smackover platform dolomites of east Texas dramatically illustrate the problems facing the modern geologist dealing with regional platform dolomitization events.

Criteria for the recognition of ancient reflux dolomites

Major criteria for the recognition of ancient reflux dolomites are outlined on Table 6.4. These criteria are arranged in a general order of decreasing reliability.

The most critical criterion that must be satisfied, if one is to interpret a dolomite as originating from the reflux of evaporated marine waters, is the close spatial relationship of the dolomite to evaporites that can reasonably be deduced as being subaqueous in origin. These evaporites must have the sedimentary structures, fabrics, mineralogical facies, and geological setting appropriate to a subaqueous origin for precipitation (Kendall, 1984). Even though reflux undoubtedly occurs in the marginal marine sabkha, the sabkha environment creates neither the hydrologic setting nor the volume of waters necessary for dolomitization of an entire platform.

The stratigraphic-tectonic and sea level setting of the associated evaporites must be appropriate for the development of a restricted or barred lagoon or basin. For example, the basin or lagoon must have a suitable sedimentologic or tectonic barrier such as a reef, grainstone complex, or horst block to restrict the interchange of marine waters between the sea and basin or lagoon. It seems that a rising sea level is the most favorable setting for the development of a sedimentologic barrier that will allow the accumulation of significant evaporites in the basin or lagoon.

The dolomitized reflux receptor beds must have exhibited high porosity and permeability, and suitable hydrologic continuity at the time of dolomitization. Platform grainstone

Table 6.4

1. Close spatial relationship of dolomite to evaporite sequences that can reasonably be deduced as being subaqueous in origin.

2. The stratigraphic, tectonic, and sea level setting of the associated evaporites must be appropriate for the development of a barred base or lagoon.

3. The dolomitized reflux receptor units must have had high porosity-permeability, and suitable hydrologic continuity at the time of dolomitization.

4. Reflux dolomitized units should exhibit dolomitization gradients, with increasing volumes of dolomite toward the source of refluxing brines.

5. Reflux dolomite crystal size are generally coarser than those found associated with sabkha sequences, because coarse receptor beds dictate fewer nucleation sites.

6. While geochemical information relating reflux dolomite to evaporative pore fluids might be lost by recrystallization, gross vertical geochemical gradients, such as increasing Sr, Na trace elements, and stable isotopes enriched in ^{18}O toward the evaporative brine source might be preserved.

Table 6.4. Criteria for the recognition of ancient evaporative reflux dolomites, listed in decreasing order of reliability.

blankets, coarse high-energy shoreline sequences, and reefs are all favorable conduits for the regional dispersal of reflux brines.

Sequences thought to have been dolomitized by evaporative brine reflux should exhibit dolomitization gradients, with increasing volumes of dolomite toward the source of refluxing brines. See for example the dolomitization gradients established in the platform dolomites of the Gulf Coast Upper Jurassic Smackover Formation (Fig. 6.35). These gradients can be modified by porosity and permeability variations in the receptor units.

Dolomite crystal sizes may be considerably coarser than those found in modern sabkha environments because coarse receptor beds dictate fewer nucleation sites than found in the mud-dominated sabkha. However, the recrystallization potential of these surface-formed dolomites may negate any original crystal size differences related to original dolomitization environment.

While geochemical information specifically relating the reflux dolomite to an evaporative pore fluid might be lost by recrystallization, gross vertical geochemical gradients such as increasing Sr and Na and stable isotopes enriched in ^{18}O toward the evaporative brine source might be preserved. In the case history of the Jurassic detailed above, all original geochemical gradients were apparently wiped out and overprinted by a major meteoric water system. Climate considerations relative to the availability of meteoric water might well determine the preservability of relevant original geochemical gradients.

Regional evaporite basins and coastal salinas, their setting and diagenetic environment

The common thread connecting large evaporite basins and coastal salinas is that tectonism or a major drop in sea level, or both, have severed their connection to the sea. Evaporite deposition is initially by evaporite drawdown, and finally by seepage reflux of marine waters through seaward porous barriers. Evaporites and fine-grained carbonates seal the basin or salina floor and any reflux of potential dolomitizing brines is confined to the margins of the basin or salina. Significant deposits of halite in addition to gypsum are to be expected under these conditions. While sea level lowstands are the most common situation for salina and evaporite basin formation, they may develop during late highstands where high frequency drops in sea level periodically isolate a lagoon or basin from the sea. The following section discusses a modern example of a coastal salina, the MacLeod salt basin, and several ancient, economically important examples of large evaporite basins, the Elk Point Basin of Canada and the Michigan Basin of the US. Finally, we will explore the conditions of deposition and associated diagenesis of the Hith evaporite and the Arab sequences of the Middle East.

The MacLeod salt basin

The MacLeod is a graben basin some 30km wide and 60km long oriented approximately north south. It is separated from the Indian Ocean on the west by a relatively porous Pleistocene carbonate ridge attaining elevations of up to 60m, and to the south by a low Holocene dune field (Fig. 6.38). During the Holocene sea level rise, from approximately 8000 to 5000 years BP, the MacLeod basin was open to the sea across a narrow southern sill and acted as a silled basin or lagoon with the precipitation of both aragonite and gypsum. During this period, seepage outflow of dense brines occurred through the lagoon floor, perhaps dolomitizing the Pleistocene sequence below (Fig. 6.39A) (Logan, 1987). At approximately 5000 years BP, sea level began to drop, and a parabolic dune field sealed the southern sill. At this point, the brine surface was depressed some 10m to the lagoon floor (Fig. 6.39B). Fresh marine waters continued to be furnished to the MacLeod in the form of seepage along the margins of the lagoon through the porous barrier, ultimately filling the MacLeod with up to 10m of gypsum and halite (Logan, 1987) (Fig. 6.39C). Free reflux of evaporative brines downward was restricted by the development of a seal across most of the lagoon floor caused by gypsum precipitation

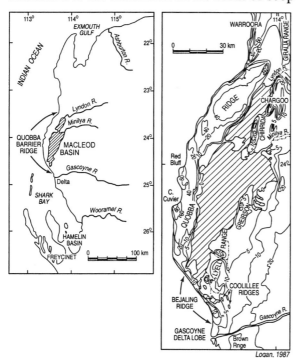

Fig. 6.38. (Left) Map showing the location of the MacLeod basin. During Quaternary marine phases the basin was a marine gulf connected southward to Shark Bay. (Right) Contour map showing the topography of the MacLeod region. The stippled area lies between sea level and -4.3 m. Used with permission of AAPG.

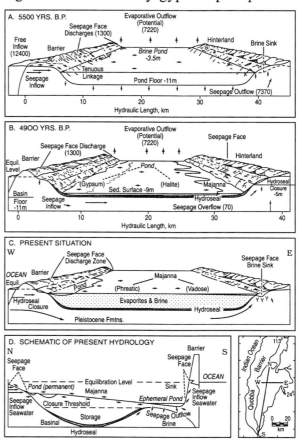

Fig. 6.39. Schematic of the MacLeod basin system from 5500 years B.P. (A) until the present (C) illustrating the effect of sea level variation on the effectiveness of the barrier, seepage reflux through the basin floor during initial transgression, the development of the hydroseal across the basin at 4900 years B.P. during the lowstand (B) and the ultimate filling of the basin with evaporites by marginal seepage reflux of marine waters (C). The present hydrology of the MacLeod basin is shown in (D). Location of cross sections shown at lower right. Used with permission of the AAPG.

associated with the initial evaporative drawdown when the southern sill was closed. Dense brines do however, reflux laterally into the adjacent sedimentary units through well-defined brine sinks developed along the margins of the lagoon (Logan, 1987) (Fig. 6.39C). The MacLeod Salt Basin is a viable modern analogue for coastal salinas as well as larger salt basins.

The Elk Point Basin of Canada

The Elk Point is an enormous middle Devonian (Givetian) basin extending some 1700 km (1100 miles) from northern Alberta, Canada, to North Dakota and Montana in the northern U.S. (Fig. 6.40). It is 700 km (450 miles) wide, extending into Saskatchewan and Manitoba on the east. The basin is rimmed with structural highs on the west (Peace River Arch, Central Montana Uplift) and the Precambrian shield and carbonate shelves on the east and south (Manitoba and Dakota shelves) (Fig. 6.40).

The Elk Point Basin is barred from the open sea to the north by the Presqu'ile barrier reef complex and is filled with a thick (up to 450m or 1500ft) evaporative sequence consisting of both anhydrite and halite, termed the Muskeg evaporites. The Presqu'ile barrier reef is part of the Keg River carbonates (Fig. 6.40). A series of economically important pinnacle reefs (Rainbow) encased in Muskeg evaporites are developed just southeast of the Presqu'ile barrier (Fig. 6.40). The carbonates closely associated with the Presqu'ile are also hosts to significant lead-zinc mineralization (the Pine Point complex) (Skall, 1975).

Shearman and Fuller (1969) interpreted the bulk of the massive evaporites of the Elk Point Basin as supratidal sabkha. Maiklem (1971) and Maiklem and Bebout (1973), however, reinterpreted the Muskeg evaporites to be

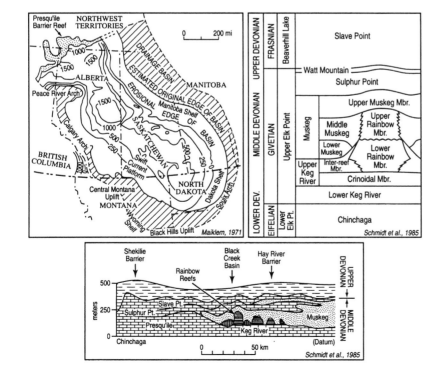

Fig. 6.40. (Upper left) Map of the Elk Point basin. Contours on the Elk Point Group in meters. (Upper right) Stratigraphic chart of Devonian formations in the Rainbow region, northwestern Elk Point basin. (Lower) Idealized regional stratigraphic cross-section of Middle Devonian formations across the Presqu'ile barrier reef in the north-western Elk Point basin. Used with permission of the Society of Canadian Petroleum Geologists.

dominantly subaqueous. They noted that most of the evaporite deposition took place during periodic drawdown episodes that resulted in the recognition of a number of sedimentologic cycles within the Muskeg and associated carbonates (Maiklem and Bebout, 1973).

Maiklem (1971), using the Elk Point Basin, presented an interesting diagenetic model based on an evaporative drawdown episode within a barred basin (Fig. 6.41). As the water in the basin evaporates, and its water level drops, a hydrologic head between the sea and the basin is established. This hydrologic head forces marine waters to reflux through the barrier into the basin as well as through platform carbonates, perhaps involving pinnacle reefs such as the Rainbow Group (Fig. 6.41). The constant reflux of large volumes of normal marine waters through the barrier, platform, and adjacent pinnacles during evaporative drawdown could lead to pervasive dolomitization of these associated limestones. Since seawater is undersaturated with respect to gypsum and halite, nearby evaporites would be dissolved as well. Schmidt and others (1985) reached a similar conclusion for early dolomitization of the Rainbow reefs southeast of the barrier (Fig. 6.41).

We shall see in later chapters, however, that the dolomitization of adjacent platforms and pinnacle reefs in the Elk Point Basin, as well as other Canadian Devonian Basins, may involve later burial fluids rather than early refluxing marine waters.

Maiklem's model of evaporative drawdown is very similar to the situation found by Logan (1987) in the MacLeod salt basin (as described above), including the dissolution of halite by refluxing marine waters.

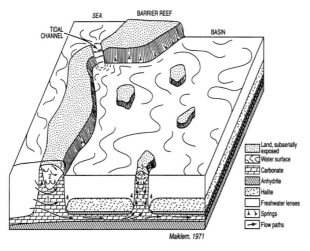

Fig. 6.41. Conceptual model of the Elk Point Basin. The sill is a barrier reef, shelfal pinnacle reefs developed behind the barrier. Diagram drawn at time of evaporative drawdown causing hydraulic head between the sea and the shelf lagoon. The numbers represent diagenetic processes and environments. 1. Normal marine waters driven through the barrier along the porous shelf sequences and up the porous pinnacle reefs, dolomitizing as it goes. 2. Floating meteoric lenses beneath barrier and pinnacles, solution and precipitation of calcite. 3. Vadose diagenesis in the upper reaches of the barrier and pinnacles, solution and precipitation. 4. Solution of evaporites by normal marine water along margins of the barrier and the pinnacles. Used with permission of the Society of Canadian Petroleum Geologists.

Michigan Basin, USA

A number of large barred basins, such as the Michigan Basin in the northern U.S., and Permian Zechstein Basin of western Europe, exhibit successions of thick post-gypsum evaporites such as halite. In some cases such as in the west Texas Permian Castile Formation, thick sequences of potash salts and bitterns are deposited in conjunction with the halite. Carbonate sequences such as the pinnacle reefs of the Michigan Basin and the Zama reefs of Alberta are often encased in these evaporites (Sears and Lucia, 1980; Schmidt et al., 1985).

In the Michigan Basin the Silurian Niagaran pinnacle reefs are partially dolomitized by reflux of brines from anhydrites forming in supratidal conditions near the crests of the reefs during evaporative drawdown. However, brines associated with the thick halites encasing the reefs tend to precipitate porosity-destructive halite in peripheral reef pore systems. This porosity-plugging halite is one of the major problems to be faced in the economic development of these reef reservoirs (Sears and Lucia, 1980).

Jurassic Hith Evaporite and the Arab Formation, Middle East

Figure 6.42 outlines the gross stratigraphic framework of the Middle and Upper Jurassic of the southern Persian Gulf and Saudi Arabia (see discussions in Droste (1990) and Alsharhan and Kendall (1986)). An Upper Jurassic 2nd order depositional sequence hosts one of the most prolific hydrocarbon systems in the world. The source rock resides in the deeper water, organic-rich carbonates of the transgressive systems tract (Hanifa Formation) (Fig. 6.42). The reservoirs and the seals occur in the highstand systems tract (Fig. 6.42). This Upper Jurassic super sequence consists of a number of shoaling upward 3rd order sequences (Fig.

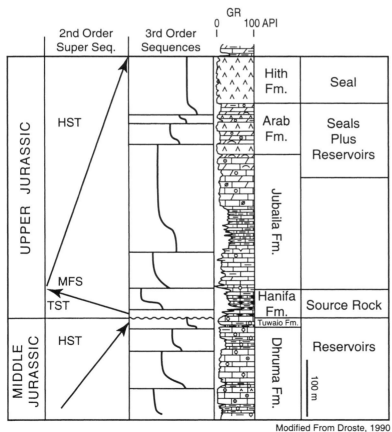

Fig. 6.42. Stratigraphic framework of the Middle to Upper Jurassic in Saudi Arabia and the Gulf Emerites. Modified in part from Droste, (1990). The Upper Jurassic 2nd order supercycle incorporates data from Haq et al. (1988). The position of the 3rd order sequences utilizes the cycle tops of Droste (1990). The shape of the coastal onlap curves are in part conceptual. Used with permission of Sedimentary Geology.

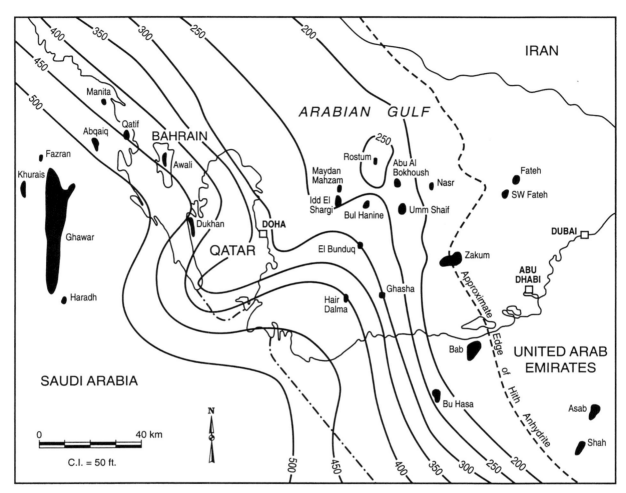

Fig.6.43. Isopach map of the Hith Anhydrite of the southern Persian Gulf and some of the major petroleum fields of the region. Used with permission of AAPG.

6.43). The terminal phase of each succeeding younger sequence becomes progressively shallower and more evaporitic culminating in the thick and laterally extensive Hith Evaporite (Formation), (Figs. 6.42, 6.43 and 6.44).

The Arab Formation hosts some of the most prolific oil reservoirs in the world. Reservoirs in the Arab are designated as the Arab A, B, C and D (Fig. 6.43). Each of these Arab reservoirs represents a single depositional sequence capped by an evaporite (Fig. 6.42 and 6.43). The Arab D is the prime reservoir at

Ghawar Field (Fig. 6.43), the largest and most prolific field in the world (Powers, 1962). The terminal anhydrites of the Arab sequences have often been ascribed to a sabkha environment because of their common nodular texture (Wilson, 1985). However, they are relatively thick (50-100 feet) and contain little interbedded carbonate toward the center of the Arabian shelf (Wilson, 1985) (Fig. 6.43). These attributes point to subaqueous lagoonal or coastal salina deposition rather than a sabkha origin. Significant portions of the Arab D

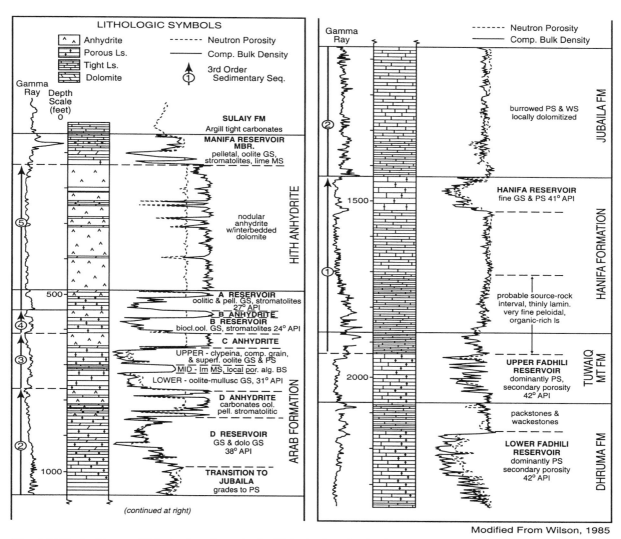

Fig. 6.44 Typical log and lithologic column in the Hith, Arab and Jubaila formation in Qatif Field (See Figure 6.43 for location of this field). Note dolomite in the D reservoir beneath the D anhydrite and limestones in the A, B and C reservoirs. At least 5 shoaling upward 3rd order sequences may be differentiated between the Dhruma and Hith formations. Reprinted with permission from Carbonate Petroleum Reservoirs, Springer-Verlag, New York.

reservoir are dolomitized at Ghawar and other fields such as Qatif (Figs. 6.43 and 6.44) (Powers, 1962 and Wilson, 1985). This dolomitization seems to be related to refluxing brines from an overlying Arab D evaporite lagoon similar to the setting of the Upper Jurassic Smackover Formation detailed in a previous section of this chapter.

The Arab A, B and C reservoirs, however, contain little dolomite although they also occur immediately below relatively thick evaporite units much like the Arab D (Powers, 1962; Wilson 1985; Saner and Abdulghani, 1995). It would seem that these evaporites (B and C) were deposited in a coastal salina, rather than a sea-linked evaporite lagoon. The base of the salina was sealed with evaporites not allowing brine reflux to dolomitize the underlying reservoirs.

These relationships suggest that the Arab B, C and D sequences have undergone a sedimentologic evolution similar to that of the MacLeod Salt Basin. Initially an evaporative lagoon with direct access to the open ocean is formed during the early highstand of the Upper Jurassic super sequence (Arab D). At this point new accommodation space is still being formed on the Arabian Shelf and the lagoon is still linked to the sea. Little new accommodation space is formed during the late highstand of the super sequence and the lagoon is isolated from the sea during a succession of higher frequency sea level lowstands (3rd order) forming successive coastal salinas (Evaporites of the Arab C and B).

The Hith Formation immediately above the Arab A is the terminal evaporite of the Upper Jurassic super sequence and forms the seal for the Arab A reservoir. Since the Hith attains a thickness in excess of 500 feet (Fig. 6.44) and the subjacent Arab A is generally not dolomitized (Fig. 6.43), it would seem that the Hith also was deposited in a salina or playa setting. Alsharhan and Kendall (1994) noted the development of a limestone barrier separating the Hith evaporite lagoon from the open Jurassic ocean to the southeast and east forming a coastal salina much like Lake MacLeod in Australia. The Hith has a number of laterally extensive interbedded dolomites, which might represent marine incursions during 4th or even 3rd order sequences.

SUMMARY

Some of the most important oil and gas fields around the world are developed in dolomite reservoirs related to diagenesis in evaporative marine environments. Evaporites seal reservoirs and occasionally destroy porosity by cementation. The precipitation of gypsum contributes significantly to dolomitization via incorporation of calcium ions into the gypsum crystals and the resultant dramatic increase in the Mg/Ca in associated fluids. Effective porosity development in sabkha-related dolomites often depends on post-dolomitization fresh water flushing as a consequence of exposure and formation of sequence boundary diagenetic terrains.

Marginal marine evaporative sequences occur in two major settings: the sabkha, where evaporites and dolomites form under dominantly subaerial conditions; and the barred evaporite lagoon or basin, where evaporites are formed subaqueously. Sabkha dolomitization is limited by a restricted hydrologic system, although adjacent porous limestones often tend to be dolomitized by refluxed brines.

There are two basic settings for the barred evaporite basin or lagoon. The first is where the barrier that restricts marine water influxing into a shelf or basin does not totally isolate the basin or lagoon from the open sea. In this case, lateral and vertical density gradients are established in the lagoon. Continued influx of marine waters allows the deposition of gypsum in the landward reaches of the lagoon and reflux of post-gypsum Mg-rich brines into the previously deposited sequence. Under these conditions both extensive reflux dolomitization of the underlying sequence and porosity enhancement are possible. The lagoonal evaporite reflux setting is commonly developed during the transgressive systems tract, where significant volumes of new accommodation

SUMMARY

space form and barring of the shelf/basin is incomplete.

The second setting is the coastal salina where the salina or lagoon is totally cut off from marine water influx by the barrier. Under these conditions evaporative drawdown is operative, and evaporites and fine-grained carbonates are deposited across the entire lagoon, effectively sealing the sequences below from brine reflux and subsequent dolomitization. Halite and even bittern salts are commonly precipitated along with gypsum and anhydrite. The coastal salina setting is commonly developed during periods of slow accommodation space formation, such as during the late highstand systems tract (HST) or the lowstand systems tract (LST). Dolomitization and porosity enhancement are less likely in such a setting, although the associated evaporites can provide an excellent seal for any pre-existing porous carbonates.

Geologic setting, timing of dolomitization, relationship of dolomite and evaporite sequences, as well as traces of leached evaporites, such as collapse breccias, are fundamental to the recognition of evaporative marine diagenesis. Although the chemical and isotopic signature of modern evaporative marine dolomites is distinctive, great care must be exercised in using dolomite geochemistry because of the possibility of metastable dolomite recrystallization.

Chapter 7

DIAGENESIS IN THE METEORIC ENVIRONMENT

INTRODUCTION

Most shallow marine carbonate sequences inevitably bear the mark of meteoric diagenesis (Land, 1986). This common diagenetic environment is also one of the most important diagenetic settings in terms of the development and evolution of carbonate porosity because of the general aggressiveness of meteoric pore fluids with respect to sedimentary carbonate minerals. This chemical aggressiveness ensures the relatively rapid dissolution of carbonate grains and matrix and the creation of secondary porosity. The carbonate derived from this dissolution can ultimately be precipitated nearby, or elsewhere in the system, as porosity-destructive carbonate cement occluding primary as well as secondary pores. Meteoric diagenetic processes, then, can lead to a dramatic and early restructuring of original depositional porosity if applied shortly after deposition to unstable mineral suites. Conversely, they can rejuvenate and enhance the porosity of older, mature carbonate sequences if applied during exposure associated with unconformities.

In this section we will review some of the fundamental geochemical, mineralogical and hydrologic aspects of the meteoric diagenetic environment relative to the development and evolution of porosity.

GEOCHEMICAL AND MINERALOGICAL CONSIDERATIONS

Geochemistry of meteoric pore fluids and precipitates

The chemical composition of meteoric waters, as compared to marine and subsurface pore fluids, is shown in Table 4.2. Meteoric waters are dilute and contain small amounts of ions important to the carbonate system such as Ca^{+2}, Mg^{+2}, Sr^{+2}, $Mn^{+2 \text{ and } +3}$ and $Fe^{+2 \text{ and } +3}$.

Meteoric waters, while exhibiting a wide range of carbonate saturation states, are often strongly aggressive (undersaturated) with respect to most carbonate mineral species. This aggressiveness is due to the fact that meteoric waters have access to large reservoirs of CO_2 present in the atmosphere and particularly in the soils of the vadose zone (Morse and Mackenzie, 1990). The P_{CO_2} of soils can often reach concentrations of 10^{-2} atm, two orders of magnitude higher than atmospheric P_{CO_2} ($10^{-3.5}$ atm) (Matthews, 1974). Meteoric waters must pass through the atmosphere in their journey to the surface of the earth. These waters often pass through the vadose zone and its attendant soils as they recharge the phreatic zone. Thus meteoric waters have ample opportunity to dissolve significant quantities of soil-derived CO_2 and to become more strongly undersaturated with respect to $CaCO_3$. The carbonate dissolved by these aggressive fluids then becomes available for later precipitation as carbonate cements as the waters evolve chemically during rock (or sediment)-water interaction.

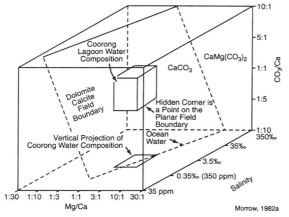

Fig.7.1. *Block diagram showing the effect on water chemistry and mineral precipitates of variation in three parameters, the Mg/Ca solution ratio, the salinity, and the CO3 / Ca ratio. The plane represents the kinetic boundary between dolomite and calcite or aragonite and it includes the hidden corner of the Coorong Lagoon waters as a point on the plane. The basal plane is after Folk and Land (1975). Note that the vertical projection of Coorong Lagoon waters falls largely on the calcite-aragonite side of the stability boundary on the basal plane. A decrease in salinity, an increase in the Mg/Ca ratio or an increase in the CO3 /Ca ratio favors the precipitation of dolomite. Used with permission of the Geological Association of Canada.*

The low Mg/Ca ratios and salinity of most meteoric waters favor the precipitation of calcite (Fig. 7.1). The obvious exception is the common occurrence of aragonite and Mg-calcite speleothems forming in caves. Gonzalez and Lohmann (1988) suggest that aragonite speleothems form as a result of high Mg/Ca ratios present in some cave waters. Cave fluids seep into large vadose cavities with lower P_{CO_2} and quickly de-gas. Calcite precipitates, and concomitant evaporation drives Mg/Ca ratios to levels that support aragonite precipitation (Fig. 7.2A). Analyses of cave waters associated with calcite and aragonite speleothems suggest that aragonite begins to precipitate at elevated CO_3^{-2} levels with Mg/Ca ratios higher

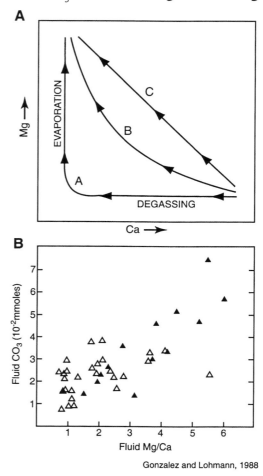

Fig.7.2. *(A). Possible changes in Ca and Mg concentration during degassing, evaporation and carbonate precipitation. Precipitation induced by fluid de-gassing results in decreasing Ca concentrations. Evaporation after de-gassing results in an increase in*

(Fig. 7.2 cont.) Mg concentrations (line A). If de-gassing and evaporation occur simultaneously and at a constant rate during carbonate precipitation, trend C will result. If de-gassing rate decreases at a constant rate of evaporation, all trends will lie under line C (e.g., line B). (B) Fluid Mg/Ca ratios and CO_2 concentrations for waters associated with monomineralic cave samples. Aragonites are solid diamonds; calcites are open diamonds. Aragonite precipitation seems to occur only at Mg/Ca ratios higher than 1.5, indicating the importance of elevated Mg concentration or Mg/Ca ratio for aragonite precipitation to occur. Used with permission from Springer-Verlag, New York.

than 1.5 (Fig. 7.2B).

Dolomites also form in continental meteoric waters associated with cave environments (González and Lohmann, 1988) as well as in evaporative lakes in the Coorong of southeastern Australia (von der Borch et al., 1975). The majority of the Coorong waters, however, fall within the calcite field of the Folk and Land (1975) calcite/dolomite stability diagram (Fig. 7.1). Morrow (1982a) suggests that the Coorong dolomites (and perhaps the spelean dolomites of Gonzalez and Lohmann) are the result of elevated CO_3/Ca ratios combined with relatively low salinity and intermediate Mg/Ca ratios (Fig. 7.1).

The calcite solubility curve does not co-vary in a linear fashion with salinity (Runnells, 1969; Plummer, 1975; Back et al., 1979; Morse and Mackenzie, 1990). When the theoretical dolomite and calcite solubility curves are plotted on a solubility/salinity diagram, a zone in which the waters should be saturated with respect to dolomite, and undersaturated with respect to calcite, is apparent (Fig. 7.3A). This relationship implies that mixed waters can be responsible for extensive carbonate dissolution and the development of significant secondary porosity. It also forms the foundation for the popular Dorag, or meteoric-marine water

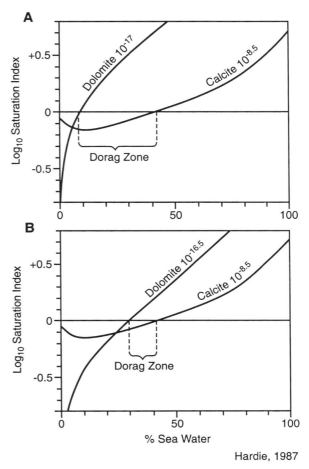

Fig. 7.3. Theoretical saturation relations of dolomite and calcite in mixtures of seawater and meteoric water at 25°C. (A) The "Dorag Zone", where mixtures are supersaturated with dolomite, undersaturated with calcite, for dolomite with $K=10^{-17}$, (B) The "Dorag Zone" for disordered dolomite with $K=10^{-16.5}$. Used with permission of SEPM.

mixing zone model of dolomitization. Hardie (1987), however, suggested that since most of our early dolomites are probably protodolomites, the K value for dolomite used in this model should be $10^{-16.5}$ rather than the 10^{-17} that was originally used. As will be stressed later, this reduces significantly the "Dorag zone" of potential dolomitization (Fig.7.3B).

Isotopic composition of meteoric waters and carbonates precipitated from meteoric waters

The isotopic composition of meteoric waters (and the calcite cements and dolomites precipitated from these waters) is highly variable. Figure 4.12 illustrates the range of isotopic compositions one might expect in natural geologic situations. The oxygen isotopic composition of meteoric waters is strongly latitude dependent and can vary in $\partial^{18}O$ from -2 to -20 per mil (SMOW) (compare the oxygen isotopic compositions of Enewetak and Barbados meteoric cements as seen on Figure 4.12) (Hudson, 1977; Anderson and Arthur, 1983). At any one geographic site, however, the latitudinal effect is zero and the initial $\partial^{18}O$ composition of meteoric water should be relatively constant. If these waters interact with limestones and sediments, however, the water's ultimate oxygen isotopic composition will be determined by the ratio O_L (number of moles oxygen from limestone) to O_W (number of moles oxygen from water) (Allan and Matthews, 1982). The solubility of calcium carbonate in surface meteoric waters is very low (1 liter of water at 25° C, $P_{CO_2} = 10^{-3.5}$, contains just 1×10^{-3} moles of total CO_2 in 55.5 moles of H_2O). Because of this low solubility, the O_L/O_W will be much less than 1, and the oxygen isotopic composition of the water should remain approximately the same regardless of the amount of rock/water interaction, or the distance down hydrologic gradient from recharge areas (Allan and Matthews, 1982). Subsequent work by Lohmann (1988) suggests that during the initial stages of mineral stabilization by meteoric water, the O_L/O_W may be great enough to buffer the oxygen isotopic composition toward more positive values.

The range of $\partial^{13}C$ values available within meteoric diagenetic environments is also enormous, ranging from -20 $\partial^{13}C$, characteristic of soil gas CO_2, to +20 $\partial^{13}C$, associated with organic fermentation processes (Fig. 4.12). In most meteoric water environments, however, soil gas and the CO_3 derived from the dissolution of marine limestones are the prime sources for the ultimate $\partial^{13}C$ composition of meteoric diagenetic waters (Allan and Mat-

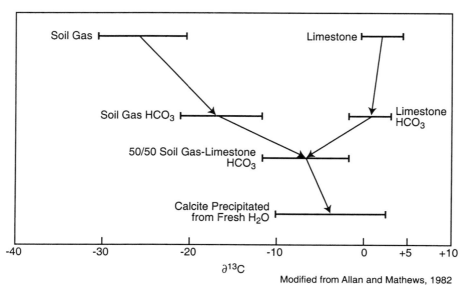

Fig. 7.4 Contribution of soil gas (CO_2) and CO_2 from dissolution of limestone to the final isotopic composition of calcite precipitated from meteoric water. It is assumed that there is a 50/50 mixture of the two sources. Fractionation factors for soil gas and limestone to HCO_3 from Emerich et al., 1970. Reprinted with permission of Sedimentology.

thews, 1982; Lohmann, 1988)(Fig. 7.4). The carbon isotopic composition of meteoric diagenetic fluids (and diagenetic carbonates precipitated from these fluids) is determined by the ratio of C_L (number of moles of carbon from limestone) to C_{SG} (number of moles of carbon from soil gas). At the surface in vadose recharge zones, the bicarbonate in meteoric waters comes equally from soil gas and limestone carbonate and the ratio is generally near 1. In this case, the meteoric waters will be depleted relative to ^{13}C because of the strongly negative carbon isotopic composition of the soil gas (Fig. 4.12 and 7.4). In the phreatic zone as meteoric waters move further from the soil gas reservoir, the contribution of heavy carbon derived from dissolution of limestone or metastable carbonate sediment is increased. Under these conditions, the C_L/C_{SG} ratio becomes larger, and hence the waters become progressively more enriched with respect to ^{13}C.

Figure 7.5 is an idealized cross plot of the variation in $\partial^{18}O$ and $\partial^{13}C$ that one might expect in the meteoric diagenetic environment based on the concepts outline above. The main variational pattern is one of constant $\partial^{18}O$ and rapidly changing $\partial^{13}C$. Lohmann (1988) terms this characteristic trend, the meteoric calcite line. During increasing rock-water interaction involving metastable carbonate mineral species at sites removed from the recharge area, one would expect the Mg and Sr compositions of diagenetic calcites to increase, reflecting the sequential stabilization of Mg-calcite and

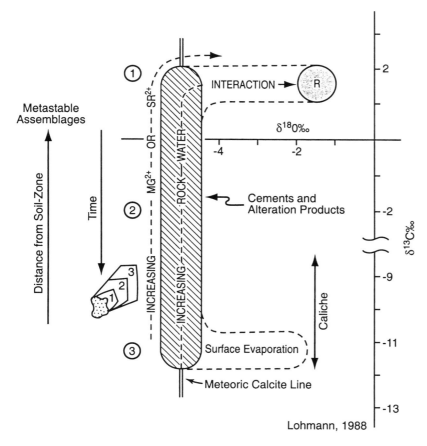

Fig. 7.5. Idealized plot of variation in $\partial^{18}O$ and $\partial^{13}C$ characteristic of meteoric vadose and phreatic carbonates. The constancy of $\partial^{18}O$ and the variable $\partial^{13}C$ define a trend termed the meteoric calcite line (2). Deviations from this line will generally take place where there is increased rock-water interaction with polymineralic suites distal from meteoric recharge and water may be buffered by rock-derived carbon, oxygen, and trace elements, such as Mg and Sr (1). Further deviation is encountered at exposure surfaces where surface evaporation may drive $\partial^{18}O$ compositions toward more positive values. Reprinted with permission of Elsevier.

aragonite (Benson and Matthews, 1971; Lohmann, 1988). Additionally, the isotopic composition of a diagenetic calcite crystal at a site removed from the recharge area should mirror the isotopic evolution of the water toward more depleted carbon isotopic compositions through time. This progressive depletion results from diminished rock-water interaction as the terrain matures and metastable components disappear (Lohmann, 1988; Morse and Mackenzie, 1990).

Local increases of $\partial^{18}O$ composition of meteoric water (or diagenetic calcites) can take place by evaporation at exposure surfaces or within caliche soil profiles (Fig. 7.5).

In zones of mixed meteoric-marine waters one would expect the carbon and oxygen isotopic composition of diagenetic calcites to exhibit positive co-variance (Fig. 4.14) (Allan and Matthews, 1982). Lohmann (1988), however, suggests that the isotopic co-variation of diagenetic cements and mineral replacements formed by mixed meteoric-marine waters will generally follow the meteoric calcite line because the mixing of marine and meteoric waters results in undersaturation and dissolution, rather than precipitation, over most of the range of mixing. The calcites, or dolomites precipitated in the mixing zone, therefore, will represent only a small segment of the range of mixing, and the resulting isotopic composition will "...define vertical trends of variation parallel to the meteoric calcite line rather than a single hyperbolic mixed-water curve"(Lohmann, 1988, p. 74).

Mineralogic drive of diagenesis within the meteoric environment

Modern carbonate sediments consist of metastable mineral suites dominated by Mg-calcite and aragonite. The solubility of aragonite is twice that of calcite, and the solubility of Mg-calcite is approximately ten times that of calcite (Morse and Mackenzie, 1990). These solubility differences provide a strong drive in the destination and intensity of meteoric diagenesis, since a water in contact with two solid phases of differing solubility simply cannot be in equilibrium with the rock (Matthews, 1974; James and Choquette, 1984; Morse and Mackenzie, 1990).

If the water is near saturation with respect to aragonite, it will be supersaturated with respect to calcite, and should precipitate calcite. If the water is near saturation with respect to calcite, it should dissolve aragonite. Ultimately equilibrium in this system can only be achieved by completely destroying the most soluble phase (Matthews, 1974). This represents a drive toward stabilization by dissolution and precipitation that plays a major role in porosity modification.

Meteoric waters interacting with a mature carbonate terrain consisting of a single stable mineral species such as calcite, will rapidly move toward equilibrium with the rock; and diagenetic processes effecting porosity modification will soon cease (Morse and Mackenzie, 1990; Budd et al., 1993).

Implications of the kinetics of the $CaCO_3$-H_2O-CO_2 system to grain stabilization and porosity evolution in meteoric diagenetic environments

If a fluid is said to be oversaturated or undersaturated with respect to a particular mineral phase, it simply means that the system is not at equilibrium and that precipitation or solution should happen. No particular time frame for the reaction is established. Reaction kinetics, on the other hand, handles what will happen to the reaction within a finite length of time (Matthews, 1974).

Chemical reactions such as the dissolution of aragonite or the precipitation of calcite can be broken into a series of steps that must occur for the reaction to achieve equilibrium. For any step in the reaction, the greater the departure from equilibrium, the faster that the step or the reaction will be able to proceed. The overall reaction cannot proceed faster than the slowest step within the reaction chain. This slowest step is termed the rate step (Matthews, 1974).

Figure 7.6 outlines the reaction steps for carbonate solution and precipitation reactions during the stabilization of a metastable mineral suite to calcite. The most significant reaction relative to porosity/permeability development is the dissolution of metastable solid phases (aragonite and Mg-calcite) and the precipitation of the stable solid phase (calcite). One of the most important rate steps in this reaction is the spontaneous nucleation of the stable calcite relative to the dissolution of the metastable aragonite. The rate of calcite growth from a solution at steady state with aragonite is at least 100 times slower than the rate at which aragonite will dissolve to replace the $CaCO_3$ lost from solution to calcite precipitation (Matthews, 1974; Morse and Mackenzie, 1990). This nucleation step can only be circumvented by precipitation of calcite onto pre-existing calcite sedimentary particles as nucleation sites.

The large kinetic difference between aragonite dissolution and calcite precipitation implies that $CaCO_3$ dissolved from aragonite is generally transported away from the site of solution. If water flux in the diagenetic system is large, and the water is strongly undersaturated with respect to aragonite, aragonite grains will undergo total dissolution, moldic voids will be formed, and all internal structure within the aragonite grains will be destroyed. The $CaCO_3$ from this grain dissolution may be deposited in the near vicinity (see for example Morse and Mackenzie, 1990), or carried a significant distance down hydrologic gradient before calcite is precipitated as a void-fill porosity occluding cement. This style of aragonite to calcite stabilization, termed by James and Choquette (1984) as macroscale dissolution, may lead to massive rearrangement of pore space and the development of cementation gradients in major meteoric phreatic water systems (Fig. 7.7 and 7.8 C and D).

If the water flux is slow and meteoric pore

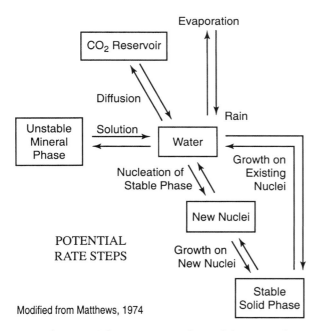

Fig.7.6. Potential rate steps in the stabilization of unstable carbonate mineral phases to stable carbonate mineral phases by dissolution and precipitation. Used with permission, Elsevier.

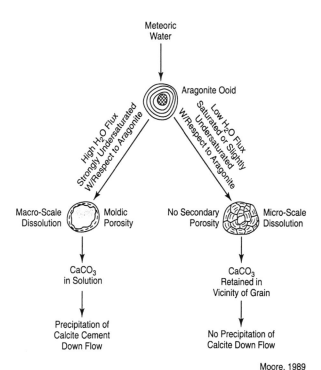

Fig. 7.7. Schematic diagram showing the two basic pathways, macroscale dissolution versus microscale dissolution, for the stabilization of aragonite grains to calcite. Major conditions controlling the path that the metastable grain will take, and the resultant porosity implications are indicated. Used with permission, Elsevier.

fluids only slightly undersaturated with respect to aragonite, stabilization of aragonite grains may proceed by microscale dissolution, or replacement (James and Choquette, 1984; Land, 1986). In this case dissolution and precipitation accomplish stabilization of aragonite across a thin reaction film within the grain (Pingitore, 1976). Saller (1992) in his studies of Pleistocene marine sediments from the Bahamas, Yucatan and Enewetak documented the delicate dissolution of aragonite allochems with later calcite cementation rather than the thin film of previous authors. The end product of both processes is a calcite mosaic with aragonite inclusions outlining elements of the original aragonite grain ultrastructure. $CaCO_3$ seems to be conserved within the grain, and microscale dissolution does not lead to appreciable secondary porosity development or to significant calcite cementation outside the grain (Fig. 7.7 and 7.8A).

The calcite precipitation rate step may be avoided if the metastable phase is Mg-calcite because appropriate nucleation sites for calcite precipitation are available within the Mg-calcite particle. In a meteoric water system undersaturated with respect to Mg-calcite, but saturated with respect to calcite, metastable $MgCO_3$ components dissolve and calcite is precipitated on readily available $CaCO_3$ nucleation sites within the grain (Matthews, 1974; James and Choquette, 1984; Land, 1986; Budd, 1992). In this case, the ultrastructure of the original Mg-calcite grain is preserved, no moldic void is developed, and no $CaCO_3$ is released to the water to be precipitated as cement elsewhere in the system. Stabilization of Mg-calcite sediments to calcite, then, generally results in only minimal modification of the original sedimentary pore system. In certain cases, however, perhaps under conditions of very high water flux and strongly undersaturated waters, Mg-calcite grains may undergo macroscale dissolution, produce moldic secondary porosity, and release $CaCO_3$ into the water for potential calcite cements elsewhere (Schroeder, 1969; Budd, 1984; James and Choquette, 1984; Saller, 1984).

Climatic effects

Climate has a pervasive effect on meteoric diagenesis. Warm temperatures promote the oxidation of organic matter making more CO_2 available to dissolve limestones and stabilize

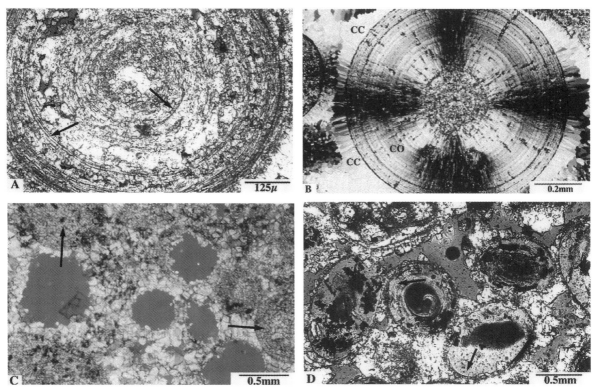

Fig.7.8. (A) Inferred aragonite Jurassic Smackover ooid that shows evidence of stabilization by microscale dissolution. Note the ghosts of original ooid lamellar structure preserved in the fine calcite crystal mosaic (arrows) (see Moore et al., 1986). Murphy Giffco #1, Miller Co. Texas, 7784' (2373m). Plain light. (B) Calcite Jurassic Smackover ooid that shows no mosaic calcite in the ooid and perfect preservation of ooid internal structure (CO) suggesting an original calcite mineralogy (see Moore et al., 1986). Note bladed to fibrous calcite cement (CC) which is in optical continuity with the internal structure of the ooid, which also suggests an original calcite mineralogy for the cements as well as the ooid. Arco Bodcaw #1, Columbia Co. Arkansas 10860' (3311m). Crossed polars. (C) Macroscale dissolution of inferred aragonite Jurassic Smackover ooids forming moldic porosity. Note that some ooids have stabilized by microscale dissolution (arrows). Same sample as (A). Plain light. (D) Macroscale dissolution of aragonite ooids (arrows) leading to oomoldic porosity, Pleistocene, Shark Bay, western Australia. Plain light.

metastable carbonate mineral suites. In addition, the rate constants for carbonate mineral dissolution increase with an increase in temperature. Of course, diagenesis is driven by water, so that it is obvious that meteoric carbonate diagenesis will be greatly enhanced by warm tropical humid climates (Fig. 7.9). In a tropical wet climate, mineral stabilization will be rapid, caves and karst features dominant, and terra rosa soils will form (Morse and Mackenzie, 1990). In arid climates, calcretes will form at the surface and mineral stabilization will be slow, allowing preservation of aragonite well into burial conditions (Fig. 7.9). Under arid conditions it will be difficult to recognize the influence of meteoric diagenesis. Under semi-arid conditions mineral stabilization will be slow in the vadose zone allowing preservation of aragonite. More complete mineral stabilization will take place in the

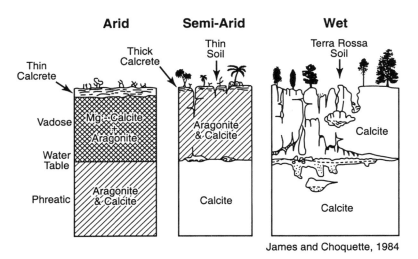

Figure 7.9. Schematic diagram illustrating the effects of climate on meteoric diagenetic pathways for coeval carbonates originally composed of a polymineralic assemblage of calcite with various amounts of magnesium and strontium-rich aragonite. Used with permission of the Geological Association of Canada.

phreatic zone. Thick calcretes and thin soils will form under these conditions (Fig. 7.9).

Hird and Tucker (1988) attributed a climatic change to the different diagenetic pathways taken by two different Carboniferous oolite formations. Both oolites were inferred to be calcite. The younger unit had little evidence of meteoric cement, while the older had abundant cement. They inferred that the younger oolite was under the influence of an arid climate, while the older unit was flushed by meteoric water under a more humid climate regimen. It would seem that the abundant fresh water allowed early onset of pressure solution with attendant intergranular cementation and porosity reduction (see also Meyers and Lohmann, 1985).

Hydrologic setting of the meteoric diagenetic environment

The meteoric diagenetic environment consists of three major hydrologic zones: 1) vadose, 2) phreatic, and 3) mixed zones (Fig. 7.10). The water table marks the boundary between the vadose and phreatic zones. Below the water table in the phreatic zone, pore spaces are saturated with water. The phreatic aquifer is often unconfined and open to the atmosphere with a well-defined vadose zone between the water table and the surface. In the vadose zone both gas and water are present in the pores, and water is generally concentrated by capillarity at the grain contacts. In the zone of infiltration near the upper surface of the vadose zone, water may be under the influence of both evaporation and plant transpiration and is actively involved in soil and caliche formation (James and Choquette, 1984). Water moves through the vadose zone in two contrasting styles. The first style is vadose seepage where water slowly trickles through a network of small pores and fractures. The second is by vadose flow, where water rapidly moves through the vadose zone through solution channels, sinkholes, joints and large fractures directly to the water table (Wright, 1991; Loucks, 1999).

Generally water moves through the phreatic zone by diffusive flow, where there is a well-defined water table and Darcy's Law can approximate flow conditions. However, free

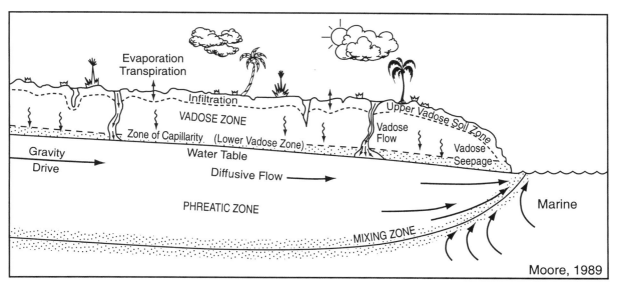

Figure 7.10. Conceptual model of the major diagenetic environments and hydrologic conditions present in the meteoric realm. Used with permission of Elsevier.

flow conditions dominate where water moves through channels and integrated conduits as a free flowing subterranean river. Water tables are difficult to recognize in free flow settings. Diffusive flow is characteristic of young carbonate sediments newly exposed and in the early stages of stabilization. Free flow is often found in mature, stabilized carbonate terrains associated with karst development (Smart and Whitaker, 1991; Wright, 1991; Loucks, 1999).

Meteoric hydrologic settings are complicated and are not static as is often portrayed in the literature (Longman, 1980). These settings often undergo an evolution forced by sedimentologic processes, climate and sea level changes. It is convenient to view these evolving hydrologic settings in terms of two end-member phases (Fig. 7.11).

The first phase, which may be termed the immature phase, is generally developed during sea level highstands and is dominated by island floating fresh-water lenses developed in mineralogically young sediments. On large isolated oceanic platforms, such as the Bahamas or Bermuda, the islands will generally be isolated, rim the platform, and exhibit some limited lateral progradation (Fig. 7.11A). In the context of continent-tied carbonate shelves, a terminal highstand coastal complex can prograde oceanward for considerable distances, such as in Quintana Roo on the Yucatan Peninsula in Mexico (Ward, 1997). Coastal progradation has produced a number of related, but individual elongate islands that may develop individual and/or overlapping floating fresh water lenses (Fig. 7.11A).

The second or *mature phase* is developed during sea level lowstands where entire carbonate platforms or shelves are exposed to meteoric water conditions on a regional scale. On large oceanic carbonate platforms such as the Bahamas, a regional unconfined meteoric lens floating on marine water with a very thick vadose zone and active marginal mixing zone may be developed (Fig. 7.11B). In continent-tied carbonate shelves, differential shelf mar-

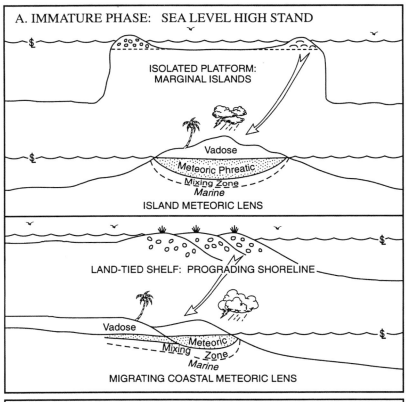

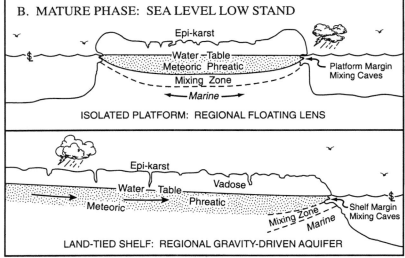

Fig. 7.11. Hydrologic settings of meteoric diagenetic environments as a function of sea level and depositional conditions (isolated platform versus land-tied shelf).

gin subsidence can lead to gravity-driven unconfined aquifers. With enough exposure, these systems can evolve into karstic free flow systems with a coastal meteoric wedge such as is seen on Barbados (Humphrey, 1997)(Fig. 7.11B).

In the ensuing transgression the unconfined aquifer may evolve into a confined regional aquifer (Moore and Heydari, 1993). Continued differential subsidence between recharge and discharge areas through time increases the difference in head between discharge and recharge areas and the system becomes a pressure driven artesian system such as is seen in peninsular Florida. Confined aquifer systems may act as efficient conduits for the distribution of meteoric waters deep into sedimentary basins as well as beneath the continental shelf (Land, 1986; Smart and Whitaker, 1991).

In the following sections we will first treat the petrographic and geochemical characteristics of the vadose diagenetic environment including porosity development and evolution. In the succeeding sections we will discuss the nature of the phreatic diagenetic environment using the framework of the two major hydrologic settings outlined above.

THE VADOSE DIAGENETIC ENVIRONMENT

Introduction

The vadose zone is important because most of the meteoric water that ultimately finds its way back to the sea must pass through the vadose environment before it is entrained into a regional meteoric aquifer system or a floating freshwater lens. The vadose environment may be divided into two zones: an *upper vadose soil* or *caliche zone* at the air-sediment or rock interface, and a *lower vadose zone* that includes the capillary fringe just above the water table (Fig. 7.10).

Upper vadose soil or caliche zone

The air-sediment interface is generally an environment of intense diagenesis in metastable carbonate terrains because infiltrating meteoric waters are generally undersaturated with respect to $CaCO_3$ due to the high P_{CO_2} usually associated with soils. In regions of high rainfall, and porous sediment, water tends to move through the zone quickly, and dissolution is probably the dominant process. In this situation, the air-sediment interface may show little evidence of subaerial exposure (Saller, 1984). The $CaCO_3$ removed by dissolution may be transported downward into the zone of capillarity, or the lower vadose zone, where calcite cements are precipitated.

In semi-arid climates, however, the potential for carbonate dissolution as well as evaporation and subsequent calcite precipitation

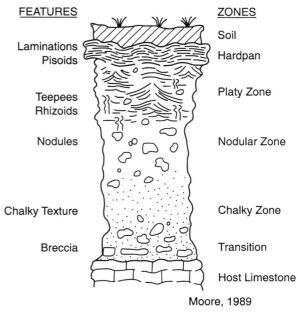

Fig. 7.12. Sketch of a typical caliche profile illustrating major zones and characteristic features. Used with permission of Elsevier.

Fig. 7.13. (A) Pisoids developed in caliche soil profiles, Permian Yates Formation on New Mexico Hwy. 7A to Carlsbad Caverns at Hairpin Curve. (B) Teepee antiform structure in Tansill Formation at the south end of the south parking lot at the entrance to Carlsbad Caverns. (C) Eastern coast of Isla Mujeres, Quintana Roo, Mexico showing latest Pleistocene eolianites. The white undulating upper surface is epikarst with caliche. Lighthouse on left for scale. (D) Rhizocretions developed in upper dune complex on Isla Mujeres suggesting plant-related stabilization during the Pleistocene. Note thin caliche crust and epikarst on upper surface. (E) Meniscus cementation pattern developed in ooid grainstones taken from the vadose zone at Joulters Cay, Bahamas. Note the rounded shape of the pores. Crossed polars. (F) Scanning electron photomicrograph of (E) showing curved crystal shapes of the meniscus cement. Meniscus photomicrographs furnished by Mitch Harris.

under near-surface conditions results in the formation of distinctive surface soil crusts variously termed caliche, calcrete or duricrust (James and Choquette, 1984) (Fig.7.9). Caliche soils have been the subject of intensive research over the past 10 years because of their potential usefulness in environmental reconstruction (Esteban and Klappa, 1983; Wright, 1991). Figure 7.12 is a sketch of a typical caliche profile showing its major characteristics.

The central features of the crust include: 1) a laminated hardpan usually developed on the top of the sequence (Fig. 7.13C and D); 2) a zone of plates and crusts; 3) a nodular, or pisolitic chalky sequence (Fig. 7.13A); and 4) a chalky transition into untouched sediments or country rock (Purvis and Wright, 1991; Wright, 1991) (Fig. 7.13).

The caliche profile is the result of intensive dissolution of original sediments, or limestones and the rapid re-precipitation of calcite, often driven by organic activity. Original textures and fabrics are generally destroyed and are replaced by a melange of distinctive fabrics, textures, and structures. These structures include nodules, pisoids (Fig. 7.13A), rhizoids (concretions around roots) (Fig. 13C and D), teepees (Fig. 7.13B) (pseudo anticlines), crystal silt (microspar), and microcodium (calcified cells of soil fungi and higher plants) (Purvis and Wright, 1991; Wright, 1991) (Fig. 7.13).

There is some modification of porosity in the upper vadose zone, with potential porosity occlusion associated with extensive hardpan development and some porosity enhancement related to crystal silt. Successive caliche profiles in a cyclical carbonate sequence could lead to vertical compartmentilization similar to that seen in marine hardground formation (Saller et al., 1999).

Lower vadose zone

The lower vadose zone is a domain of gravity percolation with water moving through on its way to the water table. Water saturation in vadose-pores increases significantly just above the water table in a region termed the zone of capillarity (Fig. 7.10). The lower vadose zone is downstream from the high organic activity and high P_{CO2} concentrations of the upper vadose soil zone and hence receives waters that are often saturated with respect to $CaCO_3$ due to carbonate dissolution in the soil zone above. This situation may result in the concentration of calcite cements immediately above, and adjacent to the water table (Longman, 1980; Saller, 1984; Budd and Land, 1990). In those sequences where vadose percolation is reduced due to climatic considerations, and where soils are not present due to sediment characteristics, climate, or evolutionary stage of development of the sedimentary sequence, vadose dissolution and cementation seem to occur in a much more localized fashion. Cements tend to precipitate immediately adjacent to the site of dissolution resulting in a rather uniform distribution of vadose cements throughout the sequence (Halley and Harris, 1979).

Porosity development in the vadose diagenetic environment

Today, most workers agree that dissolution in the vadose zone is balanced by precipitation so that gross porosity values remain the same during stabilization of an unstable mineral suite under meteoric vadose conditions (Harrison, 1975; Halley and Harris, 1979; Budd, 1984; Saller, 1984). The slow movement of diagenetic fluids responsible for solution and precipitation through the vadose zone generally precludes the development of extensive moldic porosity in aragonitic sediments in the vadose environment. Since cements are concentrated at pore throats, permeability in the vadose zone may be significantly reduced (Halley and Harris, 1979).

In a humid climate and a stable mineral suite such as will be found in an exhumed aquifer along an unconformity, fractures will be enlarged by dissolution ultimately forming large vertical shafts that allow meteoric water direct access to the water table. Cave ceilings will begin to collapse, forming large breakout domes and crackle breccias above the dome. Porosity volume as well as permeability will generally be enhanced (Fig. 7.34)(Loucks, 1999).

THE METEORIC PHREATIC DIAGENETIC ENVIRONMENT

Introduction

There is a general consensus among workers today that diagenetic processes are more intense and efficient in the meteoric phreatic environment than in the vadose zone above. Since diagenetic processes are water driven, greater diagenetic activity in the phreatic zone is the result of the greater volume of water found there (Steinen and Matthews, 1973; Matthews, 1974; Land, 1986; Dickson, 1990; Morse and Mackenzie, 1990). The discussions of the meteoric phreatic diagenetic environment that follow will utilize the two part evolutionary hydrologic framework (*immature and mature* phases) established in the previous section (Fig. 7.11). In each case we will utilize modern and ancient case histories to establish and illustrate major concepts.

The major differences between vadose and phreatic environments in these settings will be emphasized where appropriate. Karst processes and products will be discussed in a later separate section because of the involvement of both vadose and phreatic processes.

Immature hydrologic phase: island-based and land-tied floating meteoric water lenses developed in mineralogically immature sediments

During sea level transgressions and subsequent highstands, carbonate islands are established along the margins of isolated carbonate platforms such as the beach ridges of Quintana Roo (Ward et al., 1985; Ward, 1997) and the islands of the Bahamas (Carew and Mylroie, 1997; Kindler and Hearty, 1997). The sediments of these islands are dominantly composed of aragonite and magnesian calcite.

The size of the meteoric lens developed beneath these islands is generally a function of recharge, island width, and hydraulic conductivity (Budd and Vacher, 1991; Vacher, 1997). Table 7.1 shows the thickness/width ratios of a number of present day islands. Measured thickness/width ratios (percent) range from 3.6 down to 0.13 depending on heterogeneity of the hydraulic conductivity of the sediment pile. Those islands with homogenous sediment packages exhibited the largest thickness/width ratios (Busta, Tarawara Atoll). Those islands with large contrasts in hydraulic conductivity such as the contrast between Holocene grainstones and karsted Pleistocene limestones (Big Pine Key, Florida) had the smallest thickness/width ratios. These relationships can be used to pre-

Table 7.1

Island	H/a (percent)	Reference
Busta, Tarawa Atoll	3.6	Lloyd et al. 1980
Bonriki, Tarawa Atoll	1.7	Marshall and Jacobson 1985
Home Is., Cocos Islands	1.3	Jacobson 1976
Laura, Majuro Atoll	1.2	Anthony et al. 1989
Deke Is., Pingelap Atoll	1.1	Ayers and Vacher 1986
Kajalein Atoll	0.8	Hunt and Peterson 1980
Schooner Cays	0.6	Budd 1984
Bermuda	0.6	Vacher 1978
Grand Cayman	0.5	Bugg and Lloyd 1976
Cozumel, Mexico	0.29	Lesser and Weidie 1988
Tongatapu, Tonga	0.22	Hunt 1979
Bahamas		Cant and Weech 1986
Abaco	0.13	
Andros	0.19	
Bimini	0.31	
Cat	0.28	
Eleuthera	0.39	
Exuma	0.26	
Grand Bahama	0.18	
Long	0.22	
New Providence	0.17	
Big Pine Key, Florida	0.2	Hanson 1980

Budd and Vacher, 1991

Table 7.1 thickness ratios of present day island lenses. Used with permission of the SEPM.

dict the thickness of meteoric lenses in ancient island sequences as can be seen in several case histories described later (Budd and Vacher, 1991).

One of the major questions concerning these immature island lenses is the rate and effectiveness of the meteoric diagenesis supported by the lens. We have learned a great deal about this question from a series of studies of very small Holocene tropical islands such as Schooner Cay (Budd, 1984) in the Bahamas. In most modern thin floating meteoric water lenses (developed on very small islands) the most active diagenetic zone exhibiting significant cementation and dissolution is generally no more than 1.5m in thickness, centered on the water table (Moore, 1977; Budd, 1984; Saller, 1984). Budd (1984) in his study of the Schooner Cays in the Bahamas, observed from 10-40 wt. % calcite cement precipitated in intergranular pores concentrated in a 1m zone centered on the water table. The volume of dissolution of metastable grains in the zone closely matched the volume of cement observed (>87% efficiency) so that total porosity volume did not change. Grain dissolution in the floating meteoric phreatic lens at Schooner Cays and other sites such as Joulters in the Bahamas (Halley and Harris, 1979) generally consists of the delicate removal of outer ooid cortaxial layers. However, Budd (1984) observed the occasional development of moldic porosity at Schooner Cays and suggested that, given time, significant moldic porosity might evolve.

As one moves away from the water table toward the mixing zone with marine water below, the intensity of cementation and grain dissolution sharply declines. Within the mixing zone of modern, very small floating meteoric lenses, cementation is muted but significant dissolution and dolomitization has not been documented (Halley and Harris, 1979; Budd, 1984; Saller, 1984).

The stabilization of metastable mineralogies through dissolution and precipitation in the fresh water phreatic environment of a small island lens is relatively rapid, particularly along the water table. Halley and Harris (1979) estimated that the aragonitic ooid grainstones at Joulters Cay would be converted totally to calcite in 10,000-20,000 years. Budd (1988) calculated that all the aragonite in the fresh water phreatic zone at Schooner Cays would be stabilized to calcite in 4,700-15,600 years, while the aragonite in the fresh marine water mixing zone would take 8,700-60,000 years to be converted to calcite. In contrast, many authors have noted that stabilization of metastable sequences is considerably slower under vadose conditions than in the phreatic environment (Steinen and Matthews, 1973; Saller, 1984; Budd, 1984). For example, Lafon and Vacher (1975) estimated the half-life of aragonite in the vadose zone of Bermuda to be about 230,000 years.

Small island meteoric lenses, then, appear to be active diagenetic environments where mineral stabilization and porosity/permeability modification are proceeding at a relatively rapid, geologically significant, rate. However, the sizes of the lenses on these small islands are restricted and ephemeral, which in turn restricts the sizes of the active zone of diagenesis. A larger system, such as Bermuda, where the floating fresh water lens is thicker and permanent, may be a more valid model for larger ancient meteoric phreatic water systems.

Bermuda, a case study of an island with a permanent floating fresh water lens

Bermuda, a moderate sized island in the Atlantic Ocean, has been studied intensely; and its geology, geohydrology and the geochemistry of its sediments and waters are well known (Morse and Mackenzie, 1990; Vacher and Rowe, 1997). Bermuda is a relatively thin limestone cap lying on a volcanic pinnacle rising from the abyssal plain of the Atlantic some 1000km east of North Carolina, USA. The island consists of a prograding Pleistocene complex of shallow marine to shoreface-dune limestones deposited at times of high sea level. Figure 7.14 shows the general geology of Bermuda. The limestones have been divided into a number of formations capped with paleosols that can be traced around the island. The core of the island consists of the oldest limestone unit, the Walsingham Formation. The

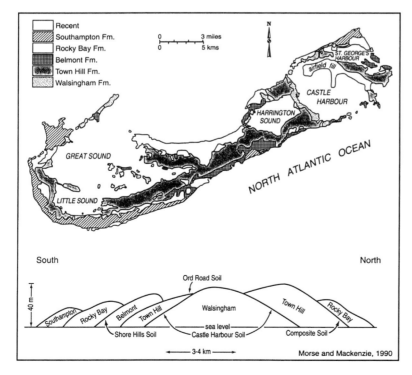

Fig.7.14. Generalized geologic map and cross-section of Bermuda. Used with permission of Elsevier.

younger limestones are arrayed in a rather symmetrical progradational pattern around the Walsingham core. The Bermuda platform has been tectonically stable throughout the Pleistocene and has been termed a "Pleistocene tidal gauge" (Land et al., 1967). Based on Land's (1967) sea level curve for Bermuda, the units deposited after 120,000 years BP (the Rocky Bay and Southampton formations) (Fig. 7.14) have spent most of their time in the vadose zone, while the older units (Belmont, Town Hill, and Walsingham formations) have been in constant contact with phreatic waters.

The limestones consist dominantly of grainstones composed of organisms found living today in the waters of the Bermuda platform. Their original mineralogy was 55% Mg calcite, 38% aragonite and 17% calcite. Figure 7.15 documents the trend in mineralogical composition of these grainstones relative to geologic time. As we might expect the mineralogy of the younger vadose-dominated units have changed very little as compared to the extremely altered older units that have had much more time in the phreatic zone. The older altered units have well-developed karst features such as caves and cavernous porosity (Morse and Mackenzie, 1990; Vacher and Rowe, 1997).

There are several distinct fresh water lenses on Bermuda where hydrology and modern rock-water interaction can be examined. The largest of these lenses is the Devonshire lens, or central lens located in an island-central position (Fig. 7.16). A hydrologic cross section (Fig. 7.16) shows a distinct lens asymmetry thought to be the response of the difference in permeability and hence hydraulic conductivity between the Rocky Bay and Town Hill formations.

The water chemistry of the Devonshire lens (Plummer et al., 1976) suggests that mineral stabilization by solution of aragonite and precipitation of calcite is still proceeding. Figure 7.17A shows the extent of the lens and location of sampled water wells. Figures 7.17C

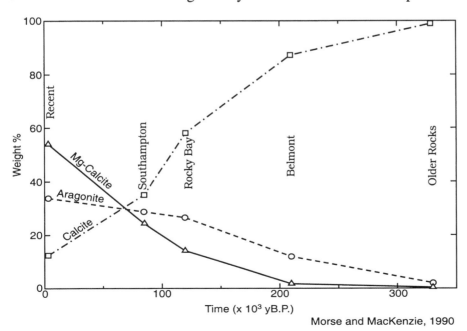

Morse and MacKenzie, 1990

Fig. 7.15. Trends in the mineralogical composition of Bermudian grainstones with geologic age. Used with permission of Elsevier.

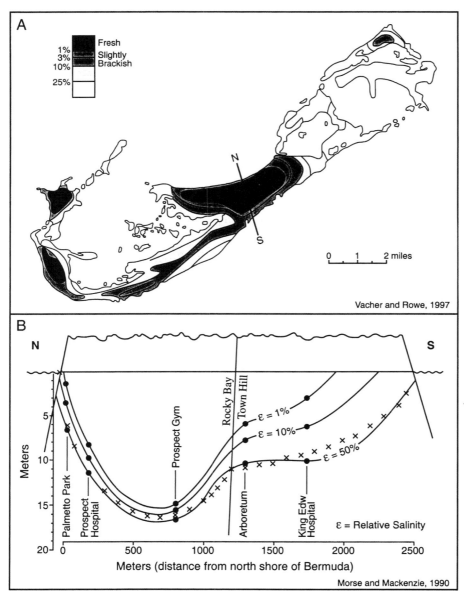

Fig. 7.16. A. Location of freshwater lenses in Bermuda. Map shows contours of percent seawater in household wells, 1972-1974. B. Hydrogeologic cross-section of the Devonshire groundwater lens. X's denote a calculated fit to the 50% isoline of relative salinity (ε) using the Gerbenn-Herzberg-Dupuit model. Used with permission of Elsevier.

and D show the saturation index for aragonite and calcite in the water of the lens. The waters are undersaturated with respect to aragonite and calcite in the center of the lens generally underlying the Devonshire marsh. Since both the Rocky Bay and Town Hill formations still contain aragonite (Fig. 7.15), aragonite should be dissolving in the center of the lens. Calcite, however, is supersaturated on the margins of the lens while aragonite is near equilibrium (Fig. 7.17C and D). In this area, calcite should be precipitating as cement. The $CaCO_3$ needed for the cement is no doubt being transported from the zone of aragonite solution in the center of the lens.

Are the permanent meteoric lenses on

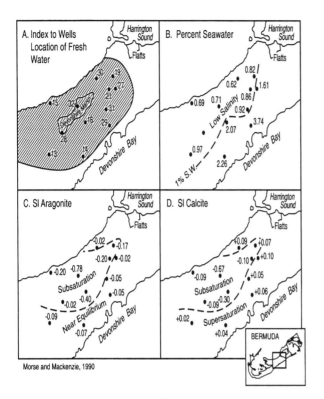

Fig.7.17. Geochemistry of the Devonshire, Bermuda groundwater lens. (A) Well samples and location of freshwater, (B) percentage of seawater in the groundwater, (C) saturation with respect to aragonite, (D) saturation with respect to calcite. Sl (=Ω) is the saturation index. Used with permission of Elsevier.

Bermuda viable models for ancient immature hydrologic phases? The rate of stabilization of aragonite in the phreatic zone of the Bermuda fresh water lens has been calculated by Vacher et al. (1990) using 4 years as the average age of the water in the lens. They suggest a half-life for aragonite in the Bermuda fresh water lens of 7,700 years, or 51,000 years for 99% of the aragonite in the phreatic zone (this excludes most of the mixing zone) to be converted to calcite. This compares favorably with estimates made by Budd (Budd, 1988) for stabilization of aragonite to calcite at Schooner Cays in the Bahamas and suggests that indeed Bermuda-type meteoric lenses in ancient sequences during terminal sea level highstands could support significant meteoric phreatic diagenesis. This would assume sediment contact with 99% fresh water for some 50,000 years, not an unreasonable assumption. While there is some evidence of $CaCO_3$ transport during aragonite stabilization, it seems that most dissolution/precipitation should be local with minimal impact on total porosity and permeability.

What about the second iteration of the immature hydrologic phase, the terminal prograding high-energy beach strandplain complex with overlapping, progressively younger meteoric floating lenses?

Meteoric diagenesis in Quintana Roo, Mexico strandplains

Ward and Brady (1979) describe an extensive Upper Pleistocene strandplain complex in northeastern Yucatan some 150km long, and 0.5-4km wide, and up to 10m thick (Fig. 7.18). The strandplain represents a rapidly prograding beach-shoreface developed during the last Pleistocene sea level highstand, or about 125,000 years ago. The sediments of the strandplain have been in the vadose zone since the last glacial fall in sea level, or at least 75,000 years (Fig. 7.18).

The beach ridges of the strandplain could have supported a substantial meteoric lens. If we use Budd and Vacher's (1991) 1% of island width to predict lens thickness, we might reasonably expect a lens thickness of 5m for a maximum of 50,000 years during highstand progradation. This relatively short period of exposure to a meteoric lens seems to have resulted in significant meteoric phreatic diagenetic overprint such as moldic porosity after

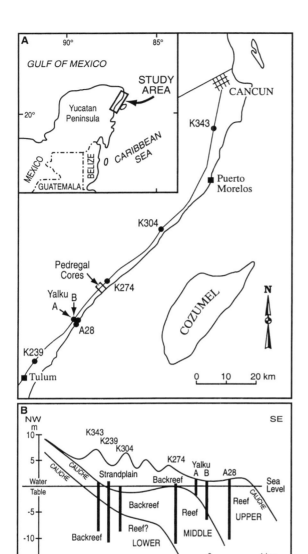

Fig. 7.18C. Outcrop photograph of the Yucatan strand plain south of Cancun. Bill Ward is pointing to diagram of tidal-dominated facies represented by the herringbone cross bedding just to the right of the diagrams.

Fig.7.18. A. Location map of the Yucatan showing outcrop of Latest Pleistocene strand plain. B. Schematic cross-section across coastal region showing the relationship of the strand plain to older Pleistocene units. Locations of cores show in Figure 7.18A. Used with permission of the New Orleans Geological Society.

aragonite dissolution, cementation by calcite circumgranular crusts, and conversion of most Mg calcite to calcite. The most intense phreatic meteoric overprint is near the Pleistocene water table. Substantial volumes of aragonite and a trace of Mg calcite still remain in spite of subsequent long exposure to vadose conditions (Ward et al., 1985). The following section discusses potential meteoric diagenesis and porosity preservation in an Upper Jurassic highstand prograding beach strandplain complex.

The Oaks Field, meteoric diagenesis in a Jurassic highstand prograding shoreline, north Louisiana, USA

Highstand systems tracts of Oxfordian to Kimmeridgian depositional sequences in north Louisiana exhibit off-lapping prograding shoreline complexes similar to the Quintana Roo beach strandplain described above (Moore and Heydari, 1993). The Oaks Field in Claiborne Parish, Louisiana, is clearly such a situation. In this field hydrocarbons are reservoired in three separate shore-parallel, off-lapping, island-shaped ooid grainstone sand bodies (Fig. 7.19) (Erwin et al., 1979). The major porous zones in each of the three reservoirs occur some 6m below the top of the reservoir and range

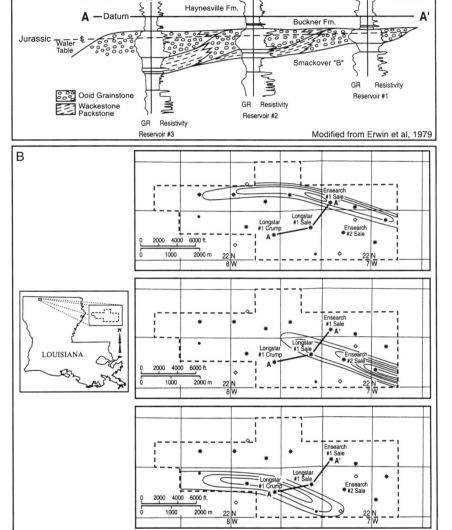

Fig.7.19. A. Dip section A-A' showing three producing reservoirs at Oaks Field (Smackover "B"). Cross-section redrawn from the original of Erwin et al., 1979 using the base of the Haynesville Formation as the datum. Separation of individual reservoirs is by tight terrigenous-rich low energy facies. B. Net pay isopachs of the three producing reservoirs at Oaks Field. Location of cross-section A-A' in Figure 7.19A is shown on each isopach. The reservoirs are the reflection of a high stand prograding shoreline complex. Each reservoir represents a 4th or 5th order cycle. Used with the permission of the GCAGS.

from 3 to 9m in thickness.

Humphrey et al. (1986) proposed that the porous zones in the Oaks Field reservoirs represented porosity preserved by early mineral stabilization within island-based meteoric water lenses. They first determined that porosity distribution was basically independent of the sedimentologic characteristics of the reservoir rock. They then noted that reduction by compaction was much higher in the core above and below the porous zone regardless of grain type (Fig. 7.20). Changes of $\partial^{13}C$ composition to more negative values immediately above the porous zone led to the conclusion that these less

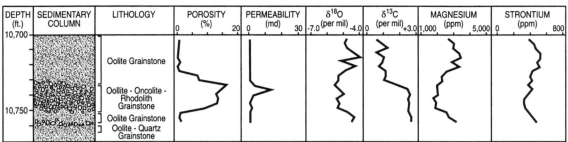

Fig. 7.20. Lithologic and geochemical data on the Ensearch #2 Sale well completed in Reservoir #2 at Oaks Field. Depleted oxygen isotopic composition and depleted magnesium and strontium content characterize the porous interval. Carbon isotopic composition shows marked shift toward lighter values at the top of the porous zone. Used with permission of AAPG.

porous rocks represented the vadose zone, and hence the porous section below represented the meteoric phreatic zone beneath the water table. Slightly lighter $\partial^{18}O$ values and lower Mg and Sr compositions of the rocks within the porous zone tend to support this conclusion (Fig. 7.20).

The Humphrey et al. (1986) model is that mineral stabilization was virtually complete in the meteoric lens prior to burial, while mineralogically unstable grains were still present in the vadose zone at the time of burial. Mineralogically unstable grains will pressure dissolve faster than those composed of more soluble minerals (Moore, 1989), hence the loss of porosity due to greater compaction in the vadose zone, and preservation of porosity in the meteoric lens below. Upper Jurassic ooids in the vicinity of Oaks Field are generally magnesian calcite (Moore et al., 1986; Moore and Heydari, 1993) which may explain the lack of early cement, oomoldic porosity and apparent total stabilization of ooids in the meteoric phreatic zone at Oaks Field.

Budd and Vacher (1991), using the data of Erwin et al. (1979) and Humphrey et al. (1986) applied their predictive island-based meteoric lens thickness model to Oaks Field (Fig. 7.21). They concluded that given the

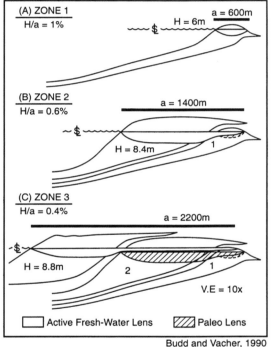

Fig. 7.21. Model for the development of Ghyben-Herzberg lenses in the Oaks Field reservoir zones 1 (A), 2 (B) and 3 (C). In this model, paleo-islands 2 and 3 were amalgamations of successively off-lapping shoal complexes, thus producing a successively larger island with lateral variations in lithology, diagenetic history and hydraulic conductivity. Note that successive freshwater lenses may have been superimposed in zones 1 and 2 with time. Used with permission of SEPM.

widths of the individual shoreline complexes that each reservoir should exhibit at least one porosity zone of 6-9m in thickness.

This predicted porosity zone thickness was within the range of porosity zone thickness observed at Oaks by Erwin et al (1979). These results validate the applicability of the Budd/Vacher model to ancient prograding shoreline complexes and strengthen the case for meaningful diagenetic processes operating in an island-based meteoric lens during a sea level highstand.

The mature hydrologic phase: regional meteoric water systems developed during sea level lowstands.

Carbonate shelves and isolated platforms generally tend to be steep sided and relatively flat topped (Chapter 2). A relative drop in sea level (eustatic fall or tectonic uplift) may lead to the exposure of the entire shelf or platform to meteoric water conditions and to the establishment of regional meteoric water aquifer systems. The ultimate position of the water table and the thickness of the vadose zone will be determined by the magnitude of the sea level drop (or tectonic uplift). During icehouse intervals, eustatic sea level drops can measure 100m or more, so one could expect to see thick vadose zones with water tables well below the exposure surface. During greenhouse intervals, sea level drops are measured in 10's of meters so vadose zones are correspondingly thinner and water tables are close to the exposure surface.

Of course climate plays a major role in the type of hydrologic system that will be developed in this setting. During ice-house conditions the climate tends to be cool and more arid, so less water will be available for the system; and resulting diagenetic potential will be muted. During greenhouse times, climate tends to be warmer and more humid so more water is available to the system; and diagenetic potential will be enhanced.

Finally, the time factor, or duration of sea level lowstand and its corresponding period of exposure is critical. Cyclic glacio-eustatic lowstands tend to be relatively brief. The late Pleistocene sea level curve for Bermuda (Vacher and Rowe, 1997) suggests that major lowstands lasted for 50,000 years or less. In a setting with a depressed water table, thick vadose zone and relatively short exposure, one would not expect to see the development of a classic karst system (Loucks, 1999). If, however, the duration of exposure is extended, possibly by attendant uplift, the development of extensive karst features can proceed if enough meteoric water is available to the system.

In the following sections we will first discuss the diagenesis and porosity evolution of an isolated carbonate platform during brief platform-wide exposure at the lowstand using the Bahamas as a case study. Our second discussion will focus on diagenesis and porosity evolution during brief lowstand exposure on a land-tied carbonate shelf. Barbados will be our "modern" example and the Mississippian shelf of New Mexico our ancient analogue. While the Bahamas and Barbados both show significant fresh-marine water mixing zones, we will discuss mixing zone dolomitization separately in a later section.

Mature hydrologic system: regional meteoric water system developed during relative sea level lowstand on an isolated carbonate platform, the Great Bahama Bank

The present Great Bahama platform is an

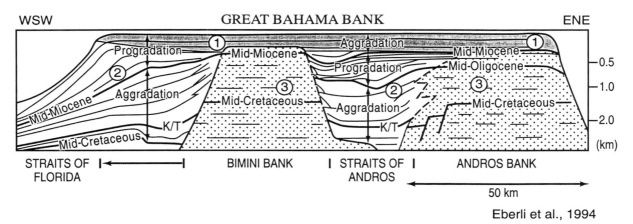

Fig.7.22. Schematic cross-section over northwestern Great Bahama Bank showing evolution of the bank and the approximate distribution of the seismic facies. Seismic facies 1 is the modern surface of the bank and the underlying high-amplitude horizontal reflections. Seismic facies 2 includes the Straits of Andros and the prograding western margin, both with high-amplitude inclined reflections. Seismic facies 3 includes the chaotic reflections of the buried Bimini and Andros banks. Used with permission of the AAPG.

isolated oceanic carbonate platform that has seen a complex geologic history (Fig. 7.22). Starting as two isolated platforms in the Cretaceous, it has coalesced by aggradation and progradation into the single large flat-topped platform seen today. Hydrogeologic studies and the results of several coring programs indicate that the interior of the Great Bahama Platform is basically open to sea water circulation from its oceanic margins, suggesting relatively high lateral porosities and permeabilities for the limestones underlying the present platform (Melim and Masaferro, 1997; Whitaker and Smart, 1997). The Pleistocene and upper Pliocene limestones beneath the platform have seen numerous glacio-eustatic sea level falls and rises. Amplitudes range from 10m to over 100m, with the last glacial maximum lowstand of 120m below present sea level having been reached some 17,000 years ago (Beach, 1995; Mylroie and Carew, 1995; Melim and Masaferro, 1997).

During sea level lowstands the Great Bahama Platform would be totally exposed to meteoric conditions. At the exposure surface, epikarst as described by Mylroie and Carew (1995) would be established, including pit caves and vadose crusts that would ultimately develop into paleosols (Fig. 7.23). The vadose solution pits provide fast access to the water table below, while the vadose crusts tend to be relatively impermeable and may channel meteoric waters laterally until a pit is encountered. A regional meteoric lens is developed across the entire platform and floats on relatively normal marine waters below. Thickness of the lens and its associated mixing zone depend on hydraulic conductivity of the limestones and meteoric water budget. With a positive water budget and relatively low conductivity, a very thick lens with a rather thin mixing zone will result. With a negative water budget and relatively high conductivity, the resulting lens will be thin and mixing zone thick (Whitaker and Smart, 1997).

In the meteoric phreatic zone of the lens, the area in the vicinity of the water table seems to be the most chemically and diagenetically

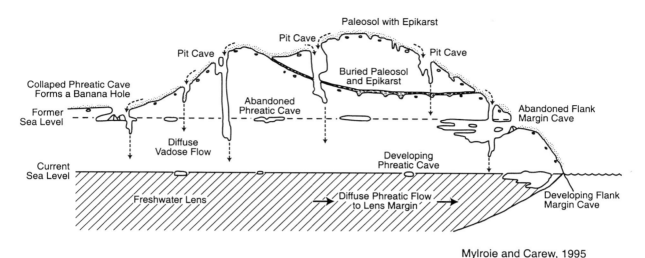

Fig.7.23. Diagrammatic representation of the main dissolutional features found on carbonate islands: epikarst (with paleosol), pit caves, banana holes and phreatic caves, and flank margin caves. Also shown are the positions of the features relative to a freshwater lens and halocline. Changes in sea level move the position of the karst features. In the Quaternary, these sea level changes led to overprinting of dissolutional environments. Reprinted by permission of the AAPG.

active with dissolution of aragonite and precipitation of calcite. When stabilization is accomplished, phreatic cave formation along and just beneath the water table is the result (Fig. 7.23). These phreatic caves and not the exposure surfaces mark the sea lowstand positions. Along the platform margins where the lowstand meteoric lens intersects the sea, excess meteoric water from the lens actively mixes with sea water and significant dissolution occurs. Sea margin caves also occur in Mexico (Ward et al., 1985). The result is a sea margin (flank margin) cave system that can attain significant proportions (Fig. 7.23).

What has been the impact of the Pleistocene-Pliocene sea level history on the diagenetic-porosity evolution of the limestones underlying the platform? There are a number of stacked exposure surfaces in the upper 100m of the record beneath the platform (Fig. 7.24C). These unconformities represent the surface epikarst developed during sea level lowstands and platform emergence. The section between the surfaces represents sedimentation during the subsequent interglacial highstand. In each case the top of the water table (near sea level) was located some distance beneath the unconformity depending on the amplitude of the glacial lowstand. Porosity is reduced significantly at each exposure surface (Beach, 1995). With each successive sea level lowstand the limestone section becomes more vertically compartmentalized, enhancing lateral fluid movement. Mylroie and Carew (1995) report several marginal caves observed during submersible operations in the upper 200m of the water column along the margins of the Great Bahama Bank. These caves no doubt mark lowstand mixing zones and positions of glacial sea level maxima.

Two deep cores acquired in 1990 on the northwestern margin of the Great Bahama Bank (Melim and Masaferro, 1997) beautifully illustrate effects of 800,000 plus years of large-

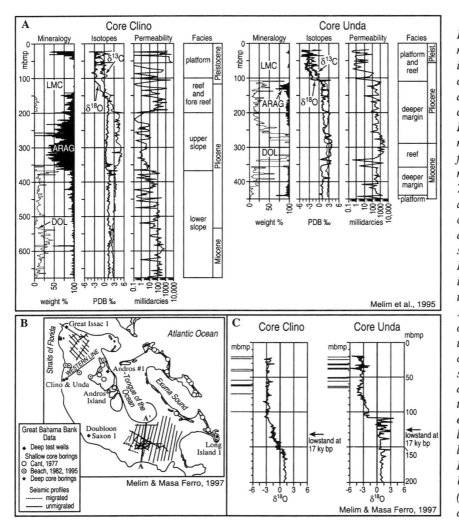

Fig.7.24A. X-ray diffraction mineralogy, bulk-rock stable isotopic data, permeability, facies, and ages for cores Clino and Unda. Key: LMC, low-Mg calcite; ARAG, aragonite; DOL, dolomite. Depths are meters below mud pit (mbmp); for Unda, sea level was 5.2 mbmp; for Clino, sea level was 7.3 mbmp. B. Bahamas regional map with detail showing location of deep test wells, deep and shallow core borings, and seismic profiles from Great Bahama Bank. A-A' is approximate line of section for Figure 7.22. Bank outlines follow 100-m contour. C. Bulk-rock oxygen isotopic data for the upper 200 m of Bahamian cores Clino and Unda. Also shown are the positions of subaerial exposure surfaces (line to the left of each plot) and the elevation in each core of the latest Pleistocene sea-level lowstand (Fairbanks, 1989). Depths in cores Clino and Unda are meters below mud pit (mbmp). Used with permission of Elsevier.

scale sea level changes resulting in entrained meteoric lenses sweeping across the upper 200m of the platform limestones (Fig. 7.24). The upper 150m are essentially stabilized to low magnesian calcite (with the exception of one 10m section of aragonite in the upper part of the Clino core). Strongly negative $\partial^{18}O$ values coupled with $\partial^{13}C$ near 0 across this 200m interval suggest that most stabilization occurred in the meteoric phreatic environment. Wide spread moldic and cavernous porosity, coupled with extensive circumgranular crust cements in this interval, support this conclusion (Beach, 1995; Melim and Masaferro, 1997). Vadose meniscus cements are restricted to the uppermost section and vadose needle fiber cements are closely associated with exposure surfaces (Beach, 1995).

Mean porosity for rocks converted to calcite in the upper 200m averages 35%; while in the youngest sequences still containing aragonite, the porosity is marginally higher, averaging some 45% (Beach, 1995). However, permeability across this upper 200m section, however, is complex and highly variable ranging from 0 to 10,000 millidarcies (Fig. 7.24A). This variability is no doubt strongly affected by cementation at unconformities and water tables as well as vadose (vertical) and phreatic (horizontal) channelized flow.

What has been learned from the Bahama story that can be applied to ancient isolated carbonate platforms? In an ice-house setting such as seen during the Pleistocene-Pliocene sequence of the Great Bahama Bank, large amplitude sea level fluctuations, coupled with floating meteoric lenses, can stabilize and karstify a relatively thick platform limestone. However, this sedimentologic, mineralogic and porosity maturation, is not a single sea level-related event. It is the result of numerous overlapping 4th and 5th order glacio-eustatic sea level fluctuations spread over a period of some million years (3rd order depositional sequence). Reservoir heterogeneity in this setting is directly controlled by exposure surfaces, paleo-water table positions and mixing zone marginal cave systems. Finally, the meteoric lens may not be diagenetically active when a large sea level fall results in an exceptionally thick vadose zone and the phreatic zone is de-coupled from soil CO_2 (Melim and Masaferro, 1997).

In a greenhouse setting the amplitude of sea level fluctuations would be considerably lower, and therefore the thickness of the platform sequence affected by meteoric water should also be smaller. Since the vadose zone under these conditions would always be relatively thin, the meteoric lens should be more diagenetically active because the water table would be closer to the soil zone. Finally, the climate should be more humid in a greenhouse setting, leading to more effective karstification of exposure surfaces and unconformities.

Mature hydrologic system: regional gravity-driven meteoric water system developed during relative sea level lowstand on a land-tied carbonate shelf

Gravity-driven regional aquifer systems are certainly important active diagenetic environments relative to porosity modification. The topography of an exposed land-tied carbonate shelf during a sea level lowstand generally exhibits a gentle slope toward the sea with the presence of inboard higher topography related to differential subsidence through time. The relatively high conductivity of the mineralogically immature high-energy terminal highstand grainstones of the exposed shelf sequence ensures the development of a water table and meteoric phreatic system if there is a positive water balance. Under these conditions the potentiometric surface should rise toward the interior of the shelf and a gravity flow system with water movement toward the sea should be established (Fig. 7.11B).

Coastal weather patterns should dictate that in a tropical humid climate the higher elevations of the shelf interior should receive more precipitation than the coastal areas. In this setting, then, the exposed shelf interior would act as a recharge area with water flow down toward the sea where excess meteoric water would discharge at shoreline springs forming a mixing zone (Fig. 7.11B). If meteoric recharge is great enough and conductivity of the sediments low enough, seawater will ultimately be displaced seaward to the vicinity of the shoreline (Back et al., 1984).

There are no modern examples of an exposed mineralogically immature land-tied, high-energy carbonate shelf where the hydrogeologic and diagenetic settings are well known. While the setting of the island of Barbados is one of continuous uplift, it is one of the few modern well-characterized, gravity-driven aquifer systems of any size involving a relatively young, mineralogically metastable carbonate terrain. Our knowledge of the Barbados system is based on a long-term research program at Brown University under the guidance of R. K. Matthews (Matthews, 1967; Harris and Matthews, 1968; Benson et al., 1972; Steinen and Matthews, 1973; Allan and Matthews, 1982; Humphrey and Matthews, 1986; Humphrey and Kimbell, 1990; Humphrey, 1997).

The general Barbados setting is shown in Figure 7.25. The main attributes of the system are an upraised Pleistocene coral cap up to 90m thick resting on a tilted Pre-Pleistocene siliciclastic aquaclude. Elevations of over 300m are attained in the center of the island. Most of the recharge waters falling on the coral cap pass through the vadose zone in sinks and fissures directly to the phreatic zone where gravity-driven ground water movement is by stream flow along the base of the coral cap. Where the stream drainage intersects sea level in the south, a 30m thick floating phreatic lens is developed that extends some 1.6km to the island coast. Water movement through this lens is by sheet flow with fresh meteoric spring outflow and a very active mixing zone along the coastline (Harris and Matthews, 1968; Humphrey, 1997). The development of the regional meteoric aquifer model discussed below is based primarily on the Barbados work of Matthews and his students.

The gravity drive is developed by differential subsidence of the shelf margin. Although open to the atmosphere along the entire length of the shelf, the main area of meteoric water

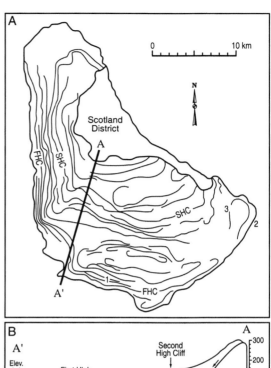

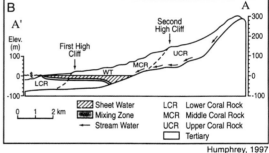

Fig.7.25A. Map of Barbados showing trends of Pleistocene reef tracts. Shaded area represents the erosional window exposing rocks of the Tertiary accretionary prism, upon which Pleistocene limestones (unshaded) unconformably lie. Reef tracts generally conform to the outline of the island and increase in age and topographic elevation toward the interior of the island. Key: FHC, First High Cliff; SHC, Second High Cliff; 1, Christ Church region; 2, Bottom Bay; 3, Golden Grove. B. Hydrogeologic cross section A-A' (see Figure 7.25A). Meteoric groundwater recharges through the Coral Rock Formation flow either as stream-water along the contact with the underlying Tertiary aquaclude, or as sheet-water that forms the coastal freshwater lens. Where the freshwater lens interfaces with marine pore fluids, a freshwater-saltwater mixing zone is formed. Used with permission of Elsevier.

recharge is at the higher elevations found in the interior of the shelf driven by coastal climatic patterns.

The diagenetic drive for regional meteoric aquifers is generally furnished by the solubility contrast between aragonite and calcite, and the rapid movement of rather chemically simple water through the aquifer (Matthews, 1974). The aquifer is recharged with water undersaturated with respect to both aragonite and calcite. As long as aragonite is present, and the water is undersaturated with respect to calcite, solution of aragonite will be the dominant process (Fig. 7.26A). As $CaCO_3$ is added to the water by dissolution of aragonite, the water will ultimately reach supersaturation with respect to calcite, and calcite precipitation will occur. Because of the calcite nucleation rate step, and the relatively rapid movement of water through the aquifer, a net transfer of $CaCO_3$ down the water flow path takes place (Fig. 7.26A). When the meteoric water flow reaches sea level, the rate of water flux slows and rapid calcite precipitation can then commence. With continued rock-water interaction down the flow path, the water becomes supersaturated with respect to calcite and is in equilibrium with respect to aragonite. Under these conditions, calcite cements can precipitate directly onto aragonite substrates without substrate dissolution.

Given enough time, the recharge area is stabilized to calcite, and calcite dissolution commences until equilibrium with the calcitized sequence is attained. Water emerging from this stabilized calcite terrain, while saturated with respect to calcite, is still undersaturated with respect to aragonite, and will dissolve any available aragonite down the flow path. The tendency, therefore, is for expansion of the stabilized calcite terrain down the flow path until all available aragonite is converted to calcite (Fig. 7.26B).

Porosity development and predictability in regional meteoric aquifer environments

During mineralogical stabilization the transfer of significant volumes of $CaCO_3$ down the flow path in regional meteoric aquifer settings infers that total porosity volume may be significantly changed by solution or precipitation at any single site. Further, given a starting metastable mineralogy dominated by aragonite, the diagenetic facies outlined in the model above, including porosity characteristics, should always be developed in a predictable sequence along the flow path during the stabilization process. If sea level rise or subsidence interrupts stabilization, these distinctive diagenetic and porosity facies will be frozen and buried into the geologic record.

Figure 7.26B schematically illustrates the up-dip to down-dip changes in the nature of carbonate pore systems that one can expect in a meteoric flow system. In zone 1, along the up-dip margin, moldic porosity is solution enlarged into vugs because calcite is undergoing dissolution. Permeability can be extremely low in areas of complete intergranular cementation. Grain neomorphism should not be common. Ubiquitous moldic porosity and almost total occlusion of intergranular pores by calcite cement characterize zone 2 in the upflow areas of the zone. Calcite cements are generally not present in aragonite molds in this zone. Zone 3, where water is supersaturated with calcite and at equilibrium with aragonite, will generally be characterized by neomorphosed aragonite grains and abundant intergranular cement. The down flow portion of zone 3 is probably the tightest

216   DIAGENESIS IN THE METEORIC ENVIRONMENT

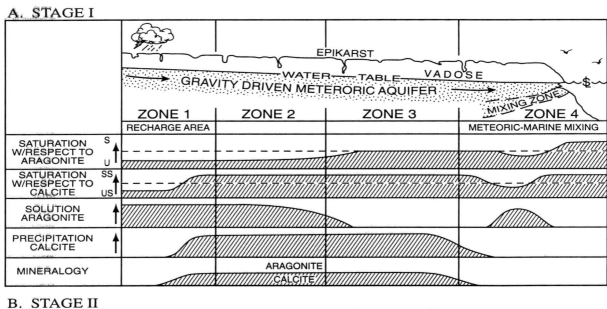

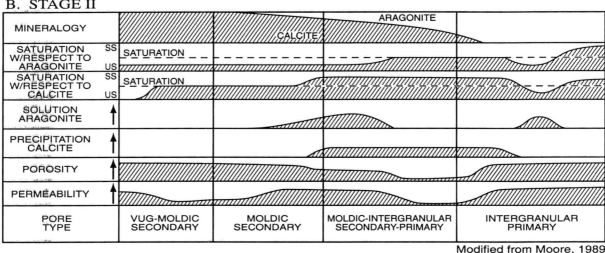

Modified from Moore, 1989

Fig. 7.26. Diagenetic model for a gravity-driven meteoric aquifer system. The hydrologic setting is shown at the top of the figure. The aquifer is divided into 4 zones, with the recharge area zone 1 and the area of meteoric-marine water mixing, zone 4. The model is divided into two stages. (A) Stage I at the initiation of recharge. Aragonite is still present in the recharge area. (B) Stage II, aragonite has been destroyed by dissolution and only calcite is present in the recharge area.

porosity zone of the system. Intergranular calcite cement in zone 4 becomes sparser in a down-dip direction, and porosity volume increases. Primary porosity preservation, little calcite cement, and common grain neomorphism generally characterize zone 4.

Thus, young metastable sedimentary sequences suffering an early meteoric diagenetic overprint in a regional meteoric aquifer environment should exhibit a predictable diagenetic and porosity facies tract oriented normal to hydrologic flow direction. This diagenetic facies tract can be extremely useful in predicting the distribution and nature of porosity on a regional basis during hydrocarbon exploration and production operations (Moore, 1984; 1989).

Geochemical trends characteristic of a regional meteoric aquifer system

Any regional geochemical trend established in a meteoric aquifer system will generally parallel the meteoric flow paths. If the aquifer flows through a siliciclastic terrain before entering the carbonate system (not an un-

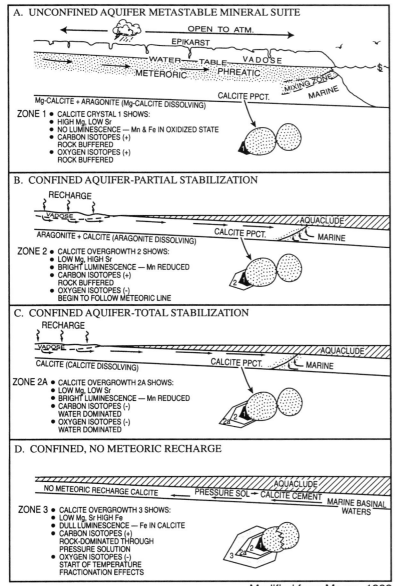

Fig.7.27. Schematic diagram showing the geochemical trends characteristic of a regional meteoric aquifer system. There are four stages, A-D, representing the evolution of the aquifer system from an initial unconfined, to a final subsurface system cut off from meteoric recharge. Cements precipitated down-flow from aquifer waters are shown diagrammatically on the right side of the diagram. Non-luminescent cement shown in black, brightly luminescent in white, and dully luminescent in a stippled pattern. Geochemical characteristics of each cement zone are shown on the left side of the diagram.

usual situation in a land-tied shelf), carbonate cements may reflect the siliciclastic signature of precipitational waters by increased contents of radiogenic Sr and trace elements Fe and Mn in an up-dip direction. In an unconfined aquifer, the full extent of the aquifer should be oxygenated by recharge through the vadose zone. Any calcite cements precipitated in the aquifer should show no luminescence (Fig. 7.27A). If subsequent transgressive deposits cut off the aquifer from vadose recharge, it becomes a confined aquifer. In such a setting, calcite overgrowths on the initial non-luminescent cement will tend to be brightly luminescent (Fig. 7.27B and C). If vadose recharge is maintained in the up-dip areas, a pH-eH gradient may be established down aquifer flow, with non-luminescent cements precipitated up-dip while luminescent phases are precipitated down-dip. As the aquifer system is buried in the subsurface, meteoric recharge is cut off and marine-related waters begin to dominate. Calcite cement overgrowths associated with pressure solution that precipitated from these subsurface fluids are generally dully luminescent. Therefore, a calcite crystal precipitated at any one site might well reflect the long-term pore fluid evolution of the aquifer by a complex cathodoluminescence zonation such as depicted in Figure 7.27D. (Meyers, 1988; Niemann and Read, 1988; Smalley et al., 1994).

In addition, the trace element composition of individual calcite crystals should also reflect the progressive mineral stabilization of the carbonate aquifer as it matures diagenetically through time. As Mg-calcites are recrystallized to calcites, the Mg to Ca ratio in the diagenetic water will increase, and increased Mg will be incorporated in the calcite cements down flow as a Mg spike (Fig 7.27A). Aragonite will begin to be dissolved as soon as all Mg-calcite is converted to calcite, increasing the Sr to Ca ratio in the water and resulting in a Sr spike, offsetting the earlier formed Mg spike in the cements precipitated down flow (Fig. 7.27B) (Benson and Matthews, 1971).

The stable isotopic composition of calcite cements precipitated in a regional meteoric aquifer system will generally follow Lohmann's (1988) meteoric water line (Fig. 7.5). Cement isotopic compositions may, however, exhibit some gross down flow regional trends particularly as the aquifer matures diagenetically. Initially, during the early stages of stabilization, the isotopic signature will be dominated by the isotopic composition of the marine sediments being stabilized. Thus both oxygen and carbon compositions may tend to be heavy (Fig. 7.27A). As diagenesis progresses and the volume of metastable minerals available decreases, phreatic water isotope signatures begin to dominate and both oxygen and carbon trend toward more depleted values (Fig. 7.27B and C). As the sequence is buried into the subsurface, carbon isotopes will again be rock dominated during pressure solution and become more positive. During burial, however oxygen isotopes will tend toward more negative values due to increasing temperature (Fig. 7.27D).

These trends will usually be manifested within a single crystal, will parallel the luminescent and trace element trends developed above, and will reflect the progressive evolution of the phreatic waters as driven by the mineral stabilization process (Fig. 7.27D) (Moore, 1989).

The following section details the diagenesis and porosity evolution of a Mississippian grainstone sequence in New Mexico as a result of its interaction with a regional gravity-driven meteoric aquifer system developed during a sea level cycle.

Mississippian grainstones of southwestern New Mexico, USA: a case history of porosity destruction in a regional meteoric aquifer system

Upper Mississippian carbonates, from Kinderhookian to Chesterian in age, form a classic transgressive-regressive shoaling-upward sedimentary sequence in the Sacramento and San Andres Mountains and the Rio Grande River valley of southwest New Mexico. A relatively thick crinoidal lime grainstone blanket termed in part the Lake Valley Formation of Osagean Age (Fig. 7.28) represents the regressive leg (highstand systems tract) of this cycle. Meyers and his co-workers (Meyers, 1988) have established that the cementation and attendant porosity occlusion of these crinoidal sands were accomplished under the influence of a regional meteoric aquifer system during the early burial stages of the sequence.

The Lake Valley grainstones consist almost exclusively of crinoid grains that were originally composed of 10 mole % $MgCO_3$ (Meyers and Lohmann, 1985). These grainstones have been cemented by four zones of complexly luminescent-zoned calcite cements that can be correlated regionally (Meyers, 1974) (Fig. 7.29). The early-most zones (1-3) are thought to be related to the regional meteoric water system. Meyers and Lohmann (1985) suggest that non-luminescent zone 1 represents a non-confined aquifer system shortly after deposition of the Lake Valley (Fig. 7.30A). Brightly luminescent zone two is thought to have been precipitated after the aqui-

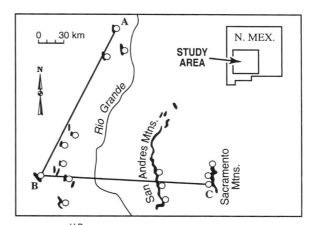
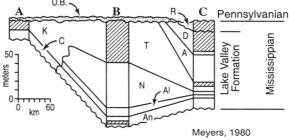

Fig.7.28. Stratigraphic setting for the Mississippian Lake Valley Formation, southern New Mexico, U.S.A. Letters in lower diagram (cf. K) represent abbreviations of laterally equivalent formations in isolated outcrops along the Rio Grande River valley. Used with permission, SEPM.

fer was shut off from extensive vadose recharge (during a sea level stillstand) (Fig. 7.30B), while non-luminescent zone 3 represents a second stage of vadose recharge developed during a subsequent sea level rise (Meyers, 1988) (Fig. 7.30C). Zone 5 was thought to represent later post-Mississippian burial cementation (Fig. 7.29)(Meyers and Lohmann, 1985).

The stable isotopic composition of these zones clearly reflects the progressive evolution of the regional meteoric aquifer water system during, and shortly after stabilization of Lake Valley crinoidal sands (Fig. 7.31). The isotopic composition of zone 1 is near the estimated original marine values for the Mississippian,

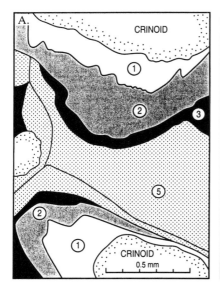
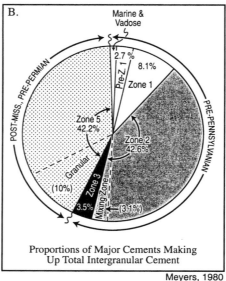

Fig. 7.29. (A) Schematic diagram showing luminescent zoned calcite cements of the Lake Valley Formation. Zone 3, shown in black is quenched. (B) Percentage of these zones found in total intergranular cement in the Lake Valley. Stippled patterns are the same as found in (B). Used with permission, SEPM.

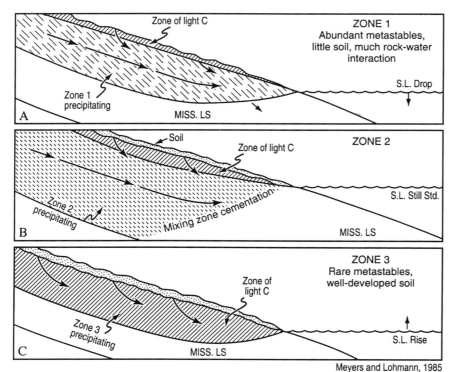

Fig. 7.30. Schematic diagram illustrating the inferred timing of calcite cement zones (1-3) of the Lake Valley with sea level fluctuations, based in part on isotopic composition of each zone. Zone 1 was precipitated during a sea level drop, zone 2 during a sea level stand, and zone 3 during an ensuing sea level rise. Used with permission, SEPM.

indicating overwhelming rock buffering of the isotopic signal during initial mineral stabilization. Zones 2 and 3 occur along the meteoric water line suggesting increasing dominance of the meteoric water signal as mineral stabilization proceeds.

All four cement zones comprise 26% of the volume of the grainstones and packstones sediments, and minimal micrite matrix, then compaction (both chemical and physical) is responsible for only 35% of the porosity loss (Meyers, 1980), while cementation is responsible for 65% of the porosity loss (Fig. 7.32). Of this 65%, approximately 50% of the cement was precipitated before significant burial (zones 1-3) and dominantly in conjunction with the meteoric aquifer (Fig. 7.32). If one assumes that compaction was accelerated by the Mississippian meteoric water system (as suggested by Meyers, 1980), it would follow that much of the 35% compaction documented for the sequence took place in the Mississippian. Assuming this figure to be 75%, the porosity of the Lake Valley at the time of burial would have been reduced to approximately 18% (Fig. 7.32). The porosity preserved at the time of the post-Mississippian unconformity (roughly 18%) consisted entirely of intergranular porosity and seems to have been distributed uniformly throughout the sequence. This porosity was lost by cement in-fill (zone 5) and the remaining 25% compaction during subsequent burial of the sequence to depths near 1000m.

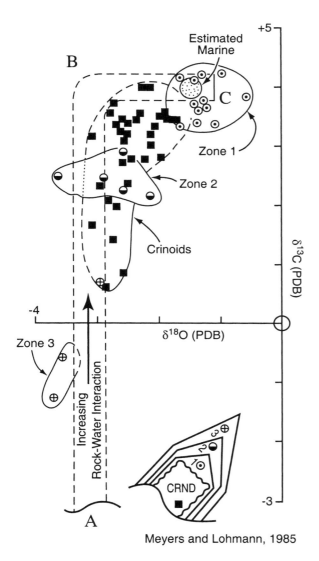

Fig.7.31. Stable isotopic composition of calcite cement zones 1-3 of the Lake Valley. The stippled area is the estimated Mississippian marine stable isotopic composition, based on the composition of crinoids. Zone 1 cement clusters around this marine value indicating a rock-buffered system, probably due to stabilization of unstable components. Zones 2 and 3 lie along the meteoric water line, with Zone 3 reflecting the least amount of rock-water interaction, indicating precipitation after most stabilization was completed. Used with permission, SEPM.

of the Lake Valley (Meyers, 1974). If one assumes an original porosity of 40% for these

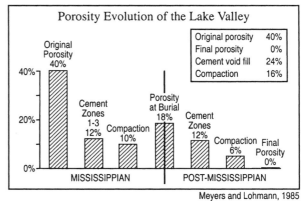

Fig.7.32. Porosity evolution of the Lake Valley Formation. Initial porosity is assumed to be 40% with minimal micrite matrix. Final porosity is 0%. Used with permission, SEPM.

The source of calcium carbonate that formed the Mississippian cements of zones 1-3 is thought to be early grain-to-grain pressure solution (occurring at a few hundred meters burial) enhanced by the high solubility of the Mg-calcite echinoid grains in an active meteoric phreatic aquifer (Meyers, 1974; Meyers and Lohmann, 1985). In such a system, porosity occlusion would be particularly effective, because no secondary porosity would be developed, and kinetic factors would be muted because of the presence of numerous calcite nucleation sites. This situation favors local precipitation of $CaCO_3$ gained from intragranular solution of the echinoid fragments. Since little $CaCO_3$ transport is necessary, the porosity occlusion rate due to cementation would be uniform down the flow path, and no significant zones of porosity enhancement or occlusion would be developed.

The source of the calcium carbonate for the post-Mississippian cements is thought to be primarily intraformational by pressure solution (although extraformational sources also driven by pressure solution cannot be ruled out). These cements are derived mainly by stylolitization during the progressive burial of the unit down to a maximum burial depth of 1000m (Meyers and Lohmann, 1985). Obviously, all porosity in this unit would have been destroyed before hydrocarbon maturation could have taken place.

In contrast, Moore and Druckman (1981) report that porosity within Jurassic calcitic ooid grainstones in the central Gulf Basin, USA, at the time of burial into the subsurface averaged some 40% with no early cement. These grainstones were exposed to meteoric water phreatic conditions along a sequence-bounding unconformity (Moore and Heydari, 1993). However, since these grainstones were laterally restricted, a regional meteoric water system comparable to the Lake Valley was not developed. If one were to assume a comparable porosity loss by cementation and compaction as seen in the Mississippian Lake Valley, it is obvious that significant porosity (perhaps as much as 20%) would have been present in the Smackover at the time of hydrocarbon maturation and migration. Indeed, Moore and Druckman (1981) report an average of 10% late subsurface cement present in upper Smackover grainstones, and between 15 to 20% preserved primary porosity in reservoirs of the sequence.

In the following section, diagenesis in the karst system will be considered.

KARST PROCESSES, PRODUCTS AND RELATED POROSITY

Introduction

Figure 7.33 schematically illustrates the active processes and some of the products produced during the development of a karst profile in a relatively humid climate. Dissolution, in both the vadose and phreatic zones, is the dominant active process (Palmer, 1995; Loucks, 1999). Karst-related dissolution is generally driven by CO_2 derived from the atmosphere and soil gasses. In addition, as noted above, the mixing of dissimilar meteoric and marine waters at the shoreline can result in significant dissolution in either an unstable or stable carbonate sequence (Smart and Whitaker, 1991; Palmer, 1995; Whitaker and Smart, 1997).

Solution cavities of all sizes from tiny vugs to gigantic caverns can be developed. Water movement in the vadose is often by conduit

SOLUTION, CEMENTATION AND POROSITY IN A KARST SYSTEM 223

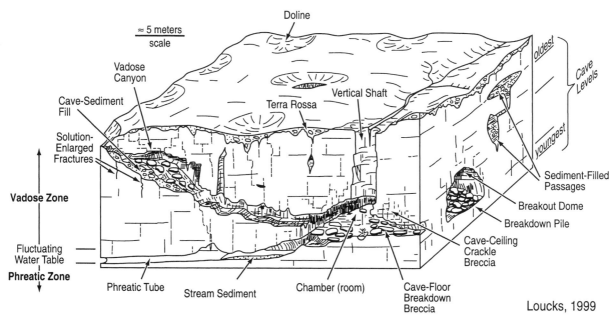

Fig.7.33. Block diagram of near-surface karst terrain including phreatic and vadose cave features. The diagram depicts four levels of cave development in a falling base-level system, with some older passages (shallowest) having sediment fill and chaotic breakdown breccias. Used with permission, AAPG.

flow through fractures that with time can be enlarged into large vertically oriented pipes and sinks (Fig. 7.33). Water movement in the phreatic zone is generally horizontal, and large horizontally oriented cavity and cave systems are often developed (Fig. 7.33) (Palmer, 1995; Loucks, 1999). If the climate is humid enough, and exposure long enough, a mature karst landscape featuring sinks, dolines, solution towers, internal drainage, and considerable relief can develop (Esteban and Klappa, 1983; Choquette and James, 1988; Purdy and Waltham, 1999). Buried karst landscapes or paleokarsts are often developed in association with important regional unconformities. Such paleokarsts can lead to significantly enhanced porosity and the development of major hydrocarbon reservoirs in the carbonate units below the unconformity (Esteban, 1991; Wright, 1991; Loucks, 1999). Esteban (1991)

described a number of paleokarst-related fields from around the world ranging in age from Ordovician to Miocene. Loucks (1999) detailed some 35 major oil and gas fields mostly in North America as paleokarst reservoirs.

The rapid and pervasive movement of water through materials associated with the vadose zone often carries soils and other sediments deep within the karst-related solution network to be deposited as internal sediments within caves and cavities. These sediments may either be siliciclastic (terra rosa), carbonate, or both. If soil derived and siliciclastic, they can significantly occlude solution porosity at even the cave scale (Fig. 7.33).

Finally, intense dissolution often results in solution-collapse features and the formation of extensive breccias that can be significantly more porous than the dissolved ground mass (Fig. 7.34). It should be noted that dissolution

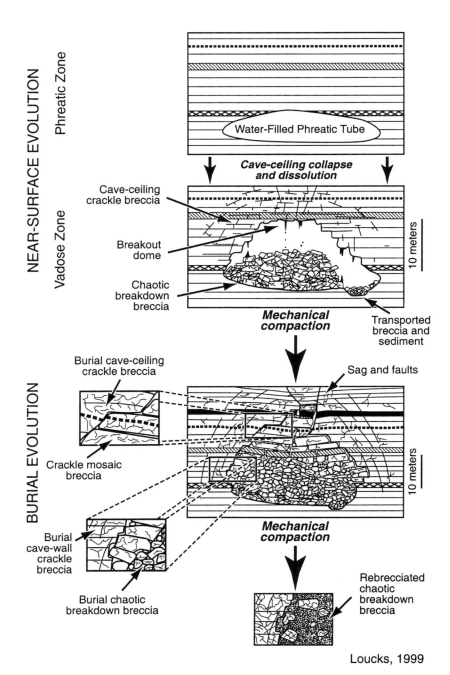

Fig.7.34. Schematic diagram showing evolution of a single cave passage from its formation in the phreatic zone of a near-surface karst environment to burial in the deeper subsurface where collapse and extensive brecciation occur. Used with permission, AAPG

of evaporites including halite and gypsum/anhydrite also result in major solution-collapse features. These evaporite solution breccias, while commonly associated with meteoric water influx through unconformities, are general- ly stratigraphically controlled and can be regional in extent (Palmer, 1995; Moore, 1996). During burial the volume of both limestone cave collapse and evaporite solution-collapse breccias with their associated fracturing, expand upon mechanical compaction (Fig. 7.34)

Collapse breccias serve as the receptor beds for extensive Paleozoic hydrothermal ore mineralization on a world- wide basis and are reservoir rocks for oil and gas (Roehl and Choquette, 1985; Loucks, 1999).

Solution, cementation, and porosity evolution in a diagenetically mature karst system

In the regional meteoric aquifer model developed in immature metastable carbonate terrains that was described in the preceding section, contrasts in mineral stability drove the diagenetic system and its porosity evolution by balancing dissolution with precipitation. As long as the mineralogical and solubility differences were present, then dissolution and precipitation proceeded, and regional transport of $CaCO_3$ to remote cementation sites was possible (Matthews, 1968; Matthews, 1974).

In the system under consideration in this section, however, there is generally no mineralogical solubility contrast because the sequence is already stabilized. Recharge waters undersaturated with respect to calcite will dissolve the calcite until the waters reach equilibrium with respect to the rock mass and little further solution or precipitation can take place. **The mature meteoric diagenetic system, therefore, is one driven primarily by solution; and one should not expect significant downstream cementation to be a major factor in porosity evolution (Matthews, 1968; 1974; James and Choquette, 1984).**

Recharge waters will generally reach calcite equilibrium relatively quickly in a stabilized diagenetic terrain (Matthews, 1968), which will tend to constrain the depth to which porosity enhancement can be accomplished beneath regional subaerial unconformities.

Unconformity-related porosity development can be localized by favorable (more porous) diagenetic and depositional facies present in the carbonates beneath the unconformity that will tend to channelize the flow of meteoric water into and through the aquifer. In addition fracture patterns associated with local structural features such as anticlines and faults can localize the development of karst related to solution porosity (Craig, 1988) (Tinker and Mruk, 1995).

The Cretaceous Golden Lane of Mexico (Boyd, 1963; Coogan et al., 1972), the Cretaceous Fateh Field, Dubai, UAE (Jordan et al., 1985) and the Permian Yates Field of west Texas, USA (Craig, 1988; Tinker and Mruk, 1995) are each giant fields with reserves in excess of 1 billion barrels of oil. These reservoirs are examples of porosity developed in mature stable carbonate sequences related to karsting along regional to interregional unconformities. The Yates field will be discussed in detail in the next section.

The Yates Field, west Texas, USA: a case history of an unconformity-related karsted reservoir complex

The Yates Field is located on the southeastern margin of the Central Basin Platform in the Permian Basin of west Texas (Fig. 7.35). Discovered in 1926, it has produced over 1.2 billion barrels of oil (as of 1995) from cavernous dolomite (Tinker and Mruk, 1995). Although oil is produced from a number of Permian formations at Yates, the main reservoirs are found in the Guadalupian San Andres Formation (Fig. 7.35). The San Andres is overlain unconformably by the Grayburg Forma-

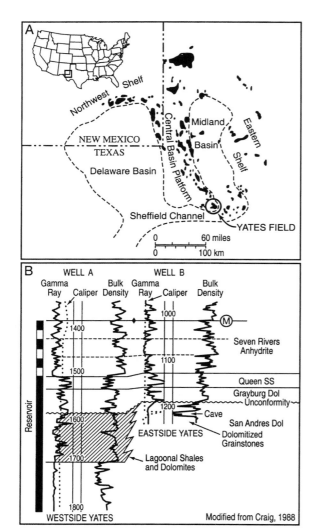

Fig.7.35. (A). Geologic setting of the Yates Field west Texas, U.S.A. (B) Log profiles of two wells showing stratigraphic setting across the field, west to east. Note the unconformity at the top of the San Andres dolomite with caves. "M" is a marker used in the reconstruction of the paleogeography at Yates. Used with permission, Elsevier.

tion. This unconformity may be an interregional unconformity of some importance particularly as it relates to the development of porosity in the underlying San Andres. The siliciclastic Queen Formation and the salina/sabkha complex of the Seven Rivers Formation overlie the Grayburg at Yates (Fig. 7.35).

The San Andres at Yates consists of two 3rd order depositional sequences (Fig. 7.36). Most of the important reservoir rock is located in sequence 2. Four higher-frequency transgressive-regressive cycles (4th order?) can be recognized within depositional sequence 2. These cycles represent discrete shelves and shelf margins, which are strongly prograding into the adjacent Midland basin to the northeast and east and presumably into the Sheffield Channel to the southeast (Fig. 7.35 and 7.36). Lateral facies within each cycle are similar. Grainstones and packstones mark the shelf margin. Open marine mudstones are found on the slope into the basin. Lagoonal, salina, and sabkha deposits were deposited landward (westward) of the shelf margin onto the Central Basin Platform (Fig. 7. 36).

These facies in all four cycles of sequence 2 were dolomitized by evaporative reflux from the adjacent lagoons and sabkhas during or shortly after deposition (Chapter 6 and Saller, 1998). Early landward to seaward fluid movement with increasing porosity toward the margin seems to have been established in each cycle.

The unconformity at the top of the San Andres is, at least in part, related to a glacial eustatic sea level fall. It is not certain that the entire Central Basin Platform was exposed at this time; but since the Permian was a time of waning icehouse conditions with high amplitude sea level changes, it can be assumed that at least a significant part of the platform was exposed. Most of cycle 4 and part of cycle 3 were truncated along the unconformity at the top of the San Andres.

Abundant drillers' logs, core, geophysical logs, and production evidence of karst-related

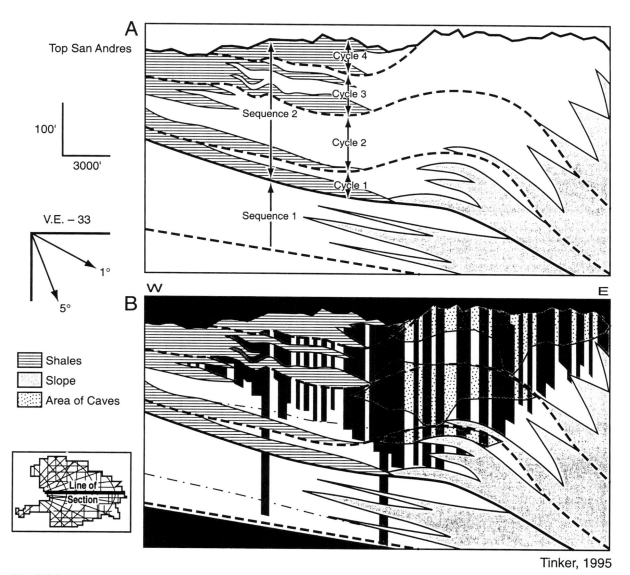

Fig. 7.36. West-to-east stratigraphic sections, with a Seven Rivers 'M' datum, (A) Lithofacies superimposed on the sequence-stratigraphic interpretation, with sequence boundaries shown in black and major cycle boundaries in dashed line. (B) Position of caves in the sequence stratigraphic framework. Used with permission of AAPG.

cave porosity exist at Yates (Craig, 1988; Tinker and Mruk, 1995). Figure 7.37 illustrates the characteristics of a 4-5 foot cave in a log suite consisting of gamma ray, caliper and formation density. These types of logs formed the backbone of the data set at Yates used by Craig (1988) and Tinker (1995).

Craig (1988) and Tinker (1995) reconstructed the paleotopography on the San Andres unconformity by constructing isopach maps of the interval between the Seven Rivers "M" datum to the top of the San Andres (Fig. 7.35 and 7.38). The maps were very similar and showed rugged topography reminiscent of a

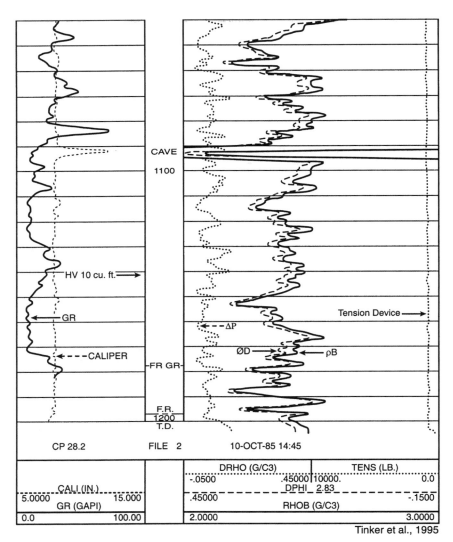

Fig. 7.37. FDC log with GR illustrating the parameters used for cave identification. At a depth of about 1090 feet note the (1) clean GR, (2) off-scale Rhob, and (3) caliper excursion of 4 inches (10cm). This is a typical signature for a cave 4-5 feet (1.2-1.5m) high. Used with permission of AAPG.

karst terrain (Fig. 7.38). Craig (1988) then plotted cave feet versus the first 500 feet below the "M" datum, separating the porous east-side shelf margin sequences from the less porous west-side lagoonal sequences in his plot. On the eastern shelf margin cave feet distribution shows a mode near 200 feet below the "M" datum (Fig. 7.39) which Craig (1988) interpreted as a paleo-water table. Note that caves are present in the landward-lagoonal side (westward) of the shelf margin in lesser abundance (Figs. 7.39 and 7.40).

Assuming that the 200 feet water table position represented sea level during the 3rd San Andres cycle (a valid assumption in a coastal situation), Craig (1988) postulated that a series of islands were present at the top of the San Andres (hachured areas on Figure 7.38). Assuming that these islands contained floating fresh water lenses (most of the islands were large enough), Craig (1988) proposed that caves in the San Andres were associated with these islands. Cave dissolution occurred in island fresh-marine water mixing zones similar

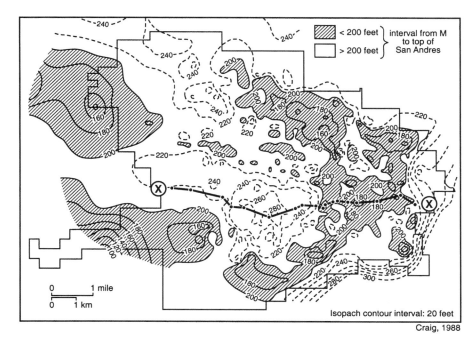

Fig.7.38. Paleotopography on the San Andres unconformity expressed in terms of the isopach interval from the Seven Rivers "M" marker to the top of the San Andres. Thins represent topographic highs and thicks are topographic lows. Hachured areas are interpreted to be San Andres islands which were created when the unconformity surface was drained or flooded to the level of 200 feet (61 m) below the "M" marker, that is, to the elevation of the Late Permian (Guadalupian) sea. Used with permission, Elsevier.

to those discussed in previous sections of this chapter.

There are several problems with Craig's (1988) model (see also Moore, 1989, pages 214-215). Perhaps the most serious is his inference that there were islands in the San Andres at the time of cave formation. After Craig (1988) decompacted the shale interbeds in the lagoonal sequence behind the grainstones of the shelf margin, most of the topographic differences between the islands and lagoon disappeared (Fig. 7.41). Tinker (1995) noted significant truncation of section along the unconformity (Fig. 7.36). Since the lagoonal facies contains shales and silts, it is obvious that these lithologies would be preferentially removed during erosion. This erosion, coupled with shale compaction, results in the raised rim and island topography seen on the isopach of Figure 7.38. It is entirely possible, therefore, that there was little or no significant relief or island development at the time of cave formation in the San Andres.

A more appropriate model might be a regional meteoric water system developed during the terminal San Andres lowstand. Since there is no doubt that platform-to-margin wa-

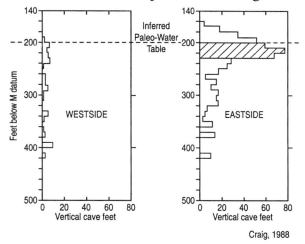

Fig.7.39. Vertical cave feet (feet of cave) as a function of depth below the "M" marker. Cave abundance in eastside Yates is focused in a relatively thin zone of 30 feet (9 m) which is thought to represent a Permian paleo-water table coinciding with a Permian sea level position (-200 feet below the "M" marker). Used with permission, Elsevier.

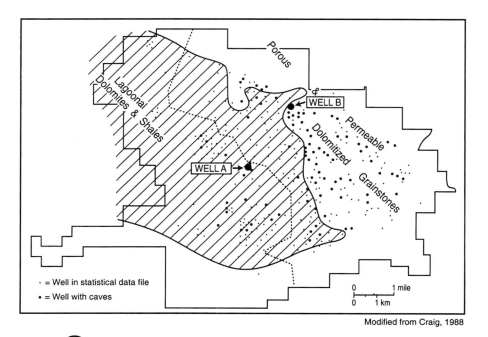

Fig.7.40. Wells with caves at Yates Field shown in relation to the paleo-geographic distribution of major lithofacies exposed at the unconformity surface of the San Andres dolomite as shown in wells A and B in Figure7.35B. Used with permission, Elsevier.

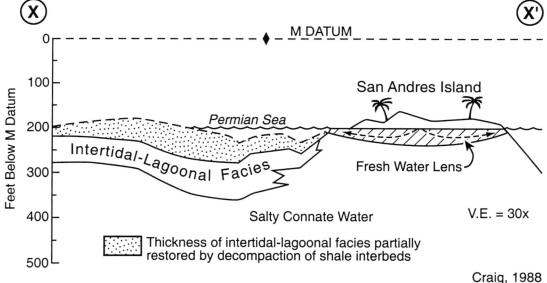

Fig.7.41.West to east cross-section X-X' in Figure 7.38 showing island and seafloor paleo-topographic profile, west-side stratigraphy, and sea level and island hydrology during San Andres island time. The key paleo-elevation, 200 feet (61 m) below M, is related to the principal cave population of Figure 7.39 and it defines the shoreline and freshwater table of the San Andres island. Estimated vertical expansion of the intertidal-lagoonal facies (stipple pattern) by de-compaction of shale interbeds suggests the possibility that the sediments of the west-side area were awash or only slightly submerged during island time rather than in water depths of up to 80 feet (24.1 m) as might be interpreted from Figure 7.38. If we take into account possible differential erosion along the top San Andres unconformity, there would be no island topography (see text and Figure 7.42). Used with permission of Springer-Verlag, New York.

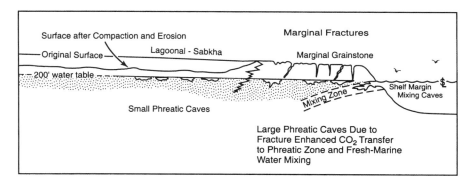

FIG. 7.42. Conceptual model of upper San Andres after de-compaction and possible preferential erosion of shelf-ward lagoonal/sabkha facies. The minus 200 feet level relates to the principal cave population in cycle #3, Figure 7.36.

ter movement during the highstand was responsible for the pervasive dolomitization of the San Andres, it is not a stretch to propose that a meteoric system with flow from platform interior to platform margin can be established during the lowstand (7.42). This water table would extend across the lagoonal sabkha sequences into the marginal dolomitized grainstones where its meteoric waters would actively mix with marine waters at the shoreline on the seaward side of the shelf margin, forming a shelf marginal cave system similar to that seen in Yucatan and discussed above. To the west of the margin in the finer-grained lagoonal sequences small phreatic caves would have developed along the water table (Figs. 7.39 and 7.42).

Tinker (1995) extended Craig's (1988) observations and utilized a 3-D seismic grid to construct a modern 3-D reservoir model for Yates Field. His main contribution (in terms of this discussion) was the recognition and mapping of the three 4th order progradational cycles (Fig. 7.36) each with its own overlapping but isolated cave system (Fig. 7.36 and 7.43). The cycle 3 cave system is more extensive and better developed than the two earlier cycles because of longer exposure at the sequence bounding unconformity at the top of the San Andres. Fracturing has played a major role in

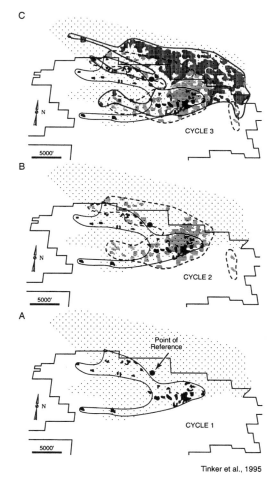

Fig. 7.43. Position of caves in the Tinker et al. 1995 3-D model. The large dot is a geographic reference point. The shift from west to east relates to progradation of successive 4th order cycles in the upper San Andres. Used with permission of AAPG.

vertical cave formation and ultimately vertical fluid conductivity in all three cycles (Craig, 1988; Tinker, 1995). The fact that karst processes were controlled by these fractures suggests that the fractures were almost syndepositional and probably associated with shelf margin instability; analogous to the neptunian dikes of Playford (1980). The observations of Craig (1988) and Tinker (1995) detailed above have made an enormous impact on the resultant development of the Yates Field.

In the following sections, the conundrum of dolomitization in the mixed meteoric-marine and meteoric-diagenetic environments will be considered.

DOLOMITIZATION ASSOCIATED WITH METEORIC AND MIXED METEORIC-MARINE WATERS

Introduction

One of the most popular and widely used dolomitization models during the 70's and 80's was the mixed meteoric-marine model, more commonly termed the mixing zone or Dorag model (Badiozamani, 1973; Land, 1986; Morse and Mackenzie, 1990; Purser et al., 1994). This model was commonly applied to ancient dolomitized sequences that were thought to have been under the influence of mixed meteoric-marine waters in settings such as shorelines (Quintana Roo, Yucatan peninsula) and shelf margin mixing zones (the Bahamas). Dolomites attributed to the mixing environment may be important oil and gas reservoirs (Roehl and Choquette, 1985).

Another meteoric diagenetic setting in which dolomite is presently accumulating involves evaporated continental alkaline groundwaters in coastal lakes and lagoons of the Coorong region of south Australia (von der Borch et al., 1975). The Coorong model has seldom been applied to ancient sequences and may not have wide applicability or economic importance (Morrow, 1982b; Moore, 1989).

The following paragraphs limit our discussion to the mixing zone model with particular emphasis on the validity and constraints of the model and its ultimate porosity potential.

Meteoric-marine mixing, or Dorag model of dolomitization

Figure 7.3A illustrates the original thermodynamic basis for the mixing model of dolomitization. When meteoric water saturated with respect to calcite and undersaturated with respect to dolomite is mixed with marine water strongly supersaturated with respect to both phases, intermediate mixtures are undersaturated with respect to calcite and strongly supersaturated with respect to dolomite. In the 10% meteoric 40% marine mixture, dolomitization should be favored (Badiozamani, 1973).

The model was published almost simultaneously by Badiozamani (1973) and Land (1973). Badiozamani (1973) based his Dorag model on Middle Ordovician sequences of Wisconsin. Land (1973) based his version of the mixed model on the dolomitization of Holocene and Pleistocene reef sequences on the north coast of Jamaica (see Figure 44A). Both workers drew heavily on the report of Hanshaw and others (1971) concerning dolomitization in the Floridian meteoric aquifer, and Runnells' (1969) work on mixing of natural waters.

The mixing zone model has been widely applied to ancient dolomitized sequences of all ages (Zenger and Dunham, 1980; Purser et al., 1994). It had particular appeal because of the inevitability of meteoric influence on shallow marine carbonate sequences and the fact that these meteoric waters replace and mix with original marine pore fluids in a variety of hydrologic situations. In each instance in which a paleo-floating meteoric water lens is identified in the record, there is the possibility of dolomitization in the mixing zone on its flanks. For every regional meteoric aquifer system, there is the potential for dolomitization in the mixing zone at the coastal terminus of the aquifer. If the sequence is a prograding shoreline sequence behind which a regional meteoric aquifer is developed, or within which a meteoric lens is supported, the resulting coastal mixing zone will migrate with the shoreline and ultimately affect a wide swath of the carbonate progradational sequence. The model has been used to explain local patchy dolomitization (Choquette and Steinen, 1985, see discussion below) as well as regional platform dolomitization (Budd and Loucks, 1981).

The recognition of mixed water dolomites in the geologic record is generally based on their association with petrographic fabrics and textures that can be related to the meteoric diagenetic environment. These fabrics include moldic porosity, early circumgranular crust equant calcite cements, and associated meteoric vadose fabrics such as microstalititic and meniscus cements within a compatible geologic setting. In addition, because of the strong contrast between the isotopic and trace element compositions of marine, meteoric and evaporative waters (see chapter 4 and introduction to this chapter), isotopic and trace element composition of the dolomite itself is generally used as a critical line of evidence. In practice, in those sequences that have been totally dolomitized, dolomite geochemistry is usually used as the prime interpretive criteria.

Lohmann (1988) suggests that mixing zone dolomites should fall in a stable isotopic population parallel to the meteoric calcite line but offset toward more enriched ^{18}O values because of dolomite fractionation. Dilution by meteoric waters should result in reduced levels of Sr, and Na in mixing zone dolomites relative to dolomites forming in seawater or evaporated seawater (Land, 1985).

Concerns about the validity of the mixing model of dolomitization

There have long been concerns over the validity of the Dorag model. Carpenter (1976) attacked the model on thermodynamic grounds, concluding that mixing seawater and meteoric water should actually inhibit dolomitization. These arguments were subsequently reiterated by Machel and Mountjoy (1986). Hardie (Hardie, 1987) noted that Badiozamani (1973) used the solubility constant for partially ordered dolomite (10^{-17}) in his meteoric-marine water saturation calculations. Hardie contended that the solubility constant for disordered dolomite ($10^{-16.52}$) should be used since most modern dolomites forming on the surface today are disordered. Use of the disordered dolomite solubility constant to calculate dolomite saturation in a mixture of meteoric and seawater will result in a much smaller dolomite window (between 30 to 40% seawater, vs. 5-50%) (Fig. 7.3B).

Perhaps the most serious concern is related to the potential for early proto-dolomites to re-crystallize and hence re-set dolomite geochemistry during the early diagenetic history of the sediments. Dolomite recrystallization was discussed earlier in Chapter 4 and in the papers of Land (1981, 1985) and Hardy (1987). This is a particularly difficult problem because dolomitization often obliterates original textures so that geochemical data becomes critical in the determination of the nature of the original dolomitizing fluids.

For the 27 years that this dolomitization model has been on the scene, many geologists have been worried over the dearth of unequivocal examples of mixing zone dolomites being formed in modern sediments. The concept of mixing zone dolomitization had its start with observations of dolomite apparently forming in the Floridian aquifer by Hanshaw and others (1971) and by Land's (1973a and b) assignment of Holocene and Pleistocene dolomites in reef rocks of Jamaica to a mixing origin.

Modern mixing zone dolomitization

One of the most compelling cases of modern mixing zone dolomitization remains Land's original study on the Pleistocene Falmouth Formation along the north coast of Jamaica. The Falmouth represents a 120,000 year highstand and consists of reef-related sediments in

Fig. 7.44. Photographs of sites of Holocene-Pleistocene mixed water dolomitization. A. Northern Jamaica near Discovery Bay. HGF is the Pleistocene Hope Gate Formation 200,000+ years BP. WCB is a wave-cut bench representing the last interglacial highstand. FF is the Pleistocene Falmouth Formation (120,000 BP) which hosts the Holocene dolomites found by Land. The arrow marks the location of the present mixing zone and is the site of one of Land's Holocene dolomite-bearing cores. B. Aerial shot of Akumal along the Yucatan coast south of Cancun. The whitish area is the zone of meteoric-marine water mixing. Photo courtesy Bill Ward. Note the extensive cave collapse in the mixing zone controlled by fractures. C. Collapsed cave system in the mixing zone at Xel-Ha south of Cancun.

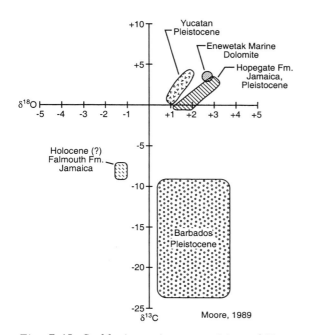

Fig. 7.45. Stable isotopic composition of Yucatan Pleistocene dolomites compared to other dolomite occurrences. Data sources: Jamaica Holocene and Pleistocene (Land, 1973a and b); Yucatan Pleistocene (Ward and Halley, 1985); Enewetak marine dolomite (Saller, 1984b); Barbados Pleistocene (Humphrey, 1988). Used with permission, Elsevier.

the process of being mineralogically stabilized (Fig. 44A). Land (1973b) took a series of cores across the Falmouth terrace just west of Discovery Bay (Fig. 44A). He discovered Holocene replacement dolomite and dolomite cement within a Falmouth cavernous zone developed along the present water table. The isotopic composition of the dolomite seems to be in equilibrium with the waters of the present mixing zone; and like them, the dolomite contains low levels of Sr and Na (Land 1973b) (Fig. 45).

Kimball et al. (1990) and Humphrey (1997) reported significant Holocene mixed water dolomites in the modern mixing zone along the southeast coast of Barbados. While it is not absolutely certain that these dolomites are actually Holocene, the waters of the mixing zone are undersaturated with respect to calcite and supersaturated with respect to dolomite (Kimball et al. 1990).

Gebelein (1977) initially reported massive dolomitization beneath palm hammocks that

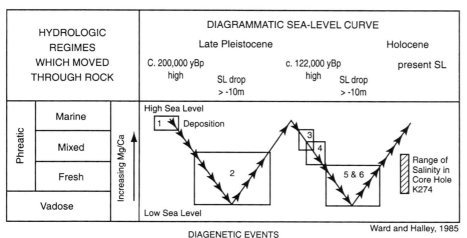

Fig. 7.46. Diagenetic events affecting Pleistocene carbonates in the Yucatan related to pore-water regimes as controlled by changes in sea level. The sea-level curve is not to scale in time or space. These rocks spent most of their history in the vadose zone. Used with permission, Elsevier.

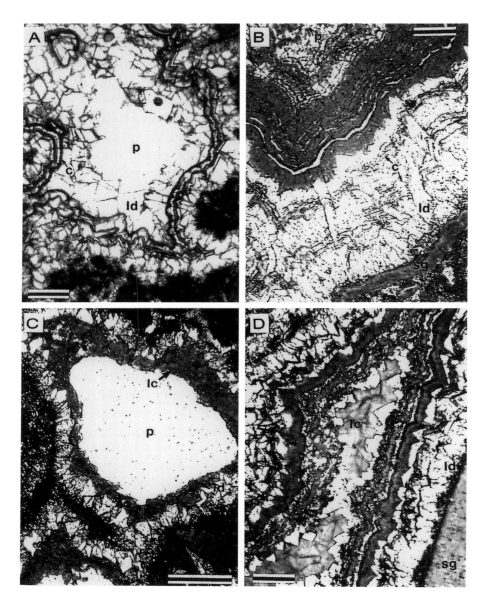

Fig. 7.47. Photomicrographs of Pleistocene mixed water dolomites from Golden Grove, Barbados. (A) Predominantly limpid dolomite cement (Id, unstained) with 2-3 syntaxial bands of calcite (c, stained red from Alizarin Red S). Note lateral continuity of the bands. Pore space (p) occupies the center of the photomicrograph. Scale bar equals 50 μm. Plain light. Sample XE-8. (B) Multiple alternating bands of limpid dolomite cement (Id, unstained) and syntaxial calcite (c, stained dark). Cement substrate is a skeletal grain (sg) in the lower right and pore space (p) is out of the field of view to the upper left. Over 20 calcite bands can be counted. Scale bar equals 50 μm. Plain light. (C) Isopachous layer of late-stage blocky calcite spar (lc, dark band overlying limpid dolomite (Id) cement. Pore space (p) in center. Scale bar equals 50 μm. Plain light. (D) Complete cement sequence in a single pore. Limpid dolomite cement (Id) growing on skeletal grain (sg), alternating bands of calcite and dolomite, and late stage calcite (lc). Scale bar is 75 μm. Plain light. Photos reproduced with permission from IAS.

supported meteoric lenses on the tidal flats of Andros Island. Subsequent investigation, however, leads to the conclusion that there was no significant potential for dolomitization in these mixing zones (Gebelein et al., 1980).

Other workers have failed to find mixing zone dolomites in modern beaches; Grand Cayman (Moore, 1973); St. Croix (Moore, 1977) or in modern island floating meteoric lenses of various sizes and hydrologic characteristics; Joulters Cay (Halley and Harris, 1979); Schooner Cay (Budd, 1984); Enewetak (Saller, 1984).

Pleistocene to Miocene mixing zone dolomitization

There have been a number of well-documented case histories of "mixing zone" dolomitization in Pleistocene and Tertiary carbonates. These studies include the Pliocene-Miocene Seroe Domi dolomites of Bonaire (Sibley, 1980) and Curacao (Fouke et al., 1996) and the Miocene Isla de Mona Dolomite of Puerto Rico (Gonzalez et al., 1997).

Some dolomite has been found in Pleistocene rocks of the Yucatan in Mexico (Ward and Halley, 1985). They occur along the coast in association with the modern regional meteoric/marine mixing zone described by Back and others (1979) (Fig. 7.44B and C). The dolomites were recovered from cores taken from the Yucatan strand plain which was discussed earlier (Fig. 7.18). However, these dolomites did not form in the present mixing zone, but are thought to have been formed in a paleo-mixing zone some 120, 000 years ago during a time of falling sea level (Fig. 7.46). Stable isotopes suggest formation in waters ranging in composition from 100% seawater to 75% seawater (Fig. 7.45). The dolomites and calcites were precipitated in a complex zonal arrangement, with no evidence of solution of the underlying calcite prior to precipitation of the dolomite (Ward and Halley, 1985) (See similar fabrics from Barbados mixing zone dolomites in Figure 7.47).

Humphrey (1988) makes a compelling case for mixed water dolomitization in one of the raised Pleistocene terraces of Barbados. These dolomites occur as limpid cements (Fig. 7.47) and mimitic replacements in reef-related sediments of the 212,000 year BP sea level highstand. Apparently the dolomitization occurred during the falling sea level immediately after the highstand as meteoric phreatic waters mixed with marine waters. It should be noted that this setting is very similar to the Yucatan mixed dolomites, only older (Fig. 7.45). Humphrey (1988) noted that the dolomitization event might only have encompassed some 5000 years. Stable isotope (Fig. 7.45) and trace element compositions as well as petrographic relationships suggest that this was a meteoric water dominated system with as little as 5% seawater involved. The anomalously depleted carbon isotope values reported by Humphrey (1988) were subsequently thought to have been the result of oxidation of thermogenetic methane derived from the underlying Barbados accretionary prism (Humphrey and Radjef, 1991; Radjef, 1992).

The mixing zone dolomites described above have a number of attributes in common: 1) Volume of dolomite is low and distribution is usually spotty; 2) The dolomite often occurs as several discrete zones of limpid dolomite cement sandwiched between zones of limpid calcite cement; 3) Waters responsible for the dolomites have a wide range of composition, from almost pure meteoric water to almost pure marine water. The resulting isotopic composition of the dolomite also shows a wide range (Fig. 7.45); 4) Independent petrographic evidence of meteoric water influences such as moldic porosity and meteoric cement textures are always present.

The following discussion of a Mississippian reservoir in the Illinois Basin presents a case history of dolomitization where setting, geologic history, and dolomite geochemistry are compatible with a mixing zone origin and reasonable alternative dolomitization models cannot be applied.

Mississippian North Bridgeport Field, Illinois Basin, USA: mixed water dolomite reservoirs

The North Bridgeport Field is a shallow (490m, 1600ft), relatively small accumulation that was discovered in the Mississippian Ste. Genevieve Limestone in 1909 in the northeastern sector of the Illinois Basin, along the LaSalle Arch (Fig. 7.48). The field has undergone a series of water flood projects since 1959 and had produced some 3.1 million barrels of 39° API gravity oil from a productive area of 1100 acres through March 1983. Ultimate recovery is estimated as 4.0 million barrels (Choquette and Steinen, 1985).

The North Bridgeport reservoir is developed in Ste. Genevieve ooid grainstones and dolomitized lime mudstones and wackestones. The ooid shoals, localized along the crest of the LaSalle Arch, occur as elongate tidal bars and channels cutting the arch at a high angle (Fig. 7.49). They are thought to be analogous to the ooid shoals forming along the northwestern margin of the present Great Bahama Platform. The dolomitized mudstones and wackestones probably represent protected lagoonal environments associated with the tidal

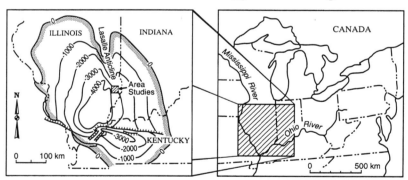

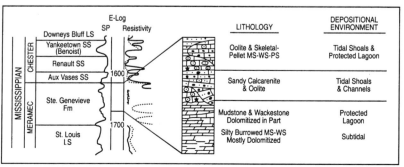

Fig. 7.48. *Location maps and generalized Ste. Genevieve lithologic column. Map at left shows regional structure on top of the New Albany Shale (Upper Devonian-Lower Mississippian). Lithologic column is a conceptualized version of a cored sequence in a Bridgeport field well. Used with permission, Elsevier.*

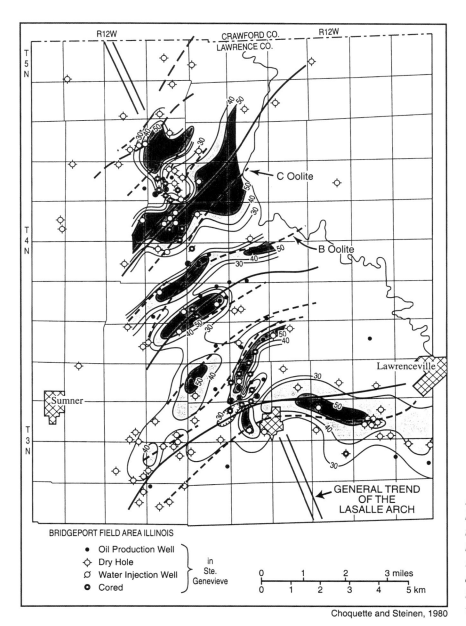

Fig.7.49. Map of percent dolomite with carbonate-sand trends superimposed, Solid lines are channel fill calcarenite trends, long and short dashed lines are B and C oolite trends respectively. Used with permission, Elsevier.

channels and shoals (Choquette and Steinen, 1985).

The ooid and dolomite reservoirs generally occur closely associated with one another, with the dolomitized sequence usually appearing beneath and/or laterally adjacent to the ooid reservoirs and allied sandy calcarenite channel sequences (Fig. 7.48). Table 7.2 and Figure 7.50 outline the differences in the pore systems supported by these two reservoir types. The ooid grainstones contain cement-reduced intergranular porosity averaging some 12% with 275 md permeability. These rocks have no secondary porosity development. The dolomite reservoirs show an average porosity of 27%, with 12 md permeability developed in

Table 7.2

	Oolite	Dolomite
Porosity types	Interparticle (BP) Intraparticle (WP)	Microintercrystal (mBC) Moldic (MO)
Inferred origin	Primary, reduced by cementation	Dissolution of $CaCO_3$
Pore size range (apparent)	10^{-2}—10^{0} mm	10^{-4}—10^{-2} mm (BC) 10^{-1}—10^{0} mm (MO)
Porosity average range n	12% 2-22% 117	27% 13-40% 90
Permeability average range n	250 md 0.1-9500 md 115	12 md 0.7-130 md 87

[1]Porosity types follow Choquette and Pray (1970). Porosity and single-point gas permeability were determined by standard core analysis of 3/4-inch (1.9-cm) diameter plugs drilled from cones.

Choquette and Steinen, 1985

Table 7.2. Comparison of some reservoir properties in oolite and dolomite reservoirs in Bridgeport Field. Used with permission, Elsevier

intercrystalline and moldic pore types. The molds represent dissolution of various bioclasts and scattered ooids. Comparative porosity/permeability plots clearly illustrate the high-porosity/low-permeability characteristics of the dolomite sequences versus the low-porosity/high-permeability attributes of the ooid grainstone reservoirs (Fig. 7.50) (Choquette and Steinen, 1985).

The ooid grainstones, while not possessing secondary porosity, do show some evidence of subaerial exposure and meteoric water influence based on the occurrence of vadose meniscus and pendent cements in the upper parts of the sand-bodies. Meteoric phreatic cements, however, are not common in the ooid grainstones. The common moldic porosity present in association with the dolomites is thought to represent significant meteoric water influence early in the diagenetic history of the Ste. Genevieve (Choquette and Steinen, 1980).

Choquette and Steinen (1980) suggested that the dolomitization of the lagoonal mudstones took place in a meteoric-marine mixing zone formed along the margins of the ooid shoals and sandy calcarenite channels. The meteoric waters were thought to be distributed

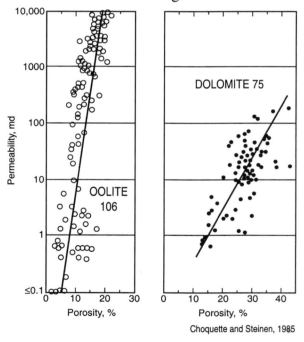

Fig.7.50. Semi-log plots of percent porosity vs air permeability in an oolite reservoir (left) and a dolomite reservoir (right) at Bridgeport Field. Used with permission, Elsevier.

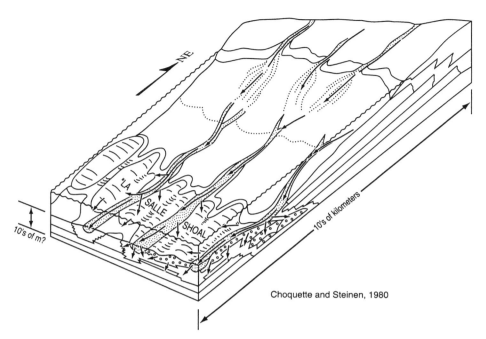

Fig.7.51. Suggested hydrologic model for dolomitization of lime-mud sediments in upper Ste. Genevieve Limestone. Mixed meteoric-marine groundwaters fed from a recharge area northeast of the LaSalle paleo-shoal, gained access to the lime muds via carbonate-sand conduits. Used with permission, Elsevier

by a gravity-driven regional meteoric aquifer system that used the porous channel and shoal systems as a distributional network (Fig. 7.51). Evaporative waters were not invoked because of the absence of any evidence for supratidal-sabkha conditions or the presence of stratigraphically associated evaporative sequences.

Stable isotope and trace element composition of the dolomite and associated calcite cements were used (Choquette and Steinen, 1980) to support a mixing zone origin for the dolomites (Fig. 7.52). But as noted by Choquette and Steinen (1980), and emphasized by Hardie (1987), the data does not unequivocally separate these dolomites from other samples known to have formed from marine and marine evaporative waters. The presence of intricate zoning in the dolomites, however, seems to preclude recrystallization and chemical re-equilibration of an earlier, more unstable phase.

As illustrated in Choquette and Steinen's (1980) photomicrographs, the Mississippian ooids at North Bridgeport Field are exceptionally well preserved, suggesting that they may have originally been composed of calcite. Sandberg (1983) indicates that the Mississippian was a time of calcite seas implying that ooids precipitated during this interval should be composed of calcite. If, indeed, a major meteoric water aquifer had been developed in these grainstones up-dip, the water would have quickly come to equilibrium with respect to calcite and little carbonate cementation would be expected as the water moved through the shoals. Where these waters began to mix with marine waters, however, a zone of calcite dissolution would be expected, as is seen in the dolomite-associated sequences below and adjacent to the shoals (Matthews, 1974; Choquette and Steinen, 1985).

Geological evidence would seem to favor mixing zone dolomitization while geochemical evidence certainly doesn't seem to preclude the use of the mixed water model for the dolomite reservoirs at North Bridgeport.

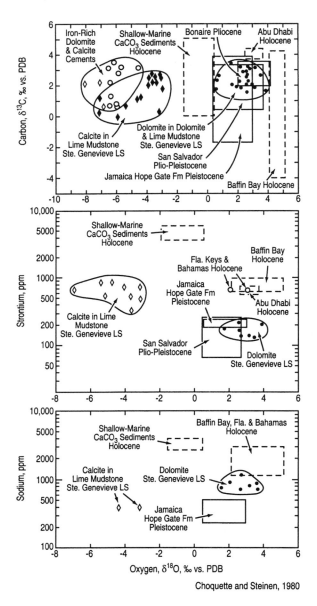

Fig.7.52. Comparison of the stable isotopic and trace element composition of Ste. Genevieve dolomites and calcite cements with dolomites from various settings. Sources of data may be found in Choquette and Steinen (1980). Used with permission, Elsevier

Where do we stand on mixing zone dolomitization today?

In 1991 there was a conference in Ortisei, Italy to honor Deodat de Dolomieu on the 200th anniversary of his classic paper describing dolomite in detail for the first time. Papers given at the Dolomieu conference were published in 1994 as special publication 21 of the International Association of Sedimentologists. Bruce Purser, Maurice Tucker and Don Zenger, the co-conveners of the conference, edited this important publication. In their introduction to this volume they highlighted and reviewed some specific concepts and problems discussed during the conference (pages 3-20). One of the concepts that elicited a great deal of discussion was mixing zone dolomitization. Their commentary as quoted below gives the reader an indication of the current status of this dolomitization model among carbonate workers.

Most ancient dolomites exhibit clear traces of dissolution, leached fossils being typical (Fairchild et al., 1991). However, dissolution may precede, postdate or be contemporaneous with dolomitization. Both meteoric and diluted seawater, unlike normal seawater, are generally undersaturated with respect to aragonite and calcite, and thus have the potential to dissolve. All three editors of this volume consider that mixed waters are potentially capable of dolomitizing. However, as noted above, the significance of the mixing zone may be more in its role of inducing fluid movement in the marine ground water below.

Perhaps some of us have missed the point. As Duncan Sibley (questionnaire) has pointed out, the word 'origin' is not very precise. The fact that dolomite may indeed form from normal or mixed sea waters is not proof that the

chemical properties we normally associate with these waters (temperature, salinity etc.) are prerequisites for dolomite formation. Clearly, dolomites may form from many different types of waters; and, as already noted, there is no unique fluid or model. The current popularity of one or other system is ephemeral.

SUMMARY

The meteoric diagenetic environment is one of the most important diagenetic settings relative to the development and evolution of carbonate porosity. Meteoric waters are chemically dilute, though highly aggressive in respect to most carbonate mineral phases because of the ready availability of CO_2. The low Mg/Ca ratios and salinities found in most meteoric waters generally favor the precipitation of calcite. Mixed meteoric and marine waters while often undersaturated with respect to calcite may remain saturated with respect to dolomite. This relationship results in extensive calcite dissolution, and the possibility of dolomitization.

The stable isotopic composition of meteoric waters, and the carbonates precipitated from these waters, is highly variable. In general, however, carbonates precipitated from meteoric waters have light stable isotopic signatures and follow a meteoric trend marked by large variations in $\partial^{13}C$ and locally invariant $\partial^{18}O$.

The solubility differences commonly present in metastable carbonate sediments provide a strong drive for the nature and intensity of meteoric diagenesis. Since mineralogical stabilization usually is accomplished by dissolution and precipitation, stabilization has profound consequences relative to porosity evolution.

During mineral stabilization by solution of aragonite and precipitation of calcite, the relatively slow calcite precipitation rate step may result in transport of $CaCO_3$ away from the site of aragonite solution. Transported $CaCO_3$ may then be precipitated down flow as void-fill calcite cement.

The meteoric diagenetic environment consists of two important sub-environments, the vadose zone and the phreatic zone.

The meteoric vadose diagenetic environment is unique because it is a two-phase system—air and water. This system is reflected in the unequal distributional patterns and unusual morphology of the cements precipitated in the environment. Such cements are most common at grain contacts; and they exhibit microstalictitic and meniscus fabrics.

Diagenesis associated with mineral stabilization is generally slow in the vadose zone. Therefore, it is possible that a metastable sequence that remains in the vadose zone will retain its unstable mineralogy (aragonite, but perhaps not Mg calcite because of its higher solubility) and significant depositional porosity upon its burial into the subsurface. Subsequently, the metastable sequence might well lose its porosity more quickly under overburden pressure during burial than will a unit containing stabilized calcite because of the higher susceptibility of the metastable mineral phases to grain-to-grain pressure solution.

The meteoric phreatic diagenetic environment is the most important meteoric environment relative to porosity modification because of the large volume of water available for dissolution and precipitation. Moldic porosity and abundant pore-fill circumgranular calcite cements are common in the phreatic zone.

The meteoric phreatic zone may occur in two basic hydrologic settings inextricably tied to sea level: 1) The immature hydrologic phase consisting of island based floating fresh water lenses formed during sea level highstand; and 2) The mature hydrologic phase consisting of regional meteoric aquifers developed during sea level lowstands.

The immature phase during sea level highstands is dominated by mixing zone dissolution along the margins of floating fresh water lenses and by rather thin water table dissolution and cement precipitation. High frequency sea level cyclicity and shoreline progradation can expand the zone of diagenesis vertically and laterally.

Thick regional meteoric water lenses floating on marine water characterize the mature phase during sea level lowstands associated with isolated platforms. High frequency sea level cycles can result in relatively thick mineralogically stable sequences. Large-scale shelf margin cave systems can develop in the lateral meteoric-marine mixing zones. Small phreatic caves are common along the water table. Karsting of the platform surface is common. The character of this karst is dependent on climate and length of exposure.

Regional open gravity-driven meteoric aquifers may be developed in land-tied carbonate shelves during lowstands because of preferential subsidence of the shelf margin. Karsting of the upper surface of the shelf is common. Mineral stabilization by dissolution of aragonite and precipitation of calcite may result in large-scale transport of $CaCO_3$ down flow and the establishment of predictable diagenetic and porosity gradients. These gradients are reflected in the geochemical characteristics of the cements. Subsequent sea level rises may result in the evolution of the open aquifer into a temporary confined aquifer. This hydrologic situation may also be reflected in the geochemistry of the cements.

The interaction of meteoric waters with mineralogically stable carbonate sequences along long-lived or tectonically enhanced unconformities results in a mature karst system. Karst processes are driven in part by climate, with humid conditions favoring the development of important porosity enhancement in potential reservoirs.

Fractures play a major role in moving aggressive meteoric waters through the vadose into the phreatic zone. Fracture dissolution enlargement enhances vertical fluid conductivity. Large-scale phreatic caverns are generally formed along and just below the water table. Fractures associated with cavern collapse enhance permeability and involve the vadose zone. Collapse breccias and related karst features such as caves are important reservoirs in the geologic record involving a number of giant fields.

Meteoric-marine mixing zone dolomitization, once the favored dolomitization model of carbonate workers has retreated in the face of widespread criticism fueled by thermodynamic concerns, dependence on geochemical criteria, and the paucity of viable Holocene examples. A much clearer understanding of the system has been developed by well-constrained studies of Pleistocene, Pliocene, and Miocene sequences in the Caribbean. Conventional wisdom suggests that while there is no doubt that mixing zone dolomitization does take place, it is not an appropriate model for pervasive shelf or platform dolomitization.

Chapter 8

SUMMARY OF EARLY DIAGENESIS AND POROSITY MODIFICATION OF CARBONATE RESERVOIRS IN A SEQUENCE STRATIGRAPHIC AND CLIMATIC FRAMEWORK

INTRODUCTION

In the preceding chapters we have discussed basic characteristics and controls over early diagenesis and porosity evolution in a variety of diagenetic environments. In this chapter by way of summary, a series of predictive, diagenesis/porosity models will be presented that should be useful in exploration, as well as exploitation, of carbonate reservoirs. Tucker (1993) was one of the first workers to present detailed stratigraphic-based diagenetic models. The discussions in this chapter draw particularly on his earlier work and the subsequent studies of Saller et al. (1994) and Harris et al. (1999).

The models presented here are couched in a sequence stratigraphic framework. See Chapter 2 for a short review of the concepts of carbonate sequence stratigraphy and Loucks and Handford (1993) and Emery and Meyers (1996) for in-depth treatments.

Changes in relative sea level drive changes in pore fluid chemical composition, which in turn control early mineralogical stabilization and porosity evolution. If, as discussed in Chapter 2, cyclical changes in relative sea level follow a predictable climate/tectonic-driven pattern, our sequence stratigraphic-based diagenetic models should be useful in predicting porosity evolution in ancient carbonate sequences.

Three cycle scales will be considered: the standard 3rd order glacio-eustatic depositional sequence, high-frequency parasequences, which comprise the 3rd order sequences, and the tectonically driven 2nd order or super sequence.

Short-term climatic conditions play a major role in any consideration of diagenesis because of the importance of the volume of water available to drive diagenetic processes (Chapter 7). Each model, therefore, will be developed for arid versus humid climates.

Long-term climatic cycles such as greenhouse versus icehouse conditions are driven in part by global tectonics (Chapter 2 and Fig. 2.14). These cycles affect amplitudes of glacio-eustatic sea level changes and mineralogy of abiotic carbonates. General patterns of diagenesis and related porosity evolution in response to these long-term climate cycles will be considered in a separate discussion.

Finally, several appropriate case histories will be presented which together will illustrate many of the major concepts developed in this chapter.

RESERVOIR DIAGENESIS AND POROSITY EVOLUTION DURING 3RD ORDER SEA LEVEL LOWSTANDS (LST)

Introduction

The lowstand systems tract is one of the most important settings for early diagenesis because mineralogically unstable marine sediments are potentially exposed to large volumes of undersaturated meteoric waters under humid conditions that have a high potential to modify porosity and permeability (Fig. 7.11B). There are a large number of potential settings that could be considered (Chapter 1). However, in order for the models to be as useful as possible, only three basic end-member settings will be utilized: the ramp, the rimmed shelf, and the rimmed isolated platform.

Diagenesis/porosity model of a carbonate ramp during sea level lowstand (LST)

Figure 8.1A illustrates the summary of diagenetic and porosity characteristics of a carbonate ramp at a 3rd order sea level lowstand (LST) under humid climatic conditions. Given the relative low relief of the ramp to basin slope, lowstand prograding complexes (LSPC) featuring prograding high-energy grainstones can commonly form seaward of the lowstand shoreline (Fig. 8.1B). There is a discussion of the ramp sequence stratigraphic model in Chapter 2).

The exposed upper ramp surface will commonly exhibit superficial epikarst features such as solution-enlarged fractures, vertical solution shafts, and pits (Fig. 7.33). The development of the karst surface is directly related to time of exposure and water availability: the more water and more exposure time, the more intense the karstification processes and products. Intense karstification will lead to more highly developed vertical conductivity. At the 3rd order sedimentary sequence level, without tectonic enhancement, lowstands generally persist for some 50,000 years (Vacher and Rowe, 1997). Under these conditions karst development could be intense and relatively deep but without the classic mature karst features such as towers, sinkholes, and major cave systems (Loucks, 1999).

In the vadose zone, cementation will be confined to grain contacts. Most primary intergranular porosity will be preserved, and little moldic porosity will be generated. Aragonite, if present, can be preserved into burial conditions while magnesian calcite is stabilized to low magnesian calcite with little porosity modification. Permeability can be negatively impacted by meniscus cement, while karstification extending through the vadose zone enhances vertical fluid conductivity (see discussion of the vadose zone, Chapter 7).

The phreatic meteoric aquifer is open to the atmosphere from ramp interior to the lowstand shoreline. Differential subsidence of basin margin relative to ramp interior can set up an open gravity-driven meteoric aquifer. Coastal climatic patterns suggest that recharge will be dominant in the ramp interior (Vacher and Quinn, 1997). Movement of aggressive

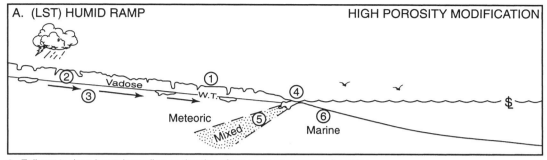

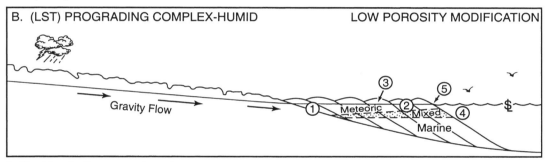

Figure 8.1 LST diagenetic/porosity model for the ramp and low stand prograding complex under humid climatic conditions.

meteoric water from recharge area toward the coast will set up distinct diagenetic and porosity gradients (Fig. 7.26). $CaCO_3$ derived from dissolution of aragonite grains in the recharge area will be transported down-stream to be precipitated as intergranular calcite cement. Vuggy and moldic porosity will form in the upstream recharge area. Aragonite grain stabilization to calcite by neomorphisim (micro-solution and precipitation) is common down stream with no moldic porosity development. Lack of secondary porosity and precipitation of transported $CaCO_3$ as calcite cement in the down stream phreatic zone can destroy most porosity and permeability toward the lowstand coastline (Fig. 7.26).

In the absence of a lowstand prograding complex (LSPC) an active meteoric-marine mixing zone will be present at the lowstand coastline. These mixed waters are generally undersaturated with respect to calcite and aragonite, leading to a high porosity zone of dissolution with moldic porosity, no intergranular cement, and potentially flank margin caves and minor dolomitization (Figs. 7.3, 7.23 and 7.46). This zone is located

SUMMARY OF EARLY DIAGENESIS AND POROSITY MODIFICATION

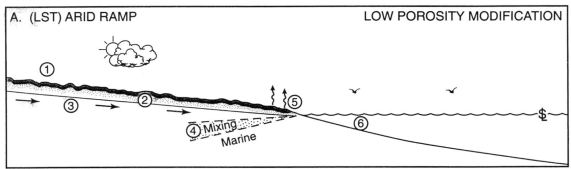

1. Impermeable caliche crusts, minor karsting.
2. Vadose zone unstable mineralogy preserved, minor cementation, high porosity and permeability.
3. Thin meteoric phreatic zone, reduced flux of water to coast little transport of CaCO₃. Diagenesis centered on water table. Reduced permeability. No diagenetic/porosity gradients established. Aragonite preserved in lower phreatic.
4. Minor solution and dolomitization in mixing zone because of reduced water flux.
5. Minor dolomitization by evaporative reflux along ramp margin.
6. Marine hardgrounds in shallow offshore.

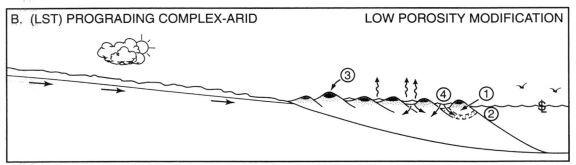

1. Ephemeral island-based meteoric lens. Minor porosity modification along water table. Preservation of aragonite.
2. Beach rock marine cements along coastline.
3. Thick impermeable caliche crusts on beach ridges.
4. Coastal salinas in swales with tee pee structures and reflux dolomitization.

Figure 8.2 LST diagenetic/porosity model for the ramp and low stand prograding complex under arid climatic conditions.

seaward of the low porosity, highly cemented zone mentioned above.

In the presence of a grainstone-dominated lowstand prograding complex (LSPC) (Fig. 8.1B), the gravity drive of the aquifer is lost because the water table is at sea level. In this situation the fresh water rides out over marine interstitial water as a thin sheet flow, much as is seen in Barbados (Fig. 7.25). The development of the floating meteoric sheet flow is progressive, following the progradation of the shoreline. The main porosity-related diagenetic modifications are concentrated at the water table with dissolution of aragonite grains and precipitation of calcite cement. The vadose zone is generally thin, and epikarst development is minor. Minor dolomitization may take place in the mixing zone.

Figure 8.2A shows the summary of diagenesis/porosity characteristics of a carbonate ramp under arid conditions without a lowstand prograding complex. All features result from the scarcity of water to drive porosity-modifying diagenetic processes.

Impermeable caliche crusts and minor epikarst mark the exposed surface of the ramp (Figs. 7.9 and 7.12). In the vadose zone, there is minor cementation; and both aragonite and magnesian calcite may be preserved into burial conditions. Along the lowstand shoreline there may be some reflux dolomitization associated with evaporative reflux of modified marine waters.

The meteoric phreatic zone will generally be thin with reduced rate of flux of meteoric water from recharge area to the sea. This reduced water flux results in less transport of $CaCO_3$ away from the site of solution. Calcite precipitation in adjacent pore spaces impacts permeability, but not total porosity. Under these conditions, diagenesis/porosity gradients are muted, porosity is enhanced, but permeability may be dramatically reduced along the water table. Away from the water table in the lower phreatic zone, aragonite mineralogy may be preserved into the subsurface.

The meteoric-marine mixing zone will be much less active because of reduced meteoric water flux into the coastal area. We might expect to see minor dissolution and dolomitization along the coastal mixing zone. If porous facies are present seaward of the lowstand shoreline, some marine hardgrounds might be formed because of higher salinities in coastal marine waters.

Figure 8.2B illustrates the summary of diagenesis and porosity characteristics of a lowstand prograding complex under arid climatic conditions. Because of reduced meteoric flow to the shoreline, island-based floating meteoric lenses would be developed during each progradational event. The lens would be thin and perhaps ephemeral. Most diagenesis and porosity modification would be minor and isolated along a narrow zone centered on the water table. Preservation of aragonite into the subsurface would be common. Marine cements precipitated as beach rock would be common along the coastline of each progradational event. Impermeable caliche crusts will cap each beach ridge, or coastal dune (Fig. 7.18 Yucatan dune). Coastal salinas can develop in swales between beach ridges and dunes leading to minor reflux dolomitization of subjacent grainstones and perhaps the occasional tee pee structure (Fig. 7.13).

Diagenesis/porosity model of a steep-rimmed carbonate shelf during sea level lowstand (LST)

Figure 8.3A shows the summary of diagenesis and porosity characteristics of a rimmed carbonate shelf during a sea level lowstand under humid climatic conditions. This model has many similarities with the humid ramp shown in Figure 8.1A. This discussion will center on the basic difference between the two models. Because of the geometry involved, most steep-rimmed shelves will lack the lowstand prograding complex so characteristic of the ramp. Instead, mega-breccia debris flows are common at the escarpment-slope contact (Fig. 2.9). Patterns of karsting, vadose and phreatic diagenesis will be similar. An important aspect of the steep margin is the development of solution-enhanced fractures marking potential sites for gravity collapse of the margin. These fractures enhance vertical conductivity of the vadose zone parallel to the shelf margin (Fig. 7.42). Marine waters are entrained by strong meteoric water flux into the mixing zone. This ocean to shelf water movement is responsible for pervasive marine

250 SUMMARY OF EARLY DIAGENESIS AND POROSITY MODIFICATION

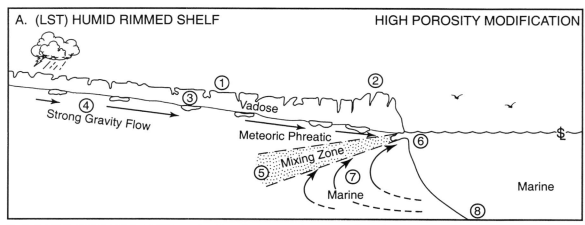

1. Epikarst to deep karst depending on duration of exposure at sequence boundary.
2. Solution enhanced fractures parallel to shelf margin related to margin collapse.
3. Preservation of intergranular porosity in vadose zone, permeability reduced, vertical conductivity may be enhanced along shelf margins by fractures.
4. Moldic to vuggy porosity, extensive cements, phreatic caves along water table, potential diagenetic/porosity gradients.
5. Major secondary porosity development, minor mixing zone dolomitization.
6. Major coastal flank cavern porosity.
7. Pervasive marine-water dolomitization of shelf margin interior driven by hydrodynamics of the mixing zone. Development of secondary porosity, perhaps some porosity destruction by marine dolomite and calcite cements.
8. Aragonite dissolution, calcite precipitation below aragonite lysocline. Net loss in porosity. Marine water influx into platform driven by mixing zone hydrodynamics and geothermal convection.

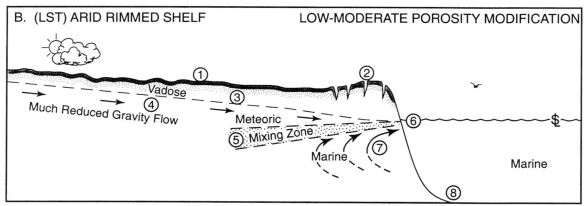

1. Impermeable caliche crusts, minor karsting.
2. Fractures parallel to shelf margin related to margin collapse.
3. Preservation of unstable mineralogy, minor cementation, high porosity and permeability. Shelf margin fracture zone increases vertical conductivity.
4. Thin meteoric phreatic zone. Much reduced flux of water to coast, little transport of $CaCO_3$. Diagenesis centered on water table. No diagenetic/porosity gradients established. Aragonite may be preserved in the lower phreatic.
5. Minor secondary porosity and dolomitization in mixing zone.
6. Flank caverns muted as result of reduced flow.
7. Minor marine water dolomitization of shelf interior due to reduced flow in meteoric water system.
8. Aragonite dissolution, calcite precipitation below aragonite lysocline. Net loss in porosity. Marine water influx into platform driven by geothermal convection.

Figure 8.3. LST diagenetic/porosity model for the rimmed shelf under humid climatic conditions (A) and under arid climatic conditions (B).

water dolomitization of the shelf interior. In addition, geothermal convection may also enhance the flux of marine water into the shelf (Fig. 5.44B). Porosity modification by marine water dolomitization may enhance porosity if the waters are undersaturated with respect to calcite or reduce porosity if the dolomite is precipitated as a cement.

Further down the slope or escarpment, below the aragonite lysocline, major dissolution of aragonite and precipitation of calcite cement can significantly reduce porosity of shelf margin limestones (Fig. 5.37 and 5.38). These diagenetic processes depend on significant flux of marine water moving into the platform either associated with mixing zone hydrodynamics or by thermal convection.

The summary of diagenesis/porosity characteristics in an arid steep-rimmed carbonate shelf during a sea level lowstand is shown in Figure 8.3B. As with the humid shelf above, the model for the arid shelf shares many characteristics with the arid ramp model shown in Figure 8.2B. Patterns of karsting, vadose, and phreatic diagenesis are similar. Shelf margin collapse fractures are generally not enhanced by solution and flank caverns are muted because of reduced activity of the mixing zone. Marine water dolomitization driven by oceanic water entrainment associated with mixing zone discharge will be significantly reduced, as will mixing zone dolomitization and associated secondary porosity development. Deep margin diagenesis and porosity loss associated with the aragonite lysocline should continue to operate driven by geothermal convection. Finally, there will generally be little evaporite-related reflux dolomitization on the steep-rimmed shelf if sea level falls below the level of the shelf interior.

Diagenesis/porosity model of a rimmed isolated carbonate platform during sea level lowstand (LST)

Figure 8.4A shows the general diagenesis/porosity characteristics of a rimmed isolated carbonate platform in a humid climate. One of the main differences between the humid rimmed shelf and isolated platform models is the absence of an active gravity-driven aquifer beneath the platform. While the shelf generally exhibits differential subsidence of the shelf margin and the subsequent development of an active open gravity-driven aquifer, the platform generally subsides as a unit resulting in a platform-wide meteoric lens floating on marine water at a sea level lowstand. Water flux in the lens is toward the platform margins. Rate of flux is slow compared to a gravity-driven aquifer and is controlled principally by the volume of excess meteoric water in the system (Whitaker and Smart, 1997). Rates of porosity-modifying diagenesis in the phreatic zone are generally slower as compared to the shelf, with diagenesis concentrated along the water table and minimal lateral transport of $CaCO_3$. Moldic porosity and calcite cementation will impact permeability negatively along the water table while porosity should remain essentially the same. If the lowstand persists long enough water table phreatic caves might ultimately form (Fig. 7.23). The mixing zone beneath the meteoric phreatic lens in the center of the platform is essentially dead relative to diagenesis. It becomes diagenetically active at the platform margins where the waters have more contact with soil and atmospheric CO_2 (Moore, 1989).

Flank margin cavernous porosity will be developed along the entire periphery of the

252 SUMMARY OF EARLY DIAGENESIS AND POROSITY MODIFICATION

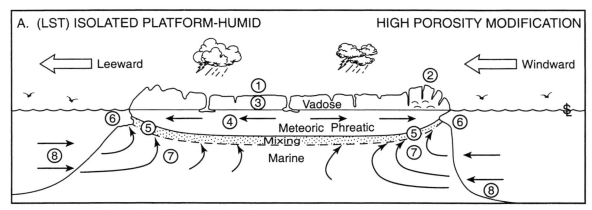

1. Epikarst to deep karst depending on duration of exposure at the sequence boundary.
2. Solution enhanced fractures parallel to windward shelf margin related to margin collapse.
3. Preservation of intergranular porosity in vadose zone, permeability reduced, vertical conductivity may be enhanced along windward shelf margin by fractures.
4. Thick meteoric lens floating on marine water. Slow water flux toward platform margin with little lateral transport of $CaCO_3$. Moldic to vuggy porosity along water table with cements precipitated in adjacent pores. There is degradation of permeability and little gain or loss in porosity. No diagenetic/porosity gradients. Phreatic caves at the water table.
5. Major secondary porosity development and minor dolomitization in the mixing zone.
6. Major coastal flank cavern porosity.
7. Pervasive marine-water dolomitization of shelf margin driven by hydrodynamics of the mixing zone.
8. Aragonite dissolution, calcite precipitation below aragonite lysocline. Net loss in porosity. Marine water influx into platform driven by mixing zone hydrodynamics and geothermal convection.

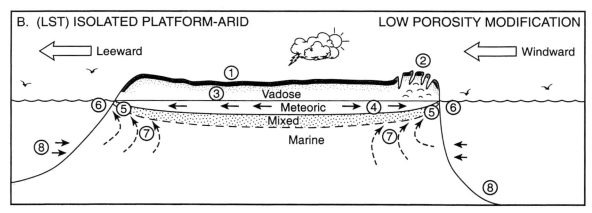

1. Impermeable caliche crusts, minor karsting.
2. Fractures parallel to windward shelf margin related to margin collapse.
3. Preservation of unstable mineralogy, minor cementation, high porosity and permeability. Shelf margin fracture zone increases vertical conductivity.
4. Very thin meteoric lens. Much reduced meteoric flux to platform margins. Minor water table caves. Minor porosity/permeability modification. Unstable mineralogy preserved. No diagenetic/porosity gradients.
5. Minor dissolution and mixed water dolomitization.
6. Minor flank cave development
7. Minor marine water dolomitization, mostly centered around margins.
8. Aragonite dissolution and calcite precipitation below aragonite lysocline driven by geothermal convection. Net porosity loss.

Figure 8.4. LST diagenetic/porosity model for the isolated platform under humid climatic conditions (A) and under arid climatic conditions (B).

platform. The solution-enhanced margin collapse jointing that increases vertical conductivity through the vadose zone will preferentially form on the reef-bound windward shelf margin where slopes are steeper.

The entrainment of marine phreatic water and its movement through the platform is controlled by the flux of meteoric water toward the platform margins. The higher the flux of meteoric waters the more marine water flows into and across the platform from the entire periphery of the platform, pervasively dolomitizing the core of the platform beneath the mixing zone (Figs.5.42 and 5.44C). In the aragonite lysocline, the net flow of marine water into the platform driven by the mixing zone as well as geothermal convection enables massive dissolution of aragonite and precipitation of marine calcite cements along the margin. Thus a net drop in porosity will occur along the margin.

The diagenesis/porosity model for an isolated platform under an arid climate is shown in Figure 8.4B. The key word for this model is *reduction*, relative to the (Fig. 8.4A) humid model : reduction in volume of the meteoric lens, reduction in porosity-modifying diagenesis associated with the lens proper, as well as a reduction in diagenesis associated with the activity of the mixing zones at the platform margins. A reduction in the rate of meteoric water flux toward the platform margins results in less movement in the underlying marine phreatic zone and consequently much less platform interior marine water dolomitization. Aragonite lysocline-related diagenesis with porosity reduction might continue if geothermal convection is present in the platform.

RESERVOIR DIAGENESIS AND POROSITY EVOLUTION DURING 3RD ORDER SEA LEVEL RISES (TST)

Introduction

Diagenesis during a rising sea level should be dominated by marine water. The sedimentological setting, whether the site is a ramp, land-tied shelf, or isolated platform, is critical to the style of diagenesis and resulting porosity modification that might ensue. Again climate plays a major role in modifying the water responsible for diagenesis. During transgression under arid conditions, marine waters are evaporated and may become a very active diagenetic fluid with major consequences for porosity modification. In the following discussion, we will erect diagenesis/porosity models for each of the three end member sedimentological settings under both humid and arid climatic conditions.

Diagenesis/porosity model for a ramp in a rising sea level (TST)

Conceptually, if we have a constant 3rd order sea level rise across a ramp, one would expect high-energy shoreface conditions to sweep across the unconformity marking the previous depositional sequence. The low-energy, fine-grained off shore deposits would progressively onlap the sequence boundary, effectively sealing a portion of the unconformity and highstand deposits of the previous sequence. During transgression and onlap then, we can effectively produce a confined gravity-driven aquifer in the highstand deposits of the previous sequence (Fig. 8.5). However, we know that there are higher order glacio-eustatic cycles that would cause a 3rd order rise to consist of a number of short-term rises and highstands. The sedimentologic pattern, then, is an on-lapping series of flooding surfaces marked by impermeable low-energy sediments followed by a relatively thin, high-

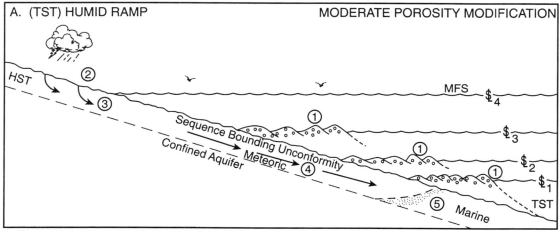

1. Floating meteoric lenses. Minor dissolution and cementation along water table. Little gain in porosity but a decrease in permeability. Minor mixed water dolomitization and beach rock cementation.
2. Recharge area for confined aquifer progressively reduced during sea level rise.
3. Stabilization of residual aragonite is by dissolution with transport of significant $CaCO_3$ in a down dip direction. Moldic and vuggy pore types give a net gain in porosity. Water flux and hence diagenesis reduced as transgression progresses.
4. Precipitation of calcite cement with a net loss in porosity. Strong diagenetic/porosity/geochemical gradient.
5. Little diagenesis or porosity modification in mixing zone because of isolation from atmosphere and CO_2.

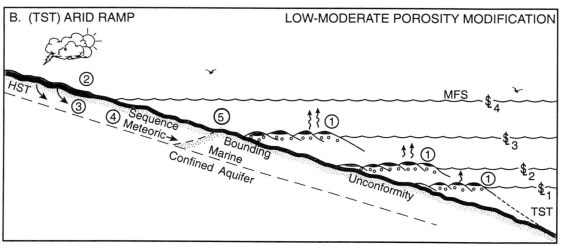

1. Small to ephemeral floating meteoric lenses. Little porosity modification associated with the lenses. Minor reflux dolomitization beneath ephemeral coastal salinas. Some porosity destruction by beach rock centered on sea level.
2. Impermeable caliche crusts further reduce minor meteoric recharge.
3-4. Low water flux reduces porosity modifying diagenesis and shuts down lateral transport of $CaCO_3$. No distinct diagenetic/porosity/geochemical gradients are established.
5. Marine-meteoric mixing zone moves progressively up-dip with transgression.

Figure 8.5. TST diagenetic/porosity model for the ramp under humid climatic conditions (A) and arid conditions (B).

energy, prograding shoreline which in turn is overwhelmed by the next high-frequency rise (Figs. 2.5 and 8.5). As the rate of sea level rise accelerates toward the maximum flooding surface, the low-energy, impermeable sediments become dominant.

Figure 8.5A illustrates the diagenesis/porosity model for a rising sea level transgressing across a ramp surface (sequence bounding unconformity of the previous depositional sequence) under humid climatic conditions. The prograding shoreline complex developed during the still stand of each high-frequency cycle may develop a floating meteoric lens. Stabilization of aragonite along the water table by dissolution and precipitation of calcite will impact permeability, but not porosity. There may be minor mixed water dolomitization and minor beach rock cementation centered on sea level. Porosity modification will be minor.

The area of up-dip meteoric recharge for the confined aquifer developing in the highstand of the previous depositional sequence will be progressively reduced as transgression proceeds. In the recharge area, any residual aragonite is quickly dissolved; and because of high water flux, there may be a considerable transport of $CaCO_3$ toward the sea. This results in a net porosity gain in the recharge area and a net porosity loss by calcite cementation in the down flow area (Fig. 7.26). There should be a distinct diagenesis/porosity/geochemical gradient established, extending from the recharge area to the mixing zone (Fig. 7.27). The mixing zone in the confined aquifer is basically inactive because it is isolated from the atmosphere and any source of CO_2. If the meteoric recharge and subsidence is great enough, the mixing zone may actually move down dip during transgression.

The diagenesis/porosity model for a rising sea level transgressing across a ramp under arid climatic conditions is shown in Figure 8.5B. The floating meteoric lenses in the high-frequency prograding shoreline complexes will be small and even ephemeral, resulting in little porosity modification. Porosity destruction by beach rock will be common, although porosity reduced zones will be relatively thin and centered on sea level. Minor reflux dolomitization may occur under ephemeral coastal salinas.

Impermeable caliche crusts developed on the sequence bounding unconformity will further reduce low meteoric recharge into the confined aquifer. Porosity-modifying diagenesis in the aquifer (both cementation and solution) is greatly slowed by reduced recharge, resulting in a lack of diagenesis/porosity/geochemical gradients and preservation of aragonite. The mixing zone in the aquifer will move progressively up dip with marine transgression.

Diagenesis/porosity model for a rimmed shelf during a rising sea level (TST)

As sea level rises on a rimmed carbonate shelf, it first encounters the raised rim of the previous sedimentary sequence, which provides an ideal substrate for reef growth. New reef growth on the previous rim tracks further sea level rise forming a new rimmed shelf margin (Fig. 8.6). In a humid setting, the lagoon behind the rim is progressively filled with mud-dominated in situ sediments progressively sealing the surface of the HST of the previous sequence (Fig. 1.14). A confined aquifer much like that seen during transgression over a ramp is formed (Fig. 8.6). The diagenesis/porosity model for this setting is shown in Figure 8.6A.

In the aggradational shelf margin reef complex, most porosity is open framework degraded by some internal sediment (Fig. 5.11F). Porosity destruction by marine cementation is concentrated on the seaward margin. However, marine cementation may be minor because of rapid reef growth leading to

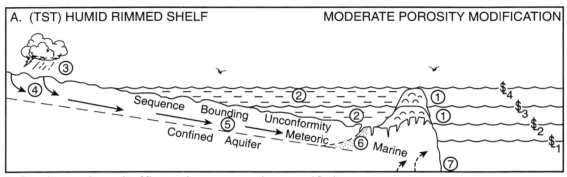

1. Open framework porosity. Minor marine cements on the seaward flank.
2. Mud dominated biotic sediments formed in situ in lagoon. No porosity modifying diagenesis.
3. Recharge area for confined aquifer progressively reduced during sea level rise.
4. Stabilization of residual aragonite is by dissolution with transport of significant $CaCO_3$ in a down dip direction. Moldic and vuggy pore types give a net gain in porosity. Water flux and hence diagenesis reduced as transgression progresses.
5. Precipitation of calcite cement with a net loss in porosity. Strong diagenetic/porosity/geochemical gradient.
6. Little diagenesis or porosity modification in mixing zone because of isolation from atmosphere and CO_2.
7. Minor marine water dolomitization. Aragonite dissolution, calcite precipitation below aragonite lysocline. Net loss in porosity. Marine water flux into shelf driven by geothermal convection. This zone may rise up shelf during transgression.

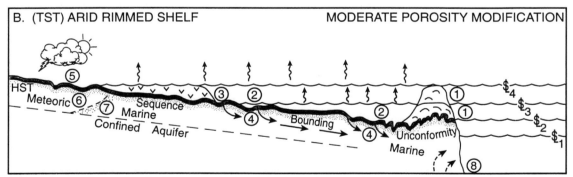

1. Open framework porosity. Significant porosity destruction by marine cements on seaward margin.
2. Hypersaline lagoonal waters depress biotic carbonate production. Significant abiotic carbonate production is concentrated just behind the shelf margin.
3. Evaporite precipitation as gypsum.
4. Reflux dolomitization of the HST of the previous depositional sequence.
5. Impermeable caliche crusts further reduce minor meteoric recharge.
6. Low water flux reduces porosity modifying diagenesis and shuts down lateral transport of $CaCO_3$. No distinct diagenetic/porosity/geochemical gradients are established.
7. Marine-meteoric mixing zone moves progressively up-dip with transgression.
8. Minor marine water dolomitization. Aragonite dissolution, calcite precipitation below aragonite lysocline. Net loss in porosity. Marine water flux into shelf driven by geothermal convection. This zone may rise up shelf during transgression.

Figure 8.6. TST diagenetic/porosity model for the rimmed shelf under humid climatic conditions (A) and under arid climatic conditions (B).

a reduction in time that the substrate is in contact with supersaturated marine waters. *This is a good setting for major framework porosity preservation in a shelf margin reef.* There is little diagenetic porosity modification in the fine-grained lagoonal sediments. Most of their primary porosity will be lost during early compaction (Moore, 1989).

The recharge area for the confined aquifer developed in the previous sedimentary sequence will be progressively reduced during the sea level rise. Dissolution will quickly destroy any residual aragonite in the recharge area. Moldic and vuggy porosity will be dominant. The resulting $CaCO_3$ will be transported down dip with a net porosity gain in the recharge area (Fig. 7.26). Calcite cements will precipitate down dip with a net loss in porosity. Strong diagenesis/porosity/geochemical gradients will develop from the recharge area down dip (Fig. 7.27). Little diagenesis or porosity modification takes place in the mixing zone. The mixing zone may move down dip during the early stages of transgression if a large volume of meteoric water is provided to the recharge area. The aragonite lysocline may transgress during sea level rise moving the locus of deep marine shelf margin diagenesis/porosity up-slope.

Figure 8.6B shows the diagenesis/porosity model for a rimmed shelf during a rising sea level under arid climatic conditions. Because of potential higher carbonate saturation states of marine waters, there may be significant destruction of porosity due to marine cementation along the shelf margin. Higher salinity of lagoon waters driven by evaporation will slow or stop biotic carbonate production while abiotic carbonates are produced and deposited near the shelf margin (Moore, 1989).

As the sea level rise proceeds, the lagoonal waters become progressively more saline as the result of evaporation until gypsum begins to be deposited in the shelf interior. Heavy post-gypsum brines flow under more normal marine waters and reflux down into and dolomitize the highstand systems tract of the previous sequence (Figs. 6.20 and 6.36). The dolomitized HST of the previous sequence will have enhanced porosity as well as permeability because the refluxing waters will be Mg rich and Ca and CO_3 poor (Chapter 6 and Moore, 1989). In order to dolomitize these rocks, calcite and aragonite must be dissolved before dolomitization can take place. Gypsum will ultimately fill the lagoon and shut down reflux as the rate of formation of new accommodation space slows at the start of the highstand (at the maximum flooding surface, MFS).

Because of reduced meteoric recharge there will be a small, even ephemeral, confined meteoric aquifer developed in the HST of the previous sequence with little porosity-modifying diagenesis. The mixing zone moves progressively up-dip during the sea level rise. Deep marine shelf margin diagenesis will be identical to that seen in the humid model.

Diagenesis/porosity model of a rimmed isolated carbonate platform during sea level rise (TST)

The interior of an isolated carbonate platform during a sea level rise will generally be open to normal marine waters, particularly on the leeward flank where sediments are dominated by subtidal grainstones. On the windward side of the platform, however, framework reefs will generally track the sea level rise, giving the platform a decidedly asymmetrical aspect (Fig. 8.7).

Figure 8.7A is a diagenesis/porosity model for an isolated carbonate platform during rising sea level under humid conditions. There is little meteoric water influence on the platform, other than the likelihood that salinity of the waters in the platform interior are maintained near sea water values by rainwater replacing

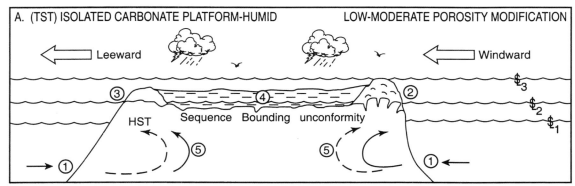

1. Aragonite dissolution and calcite cementation below aragonite lysocline driven by thermal convection. Diagenetic zone may rise up shelf during transgression.
2. Open framework porosity on windward flank. Minor porosity destruction by marine cementation.
3. Intergranular porosity in grainstones on leeward margin. Minor marine cements in hardgrounds. While there is no impact on porosity, vertical conductivity may be affected.
4. Significant biotic in situ carbonate production in the platform interior dominated by muds. No porosity modifying diagenesis.
5. Minor dolomitization of platform margins by marine waters driven by thermal convection. Net loss of porosity because Mg, Ca and CO_3 are imported with the marine water and dolomite generally precipitated as cement.

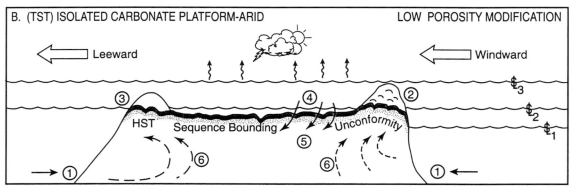

1. Aragonite dissolution and calcite cementation below aragonite lysocline driven by thermal convection. Diagenetic zone may rise up shelf during transgression.
2. Open framework porosity on windward flank. Minor porosity destruction by marine cementation.
3. Intergranular porosity in grainstones on leeward margin. Some marine cements in hardgrounds. While there is little impact on porosity, vertical conductivity may be affected.
4. Salinity increases progressively into the platform interior, particularly in the sheltered zone behind the windward margin. Heavier, more saline waters reflux into HST of previous sequence.
5. Minor reflux dolomitization. Net loss in porosity because Mg, Ca and CO_3 are imported with the marine water and dolomite generally precipitated as cement.
6. Minor dolomitization of platform margins by marine waters driven by thermal convection. Net loss of porosity because Mg, Ca and CO_3 are imported with the marine water and dolomite generally precipitated as cement.

Figure 8.7. TST diagenetic/porosity model for the isolated platform under humid climatic conditions (A) and under arid climatic conditions (B).

evaporation volume loss. On the deep platform flanks below the aragonite lysocline, there is minor aragonite dissolution and calcite cementation. This diagenetic zone may climb the margin during sea level rise. Water flux into the platform may be driven by thermal convection if the platform is large enough. Net porosity loss is to be expected. This same thermal convection may cause marine water dolomitization along the flanks of the platform; and, if efficient enough, may even affect the platform core. We might expect to see a net porosity reduction as a result of dolomitization, since the marine waters are importing Mg as well as Ca and CO_3 and the dolomite may be precipitated as cement.

On the windward side, there will be some porosity reduction by marine cements, but the rising sea level will preclude massive marine cementation. On the leeward flank, the active grainstone shoals will exhibit minor scattered marine hardgrounds; which, while not impacting net porosity, may influence vertical conductivity. The platform interior is dominated by fine-grained biotic carbonates and little porosity-modifying diagenesis occurs.

Under arid conditions the model for an isolated carbonate platform will be much the same as a humid platform (Fig. 8.7B). The main area of departure will be the situation in the platform interior where marine waters transiting from the leeward margin are progressively evaporated to higher salinities. There should be less fresh marine water entering the platform from the windward margin because framework reefs accrete to sea level. While evaporites are seldom precipitated, the evaporated waters are heavy relative to normal seawater, and can reflux into and dolomitize the HST of the sequence below.

There will probably be a net loss of porosity because again the dolomites will generally be precipitated as cements.

RESERVOIR DIAGENESIS AND POROSITY EVOLUTION DURING 3RD ORDER SEA LEVEL HIGHSTANDS (HST)

Introduction

Diagenesis and porosity modification during a sea level highstand are extremely variable and strongly influenced by sedimentological and climatic setting as well as the rate at which new accommodation space is being generated. During the lowstand, little new accommodation space is generated except on the ramp. During a rising sea level the generation of new accommodation space accelerates leading to strong onlap onto the ramp and rapid marginal accretion on the shelf and platform. In the case of the highstand, the rate of formation of new accommodation space begins to decelerate, leading to strong progradation on the ramp (if sediment production is sufficient) and slow accretion on steep shelf and platform margins. These are all-important factors in determining patterns of diagenesis and porosity modification. In the following discussion we will again use our three end-member settings, the ramp, land-tied rimmed shelf, and isolated platform. It should again be emphasized that these settings are end-members and that one may grade into another.

Diagenesis/porosity model for a ramp during a sea level highstand (HST)

In a ramp setting the rate of formation of

new accommodation space decelerates after maximum flooding (MFS) and initiation of sea level highstand. If there is sufficient carbonate production, the high-energy shoreline complex of the ramp will be strongly progradational, ultimately evolving into a shelf (Wilson, 1975) (Fig. 8.8).

Figure 8.8A illustrates the diagenesis/porosity model for a ramp during a sea level highstand under humid climatic conditions. During the maximum flood, at the point where the highstand commences, the shoreline-

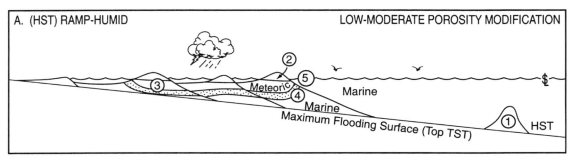

1. Microbial mound with organic processes driving dissolution and precipitation. Net relative increase of both primary and secondary porosity.
2. Vadose meniscus cement, minor dissolution and no net gain in porosity. Permeability may decrease.
3. Overlapping meteoric lenses floating on marine water. Dissolution of aragonite and precipitation of calcite centered on the water table. No net gain in porosity. Permeability is impacted at the water table. Some aragonite preserved in the lower phreatic.
4. Minor dissolution of aragonite and minor dolomitization in the active mixing zone at shoreline. No diagenesis in the mixing zone beneath the lenses.
5. Minor beach rock cementation at the shoreline.

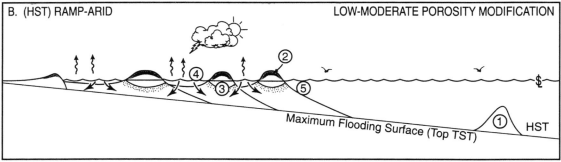

1. Microbial mound with organic processes driving dissolution and precipitation. Net relative increase of both primary and secondary porosity.
2. Caliche crusts formed on coastal dunes. Vadose meniscus cement minor, dissolution. No net gain in porosity, permeability reduced.
3. Ephemeral meteoric lenses isolated by saline ponds. Minor dissolution and precipitation at the water table with little porosity modification. There will be significant preservation of aragonite.
4. Heavy brines reflux downward from coastal salinas and dolomitize underlying coastal sediments. Porosity may be enhanced if evaporites are present.
5. Some porosity destruction by beach rock cementation along the shoreline centered on sea level. Major effect will be reduced vertical conductivity.

Figure 8.8. HST diagenetic/porosity model for the ramp under humid climatic conditions (A) and arid conditions (B).

associated siliciclastics and sources of fine-grained terrigenous sediment are at their maximum distance from the deeper waters of the slope and basin. The slope and basin receive minimal sediment input stress permitting deep-water microbial mounds to form. Mound building is perhaps fueled by hydrocarbon or hydrothermal fluid leakage from the deep basin (Figs. 5.26 and 5.36). Within these mounds, all diagenetic products and processes are ultimately fueled by microbial activity, including porosity-enhancing dissolution as well as porosity-destructive marine cementation (Figs. 5.30 and 5.31). Once formed, these mounds will continue to grow until buried by carbonate sediment from the prograding ramp.

The strongly progradational shoreline complex of the ramp highstand systems tract will support a series of progressively overlapping floating fresh water lenses (Fig. 7.21). Barrier size and rate of meteoric recharge (Table 7.1) will determine lense size. The lenses can be permanent. A low coastal dune complex may be present with vadose processes and products dominant (Figs. 7.18 and 7.41). Porosity modification in the vadose zone will be minor while permeability might be impacted by meniscus cement. Most porosity modification will center on the water table. There will be little net porosity gain, but permeability could decrease significantly. Some moldic porosity may be developed, but primary intergranular porosity will be dominant. Aragonite and magnesian calcite may be stabilized to calcite through the entire meteoric lens if progradation is slow enough (Vacher and Quinn, 1997). There may be minor dissolution and dolomitization in the seaward margins of the mixing zone. This zone may move seaward with progradation. Porosity destruction by beach rock cementation will be minor on the seaward edges of the shoreline complex and not present where meteoric water mixes with marine water at the shoreline (Chapter 5). The mixing zone beneath the meteoric lens will be essentially inactive because of isolation from the atmosphere and CO_2.

The major difference between the arid ramp and the humid ramp during a sea level highstand relates to the volume of meteoric water available and evaporation of seawater in salinas between coastal dune fields (Fig. 8.8B). While caliche crusts will develop on the crests of coastal dunes, the ponds between these dunes become hypersaline by evaporation. Evaporites may or may not form depending on climate. The heavy evaporated seawater refluxes down and dolomitizes the underlying carbonate coastal sands. If gypsum forms in the overlying pond the porosity in the underlying sediments may be enhanced by dolomitization because the Ca and CO_3 necessary for dolomitization must be furnished by dissolution of the underlying sediments. If, however, gypsum does not form in the pond above, all Ca and CO_3 necessary for dolomitization are transported into the underlying sediment and the dolomite is precipitated as dolomite cement reducing porosity.

Reduced meteoric recharge results in thin to ephemeral meteoric lenses. The salinas between the dune fields will fragment and separate the individual lenses (Whitaker and Smart, 1997). Minor solution of aragonite and precipitation of calcite with minor impact on porosity will be centered on the water table of each lens. Aragonite preservation is common. The mixing zone will be reduced and diagenetic

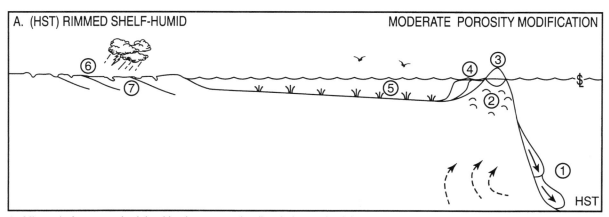

1. Minor platform margin dolomitization, aragonite dissolution and calcite cementation. May be driven by thermal convection. Fine mud-dominated turbidites. Diagenesis rarely alters porosity.
2. Extensive porosity destructive marine cementation.
3. Local floating meteoric lenses. Porosity modification centered on water table. No net gain in porosity. Minor flank caves and mixing zone dolomitization.
4. Island-tied tidal flats/sabkhas. Minor epikarst and dolomitization. Little porosity modification.
5. Mud-dominated in situ carbonate production. No porosity modifying diagenesis.
6. Minor surficial epikarst. Becomes more intense landward.
7. Upper layers of coastal zone stabilized to calcite without appreciable impact on porosity. Minor dolomitization.

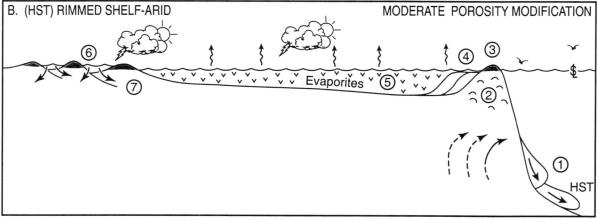

1. Minor platform margin dolomitization, aragonite dissolution and calcite cementation. May be driven by thermal convection. Fine mud-dominated turbidites. Diagenesis rarely alters porosity.
2. Extensive porosity destructive marine cementation.
3. Caliche crusts develop on island tied sabkhas. Dolomitization beneath surface of tidal flats and coastal salinas. May be porosity enhancing or destructive.
4. Shelf margin reefs and islands form barriers to inflow of normal marine water during high frequency low stands.
5. Evaporative drawdown leads to precipitation of gypsum and halite across lagoon. Evaporites seal lagoon floor precluding reflux and dolomitization of sediments below lagoon.
6. Low permeability caliche crusts occur inland from the coastal zone.
7. Evaporative brine reflux dolomitization associated with sabkhas and coastal salinas. Porosity/permeability may be enhanced depending on presence/absence of evaporites.

Figure 8.9. HST diagenetic/porosity model for the rimmed shelf under humid climatic conditions (A) and under arid climatic conditions (B).

processes muted. Porosity destruction by beach rock cementation will be common along the shoreline. Major impact will be reduced vertical conductivity.

Diagenesis/porosity model for a rimmed shelf during a sea level highstand (HST)

During the late highstand the rate of generation of new accommodation space slows perceptibly (Chapter 2). This slow-down allows the accreting shelf margin reef to more easily track sea level and to form storm-related sandy islands and occasionally to form a lagoonal barrier during periods of high-frequency sea level lowstands (Fig. 8.9). In the shelf interior the reduced rate of accommodation space formation allows the rapid progradation of mud-dominated coastal tidal flats (Fig. 6.17), (Tucker and Wright, 1990). The sediment for the coastal zone is sourced from biotic carbonate production in the lagoon (Fig. 8.9).

Given this setting Figure 8.9A illustrates the diagenesis/porosity model for a rimmed shelf during a sea level highstand under humid climatic conditions. During the highstand the adjacent deep-water slopes receive fine-grained turbidites transported out of the lagoon and off the shelf. If these deposits intersect the aragonite lysocline, there should be a net loss of aragonite and a temporary gain in porosity. Considerable porosity may be lost to calcite cementation if the turbidite surface is exposed to marine waters for a significant time. This is not a likely scenario during the highstand because of the high volume of shelf and lagoonal sediments transported onto and across the slope (Figs. 1.11 and 1.12). Some dolomitization of the margin may occur if a thermal convection cell is generated within the shelf. Some porosity loss would be expected.

The slow accretion of shelf margin reefs as they track sea level during the late highstand leads to high volumes of internal sediment and pervasive marine cementation. There is extensive destruction of porosity along the shelf margin. ***Thus highstand reefs that are not dolomitized or leached by meteoric water may be largely tight and poor exploration targets.*** Storm-related sand islands along the margin are associated with the reefs. These islands can support local floating meteoric lenses. Porosity-modifying diagenesis is centered on the water table and no net porosity loss or gain is to be expected. Some minor mixed water dolomitization and dissolution may be found in the shore mixing zone. In the absence of a mixing zone one could expect to see minor beach rock cementation centered on sea level.

Carbonate production in the lagoon is generally biological and mud dominated. There is little porosity-modifying diagenesis except for marine hardgrounds developed on the coarse lagoonal reef aprons immediately behind the shelf margin reefs. Muds from the lagoon source the inner shelf prograding tidal flats. Epikarst may develop on the surface of the tidal flats. Karsting becomes more intense inland where the surface has been exposed longer. Small meteoric lenses can develop within the tidal flat complex but the lack of hydraulic head and low permeability of the sediments restrict the potential for porosity-modifying diagenesis. Some aragonite stabilization to calcite can occur in the meteoric lens. Some minor porosity-neutral dolomitization may be seen if the climatic pattern is tropical (distinct wet and dry seasons).

The major differences between the diagenesis on arid shelves versus humid shelves during sea level highstands centers on evaporation and the lagoon (Fig. 8.9B) The meteoric lenses in the shelf margin islands and reefs become tiny and ephemeral, basically shutting down most porosity-modifying diagenesis. Beach rock becomes more common along the shelf margin shorelines and impacts vertical conductivity (Fig. 5.8A). In the late

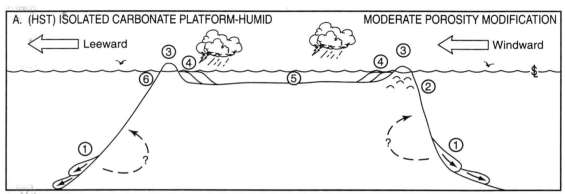

1. Minor platform margin dolomitization, aragonite dissolution and calcite cementation. May be driven by thermal convection. Fine mud-dominated turbidites. Diagenesis rarely alters porosity.
2. Extensive porosity destructive marine cementation.
3. Local floating meteoric lenses. Porosity modification centered on water table. No net gain in porosity. Minor flank caves and mixing zone dolomitization.
4. Island-tied tidal flats/sabkhas. Minor epikarst and dolomitization. Little porosity modification.
5. Mud-dominated in situ carbonate production. No porosity modifying diagenesis.
6. Minor marine cementation as beach rock and marine hardgrounds. Little porosity impact.

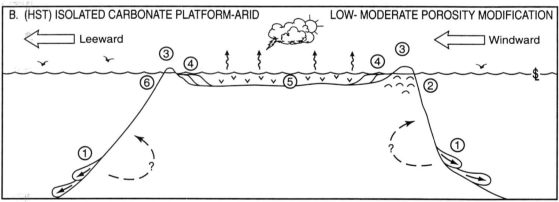

1. Minor platfrom margin dolomitization, aragonite dissolution and calcite cementation. May be driven by thermal convection. Fine mud-dominated turbidites. Diagenesis rarely alters porosity.
2. Extensive porosity destructive marine cementation.
3. Ephemeral meteoric lenses. Little porosity modifying diagenesis except for caliche crusts on the islands.
4. Caliche crusts develop on island tied sabkhas. Dolomitization beneath surface of tidal flats and coastal salinas. May be porosity enhancing or destructive.
5. Shelf margin reefs and islands form barriers to inflow of normal marine water during high frequency low stands. Evaporative drawdown leads to precipitation of gypsum and halite across lagoon. Evaporites seal lagoon floor precluding reflux and dolomitization of sediments below lagoon.
6. Moderate marine cementation as beach rock and marine hardgrounds. Little porosity loss, but vertical conductivity is reduced.

Figure 8.10. HST diagenetic/porosity model for the isolated platform under humid climatic conditions (A) and under arid climatic conditions (B).

highstand, shelf margin reefs and islands may form barriers to free inflow of marine waters into the lagoon and salinity begins to rise in the arid climate. High-frequency lowstands may effectively isolate the lagoon and evaporative drawdown results in widespread gypsum and halite precipitation across the floor of the lagoon (Figs. 6.42 and 6.44). These evaporites effectively seal the lagoon floor, and reflux dolomitization of the sediments beneath the lagoon is precluded. The end result is a relatively thick and extensive lagoonal evaporite deposit (Fig. 6.43).

Progradation of the mud-dominated coastal complex is reduced and ultimately stopped as the source of its sediment is covered by the evaporites. Some brine reflux dolomitization may be seen along the interior coastal margin associated with sabkhas and salinas that parallel the coast. If gypsum and anhydrite are present, dolomitization will be by replacement and porosity/permeability will be enhanced. If no evaporite is present, dolomitization is by cementation and porosity is reduced.

Diagenesis/porosity model for an isolated carbonate platform during a sea level highstand (HST)

The diagenesis/porosity model for an isolated platform during a sea level highstand under humid climatic conditions (Fig. 8.10A) is similar to the model for a land-tied shelf discussed in the last section (Fig. 8.9A). As the rate of generation of new accommodation space slows down in the late highstand, reefs on the windward side of the platform build to sea level with significant porosity destruction by marine cementation. Reef-associated islands can build above sea level, supporting meteoric lenses with porosity-modifying diagenetic processes that are identical to those found on the shelf margin. On the isolated platform, however, there is a leeward flank where high-energy grainstones generally accumulate. During late highstands, islands can often form much as they do on the windward margin with identical diagenetic consequences. Marine cementation, however, is limited to beach rock and marine hardgrounds with little porosity impact.

Fine muds generated in situ in the lagoon can accumulate as small progradational tidal flats on the lagoonward margins of both windward and leeward islands. As in the shelf model, minor epikarst may be found in the interior of the tidal flats with minor dolomitization and little porosity modification.

The arid diagenesis/porosity (HST) model for isolated platforms (Fig. 8.10B) is identical to the (HST) arid shelf model (Fig. 8.9B) except that the shelf interior progradational tidal flats/sabkhas would probably be more extensive than the island-based flats of the platform.

DIAGENESIS AND POROSITY AT THE PARASEQUENCE SCALE

Introduction

Carbonate shelf and platform sequences are generally composed of numerous carbonate cycles of various scales ranging in thickness from 1 meter to 10's of meters. These cycles, or parasequences, start with a distinct deepening event (marine flooding surface) and exhibit a shallowing to shoaling-upward tendency. Exposure surfaces or marine hardgrounds may cap the cycles (Fig. 2.13). As noted earlier, the origin of carbonate

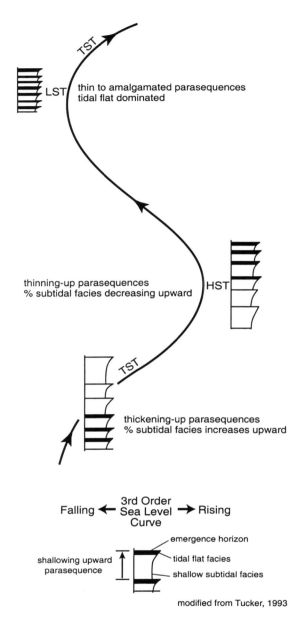

Figure 8.11. Conceptual stacking patterns for parasequences (4th to 5th orders) deposited on a carbonate platform in a background of a 3rd order lower frequency sea level cycle. As accommodation space changes during the low frequency cycle, the thickness of the parasequences change. Individual parasequences generally show a lateral facies variation across a platform as seen in figure 8.12. Used with permission of Blackwell Science.

parasequences is a matter of debate. They may be allocyclic related to orbital forcing in the Milankovitch band (10,000-100,000 years) or autocyclic such as the Ginsburg tidal flat progradation model (Tucker and Wright, 1990).

The diagenesis/porosity models developed for 3rd order sequences in the previous section generally apply to parasequences. During the marine flood back or transgression, marine diagenetic processes tend to dominate. The regressive leg of the cycle with its potential subaerial exposure is generally dominated by meteoric diagenesis. As in the 3rd order cycle, climate plays a major role at the parasequence level, particularly influencing the suite of diagenetic processes produced at the exposure surface.

In a humid climate where meteoric diagenesis is active, the cycle top may exhibit immature paleokarst, vadose cements, and moldic porosity. In an arid climate, where meteoric diagenesis is subdued, the cycle top may contain evaporites, caliche crusts, tee pee structures, pisoids and evaporite-related dolomitization. It should be noted that the period of exposure at a parasequence cycle top is generally quite short as compared to a 3rd order sequence boundary. Therefore, the effects of diagenetic processes at the individual parasequence scale are reduced accordingly.

Cumulative diagenesis associated with parasequence stacking patterns in a 3rd order sequence

Parasequences respond to changes in the rate of generation of accommodation space during the development of a 3rd order sea level cycle (Fig. 8.11). During the transgressive systems tract (TST), the rate of formation of

DIAGENESIS ASSOCIATED WITH PARASEQUENCE STACKING PATTERNS

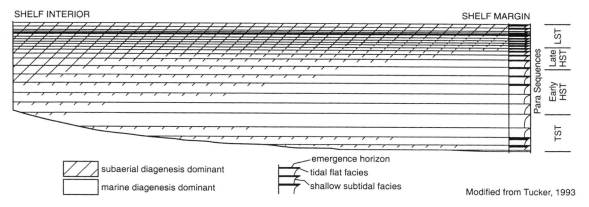

Figure 8.12. Generalized scheme indicating the lateral and vertical variations in carbonate diagenesis for parasequences of different systems tracts, depending of course on climate. Subaerial diagenesis only affects the upper parts of the platform interior parasequences deposited during the 3rd order sea level rise (TST), but it may affect all parasequences deposited during the 3rd order sea level fall (late HST-LST). Parasequences of the LST may be severely affected by subaerial diagenesis. Used with permission of Blackwell Science.

new accommodation space starts slowly and steadily accelerates. In the early TST, parasequences are generally thin with well-developed exposure at cycle tops. As the rate of formation of accommodation increases, parasequences thicken and marine diagenesis, such as the formation of marine hardgrounds, tends to dominate at cycle tops (Fig. 8.11).

During the early highstand, parasequences will tend to be thick and marine dominated. In the late highstand as the rate of formation of new accommodation space decelerates, the trend will be toward thinner parasequences with exposure at parasequence tops (Fig. 8.11).

During the lowstand, parasequences will be very thin or amalgamated into a sequence-bounding unconformity. Exposure-related diagenesis will be dominant (Fig. 8.11).

Figure 8.12 shows a schematic diagram relating 3rd order systems tracts to the lateral and vertical variations of diagenesis associated with parasequences on a carbonate shelf. In the transgressive systems tract, subaerial diagenesis only impacts cycle tops in the early transgression. As the formation rate of accommodation space accelerates, marine diagenesis tends to dominate (Fig. 8.12). In the early highstand only cycle tops in the platform interior undergo exposure diagenesis. As accommodation space is reduced in the late highstand, exposure diagenesis is dominant and all cycle tops show signs of exposure from shelf interior to shelf margin (Fig. 8.12). During the lowstand, as parasequences thin and begin to amalgamate, subaerial diagenesis becomes more intense and tends to be cumulative, impacting previously deposited parasequences. In these cases the entire parasequence exhibits the effects of subaerial diagenesis. This cumulative subaerial diagenesis and resulting porosity modification affects more parasequences in the shelf interior than on the shelf margin (Fig. 8.12).

It is important to note that carbonate reservoir scale architecture, and ultimately reservoir heterogeneity, is controlled by parasequence stacking patterns responding to higher order relative sea level changes.

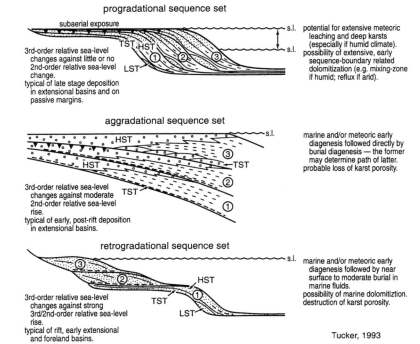

Figure 8.13. Diagenetic models of carbonate shelves/ramps in response to 3rd order sequence stacking patterns in a background of 2nd order, tectonically driven super-sequences. The progradational and retrogradational stacking patterns are illustrated with carbonate shelves, while the aggradational sequence sets shows carbonate ramps. Used with permission of Blackwell Science.

DIAGENESIS AND POROSITY AT THE SUPERSEQUENCE (2ND ORDER) SCALE: SEQUENCE STACKING PATTERNS

Introduction

Our discussion to this point has involved a single depositional sequence and the parasequences sets that comprise the sequence. In the last section it was emphasized that under conditions of low accommodation space generation, older parasequences were affected by the subaerial diagenesis being suffered by younger cycles (cumulative diagenesis). As an example, older cycles could well be dolomitized by brines generated during the deposition and exposure of a younger cycle. We know that 3rd order depositional sequences (.5 to 3my duration) may exhibit a distinct stacking pattern, much like parasequences, in response to longer-period, tectonically-driven relative sea level cycles. These cycles are the 2nd order sequence, or super sequence (3 to 50my duration) (Fig. 2.6). It behooves us to examine the diagenetic interaction and potential cumulative diagenesis of 3rd order sedimentary sequences in the context of the super sequence.

Three end-member sequence stacking patterns can be recognized (Fig. 8.13) in the context of the change in relative sea level of the super sequence. These stacking patterns are similar to parasequence sets and are also termed progradational, aggradational, and retrogradational sequence stacking sets.

Progradational sequence set and diagenesis

When the 3rd order sea level cycle occurs against a backdrop of minimal or no 2nd order relative sea level change, a progradational sequence set is formed such as the Mesozoic of the Neuquen Basin of Argentina (Mitchum and Uliana, 1985). This setting is typical of passive margins and extensional basins (Fig. 8.13).

Where sequence sets are arranged in a progradational or an off-lapping geometry, post-marine, near surface or subaerial diagenesis operates for an extended period of time until a change in the 2nd order relative sea level cycle occurs. In a humid climate, mature paleokarst can be developed involving several sequences. Extensive moldic, vuggy and cavernous porosity may be produced. An example is the Devonian Wabamun Group in the Alberta Sedimentary Basin (Saller and Yaremko, 1994).

In an arid climate, extensive sabkha and reflux dolomitization can extend across several sequences. Amalgamated pisoid and tee pee horizons are common. The zone at the shelf margin where sequences are superimposed ultimately develops the most intense, deep and mature diagenetic overprint and porosity modification.

The sequence set will be capped by an amalgamation of sequence boundaries. The early diagenesis of the sequences within the set will be related to this master sequence boundary and hence will be similar. Mechanical and chemical compaction should be delayed on burial into the subsurface because of intensive early mineral stabilization, cementation and dolomitization. The Upper Permian Capitan and related formations of the Permian Basin are examples (Mutti and Simo, 1991; Mutti and Simo, 1994).

Aggradational sequence set diagenesis

When a 3rd order sequence is formed during a moderate 2nd order sea level rise, such as during a 2nd order early highstand, an aggradational sequence-stacking pattern is produced. This setting is typical of early post-rift deposition in an extensional basin. The Lower Cretaceous of Texas is an example (Bebout et al., 1981; Bebout and Loucks, 1983; Moore, 1989; Moore and Bebout, 1989)(Fig. 13B).

Aggradational sequence sets will undergo early surficial meteoric and marine diagenesis followed immediately by burial into the subsurface. Early diagenesis will, therefore, determine the path of burial diagenesis. This is a setting where confined aquifers will likely be formed in highstand deposits between 3rd order transgressions. Under humid conditions, confined aquifers developed in 3rd order highstands should undergo extensive diagenesis and porosity modification such as the Middle Jurassic shelf margin grainstones of the Paris Basin (Purser, 1985). In this situation subsequent burial will see little mechanical or chemical compaction. Under arid conditions, however, moderate to low diagenetic and porosity modification will lead to high levels of mechanical and chemical compaction such as Tucker and Hird found in Carboniferous ooid grainstones from south Wales (1988).

Retrogradational sequence set and diagenesis

When the 3rd order sequence is formed during a 2nd order relative sea level rise, a retrogradational sequence set is produced such as the Lower and Middle Devonian of the Alberta Sedimentary Basin in Canada (Stoakes et al., 1992). This setting is characteristic of rift, extensional, and foreland basins (Fig. 8.13)

In a retrogradational sequence set, early meteoric and /or marine diagenesis is immediately followed by burial in marine pore fluids. Exposure times along sequence-bounding unconformities in this setting are relatively short, hence accompanying meteoric humid or arid diagenesis related to the unconformity is not intense. Although cementation and moldic porosity will undoubtedly be present, primary porosity may well be dominant. While early cementation will no doubt delay compaction, the possibility of preservation of unstable mineralogy into the subsurface because of superficial early diagenesis may lead to early chemical compaction. This said, the retrogradational sequence set seems to be the ideal setting for the preservation of primary porosity into the subsurface.

When the 3rd order transgressive systems tract coincides with a 2nd order sea level rise, a large volume of new accommodation space is generated. There may be two important consequences. If the climate is arid and the shelf is rimmed with reefs or high-energy grainstones, we have a setting where large volumes of lagoonal evaporites might be deposited and the highstand systems tract of the previous sequence is dolomitized by evaporative reflux. The Upper Jurassic of east Texas, USA, is an example (Moore, 1989, Chapter 10).

Secondly, as the rate of 2nd order sea level rise accelerates, earlier deposited 3rd order sequences could be overlain by deeper water, organic-rich condensed sections, forming seals as well as depositing potential source rocks for subjacent reservoirs. Important examples include major hydrocarbon provinces in mid-Upper Silurian (Europe and USA), mid-Upper Devonian (Canada) and mid-Upper Jurassic (Middle East) carbonate platforms, all of which are periods of 2nd order sea level rise (Tucker, 1993).

As Tucker (1993) so aptly wrote, *"In terms of reservoir, seal, source rock and primary porosity retention, carbonate platforms formed during second-order relative sea-level rises have the highest hydrocarbon potential."*

DIAGENESIS AND POROSITY AT THE 1ST ORDER SCALE: ICEHOUSE VERSUS GREENHOUSE

Introduction

Figure 8.14 shows two versions of the long-term (1st Order) relative sea level curve for the Phanerozoic. Both curves have two distinct relative sea level highstands centered on the Silurian and the middle Mesozoic (the youngest Vail highstand is, however, offset to the Tertiary-Cretaceous boundary). Long-term Phanerozoic climate patterns (cold versus warm) generally follow the relative sea level curve with warm greenhouse conditions occurring during times of highstands and cold icehouse conditions during times of relative sea level lowstands (with the exception of mid Ordovician and Silurian, Fig. 8.14). In addition to a causal pattern between long-term climate and relative sea level, long-term changes in the mineralogy of marine abiotic carbonates may also parallel the Phanerozoic climate and sea level curves (Fig. 8.14).

In the paragraphs that follow we will discuss the impact of temporal changes in marine abiotic carbonate mineralogy on long-term diagenetic and porosity trends. Finally, we will explore the differences in 3rd order sedimentary sequence architectures and their porosity evolution as a function of long-term climatic trends (icehouse versus greenhouse).

Long-term temporal changes in carbonate mineralogy: impact on diagenesis and porosity evolution

The carbonate mineralogic control over ultimate diagenetic pathways and porosity evolution was emphasized in Chapters 3 and 7. The mineralogy of abiotic marine carbonate components such as ooids, marine cements, and chemically precipitated mud has changed through the Phanerozoic (Fig. 8.14)(see Chapter 3 for an in-depth discussion). In general terms, marine abiotic components have a calcite composition during greenhouse and high relative sea level periods and are composed of aragonite and magnesian calcite during icehouse and low relative sea level conditions.

As noted previously, aragonite is considerably more soluble than calcite and if flushed with meteoric water will generally dissolve and produce a like-volume of calcite cement. The diagenetic potential of aragonite dominated sediments is high, and chances of significant early surficial porosity modification is good. During greenhouse periods, however, ooids would be composed of calcite, and meteoric water flushing would produce little moldic porosity and cement, leading to primary intergranular porosity preservation into the subsurface. This scenario is certainly most applicable to ooid grainstone reservoirs such as the Jurassic Arab sequence in the Middle East, the Jurassic reservoirs of the Paris Basin,

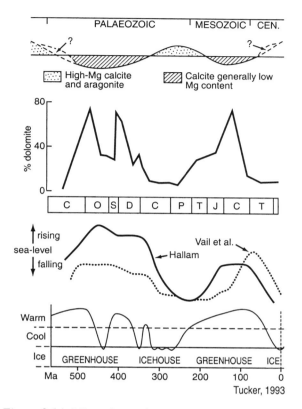

Figure 8.14. *Mineralogy of marine abiotic precipitates (ooids, cements and muds) and dolomite abundance through the Phanerozoic compared with the 1st order sea level curve and icehouse greenhouse conditions. Used with permission of Blackwell Science.*

and the Jurassic Smackover reservoirs of the Gulf of Mexico.

If greenhouse seas are saturated or undersaturated with respect to aragonite and supersaturated with respect to calcite, it follows that aragonite bioclasts may well undergo dissolution under shallow marine conditions mimicking fabrics and textures of meteoric diagenesis. While some workers have reported shallow marine dissolution of aragonite (Chapter 5) the scientific community has given little consideration to this potential problem.

The mineralogy of chemically precipitated muds (such as the aragonite muds of the

Bahamas) should track the mineralogy of ooids and marine cements. During warm greenhouse conditions, chemically precipitated marine muds should be dominantly calcite. During burial into the subsurface, individual calcite crystals will resist the onset of chemical compaction because of stable mineralogy and maintain porosities near 40%. Permeabilities, of course, will be reduced. The parallels with the burial history of chalks are striking. Lower Cretaceous reservoirs of the Middle East (Shuaiba Formation) and the Gulf Coast (Edwards Formation) are noted for the presence of extensive microporosity which impacts reservoir development (Moshier, 1989b; Moshier, 1989a). Aragonite muds, however, because of higher solubility and larger surface areas, are highly susceptible to diagenesis and quickly stabilize into an interlocking calcite mosaic, losing all porosity and permeability.

The volume of dolomite in the sedimentary record apparently also varies through time in concert with the 1st order relative sea level curve and the long-term climatic cycle. Greenhouse periods are seemingly times of dolomitization while icehouse conditions see significantly lower volumes of dolomitized sequences. Conventional wisdom suggests that higher sea levels during greenhouse times provide more opportunity for marine waters to interact with sediments and limestones, deliver more Mg, and ultimately dolomitize (Hardie, 1987). Another factor may be thermodynamic. During greenhouse times, surface marine waters are presumably just saturated or undersaturated with aragonite; and only calcite and dolomite can precipitate from marine water under these conditions. It follows that it should be significantly easier to dolomitize sediments and limestones under greenhouse conditions.

Diagenesis along shelf margins should also be impacted. During greenhouse periods, surface seawater will be undersaturated with respect to aragonite and supersaturated with respect to calcite and dolomite. Because of higher P_{CO_2} levels in the atmosphere, the calcite lysocline should be significantly shallower along shelf margins and dolomitization associated with mixing zone hydrology and thermal convection should certainly be easier.

In summary, temporal changes in the chemistry of seawater driven by CO_2 flux have an enormous impact on the early porosity evolution of carbonate reservoirs and control to a great extent the pathways of subsequent burial diagenesis.

The architecture of sedimentary sequences and their diagenesis/porosity evolution as a function of long-term climatic cycles (icehouse versus greenhouse)

There is little doubt that the nature of depositional sequences has varied throughout the Phanerozoic (Tucker, 1993). When one looks at outcrops of Lower Cretaceous platform carbonates the fine meter scale cyclicity recorded in the sequence is striking as is the general lack of deep karst features at the cycle caps. In contrast, Pennsylvanian sequences seem to exhibit fewer, but thicker high-frequency cycles, and the 4th and 5th order cycle caps often show deep karst development. These differences are driven by changes in the amplitude of high-frequency sea level perturbations as a function of ice cap development.

During greenhouse times, the ice caps are small, or absent, and high-frequency sea level

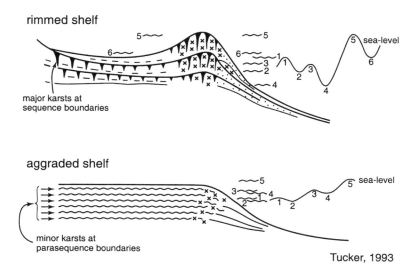

Figure 8.15. Conceptual diagenetic models for a carbonate shelf typical of icehouse and greenhouse conditions. Under humid conditions and high amplitude sea level cycles (icehouse) rimmed shelves are favored and meteoric diagenesis is very intense. Under greenhouse conditions and a humid climate, accretionary shelves are favored and meteoric diagenesis is muted and much less intense than seen on the rimmed icehouse shelves. Used with permission of Blackwell Science.

changes are relatively small, measured at the meter scale, resulting in many thin-stacked parasequences (Fig. 8.15). Even at the 3rd order scale relative sea level changes will be measured in 10's of meters, periods of nondeposition (exposure) will be of shorter duration and the depositional sequence will be relatively thick (50-100m). Sequence-bounding unconformities with significant karstification and subaerial diagenesis are probably less common. During greenhouse times, the meter-scale high-frequency sea level changes, within a framework of larger scale 3rd order change will favor the development of aggraded shelves with little topographic relief from margin to shelf interior The Upper Jurassic of the central Gulf of Mexico is an excellent example (Chapter 10).

During icehouse periods, polar ice caps are large. The waxing and waning of continental ice sheets is controlled by orbital-forcing mechanisms (Milankovitch cycles). The resulting sea level changes are enormous, often exceeding 100m (as witnessed by the well-documented history of the Pleistocene) (Morse and Mackenzie, 1990); and the periodicity of the change is fast (in the range of 4th and 5th order sequences). The amplitudes of these 4th and 5th order glacio-eustatic sea level changes are generally greater than the background 3rd order sea level change responsible for the 3rd order depositional sequence. This results in relatively thick 4th and 5th order sequences capped with distinct unconformities across the shelf, platform, or ramp. These high-frequency sequences are thicker than their greenhouse high-frequency counterparts, but thinner than greenhouse 3rd order sequences (see discussion of Bahama Platform in Chapter 7). In terms of patterns of sedimentation, icehouse conditions and their large amplitude sea level changes seem to promote shelf margin growth leading to the development of rimmed shelves and platforms (Fig. 8.15).

There are major differences in diagenetic trends, and ultimately differences in porosity evolution, between icehouse and greenhouse conditions. The high-amplitude sea level changes associated with 4th and 5th order high-frequency sequences under humid climatic conditions assure that shelves and platforms will undergo deep, intense and overlapping karst development (Fig. 8.15). During lowstands, hydrologic head in open meteoric aquifers will necessarily be large, leading to rapid meteoric water flux and intense porosity modification. Aragonite mineralogy of abiotic components will modify porosity characteristics and its distribution in the aquifers. Under arid conditions at high-amplitude lowstands, the entire platform will be exposed, and most deposition will occur in the basin (reciprocal sedimentation of siliciclastics). Such conditions can result in massive lowstand evaporite deposition in adjacent basins if sea level falls below basin sills. An excellent example is the Permian Salado Evaporite in the Delaware Basin of west Texas.

Under greenhouse conditions in a humid climate, 4th and 5th order parasequences will exhibit only surficial karst development at parasequence caps. The only significant karst will be seen along 3rd order sequence bounding unconformities that result from higher-amplitude sea level lowstands (Fig. 8.15). The only significant open gravity-driven aquifer systems will develop along sequence-bounding unconformities. The higher amplitude of the 3rd order sea level lowstand will generate sufficient hydrologic head to maintain a reasonable water flux toward the basin. Because of lower head, lower water flux, and a more calcite-rich starting sediment, porosity modification during greenhouse times will still be less intense than that seen during icehouse periods.

Finally, during arid conditions, low-amplitude, high-frequency sea level lowstands may drop sea level below the shelf rim but not below the level of the shelf interior lagoonal floor. The result may be to allow marine water reflux through the rim and evaporite deposition across the adjacent lagoon floor. The Jurassic Hith Evaporite of the Middle East may well be an example (Alsharhan and Kendall, 1994). Finally, because of the relatively low amplitude of sea level lowstands in greenhouse periods, sea level will generally not fall below the level of basin sills and lowstand basin evaporites will be less common.

CASE HISTORIES

Introduction

The following section presents two economically important case histories where early diagenesis and porosity evolution was evaluated in a sequence stratigraphic framework.

The first study, published by Art Saller, Tony Dickson, and Fumiaki Matsuda in 1999, concerns Pennsylvanian and Permian shelf reservoirs located on the Central Basin Platform in west Texas. These sequences were deposited under tropical humid conditions during a major icehouse period characterized by high-amplitude, high-frequency sea level cycles. Economic porosity distribution can be tied to the relationship between exposure duration and long-term rate of formation of new accommodation space.

Hebert Eichenseer, Frederic Walgenwitz, and Patrick J. Biondi published the second study, also in 1999. Their work considers an Aptian-Albian mixed siliciclastic/carbonate/evaporite ramp sequence deposited under semi-arid climatic conditions during a greenhouse time characterized by low-amplitude, low-frequency sea level cycles. Porosity distribution was tied to sea level-controlled sedimentologic architecture. The nature of porosity modification and preservation was related to early climate-driven diagenetic processes.

Controls over porosity evolution in Upper Paleozoic shelf limestones, Southwest Andrews area, Central Basin Platform of the Permian Basin, USA: porosity development as a function of subaerial exposure during an icehouse time

The Pennsylvanian-Permian carbonates in the Andrews area on the Central Basin Platform (Fig. 8.16) were deposited during icehouse conditions where the amplitude of sea level changes and their frequencies were high. As discussed earlier in this chapter in the section

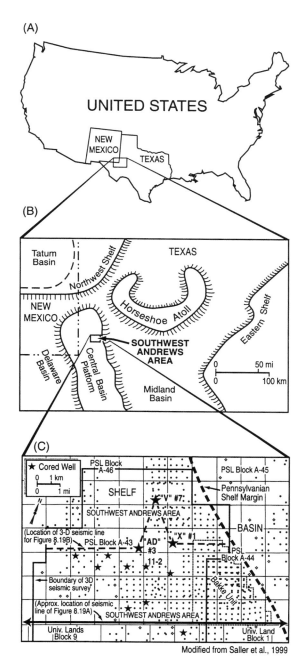

Figure 8.16. A) Location of the Central Basin Platform. B) Location of the Southwest Andrews area and the paleogeography of west Texas during the Late Pennsylvanian. C) Field map of the southwest Andrews map showing location of wells, and seismic lines shown in Figure 8.19. Used with permission of AAPG.

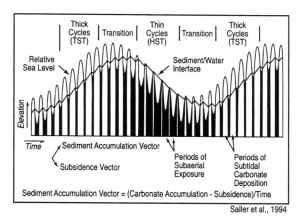

Figure 8.17. Schematic model for high frequency carbonate cycles (4th and 5th orders) developed in a background of low frequency cyclicity (3rd order). Rate of sedimentation is assumed to be constant when sediment-water interface is below sea level. No carbonate sedimentation occurs when sediment-water interface is above sea level. Rate of subsidence is constant. High frequency cycles deposited during 3rd order sea level rises (TST) exhibit long periods of carbonate deposition and short periods of subaerial exposure (shown in black), while those deposited during sea level falls (LST) have short periods of carbonate and long intense periods of subaerial exposure (shown in black). Used with permission of AAPG.

on parasequences and diagenesis during long-term sea level changes, high-frequency parasequence stacking patterns were controlled by the long-term sea level cycle (Fig. 8.11). The effects are amplified during icehouse times as can be seen in Figure 8.17. Here we have a schematic representation (Fisher plot) of cyclic carbonate sedimentation with composite sea level changes involving short-term, high-amplitude fluctuations superimposed on a long-term cycle. During a long-term rising sea level, periods of parasequence subaerial exposure would be short. However, when the rate of long-term sea level rise slows or begins to drop as in the late highstand and early lowstand, periods of parasequence subaerial exposure begin to lengthen and dominate the higher order cycles. The theme of this case history is the effect of length of subaerial exposure at the parasequence level on reservoir porosity modification as a function of the long-term sea level cycle.

The reservoir rocks in the Andrews field area involve some 500m of the Upper Pennsylvanian and Lower Permian Series (Fig. 8.18). The interval is divided into several formations, with the Pennsylvanian Strawn, Canyon and Cisco, and the Permian Wolfcamp formations of economic interest in the field area that is located on the eastern margin of the Central Basin Platform (Fig. 8.16). Limestones punctuated by minor fine-grained mud-dominated clastics comprise the interval of interest. The abundance of shales, lack of evaporites, and paucity of dolomite suggest that the Pennsylvanian/Permian of the Andrews area was deposited under humid climatic conditions near the Pennsylvanian/Permian paleoequator.

Figure 8.19A, based on a 2-D seismic line in the Andrews area, illustrates the development of a westward thinning, eastward prograding shelf from the Strawn to the Wolfcamp. A major flood-back at the end of the Wolfcamp carried the shelf margin some 17km westward onto the Central Basin Platform. Abo/Clear Fork shelves then prograded eastward over the Strawn/Wolfcamp Shelf. Figure 8.19B shows a seismic line across the margin of the Central Basin Platform at the Southwest Andrews Area.

The most porous limestones occur on structural highs near the eastern margin of the field, with stratigraphic traps associated with porosity pinch-outs to the west. Only 5-25% of the 400m reservoir interval contains economic porosity (>4%). Porosity generally

SERIES	MID-CONTINENT STAGE	WEST TEXAS FORMATION
LOWER PERMIAN	Leonardian	Leonard
LOWER PERMIAN	Wolfcampian	Woffcamp
	286 ± 6 m.y.	
PENNSYLVANIAN	Virgilian	Cisco
PENNSYLVANIAN	Missourian	Canyon
	296 ± 5 m.y.	
PENNSYLVANIAN	Des Moinesian	Strawn
PENNSYLVANIAN	Atokan	Atoka
PENNSYLVANIAN	Morrowan	Morrow
	320 ± 10 m.y.	

Saller et al., 1994

Figure 8.18. General stratigraphy of the Southwest Andrews area and comparison of west Texas stratigraphy with mid-continent stratigraphy. Used with permission of AAPG.

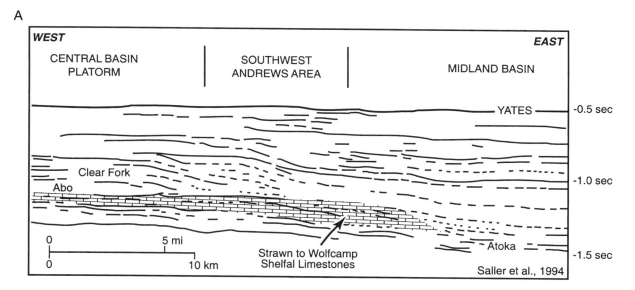
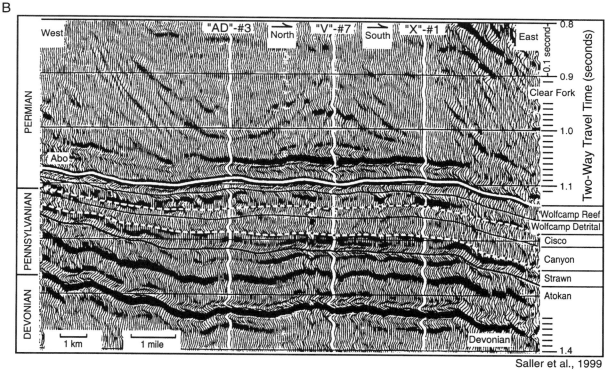

Figure 8.19A. Line drawing illustrating the general stratigraphy derived from an east-west seismic line near the Southwest Andrews area. Position of line shown on Figure 8.16C. Note that the upper Wolfcamp shelf was drowned and the shelf margin stepped back some 17km to its position in the Abo, then prograded eastward toward the Midland Basin during deposition of the Clear Fork and overlying formations. B. Composite seismic line showing detailed stratigraphy of the Southwest Andrews area. Location of line shown on Figure 16C. Used with permission of AAPG.

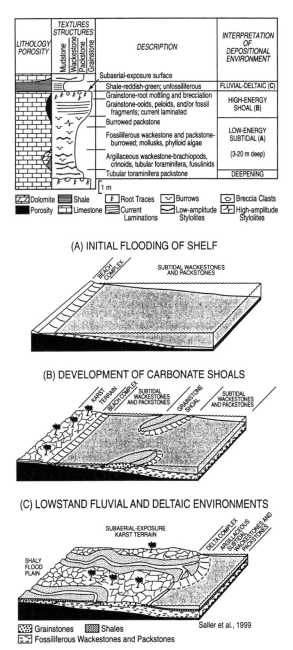

Figure 8.20. Idealized cycle from Southwest Andrews area. A-C Depositional model through a high frequency sea level cycle. A) Rapid sea level rise and flooding of the platform with widespread deposition of fossiliferous wackestones and packstones. B) Stable sea level, development of grainstone shoals. C) Sea level drop exposing the carbonate platform and causing clastics

(Fig. 8.20 cont.) to bypass the platform and be deposited in deeper water. Clastics were re-worked and re-distributed during the subsequent transgression. Used with permission of AAPG.

occurs in 1-6m zones within high-frequency sedimentary cycles. Production extension in Southwest Andrews is in its 47th year with secondary water flood projects just commencing. The most productive wells have produced up to 500,000 barrels of oil on a 40-acre spacing.

Figure 8.20 illustrates an idealized sedimentary cycle formed on the Strawn/Wolfcamp shelf in response to the high-frequency, high-amplitude sea level changes characteristic of this icehouse period. Shelf flooding (Fig. 8.20A) is characterized by argillaceous wackestones with an open marine fauna. As sea level reaches its highstand, the sequence shallows and high-energy ooid and peloidal grainstones are deposited (Fig. 8.20B). During the subsequent lowstand, siliciclastic fluvial deltaic deposits develop over subaerial exposure surfaces marked by root mottling and brecciation (Fig. 8.20C).

Within the Strawn/Wolfcamp interval, these high-frequency cycles are superimposed on long-term changes in the rate of accommodation (Fig. 8.21). Saller et al., (1999) recognized some 87 cycles in this interval and estimated that each cycle represented some 160,000 years. The Strawn and Canyon were periods of high rate of formation of accommodation space (2nd order TST-HST) with relatively thick high-frequency cycles whose characteristics suggest that they spent most of their time accumulating sediment in the subtidal environment. In contrast, the Cisco and Wolfcamp Detrital were times of reduced rate of formation of accommodation space (2nd

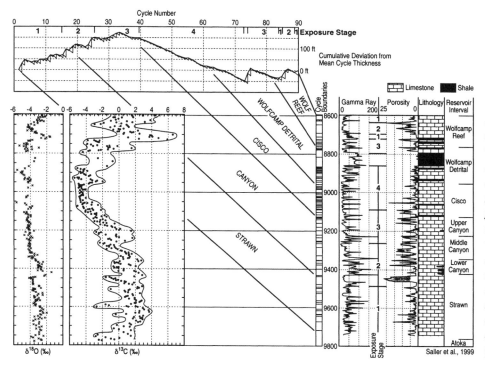

Figure 8.21. Summary of core data from the X-1 well. Location shown on Figure 8.16C. Bulk rock stable carbon and oxygen isotope profiles are plotted at left. Cycle boundaries, gamma-ray wireline log, exposure stage, core porosity (%), lithology, reservoir stratigraphic subdivisions, and a Fischer plot of cycle thickness are also shown. Used with permission of AAPG.

Table 1.1

Number of Cycles	Average Cycle Thickness (m)	Average Limestone Porosity (%)	Average Limestone Permeability (md)	Average $\delta^{13}C$ (‰, PDB)	Average $\delta^{18}O$ (‰, PDB)	Limestone >4% Porosity/Total (ft)	Ave. Porosity Limestones >4% Porosity (%)	Ave. Permeability Limestones >4% Porosity (md)
Exposure Stage 1								
22	7.93	1.63	1.33	1.05	-2.88	42/557	5.52	8.68
SD*	(5.18)	(1.37)	(17.7)	(1.77)	(0.54)	7.5%**	(1.7)	(46.6)
Exposure Stage 2								
74	5.49	4.27	10.77	0.10	-3.57	458/1313	9.54	19.4
SD*	(3.35)	(4.72)	(137.1)	(1.45)	(0.62)	35%**	(4.45)	(184.4)
Exposure Stage 3								
53	3.35	3.13	1.89	-2.46	-4.26	112/455	8.03	4.31
SD*	(1.65)	(3.26)	(6.41)	(1.46)	(0.50)	25%**	(3.07)	(9.63)
Exposure Stage 4								
92	2.38	2.24	0.18	-4.29	-4.54	61/512	6.04	0.49
SD*	(1.13)	(1.84)	(0.39)	(0.81)	(0.36)	12%**	(2.0)	(0.78)

*SD = Standard deviations for average values.
**Percentage of gross reservoir section with porosity >4%

Saller et al., 1999

Table 8.1. Characteristics of limestones by stage of subaerial exposure. Used with permission of AAPG.

order late HST-LST) with the resulting development of thin, high-frequency cycles that show abundant evidence of subaerial exposure (Fig. 8.21). Significantly more negative $\partial^{13}C$ values (Fig. 8.21) support more intense subaerial exposure of the high-frequency cycles during the Cisco and Wolfcamp Detrital interval.

Saller et al. (1999) established 4 exposure stages based on cycle thickness and nature and

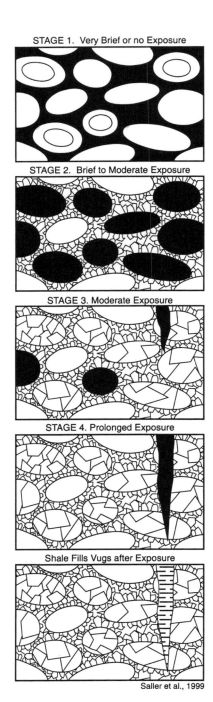

Figure 8.22 Schematic diagram showing the alteration of porosity during the four stages of subaerial exposure. Pores are shown in black. Shale filled many vugs during transgression after subaerial exposure (bottom). Used with permission of AAPG.

intensity of diagenetic overprint suffered by each cycle (Figs. 8.21 and 8.22). Stage 1 is found in thick cycles associated with the TST and shows little or no evidence of subaerial exposure. Packstones and wackestones are the dominant lithologies. While initial porosity is high, it is soon destroyed by compaction in the shallow subsurface. Stage 2 is developed in the transition between the TST and HST where accommodation rate is beginning to slow down and cycles are still relatively thick. The terminal cycle facies are grainstones that exhibit brief-to-moderate subaerial exposure. Thin caliche crusts and root mottling are seen on the cycle tops, and there is evidence that meteoric waters did not penetrate the entire cycle. Moldic and vuggy porosity are dominant, with primary porosity reduced by cement. Overall cycle exposure time was brief to moderate (Fig. 8.22). Stage 3 is found in the late HST where the rate of accommodation space formation is significantly slower and cycles are thinner. Meteoric diagenesis affects the entire cycle. Caliche crusts at cycle tops are thick and have associated rhizocretions, fractures, fissures, and isolated sinkholes. While all intergranular porosity is filled with cement, some moldic, vuggy, and fissure porosity remain unfilled (Fig. 8.22). Exposure was moderate to prolonged. The thin cycles of stage 4 are characterized by intense subaerial exposure resulting in brecciation, fracturing, and fissuring. Red shales fill much of the secondary porosity developed during dissolution. Cycle thickness may be irregular because of widespread, intense karstification including large-scale, shale-filled stream valleys (Fig. 8.22). Exposure was prolonged.

The distribution of porosity in the Strawn/Wolfcamp interval is shown on Figure 8.21 and

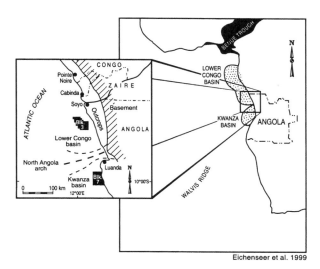

Figure 8.23. Location map showing Block 3, Lower Congo Basin and Block 7, northern Kwanza Basin. Used with permission of AAPG.

sequences to the south in Block 7 are mixed carbonates/sandstones/evaporites. As a result, the lithostratigraphic framework (and the nature of the reservoirs) changes dramatically between the two blocks (Fig. 8.24).

The Cretaceous of offshore Angola was the site of major rifting associated with the opening of the Atlantic. Organic-rich rift lake sequences provide some of the major source beds for later Cretaceous reservoirs, particularly to the south in the Kwanza Basin (Eichenseer et al., 1999). The first major marine incursion occurred as rifting began to switch to drifting, resulting in the deposition of the Loeme evaporites. The overlying Albian-

detailed in Table 8.1. Note that porous intervals coincide with cycles exhibiting exposure stages 2 and 3 with stage 1 and 4 cycles being relatively non-porous.

In summary, brief-to-moderate subaerial exposure in a humid icehouse period developed during a 2nd order HST is necessary for the formation and preservation of economic porosity. Long-term, intense subaerial exposure developed during a 2nd order LST may well be porosity destructive.

Stratigraphic controls over porosity development in an Albian ramp sequence, offshore Angola, Africa: Porosity evolution under greenhouse conditions in a mixed siliciclastic/carbonate/evaporite setting

Albian mixed ramp sequences are major oil producers in the Angola offshore Lower Congo (Block 3) and Kwanza (Block 7) basins (Fig. 8.23). The sequences in northern Block 3 are mixed sandstones/dolomites, while the

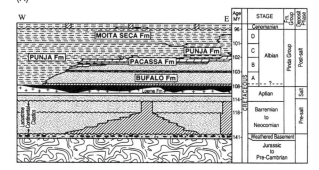

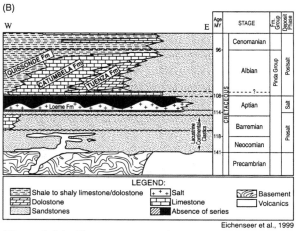

Figure 8.24. Chronostratigraphy and lithostratigraphy of the Pinda Group for A) Block 3 and B) Block 7. Used with permission of AAPG.

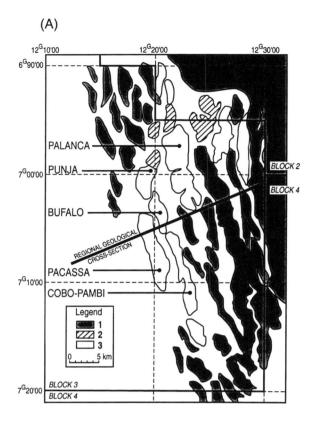

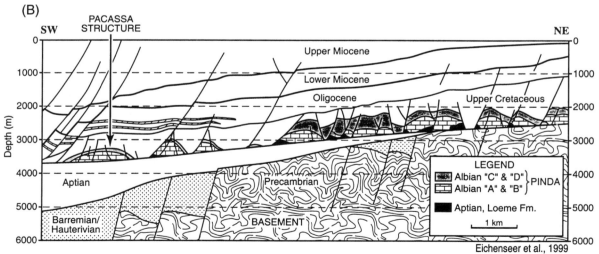

Figure 8.25. A) Schematic seismic map showing salt tectonic "turtle back" structures. 1=faulted segments of the Albian ramp, 2=oil and gas fields, 3=oil fields. B) Dip-directed geological cross section through Block 3 showing present day "turtle back" structures of the Albian ramp margin. Used with permission of AAPG.

Cenomanian sequence consists of a series of ramps deposited during an overall subsidence-driven 2nd order transgression. Each separate ramp sequence was the result of greenhouse low-amplitude 3rd order sea level cycles superimposed on the overall transgression (Fig. 8.24).

Subsequent drift-related differential

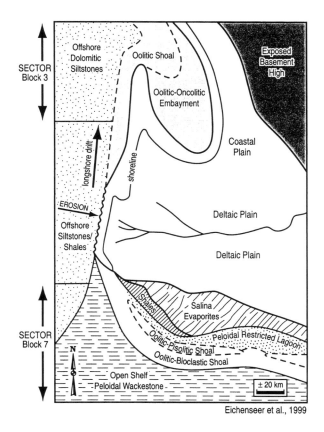

Figure 8.26. Paleogeographic re-construction showing main depositional systems in the Pinda Group, offshore Angola. Longshore drift supplied siliciclastics to Block 3 from a delta complex to the south. An oolitic-evaporitic ramp developed at the same time south of the delta. Used with permission of AAPG.

subsidence and sediment loading of the Loeme evaporites caused halokinesis which resulted in the antiform "turtle-back" structures characteristic of the Cretaceous Angola offshore (Fig. 8.25).

Figure 8.26 shows the gross paleogeography of the Albian in the Angola offshore. The North Angola Arch (Fig. 8.23) provided coarse clastics to a major west-flowing river system that terminated in a wave-dominated delta in the area between Blocks 3 and 7. Strong longshore drift carried most of the clastics northward resulting in the mixed sandstone/carbonate ramp characteristic of Block 3 in the Lower Congo Basin. To the south, in Block 7 of the Kwanza Basin, ooid grainstones, lagoonal wackestones and marginal salina evaporites dominated the shallow-water depositional sequences. The presence of evaporites to the south suggests arid climatic conditions in this area. Eichenseer et al. (1999) developed high-frequency sea level-driven sequence models for each setting in order to illustrate the sedimentary controls over porosity distribution.

A typical high-frequency sedimentary cycle as is commonly developed on a mixed sandstone/carbonate ramp in Block 3 is shown in Figure 8.27. Each cycle (4th or 5th order parasequence) represents a complete transgressive-regressive sequence in response to changes in accommodation space resulting from subsidence and eustatic sea level signals prevalent on the ramp at that time (see Chapter 2 discussions). The initial flood-back related to the transgression re-works any lowstand sandstone into coarse upper shoreface deposits. As the transgression accelerates, the upper shoreface sands are truncated and upper offshore to lower shoreface oncolitic wackestones and packstones are deposited. At maximum flood, the deepest shelf deposits consisting of silty dolomites are deposited over the higher energy "transgressive barrier"(Fig. 8.27A). In the regressive phase, offshore to lower shoreface mixed oolitic sandstones are deposited in a classic shoaling-upward sequence marked by hummocky cross-stratification. The regressive phase and the cycle are terminated by erosion at the accommodation minimum or cycle lowstand (Fig. 8.27A). Lateral facies relationships of facies within a cycle are shown in Figure 8.27B. Figure 8.27C shows 3-dimensional fa-

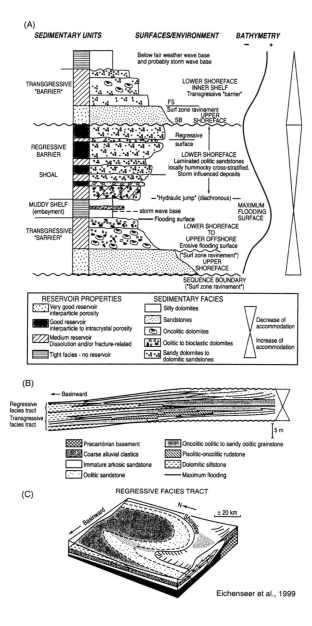

Figure 8.27. Typical model of a high frequency sequence in the oolitic-siliciclastic ramp in Block 3. A) Vertical succession of facies, paleoenvironments, and reservoir characteristics. B) Dip-directed cross-section of a high frequency sequence showing lateral facies distribution. C) Block diagram showing three-dimensional facies distribution. Used with permission of AAPG.

cies distribution.

All carbonates suffered pervasive early (mixed water?) dolomitization. Later meteoric water flushing, perhaps related to cycle and sequence boundaries, dissolved non-dolomitized grains to form significant secondary porosity (Eichenseer et al., 1999). The best reservoirs are developed in the coarse sandstones of the transgressive leg and the oolitic sandstones and sandy oolites of the upper regressive leg. The silty dolomites of the maximum flood generally form aquicludes that tend to compartmentalize the reservoir. Dolomitized wackestones, packstones, and grainstones found in the transgressive and the lower regressive legs may show moderate porosity and low permeability.

A sedimentary cycle model for the oolitic/evaporitic ramp sequence developed in the Kwanza Basin at Block 7 is shown in Figure 8.28. Again the cycle represents a complete high-frequency transgressive-regressive sequence developed in response to changing accommodation space in a background of longer (3rd order) relative sea level changes.

The transgressive leg starts with low-energy lagoonal facies which rapidly change upward into the deeper water mudstones of the shelf centered on the maximum flooding surface. The lower part of the regressive leg is marked by high-energy middle shoreface tidal-dominated ooid-rhodolite grainstones (oncoids of Eichenseer et al., 1999) (Fig. 8.28). The upper regressive leg is marked by upper shoreface to foreshore fine ooid grainstones followed by low-energy, fine-grained lagoonal to supratidal wackestones, mudstones and evaporites. Exposure and perhaps erosion prior to the commencement of the next cycle terminates the parasequence.

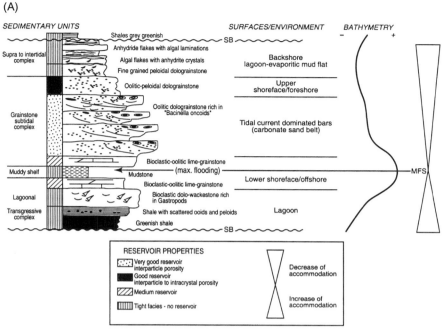

Figure 8.28. Typical model of a high frequency sequence in the oolitic-evaporitic ramp system in Block 7. A) Vertical succession of facies, paleoenvironments and reservoir characteristics. B) Dip-directed cross-section of a high frequency cycle showing lateral facies distribution in relation to early dolomitization. Used with permission of AAPG.

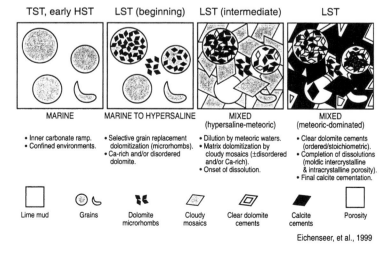

Figure 8.29. Schematic of successive stages of dolomitization, dissolution and cementation occurring in the mixing zone of fresh water with marine to hypersaline waters during relative sea level fall in Block 3 and Block 7. Used with permission of AAPG.

SUMMARY OF EARLY DIAGENESIS AND POROSITY MODIFICATION

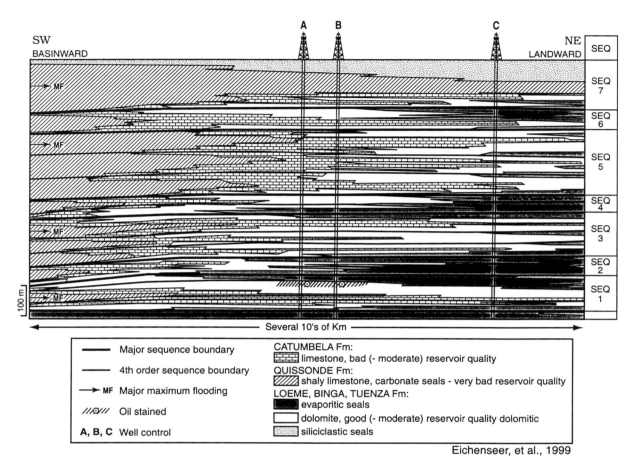

Figure 8.30. Dip-directed northwest-southeast cross-section showing stratigraphic subdivisions of the Albian ramp margin in Block 7 into eight 3rd order sequences and the distribution of main dolomitic reservoirs. Sequences 3-6 correlate with the entire reservoir section of Block 3. Used with permission of AAPG.

Figure 8.28 B illustrates the lateral facies relationships in a single high-frequency cycle. All carbonates of the middle shoreface to the foreshore show early pervasive dolomitization and moldic porosity. Eichenseer et al. (1999) theorized that the dolomitization was a continuing process starting with marine waters during the TST, evaporative waters associated with early LST, and culminating with mixed marine-meteoric waters during the late LST (Fig. 8.29). During the later phase where meteoric waters began to dominate, dissolution of undolomitized rock components resulted in the development of significant moldic and vuggy porosity that characterizes the major reservoirs of the Albian in the Kwanza Basin (Fig. 8.29).

The best reservoirs are developed in the tidal-dominated ooid/rhodolite carbonate grainstones of the regressive leg of the cycle. The fine-grained components of the transgressive leg and the lagoonal deposits of the late highstand are generally tight and form effective seals (Fig. 8.28).

The Albian of Block 7 in the Kwanza basin can be divided into eight 3rd order sequences, with each sequence consisting of a number of the high-frequency cycles described above (Fig. 8.30). Sequence stacking patterns show a distinct 2nd order transgression (retrogradational stacking pattern) toward the northeast. In this framework, the best reservoirs are concentrated in the early and late phases of the 2nd order transgressive leg, where rapid regressions during the HST of sequences 2, 4 and 7 contain volumetrically the highest percentage of reservoir rock.

In summary, Albian porosity distribution and evolution in the siliciclastic-dominated Lower Congo Basin in Angola were strictly stratigraphically controlled. Porosity distribution and evolution is a more complex interplay of stratigraphy and sea level-related early diagenesis to the south in the Kwanza Basin where the Albian is dominated by carbonates and evaporites.

SUMMARY

In this chapter we have developed a series of predictive diagenesis/porosity models to help summarize the early diagenetic concepts developed in earlier chapters. These models are based on a sequence stratigraphic framework that ties together climate and the chemistry of fluids available for diagenesis and porosity modification to a predictive sea level-driven stratigraphy.

The models represent standard sedimentary sequences (3rd order) in three basic settings, the ramp, the land-tied rimmed shelf, and the isolated platform. Arid and humid climatic conditions were considered for each model because of the importance of water to diagenetic processes. The models were taken through a complete sea level cycle, from LST to TST and finally to HST.

In terms of porosity evolution, the LST under humid climatic conditions is the most important part of the sea level cycle. In all three settings surface karsting and mixing zone dissolution at ramp, shelf, and platform margins can be extremely important. In the ramp and land-tied shelf, however, gravity-driven open aquifer systems can develop, leading to extensive water table moldic and vuggy porosity, regional diagenesis and porosity gradients and enhancement of surface karsting by collapse of phreatic caverns. In contrast, the isolated platform's phreatic system is considerably less active; and phreatic diagenesis will be muted relative to the ramp and land-tied shelf.

The steep margins of land-tied shelves and of isolated platforms are sites of significant marine diagenesis during all parts of the sea level cycle. Mixing zone marine water entrainment as well as thermal convection can drive circulation of deeper marine waters through the platform and shelf margin. Porosity-modifying dissolution, cementation and pervasive dolomitization are the result. These deeper marine diagenetic processes are essentially absent on the ramp.

During lowstands the ramp can support lowstand prograding complexes. Diagenesis and porosity modification in this setting is relatively minor and primary intergranular porosity preservation is the rule.

In an arid climate, meteoric diagenesis/porosity modification in all three settings will be muted. Some mixed water dolomites will form in a weakened mixing zone. Small-scale evaporite-related dolomitization will be common in the lowstand prograding complexes of the ramp.

SUMMARY OF EARLY DIAGENESIS AND POROSITY MODIFICATION

Diagenesis and porosity modification during the TST is complex and varied. In a ramp setting, some minor meteoric diagenesis can be expected at the ends of retrogradational parasequences. Transgression and subsidence of the ramp and the land-tied shelf will allow the establishment of a confined gravity aquifer system in the previous sequence, leading to further meteoric diagenesis, mineralogic stabilization, porosity modification, and diagenesis/porosity/geochemical gradients. Mixing zone diagenesis will not be a factor because the mixing zone is not open to the atmosphere.

In the platform setting, aggradational windward reef margins will produce significant framework porosity with minimal porosity destruction by marine cementation. Aggradational to retrogradational grainstones on the leeward margin may retain excellent intergranular porosity. The platform interior will be covered with fine-grained in-situ carbonate sediments that will lose most porosity after very shallow burial.

Under arid conditions during the transgression, most meteoric diagenesis will be muted; and confined aquifers, if present at all, will be small and porosity modification minor. Reflux dolomitization across sequence boundaries associated with transgressive lagoonal evaporites is a major porosity-modifying diagenetic event on land-tied shelves. On the isolated platform, however, more open marine circulation allows evaporation without precipitation of evaporites. Heavy marine water refluxes across the sequence boundary dolomitizing the HST of the previous sequence. Dolomitization is porosity destructive because CO_3 as well as Mg are furnished to the limestones below.

In the HST under humid climatic conditions, meteoric diagenesis is relatively minor in all settings, being centered primarily in island floating fresh water lenses. Porosity destruction by marine cementation on reef-bound platform and shelf margins becomes extremely important as rate of accommodation space formation slows during the highstand.

In an arid climate, dolomitization associated with tidal flat and sabkha evaporites becomes important on the ramp and in the interior of the land-tied shelf. In the late highstand when the rate of generation of new accommodation space is at its minimum, high-frequency sea level lowstands (4th and 5th order) begin to isolate the shelf lagoon and platform interior. Isolation and continued evaporation cause progressive evaporative drawdown and massive precipitation of evaporites (including halite) which preclude any reflux dolomitization.

At the parasequence level, diagenesis is driven by the background 3rd order sea level cycle. During the 3rd order TST into the early HST, parasequences will tend to be thick and to be dominated by marine diagenesis. During most of the HST, parasequences will be thinner and exposure at the cycle tops becomes important and meteoric influence increases. During the late HST and the LST, subaerial diagenesis (both meteoric and evaporative related) becomes dominant.

Regional diagenetic patterns and porosity evolution, developed in association with contrasting 3rd order sequence stacking patterns responding to 2nd order, tectonically-driven sea level cycles, can be useful in frontier exploration efforts.

Retrogradational sequence sets developed during a strong relative sea level rise, such as is found in rift or foreland basins, will be dominated by marine diagenesis followed by some meteoric water overprint with moderate porosity modification and finally burial diagenesis.

During early post-rift in extensional basins, aggradational sequence sets will develop. Moderate meteoric water diagenesis and porosity modification at sequence boundaries will be followed immediately by burial diagenesis.

Progradational sequence sets develop on passive margins where there is little 2nd order sea level change. This setting supports the important porosity-modifying processes such as deep and cumulative karstification, extensive phreatic meteoric diagenesis, and under arid conditions, reflux dolomitization.

At the 1st order scale, icehouse conditions see high-frequency, high-amplitude sea level changes that support rimmed carbonate shelves. Aragonite-dominated sediments deposited on these aggraded shelves tend to exhibit significantly more meteoric diagenesis and porosity modification.

In a greenhouse setting where sea level changes are of low-amplitude, carbonate ramps tend to form. Calcite-dominated sediments deposited on these ramps tend to exhibit more muted meteoric diagenesis and porosity modification.

Finally two case histories, the first an icehouse Pennsylvanian-Permian rimmed shelf in the Permian Basin of West Texas, and the second, a greenhouse Cretaceous ramp in offshore Angola, were presented to illustrate the basic concepts developed in this chapter.

Chapter 9

BURIAL DIAGENETIC ENVIRONMENT

INTRODUCTION

 Early surficial porosity modifications, such as dissolution, cementation, and dolomitization, discussed in previous chapters in a background of sea level changes, are generally accomplished in an infinitesimally small geological time frame. In point of fact, carbonate sequences spend most of their geologic history, measured in tens to hundreds of millions of years in the subsurface, in a diagenetic environment quite different from the environments encountered by newly formed carbonate sediments during the early stages of lithification and burial.

The nature of this burial environment and the diagenetic processes that affect carbonate rock sequences residing within it, have received significant attention during the '80's and '90's, as exploratory drilling expanded into progressively deeper frontiers and as new and improved analytical techniques became available. Several excellent review papers and the work of a number of investigators have focused our attention on the burial regimen and have clearly demonstrated its importance in casting the ultimate evolutionary pathways of carbonate porosity (Bathurst, 1980; Moore and Druckman, 1981; Moore, 1985; Scholle and Halley, 1985; Bathurst and Land, 1986; Halley, 1987; Moore, 1989; Choquette and James, 1990; Moore and Heydari, 1993; Heydari, 1997a; Montanez et al., 1997).

In this chapter we will attempt to present a balanced view of the present state of knowledge of this important, but still poorly understood diagenetic regimen, and to develop a reasonable assessment of the fate of carbonate porosity during the burial process.

THE BURIAL SETTING

Introduction

In all of our previously discussed diagenetic environments, we dealt primarily with surficial conditions, where temperatures remained near surface levels, pressure was not a consideration, and the rocks contained enough pore space to enable the free movement of diagenetic fluids in response to tides, waves, gravity, and evaporation. The starting materials were generally metastable, and pore fluids evolved rapidly by mixing or by chemical exchange during stabilization. The system often had access to significant, localized reservoirs of CO_2 that drove diagenetic processes at a relatively rapid pace. The major processes affecting porosity were dissolution, cementation, and dolomitization.

The burial diagenetic environment starts when sedimentary sequences are buried beneath the reach of surface-related processes. It includes the mesogenetic or deeper burial regime of Choquette and Pray (1970), and may extend into the zone of low-grade metamorphism in the deeper reaches of some sedimentary basins (Fig. 9.1). While extending into the shallow subsurface, regional confined meteoric aquifers are treated as surficial environments because of surface meteoric recharge.

In the burial environment, diagenetic fluids are cut off from free exchange with chemically active gases of the atmosphere (most particularly oxygen and CO_2). In addition, temperature and pressure progressively increase, while the exchange of diagenetic fluids diminishs due to continuous porosity reduction. Pore fluids undergo a slow compositional evolution driven by rock-water interaction and mixing of basin-derived waters. The major process affecting porosity is compaction.

In the following paragraphs we will discuss some the most important parameters of the burial setting, such as pressure, temperature, and water chemistry, in order to set the stage for consideration of burial diagenetic processes and products.

Pressure

There are three types of pressure that influence sedimentary sequences during burial: hydrostatic, lithostatic, and directed pressure. Hydrostatic pressure is transmitted only through the water column, as represented by the sediment pore system. Lithostatic pressure is transmitted through the rock framework, while directed pressure is related to tectonic stresses (Bathurst, 1975; Collins, 1975; Bathurst, 1980; Bathurst, 1987; Choquette and James, 1987). Figure 9.2A illustrates the ranges of pressures one might expect in a sedimentary basin setting. Hydrostatic pressure will vary depending on fluid salinity and temperature.

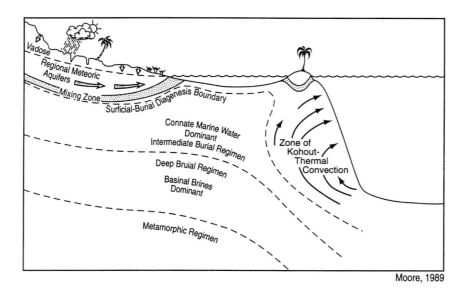

Fig.9.1. Schematic diagram showing the relationship of surficial diagenetic environments to the several zones of the burial diagenetic regimen. Used with permission of Elsevier.

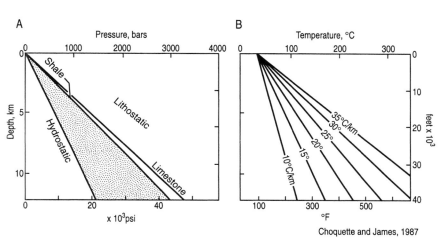

Fig 9.2. Graphs showing the general ranges of pressure and temperature in the burial diagenetic regimen (A) Static pressure variations. The position of the hydrostatic pressure curve varies depending on the concentration and density of pore waters. Most pore-fluid pressures in the subsurface would plot on curves in the stippled area. (B) Temperature ranges encountered in the subsurface assuming different geothermal gradients. Used with permission of Elsevier.

Most ambient hydrostatic pressures found during drilling will fall in the shaded area shown on Figure 9.2A. The net pressure on a sedimentary particle in the subsurface is found by subtracting the hydrostatic pressure from the lithostatic pressure. The application of this net pressure on the rock mass results in strain. This strain is relieved by dissolution and is the force that drives chemical compaction, or pressure solution.

During burial, if interstitial water is trapped in the pore spaces while the sediment is undergoing compaction, hydrostatic pressure can increase dramatically and even approach ambient lithostatic pressure, resulting in very low net pressures at rock and grain interfaces. This condition is termed overpressuring, or geopressure (Choquette and James, 1990). These low net pressures, acting at grain contacts, may slow and even stop physical and chemical compaction. In this case the high fluid pressures act as an intergranular buttress, resulting in exceptionally high porosities under relatively deep burial conditions (Scholle et al., 1983b) (see, for example, the North Sea chalk case history later in this chapter).

Geopressure may occur under conditions of rapid burial where aquacludes, such as evaporites, shales, or even thin, cemented layers, (marine hardgrounds or paleo-water tables), are present in the section. Geopressured zones tend to isolate pore fluids from surrounding diagenetic waters, preventing fluid and ion transfer and slowing or preventing diagenetic processes such as cementation (Feazel and Schatzinger, 1985). Hydrofracturing associated with geopressured zones may enhance porosity (as in the case of the Monterey Formation in California, as described by Roehl and Weinbrandt (1985), but CO_2 degassing associated with the release of pressure during fracturing can lead to massive cementation (Woronick and Land, 1985).

Temperature

The progressive increase of temperature with depth into the subsurface is termed the geothermal gradient. Figure 9.2B illustrates the range of temperatures one might expect to encounter in the subsurface assuming different geothermal gradients. Each sedimentary basin has a different thermal history, and hence, geothermal gradient, as driven by burial history, sediment type, tectonic setting, and hydrology.

Taken alone, increasing temperature speeds chemical reactions and the rate of ionic diffusion. Increasing temperature will, however, decrease the solubility of carbonates because of the influence of CO_2 on the carbonate system. These relationships favor dolomitization and the precipitation of calcite cements in the subsurface (Morse and McKenzie, 1990). The elevated temperatures found under burial conditions strongly influence the oxygen isotopic composition of subsurface calcite cements and dolomites because of the strong temperature dependency of oxygen fractionation between solid carbonate phases and water (Anderson and Arthur, 1983).

Increasing both temperature and pressure during burial triggers a series of mineral reactions and phase changes that may release water and ions that can become involved in carbonate diagenetic processes in the subsurface. For example, the conversion of gypsum to anhydrite at about 1000m releases significant water that may become involved in solution, cementation, or dolomitization (Kendall, 1984) The conversion of smectite to illite commencing at some 2000m and 60° C releases water of crystallization as well as cations such as Mg, which could be used in dolomitization (Boles and Franks, 1979; Wanless, 1979).

The application of elevated temperatures over extended time periods to sedimentary organic material leads to the catagenesis of the organic compounds and the formation of oil and gas (Tissot and Welte, 1978). Figure 9.3 outlines the general framework of organic

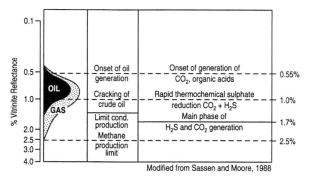

Fig. 9.3. Framework of hydrocarbon maturation, destruction, and organic diagenesis as a function of vitrinite reflectance. Used with permission of Elsevier.

diagenesis relative to vitranite reflectance, a measure of thermal maturity. Thermal maturity is a function of temperature and the time interval over which that temperature has been applied and is generally independent of lithostatic pressure (Tissot and Welte, 1978).

The by-products of the later stages of organic catagenesis can dramatically influence deep burial carbonate diagenesis. As seen on Figure 9.3, at the onset of oil generation and at a vitrinite reflectance of about .55%, CO_2 and organic acids are produced. At a vitrinite reflectance of 1.0%, rapid thermochemical sulfate reduction commences and CO_2 and H_2S are generated, reaching a peak at a vitrinite reflectance of about 1.7% (Sassen and Moore, 1988). These organic diagenetic processes can lead to the formation of aggressive subsurface fluids that may effect carbonate dissolution and produce secondary porosity (Schmidt and McDonald, 1979; Surdam et al., 1984). The $CaCO_3$ furnished to the pore fluids as a result of this dissolution phase is then available for precipitation as calcite cement or dolomite elsewhere in the system (Moore, 1985; Sassen and Moore, 1988; Heydari and Moore, 1989; Heydari, 1997b).

Deep burial pore fluids

Our knowledge of subsurface fluids is garnered from samples of oil field waters gathered during initial testing and during later production of oil and gas. Figure 9.4 illustrates the relative abundance of oil field waters of different salinities. Some 74% of these waters are classed as saline water or brine (10,000-100000 ppm dissolved salts) and over 50% of these waters are more saline than seawater. Most of the subsurface waters that have been analyzed are CaCl waters with major cations in the following proportions: Na>Ca>Mg (Collins, 1975). Those subsurface waters that are classed as brines (>100,000 ppm dissolved solids) are thought to be related to evaporites, originating either as evaporite interstitial fluids (Carpenter, 1978) or as the result of subsurface evaporite dissolution (Land and Prezbindowski, 1981). Those subsurface fluids with salinities less than seawater are thought to be mixtures of meteoric, marine, and perhaps, basinal fluids (Land and Prezbindowski, 1981; Stoessell and Moore, 1983).

In detail, the composition of subsurface fluids is complex and varies widely within and between basins, due to mixing of chemically dissimilar waters and continuous rock-water interaction (dissolution as well as precipitation) during burial (Collins, 1975; Carpenter, 1978; Land and Prezbindowski, 1981; Morse et al. 1997). Most subsurface waters, however, have very low Mg/Ca (Collins, 1975) and tend to exhibit progressively heavier oxygen isotopic values with depth due to temperature fractionation effects (light isotopes preferentially incorporated in the solid phase at elevated temperature). Land and Prezbindowski (1981) report ∂^{18O} values for Cretaceous formation waters in excess of +20 (per mil, SMOW). See Table 4.2 for the composition of Jurassic brines from the Gulf Coast of the USA

Tectonics and basin hydrology

Most of our knowledge of hydrology concerns gravity-driven systems, such as were treated in the section on meteoric diagenetic environments. Deep basin hydrology, however, has assumed a position of importance as investigators have taken a more integrated approach to basin studies and as our exploration frontiers have deepened. These studies have clarified the role of earth-scale tectonics in defining the nature and evolution of subsurface basin-wide hydrologic systems. The thermal, geochemical and flow characteristics of these basin hydrologic realms control the pathways of subsequent burial diagenesis of the basin sedimentary sequence. Three basic hydrologic/

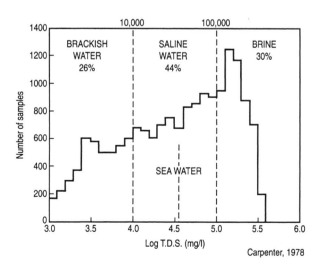

Fig. 9.4. Histogram illustrating the relative abundance of oil-field waters of different salinities. Used with permission of Elsevier.

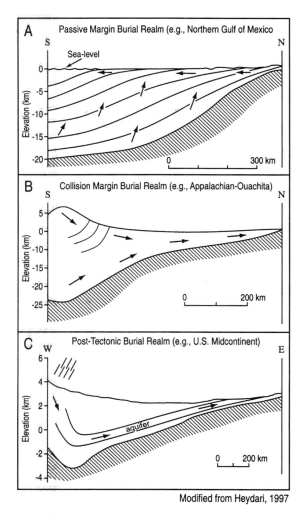

Fig. 9.5. Models of diagenetic fluid movement within platform carbonates under different hydrologic/tectonic settings. Arrows indicate fluid flow. Thin solid lines are geologic contacts. Cross hatchering represents the basement. The passive margin burial diagenetic realm (A) is characterized by tensional tectonics, growth faulting, uniform subsidence rates and slow upward flow of waters driven by compaction. The collision margin burial diagenetic realm (B) is characterized by compressional tectonics, thrust faulting, extensive fracturing, and episodic flow of tectonic waters toward the craton. The post-orogenic burial diagenetic realm (C) is characterized by a relative absence of tectonic activity, dominance of topographically-driven fluid flow, high flow rates and systematic changes in fluid compositions toward the recharge area. Used with permission of SEPM.

tectonic burial regimens can be identified: 1) the passive margin, 2) the collision or active margin, and 3) the post-tectonic regimen (Heydari, 1997a) (Fig. 9.5).

The northern Gulf of Mexico, the Mesozoic of West Africa, and eastern Brazil are all passive margin burial systems (Fig. 9.5A). The passive margin is characterized by extensional tectonics, growth faulting and relatively slow, but steady subsidence. Three hydrologic flow regimens are commonly found in this setting (Harrison and Summa, 1991). The first is a high-velocity, near-surface, gravity-driven meteoric system with flow toward the basin. The second is a deep (greater than 1km) compactional system with horizontal as well as vertical components and moderate flow velocities. The third is a thermohaline system where sediments buried less than 3km are hydrostatically pressured while sediments buried deeper may be geopressured. The geopressure zone is hydrologically isolated and often is characterized by lower salinities (Hanor, 1987). Burial diagenesis in a passive margin setting generally exhibits two phases. The first phase takes place before hydrocarbon migration (approximately 2-3km burial) and is dominated by mechanical and chemical compaction in response to steadily increasing temperatures and pressures. The second phase commences as hydrocarbons begin to migrate into reservoirs in conjunction with basinal brines moving up dip. Aggressive pore fluids associated with hydrocarbon diagenesis including thermochemical sulphate reduction (Fig. 9.3) dominate diagenesis in this setting and produce very late calcite cements and some dissolutional secondary porosity.

The Lower Paleozoic Appalachian-Ouachita tectonic belt of the eastern USA, the northern Rocky Mountain Paleozoic tectonic belt in Canada, and the Mesozoic tectonic belt of Switzerland all support typical active margin burial regimens (Oliver, 1986; Garven et al., 1993; Montañez, 1994) (Fig. 9.5B). This setting is characterized by compressional tectonics, thrust faulting, and extensive fracturing of carbonates with episodic high temperature-focused fluid flow in response to tectonic loading. The rapid movement of these hot fluids through contiguous carbonate units recrystallizes early calcites and dolomites, remobilizes early formed carbonates and evaporites, and precipitates dolomite, calcite and Mississippi Valley type ore deposits in an up-flow direction. Extensive subsurface dissolution and secondary porosity development can accompany these processes.

The post-orogenic setting is exemplified by present Upper Paleozoic Mississippian aquifers in the mid-continent of the USA and Paleozoic aquifers of the Alberta Sedimentary Basin (Fig. 9.5C). In both cases Paleozoic aquifers are exposed and recharged by meteoric water along the elevated flanks of the Rocky Mountains, setting up a topographically controlled, gravity-driven hydrologic system in the adjacent basins (Bachu, 1997). These systems are characterized by high fluid flux with fluid compositions controlled by meteoric-subsurface brine mixing and dissolution of evaporites if present. While important relative to secondary hydrocarbon migration and trapping (Bachu, 1997), this tectonic-hydrologic setting is responsible for only minor porosity-related diagenesis (Banner et al., 1989; Banner et al., 1994) centered on dissolution in the basin-flank recharge area.

The following discussion of burial diagenesis will be structured within the tectonic-hydrologic framework outlined above. We will start with the passive margin setting where mechanical compaction, chemical compaction and hydrocarbon-related dissolution and cementation is driven by increasing temperatures and pressures during slow and continuous burial.

PASSIVE MARGIN BURIAL REGIMEN

Introduction

During the early, mechanical compaction phase of sedimentary sequence subsurface burial, confined gravity driven meteoric water systems may continue to be active until terminated by subsequent transgression and closure of up-dip recharge areas. Active fluid movement is dramatically reduced and connate marine waters dominate as burial progresses, and available porosity continues to be reduced by pressure solution and cementation (chemical compaction). Initial high water-rock ratios are progressively reduced as porosity is lost during burial.

The following discussion centers on the two primary diagenetic processes active in the passive margin burial setting, mechanical compaction/de-watering and chemical compaction with attendant porosity-filling cementation.

Mechanical compaction and de-watering

As one might expect, mechanical compaction and de-watering are particularly important in mud-dominated sediments. Most of the work on mechanical compaction and de-watering has centered on outcrop, subsurface, and oceanographic studies of Mesozoic to

Holocene pelagic oozes and chalks (Scholle, 1971; Neugebauer, 1973; Schlanger and Douglas, 1974; Garrison, 1981; Kennedy, 1987; Maliva and Dickson, 1992).

Pelagic oozes commonly have a total initial porosity of up to 80%, consisting of 45% intragranular porosity and 35% intergranular porosity (Garrison, 1981). The intergranular porosity is close to the porosity that was observed by Graton and Fraser (1935) for the packed spheres used in their early porosity experiments, suggesting that most pelagic oozes have the initial porosity characteristics of a grain-supported sediment.

Matter and others (1975) observed a 10% porosity loss for these oozes in the first 50m of burial. This porosity loss approximates that observed by Graton and Fraser (1935) in their packing experiments between unstable and stable packing geometies. It follows then, that the initial gravitational porosity loss suffered by pelagic oozes is basically a mechanical rearrangement to a more stable grain geometry. This type of mechanical compaction is completed when porosity has been reduced to approximately 70%. At this point, the breakage of thin-walled foraminifera caused by loading may decrease the porosity further to some 60% (Fig. 9.6).

Mechanical compaction of muddy shelf sequences may take a different path than pelagic oozes because of the following factors: 1) the elongate needle shape of the shelf mud particles versus the more equant shapes for the pelagic oozes (Enos and Sawatsky, 1981); 2) the general lack of large volumes of intragranular porosity in the shelf muds; and 3) the common occurrence of significant (up to 40%) tightly bound water sheaths surrounding the elongate particles of shelf

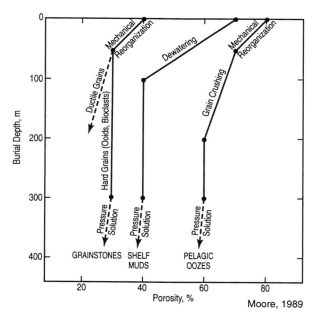

Fig. 9.6. *Schematic diagram illustrating the porosity evolution of shelf grainstones and mud versus pelagic oozes during the first 400m of burial. Curves based on data from Neugebauer (1973), Scholle (1977), Garrison (1981), and Enos and Sawatsky (1981). Used with permission of Elsevier.*

muds. Enos and Sawatsky (1981) indicated that the elongate aragonite needles in shelf muds were bi-polar, thus attracting and binding water molecules. Under these conditions, the de-watering phase is probably the most important early compaction event in muddy shelf sequences. De-watering would lead to the highest loss of porosity, (reduction to 40% or less), before a stable grain framework could be established (Fig. 9.6).

The compaction experiments of Shinn and Robbins (1983), using modern, mud-rich sediments from many of the same environments and localities as sampled by Enos and Sawatsky (1981), seem to confirm the importance of the de-watering phase in muddy shelf sediment compaction. They report a decrease of porosity from 70% to 40% under a load of 933psi (approximately equivalent to 933 feet or 298m of burial) for mud-dominated sediment cores from Florida Bay. Significant grain breakage or detectable pressure dissolution of the fine aragonitic matrix did not accompany this porosity loss. This dramatic porosity drop was noted within the first 24 hours of the experiment. No further significant porosity decline was observed regardless of the duration of time that the pressure was maintained. These results indicate that this initial porosity loss was dominantly the result of de-watering, and that apparently a stable grain framework is reached for this type of sediment at approximately 40% porosity under burial conditions of 100m or less (Shinn and Robbins, 1983).

Grain-supported shallow water sediments, such as carbonate sands, will have a slightly different compactional history than that of the mud-dominated shelf sediments discussed

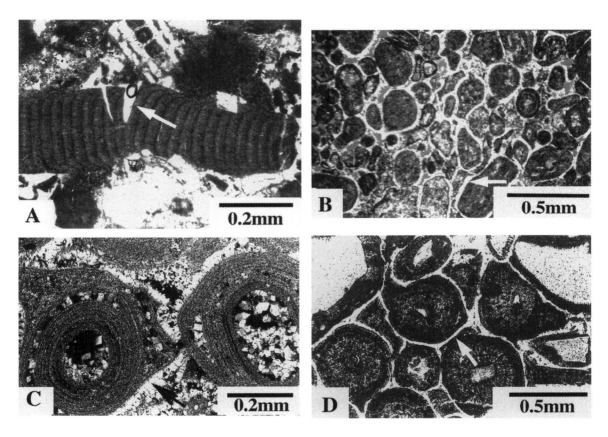

Fig. 9.7. (A) Halimeda fractured by burial compaction (arrow) under 762 m (2500') of limestone at Enewetak Atoll, western Pacific. Sample taken from Miocene section in well F-1. Plain light. (B) Compacted Jurassic Smackover Formation peloids under 9000' (2744m) of burial, southern Arkansas. The peloids were coated by early isopachous cement that is also involved in compaction (arrow). Plain light. (C) Compaction of Jurassic Smackover Formation ooids that have undergone dissolution during stabilization. Early isopachous calcite cements are also involved in compaction (arrow). Murphy Giffco #1, 7788' (2374 m) Bowie County, Texas. Crossed polars. (D) Strongly compacted Jurassic Smackover Formation calcite ooids. Again note the early isopachous cement that was also involved in compaction (arrow). Phillips Flurry A-l, 19,894' (6065 m) southern Mississippi.

above and will more nearly parallel the early burial history of pelagic oozes.

The first loss of porosity is associated with mechanical grain rearrangement to the most stable packing geometry, which, if the grains are spheres such as ooids, would involve a loss of up to 10% intergranular porosity (Graton and Fraser, 1935). The next phase of compaction for grain-supported sediments involves the mechanical failure of the grains themselves, either by fracturing, or by plastic deformation, both leading to the loss of porosity (Coogan, 1970; Bhattacharyya and Friedman, 1979) (Fig. 9.6).

The experiments of Fruth and others (1966), while easily documenting the mechanical failure of ooids and bioclasts, suggests that the actual loss of porosity due to this failure may be less than 10% at confining pressures equivalent to some 2500m overburden. Other grain types such as soft fecal pellets or bioclasts that have very thin shell walls may be particularly susceptible to mechanical compaction, failing by plastic deformation, or crushing, with an attendant loss of significant porosity (Fig. 9.7) (Rittenhouse, 1971; Enos and Sawatsky, 1981; Moore and Brock, 1982; Shinn and Robbin, 1983).

Mechanical compaction effects in grain-supported sediments can be moderated by early cementation either in the marine environment or in the meteoric diagenetic environment, where intergranular cements tend to shield the enclosed grains from increasing overburden pressure during burial (Shinn et al., 1983; Railsback, 1993). In some cases, however, if the grains were originally soft, even early cementation may not be sufficient to stop grain failure and attendant porosity loss (Fig. 9.7).

Chemical compaction

Once a carbonate sediment has been mechanically compacted, and a stable grain framework has been established (regardless of whether the grains are mud sized or sand sized), continued burial will increase the elastic strain at individual grain contacts. Increased strain leads to an increased potential for chemical reaction, reflected as an increase in solubility at the grain contact, and ultimately results in point dissolution at the contact. Away from the grain contact, elastic strain is less, the relative solubility of calcium carbonate is reduced, and ions diffusing away from the site of dissolution will tend to precipitate on the unstrained grain surfaces (Bathurst, 1991).

Pressure solution is manifested in a variety of scales from microscopic, sutured, grain contacts to the classic, macroscopic, high-amplitude stylolites (Figs. 9.8 and 9.9) (Bathurst, 1983; Koepnick, 1983).

In grainstones, initial pressure solution is concentrated at grain contacts, resulting in microscopic sutured contacts with progressive grain interpenetration (Fig. 9.9A and B). During this stage, increasing lithostatic load is accommodated by individual grain solution. Ultimately, however, the rock will begin to act as a unit, and lithostatic load will then be accommodated by macroscopic stratal solution features such as stylolites (Figs. 9.8 and 9.9C and D). It is unclear whether mud-dominated limestones undergo a similar compactional evolution. The dual process of pressure solution and cementation, described above, is commonly termed *chemical compaction* (Lloyd, 1977; Choquette and James, 1987).

The pressure solution process (ignoring cementation) involves the reduction of the bulk

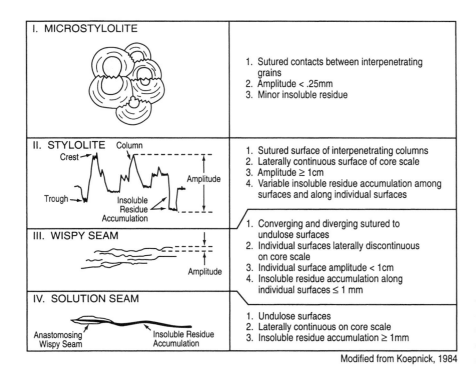

Fig 9.8. Types and characteristics of pressure solution features encountered in the subsurface. Used with permission of Elsevier.

Modified from Koepnick, 1984

volume of the rock. If one reduces the bulk volume of a rock, it follows that there will be a loss in porosity (Rittenhouse, 1971). For example, sediment composed of spheres (i.e., an oolite) with orthorhombic packing that loses 40% of its bulk volume to pressure solution will suffer a 44% reduction in its original porosity.

Pressure solution rarely occurs without cementation because the dissolved components typically precipitate nearby as cement (Bathurst, 1975; Coogan and Manus, 1975; Bathurst, 1984). Rittenhouse using a variety of packing geometries and textural arrangements (1971) calculated the volumes of cement that could be derived from grain-to-grain pressure solution. Rittenhouse's work was based on siliciclastic sands, but the principles apply equally well to carbonates.

Using Rittenhouse's method and his orthorhombic packing case, assuming an original porosity of 40% and a bulk volume decrease of 30% by pressure solution, we can calculate a decrease of 36% of the original porosity, down to 25%. With this 14% drop in original porosity we can expect the potential production of only about 4% cement, assuming that the cementation process is 100% efficient (Fig. 9.10, case 1). We can, therefore, only reduce our original porosity down to approximately 21% by chemical compaction, assuming a bulk volume decrease of 30% by pressure solution (Fig. 9.10, case 1). The 30% volume decrease used in our calculations above is a figure that has been used by a number of authors as a typical volume decrease in the geological record (Coogan and Manus, 1975; Bathurst, 1975, 1984). The 30% figure was originally derived by measuring the accumulated amplitudes of megascopic stylolites, and ignores, among other things, volume lost by grain interpenetration, solution seams, and

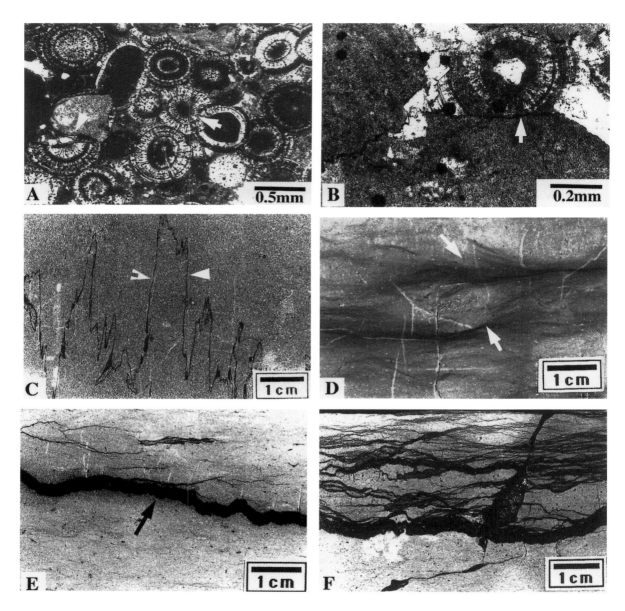

Fig. 9.9 (A) Strongly compacted Jurassic Smackover Formation ooid grainstone showing grain-to-grain pressure solution (arrows). I.P.C.#1, 9742 ' (2970 m), Columbia Co. Arkansas, USA. (B) Grain-to-grain pressure solution between ooid and larger algally coated grains (arrow), Jurassic Smackover Formation, Arco Bodcaw #1, Walker Creek Field, 10,906' (3325 m), Columbia Co., Arkansas. (C) High amplitude stylolites (arrows) in Jurassic Smackover Formation, Getty #1 Reddoch l7-15, 13,548' (4l30m), Clarke Co., Mississippi. (D) Wispy seam stylolites with "horse tails"(arrows) in the Jurassic Smackover Formation LL and E Shaw Everett, 13,657' (4164 m), Clarke Co., Mississippi. (E) Solution seam with insoluble residue accumulation (arrow) in the Jurassic Smackover. Same well as (D), 13,603' (4147 m). Anastomosing wispy seams in the Jurassic Smackover Formation. Same well as (D), 13,635' (4156 m).

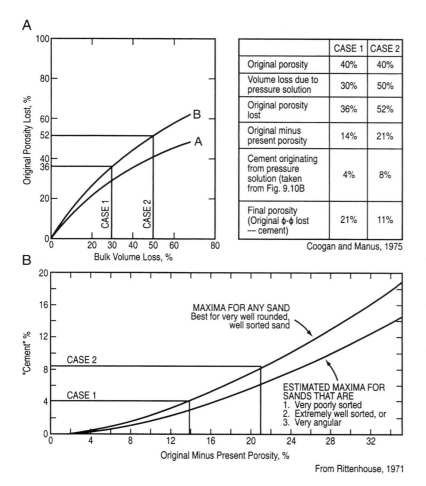

Fig. 9.10. Volume of cement, and porosiy loss that may be attributed to pressure solution. Starting porosiy in both cases is 40%. In case l, 30% of the total volume of the rock is lost by pressure solution, and the final porosiy is 21%, while in case 2, 50% of the total volume of the rock is lost by pressure solution and the final porosiy is 11%. (A) Original porosity lost plotted vs bulk volume loss. Curve B is orthorhombic packing. (B) Original minus present porosity plotted vs % cement. This is the curve used to estimate volume of cement that can be gained from pressure solution. Used with permission of Elsevier.

mechanical compaction (Bathurst, 1984). If we increase our estimate of bulk volume lost during burial to 50% (Fig. 9.10, case 2) to accommodate the possible effects of grain-to-grain pressure solution and mechanical compaction, we can reduce our original porosity down to approximately 11%. Even in this more extreme case, we are only able to produce 8% cement as a result of chemical compaction (Fig. 9.10, case 2).

Heydari (2000) recently published an interesting study on porosity loss in ooid grainstones of the Upper Jurassic Smackover Formation during deep burial in the USA, northeastern Gulf of Mexico. His results, using grain shortening measurements on ooids (Fig. 9.11) and the experimental data of Mitra and Beard (1980), parallel the results of the Rittenhouse exercise detailed in the paragraphs above. Heydari (2000) calculated an average vertical shortening in his core of 28%, a figure strikingly consistent with the general observations of previous workers mentioned above who based their estimates on measurements of stylolite amplitudes. Using his calculated average vertical shortening of 28% and the experimental data of Mitra and Beard (1980) (Fig. 9.12), Heydari (2000) estimated porosity loss by pressure solution in his core of about 15 porosity units or 37.5% of the starting porosity. This is almost identical to the results of the Rittenhouse exercise (Fig. 9.10). Pressure solution-related cements in his study core totaled 6.7% by thin section point count. This cement volume, while consistent with this author's observations in the Smackover elsewhere in the Gulf of Mexico

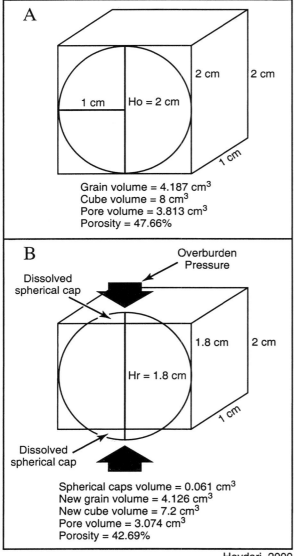

Figure 9.11. (A) Hypothetical spherical grain encased within a cube. The grain radius is 1 cm, and each limb of the cube is 2 cm. This cube is cut vertically in half to better illustrate changes in grain shape, and spherical caps dissolved. (B) Vertical shortening of 0.1 cm at both top and bottom of the grain. Used with permission of AAPG.

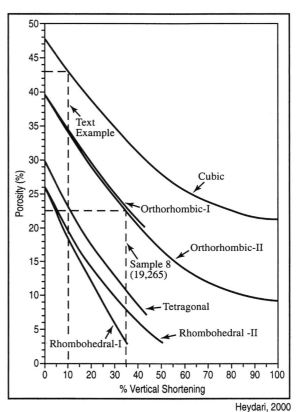

Figure 9.12. Porosity reduction in relation to vertical shortening of spherical grains in different packing arrangements (data from Mitra and Beard, 1980). The orthorhombic II packing arrangement curve was used to extrapolate porosity loss due to vertical shortening in the study. For example, sample 8 (19,265') has an average vertical shortening of 35%, which corresponds to a porosity loss by intergranular pressure solution of 18 porosity units (from 40 to 22%). The line marked "text example represents theoretical compaction and the resultant 10% vertical shortening of a sphere in cubic packing arrangement, resulting in loss of about 5 porosity units (from 47 to 42%). Used with permission of AAPG.

region, is high relative to the Rittenhouse estimates (4% versus 6.7%, Figure 9.10) and may represent a contribution from nearby high-amplitude stylolites not represented in the data set. One must draw the conclusion, as did Heydari (2000), that most porosity loss during burial is due to mechanical compaction and pressure solution and that porosity loss due to pressure solution-related cementation is strictly limited. It should be noted, however, that $CaCO_3$ may be transported in the subsurface from sites of dissolution (donor beds) to sites of precipitation (receptor beds) (Heydari, 2000).

Factors affecting the efficiency of chemical compaction

Burial depth with attendant lithostatic load is one of the most critical factors determining the onset and ultimate efficiency of chemical compaction. Dunnington, (1967) in his careful analysis of compaction effects on carbonate reservoirs, indicated that 600m was a minimum burial depth for the onset of pressure solution. Buxton and Sibley (1981) deduced from stratigraphic reconstruction that chemical compaction took place under some 1500m burial in the Devonian of the Michigan Basin. Neugebauer (1974) stated that while minor solution takes place in the first 200m of burial significant pressure solution does not occur in chalks until at least 1000m of burial. Borre and Fabricius (1998) report pressure solution-related cementation commencing at 850m in deep-sea chalk cores from the Ontong Java Plateau. Saller (1984) observed no pressure solution in uncemented grainstones in cores taken at Enewetak Atoll until some 800m of burial. However, Meyers and Hill (1983) felt

FACTORS AFFECTING CHEMICAL COMPACTION

FACTORS ENHANCING CHEMICAL COMPACTION	FACTORS RETARDING CHEMICAL COMPACTION
1. Metastable (i.e.: aragonite) mineralogy at the time of burial (Wagner and Matthews, 1982)	1. Stable (i.e.: calcite, dolomite) mineralogy at the time of burial (Moshier, 1987)
2. Magnesium poor meteoric waters in pores (Neugebauer, 1973)	2. Oil in pores (Dunnington, 1967; Feazel and Schatzinger, 1985)
3. Insolubles such as clays, quartz, and organics (Weyl, 1959)	3. Elevated pore pressures as found in geopressured zones (Harper and Shaw, 1974)
4. Tectonic stress (Bathurst, 1975)	4. Organic boundstones (Playford, 1980)

Koepnick, 1984

Table 9.1. Factors affecting chemical compaction. Used with permission of Elsevier.

that pressure solution in Mississippian grainstones started after only tens to hundreds of meters of burial. In addition, Bathurst, (1975) cited numerous occurrences of shallow burial onset of pressure solution. These widely divergent reports clearly indicate that one cannot make sweeping generalizations concerning minimum burial needed for chemical compaction to start because of the many factors, as noted in the following paragraphs, that may enhance, or retard the process (Table 9.1).

The original mineralogy of the sediment undergoing compaction is a very important control over the nature of its early chemical compactional history. A mud-bearing sediment composed of metastable mineral components (aragonite) should be much more susceptible to early chemical compaction and chemically-related porosity loss than the stable (less soluble) calcite mineralogy characteristic of deep marine pelagic oozes and chalks (Scholle, 1977a). If the sediments are buried with marine pore waters, the solubility of the various touching grains will determine the timing of the onset of pressure solution. Grains composed of metastable minerals such as aragonite or magnesian calcite should begin to dissolve well before calcite or dolomite (Scholle and Halley, 1985; Humphrey et al., 1986).

Most shallow marine mud-bearing carbonate sequences seem to lose their porosity early in their burial history (Bathurst, 1975; Wilson, 1975; Moshier, 1989). Mud matrix microporosity, however, is common and is very important in certain Mesozoic reservoirs, particularly in the Middle East (Wilson, 1980; Moshier et al., 1988). Some workers relate this microporosity to dissolution associated with unconformity exposure (Harris and Frost, 1984). Moshier (1989), however, makes a compelling case for original calcite mineralogy for these Cretaceous shallow marine muds. Original calcite mineralogy suggests that the microporosity is actually preserved depositional porosity because of the mineralogical stability of the original mud that allows the sequence to follow the characteristic porosity evolution of chalks and pelagic oozes. Moshier's thesis is consistent with Sandberg's view (1983) that the Cretaceous was a calcite sea in which most abiotic components (including mud-sized material precipitated as whitings across shallow marine shelves) would have precipitated as calcite.

The chemistry of pore fluids in contact with a sedimentary sequence during chemical compaction may have a dramatic impact on the rate of porosity loss during burial. Neugebauer (1973; 1974), in his experimental work, and Scholle (1977b), in his field and subsurface studies, both indicate that meteoric waters (Mg poor) enhance the process of chemical compaction as compared to marine pore fluids. On the other hand, the early introduction of hydrocarbons into pore space can totally stop solution transfer and hence chemical compaction (Dunnington, 1967; Feazel and Schatzinger, 1985).

Pore fluid pressures can dramatically affect the rate of chemical compaction. In overpressured settings, where hydrostatic pressures begin to approach lithostatic load, the stress at grain contacts, and hence, the strain on the mineral is reduced, and pressure solution is slowed or stopped. This effect has been demonstrated in the North Sea (Harper and Shaw, 1974; Scholle, 1977b) where chalks exhibit porosities of some 40% at 3000m burial in zones of over pressuring. Using Scholle's burial curve for chalks (1983) (Fig. 9.19), a 40% porosity would typically be expected in chalks encountered at 1000m of burial or less.

Insolubles, such as clay and organic material, seem to enhance chemical compaction in fine-grained carbonate sequences (Choquette and James, 1987). Weyl (1959) and Oldershaw and Scoffin (1967) observed the preferential formation of stylolites in clay-rich sequences, while Sassen and others (1987) related more abundant stylolites intensity to the presence of fine-grained, organic-rich, source beds in the Jurassic of the Gulf of Mexico.

Any factor that increases stress at the point of contact between carbonate grains will tend to enhance chemical compaction. Tectonic stress such as folding or faulting concentrates stress into well-defined zones leading to tectonically controlled concentration of pressure solution-related phenomena such as stylolites (Miran, 1977; Bathurst, 1983; Railsback and Hood, 1993).

Intuitively, one would expect the temporal subsidence and thermal history of a basin to affect the efficiency of the chemical compaction process. In siliciclastics, the effect of thermal regimen on porosity-depth relationships has long been known. Loucks and others (Loucks et al., 1977) documented the loss of sandstone porosity with depth of burial in the Texas and Louisiana regions of the USA Gulf Coast. They demonstrated that eqivalent porosity destruction occurred at shallower burial in Texas (versus Louisiana), due to the much higher geothermal gradient of south Texas, resulting in more efficient chemical compaction in that region. Lockridge and Scholle (1978) could detect no difference in the porosity-depth curves for chalks between North America and the North Sea, even though the geothermal gradients in the North Sea are 1.5 to 2 times higher than encountered in North America. Schmoker (1984), however, has generated a compelling case for subsurface porosity loss (presumably driven by chemical compaction) depending exponentially on temperature and linearly on time and hence operating as a function of thermal maturity in the basin of deposition. We will consider this important topic further in our summary discussion of carbonate porosity evolution later in this chapter.

Finally, boundstones, because of the presence of an integral biologic framework and intense cementation at the time of deposition, will commonly exhibit little evidence of compaction of any type. The porosities of reef-associated lagoonal, shelf, and slope sequences, however, are significantly reduced by mechanical as well as chemical compaction. Shelf margin relief relative to adjacent shelf and periplatform facies is thought to have been enhanced in the Devonian sequences of the Canning Basin by the resistance to compaction of the well-cemented shelf margin reefs while adjacent muddy and grainy facies show abundant stylotization (Playford, 1980; Playford and Cockbain, 1989). Wong and Oldershaw (1981) demonstrated a similar facies control over chemical compaction in the Devonian Kaybob reef, Swans Hills Formation, Alberta.

Subsurface cements in a passive margin setting

The introduction of cement into pore space requires a ready source of $CaCO_3$ during the precipitational event. In surficial diagenetic environments, large volumes of cement are precipitated into pore spaces prior to significant burial (James and Choquette, 1984). The $CaCO_3$ needed for these cements is provided directly from supersaturated marine and evaporated marine waters, or indirectly, with the $CaCO_3$ coming from the dissolution of metastable carbonate phases in the sediment/rock. In each case, the rate of precipitation and the ultimate volume of cement precipitated are influenced strongly by the rate of fluid flux through the pore system.

While the volume of cement is not as large as encountered in pre-burial environments, subsurface cements have been documented from a number of sedimentary basins (Meyers, 1974; Moore, 1985; Niemann and Read, 1988). Under burial conditions, however, we are cut off from supersaturated surface marine waters, a ready reservoir of atmospheric and soil gas CO_2 is not available, and most subsurface fluids are in equilibrium with $CaCO_3$. Under these conditions, neither solution nor precipitation tends to occur (Matthews, 1984; Choquette and James, 1987). Morse et al.(1997), however, noted that mixtures of diverse subsurface fluids (such as brines and meteoric water) may be both saturated as well as undersaturated with respect to $CaCO_3$.

If metastable mineral grains survive early surficial diagenesis and are buried into the subsurface, a solubility contrast is created that must be satisfied by dissolution and precipitation (Chapter 7). Mineralogical stabilization in the subsurface tends to take place by microsolution and precipitation because subsurface diagenetic fluids move slowly and are so near equilibrium with the surrounding rock. In this case, then, solution and precipitation takes place in a closed system and little $CaCO_3$ leaves the grain to be made available for cementation. While increasing temperature does favor cementation in the subsurface (Bathurst, 1986) an adequate source of $CaCO_3$ to form the cement is an obvious problem (Bathurst, 1975).

Most workers turn to pressure solution, active during chemical compaction, as the $CaCO_3$ source for the bulk of the pre-hydrocarbon burial cements encountered in the record (Meyers, 1974; Bathurst, 1975; Scholle and Halley, 1985; Moore, 1985; Choquette and James, 1987). As was seen in the discussion of chemical compaction earlier in this chapter, however, an enormous amount of section must be removed by pressure solution to form a relatively small volume of cement. In many cases it seems obvious that we must supplement the $CaCO_3$ provided by pressure solution with other sources. $CaCO_3$ may be gleaned from the dissolution of limestones and dolomites by aggressive fluids formed in association with hydrocarbon maturation and diagenesis (Fig. 9.3) (Schmidt and McDonald, 1979; Druckman and Moore, 1985; Moore, 1985; Mazzullo and Harris, 1991). In addition, $CaCO_3$ may be formed as a result of high temperature thermochemical sulfate reduction (Fig. 9.3) (Machel, 1987; Heydari and Moore, 1989; Heydari, 1997a). Finally, $CaCO_3$ may be imported from adjacent subsurface carbonate units also undergoing chemical compaction (Bathurst, 1975; Wong and Oldershaw, 1981).

We will first briefly outline those criteria useful for the recognition of subsurface cements and then evaluate the impact of subsurface cementation on reservoir porosity evolution.

Petrography of burial cements

In the following discussion, only those cements precipitated in the subsurface from waters that are isolated from surficial influences will be considered. Regional artesian meteoric aquifer systems were considered in the previous chapter and will not be discussed here (Fig. 9.1).

In general, subsurface cements are coarse, commonly poikilotopic, dully luminescent, and not as inclined to luminescent zoning as their surficial counterparts (Moore, 1985) (Choquette and James, 1990; Heydari and Moore, 1993). These burial cements are commonly either calcite or saddle dolomite.

Perhaps the most reliable indicator for the late subsurface origin of cements is the determination that the cement clearly postdates burial phenomena. Relationships of cements to such features as fractures, compacted and fractured grains, stylolites, and pyrobitumen have long been used as the basic criteria for the recognition of these late cements (Moore, 1985; Choquette and James, 1990; Heydari, 1997a) (Chapter 4) (Fig. 9.13).

While burial cements will generally exhibit two-phase fluid inclusions, the presence of two-phase fluid inclusions in cements are is not always a valid criteria for late subsurface

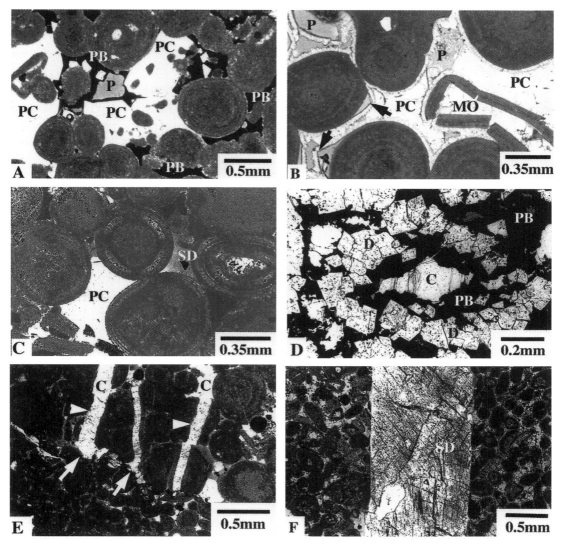

Fig. 9.13. (A) Jurassic Smackover with intergranular pores (P) filled with post-compaction calcite (PC) cement overlain by phyrobitumen (PB). Getty Masonire 18-8, 13,613' (4150 m) Clarke Co., Miss. Plain light. (B) Post-compaction poikilotopic calcite cement (PC). Note spalled, early, circumgranular calcite cement (arrows) incorporated in later cement. (MO) is an ooid coated aragonitic grain subsequently dissolved prior to compaction, crushed by compaction then incorporated into the burial poikilotopic calcite cement (PC). These relationships clearly suggest that the Jurassic ooids were originally calcite since they were not dissolved. Same well and depth as (A). Plain light (C) Post-compaction poikilotopic calcite cement (PC) and saddle dolomite (SD) filling pore space in Jurassic Smackover. Arco Bodcaw #1, 10,910' (3326 m). Crossed polars. (D) Pervasive dolomite (D) in Jurassic Smackover Formation, porosity filled with bitumen (PB). Mold filled with post-bitumen calcite(C).Phillips Flurry A-l, l9,894' (6065m), Stone Co.Miss. Sample furnished by E. Heydari. Plain light. (E) Calcite (C) filled tension gash fractures (small arrows) associated with stylolites (large arrows). Smackover Formation, Getty #1 Reddoch, Clarke Co., Mississippi, 13692' (4174 m). Sample furnished by E. Heydari. Plain light. (F) Large fracture filled with saddle dolomite (SD) in Jurassic Smackover, Evers and Rhodes #1, 11,647' (3551 m), east Texas. Plain light.

origin because of the demonstrated potential for step-wise subsurface re-equilibration (Moore and Druckman, 1981; Moore and Heydari, 1993; Eichenseer et al., 1999).

Geochemistry of burial cements

Subsurface cements are precipitated from fluids that are in the process of chemically evolving, as a result of continuous rock-water interaction in the burial diagenetic environment over extremely long periods of time. These rock-water interactions are not confined specifically to carbonates but commonly include related basinal siliciclastics and evaporites. Rock-water interaction with such a mineralogically diverse rock pile often results in the formation of progressively more concentrated brines, rich in metals such as Fe, Mn, Pb and Zn (Collins,1975; Carpenter, 1978; Land and Prezbindowski, 1981;Morse et al., 1997).

Thus burial cements are commonly Fe and Mn rich, with late burial saddle dolomites often containing over 5wt% Fe. Lead and zinc mineralization are often associated with these late Fe-rich cements and dolomites (Carpenter, 1978; Prezbindowski, 1985; Moore, 1985).

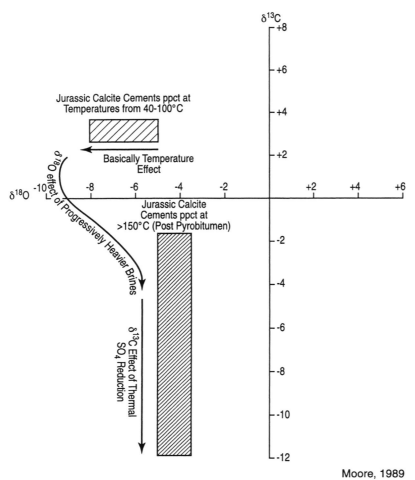

Fig. 9.14. Stable isotopic composition of burial calcite cements showing the effects of increasing temperatures, increasingly heavier brines, and thermal sulfate reduction on the oxygen and carbon isotopic composition. Data on calcites precipitated from 40 to 100°C. Used with permission of Elsevier.

However, Sr concentrations, are commonly quite low, even though subsurface brines may often exhibit high Sr/Ca ratios (Collins, 1975; Moore, 1985). Moore and Druckman (1981) and Moore (1985) suggested that the slow rate of precipitation that characterizes subsurface calcite and dolomite cements may be responsible for the apparent very small Sr distribution coefficients indicated for these cements (Chapter 4 discussion of distribution coefficients).

The oxygen isotopic composition of burial calcite and dolomite cements is clearly affected by the strongly temperature-dependant fractionation of oxygen isotopes between cement and fluid (Anderson and Arthur, 1983) (Chapter 4). As temperature increases during burial, the calcites precipitated should exhibit progressively lighter oxygen isotopic compositions (Moore and Druckman, 1981; Moore, 1985; Choquette and James, 1990). This same fractionation process during subsurface rock-water interaction, however, leads to the progressive enrichment of subsurface fluids in ^{18}O (Land and Prezbindowski, 1981). The progressive depletion of ^{18}O in the cement, therefore, may ultimately be buffered, and even reversed, by the increasing enrichment of ^{18}O in the pore fluid through time (Heydari and Moore, 1993; Moore and Heydari, 1993; Heydari, 1997a) (Fig. 9.14).

The carbon isotopic composition of most subsurface cements is generally rock buffered as a result of chemical compaction so it will show little variation and is commonly enriched with ^{13}C. There is usually a trend, however, toward slightly depleted ^{13}C compositions in those cements precipitated shortly before hydrocarbon migration. This trend is probably the result of the presence of increasing volumes of ^{12}C as a byproduct of hydrocarbon maturation (Moore, 1985) (Fig. 9.14).

Very late cementation events, particularly those involved with late sulfate reduction associated with thermal degradation of hydrocarbons at temperatures in excess of 150° C, can exhibit very light carbon isotopic compositions (Heydari and Moore, 1989; Heydari, 1997b) (Fig. 9.14).

Sr isotopes can be quite useful in the study of late burial cements. During the chemical compaction phase of burial diagenesis, before pore fluids have had the opportunity to evolve by mixing with basinal brines, the Sr isotopic composition of burial cements will generally represent the Sr isotopic composition of the compacting limestones and dolomites. The Sr isotopic composition of the cements therefore reflects the Sr isotopic composition of the marine water in which the limestones and dolomites formed. Basinal fluids commonly have the opportunity to interact with

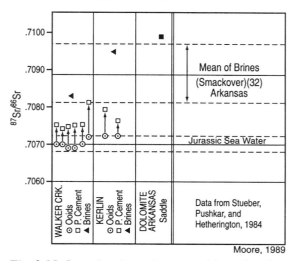

Fig. 9.15. Strontium isotopic composition of carbonate components and formation waters from oil fields in southern Arkansas, USA. Jurassic seawater value of .7070 is taken from the curve of Burke and others, 1983. Used with permission of Elsevier.

siliciclastic, clay-rich sequences at elevated temperatures (>100° C), and hence are often enriched with respect to ^{87}Sr. As these pore fluids arrive (generally shortly before or during hydrocarbon migration) and mix with marine connate waters in carbonate shelf sequences, the ^{87}Sr/^{86}Sr ratio of the pore fluid, as well as the cements precipitated from the pore fluid, increase. The ^{87}Sr/^{86}Sr ratio in the burial cement may, therefore, be a gross measure of the relative proportions of basinal brine and connate water present at the time of cement precipitation. As such, the Sr isotopic composition of the cement may reflect the stage of pore fluid evolution at the time of the cementation event (Moore, 1985; Heydari and Moore, 1989)(Banner and Hanson, 1990; Banner, 1995) (Fig. 9.15).

Impact of late subsurface cementation on reservoir porosity

Compared to pressure solution, post-compaction subsurface cements are probably of minor importance in reducing the total pore volume available for hydrocarbons in a subsurface reservoir. Several observations support this conclusion. The actual volume of late, post-compaction cements documented in hydrocarbon reservoirs is relatively small (Moore, 1985; Scholle and Halley, 1985; Prezbindowski, 1985; Heydari, 2000). These data support the conclusion of Rittenhouse (1971) that a large volume reduction by pressure solution yields a comparatively small volume of cement. This small yield is compounded by the fact that as hydrocarbon migration commences, reservoirs will generally be filled; and reservoir diagenesis may be terminated by the presence of hydrocarbons long before chemical compaction (the main source for cementation materials) is completed. Even though cementation will continue as long as chemical compaction and subsurface dissolution provides a carbonate source, these later cements will only impact that porosity not filled with oil and hence are generally economically unimportant except in those cases of multiple migration.

Subsurface dissolution in a passive margin setting

There is abundant evidence that aggressive pore fluids, derived as a result of hydrocarbon maturation and hydrocarbon thermal degradation, exist and affect rock sequences in the subsurface (Schmidt and McDonald, 1979; Moore and Druckman, 1981; Surdam et al., 1984); Mazzullo and Harris, 1991). Most workers call upon the decarboxilation of organic material during the maturation process to provide the CO_2 and organic acids necessary to significantly expand porosity by dissolution (Fig. 9.3).

The porosity enhancement related to hydrocarbon maturation will normally occur prior to the arrival of hydrocarbons in reservoirs and hence is of economic concern. Intergranular porosity may undergo solution enlargement that can ultimately evolve into vugs. Moldic porosity is seldom seen. Late stage pores are generally rounded and can be confused with meniscus vadose pores. However, unlike vadose pores, these rounded pores cut across pre-existing grains and earlier cements. They cut all textural elements including demonstrable late subsurface cements and may enlarge stylolites (Fig. 9.16).

The thermal degradation of hydrocarbons

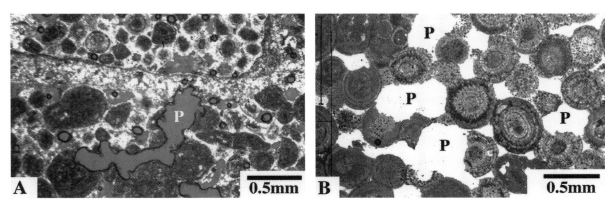

Fig. 9.16. Late subsurface dissolution fabrics of Jurassic Smackover Formation ooid grainstonc reservoir rocks. Note that all pores (P) are rounded and mimic vadose pore geometries. (A) #1 Fincher Clark at 7953'. Plain light with gypsum plate. (B) Arco Bodcaw #1 10,894' (3321 m). Plain light.

after emplacement in the reservoir (at temperatures in excess of 150°C, Figure 9.3) leads to the production of CO_2, H_2S, methane, and solid pyrobitumen (Tissot and Welte, 1978; Sassen and Moore, 1988). While the CO_2 and H_2S may combine with water to trigger porosity enhancement in the growing gas cap by dissolution, the concurrent precipitation of solid pyrobitumen can seriously degrade the porosity and permeability of the reservoir (Fig. 9.13A and D). The presence of SO_4 acts as a catalyst in the breakdown of methane and can accelerate the production of aggressive pore fluids (Sassen and Moore, 1988).

At this point we do not have enough data to fully evaluate the importance of subsurface dissolution and the impact of late organic diagenetic processes on the final stages of porosity evolution in carbonate rock sequences in a passive margin. We can, however, state with some certainty that: 1) porosity is being enhanced in the deep subsurface, 2) that dissolution does provide carbonate for further cementation, and 3) that pyrobitumen formation during thermal destruction of hydrocarbons is an important, but seldom considered, porosity-destructive process.

The North Sea Ekofisk Field: a case history of porosity preservation in chalks

The Ekofisk field is a giant oil field developed in Upper Cretaceous and Lower Tertiary chalks in the Norwegian and Danish sectors of the central graben of the North Sea

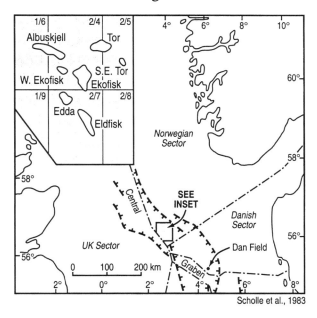

Fig. 9.17. Location of Ekofisk Field and other major Mesozoic chalk fields in the Central Graben area of the North Sea. Used with permission of Elsevier.

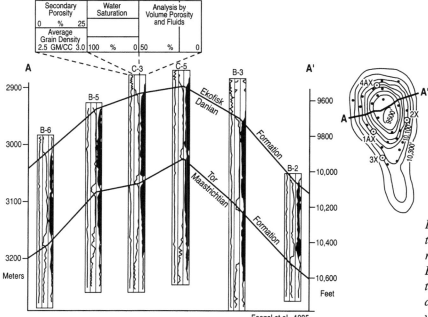

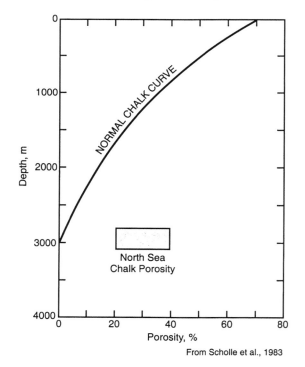

Fig. 9.18. Cross section through the Ekofisk structure. Map to the right shows structure on top of the Ekofisk Formation. Black area to the right of one set of curves shows amount of oil-filled porosity. Used with permission of Elsevier.

Basin (Fig. 9.17). Discovered in 1969, it is currently producing 280,000 barrels of oil and 580 mcf of gas per day. It had produced approximately 2 billion barrels of oil to the end of 1996, out of an estimated 6 billion barrels of oil in place. (Landa et al., 1998). The chalk fields of the central graben are anticlinal structures developed as a result of Permian salt movement. Overpressured Late Tertiary shales seal the reservoirs. At present, the reservoirs are encountered between 2930m and 3080m (Fig. 9.18).

The reservoirs are Danian and Maastrichtian chalks deposited in the central graben area in relatively deep waters. Deposition was slow, as a pelagic sediment rain, as well as rapid mass flow deposits. These mass flow deposits originated along the Lindesnes Ridge to the south (Scholle et al., 1983a; Taylor and Lapré, 1987; Herrington et al., 1991). Salt related structures were present within the graben during the Late Cretaceous

Fig. 9.19. Range of North Sea chalk porosity encountered at a depth of 3000 m compared to an average chalk burial-porosity curve. Used with permission of Elsevier.

and Early Tertiary and may have affected the deposition of the chalks (Feazel et al., 1985). Oil production is from preserved primary porosity that ranges from 25 to 42%. Permeabilities are low, averaging about 1 md. Production tests indicate effective permeabilities of over 12 md as a result of the extensive fracturing common to most of the chalk fields of the central graben. The reservoirs are sourced from the underlying Jurassic Kimmeridgian shales (Scholle et al., 1983b).

The biggest anomaly concerning production from the Ekofisk Field is the extraordinary levels of primary porosity encountered in chalks buried to depths of some 3000m. Scholle (1983), based on his studies of European and North American chalk sequences, suggests that the original porosity of most chalks should be completely destroyed by mechanical and chemical compaction when buried to this depth (Fig. 9.19).

Most workers at Ekofisk agree that reservoir overpressuring is one of the major contributory factors to porosity preservation in the greater Ekofisk area (Scholle, 1977a; Feazel et al., 1985; Taylor and Lapré, 1987; Maliva and Dickson, 1992; Caillet et al., 1996). Figure 9.20 is a generalized depth-pressure plot for a typical Ekofisk well. At 3050m, the average depth of Ekofisk reservoirs, normal pore pressures should be approximately 4300 psi and lithostatic pressure 9000 psi. However, pore fluid pressures of over 7000 psi have been reported at that depth, reducing the net lithostatic load responsible for compaction to a level of approximately 1900 psi. This pressure is equivalent to about 1000m burial depth, and using Scholle's (1983) chalk burial curve (Fig. 9.19) one would expect to encounter porosities in the range found at Ekofisk.

Another contributory factor affecting porosity preservation at Ekofisk may have been the early migration of oil into the reservoirs. The geothermal gradients coupled with the burial history of the Jurassic source rocks point toward early maturation and migration in the region (Harper, 1971; Van den Bark and Thomas, 1981). The general trend of porosity reduction coupled with depth and lateral distance from structural closure has been documented in many of the Greater Ekofisk fields including Ekofisk itself. These observations certainly support some functional relationship between hydrocarbons and porosity preservation. As was noted by Feazel and Schatzinger (1985), when oil displaces water in the pores, the chemical compaction process will be shut down because solution and

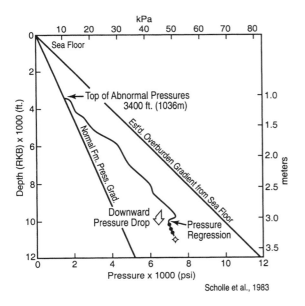

Fig. 9.20. Generalized pressure-depth plot for a typical Ekofisk well in the Central Graben of the North Sea compared to expected formation pressures. Abnormal pressures are encountered in most of the Lower Tertiary and Upper Cretaceous section including the chalk reservoirs at 3 to 3.5km depth. Used with permission of Elsevier.

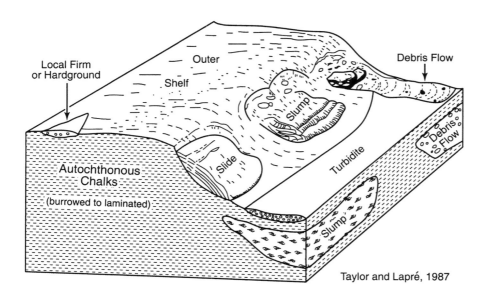

Fig. 9.21. Depositional model for chalks emphasizing slope failure and transport of allochthonous chalks in the moderately deep marine environment of the Central Graben area of the North Sea. Used with permission of Elsevier.

precipitation of carbonate requires a film of water.

Many workers, such as Taylor and Lapre´ (1987), feel that extensive mass-movement of the chalks destroyed early sea floor cements and deposited poorly packed sediments rapidly across future reservoir areas. These rapidly deposited allochthonous chalks were more porous than the slowly deposited in-situ chalks because they were not subjected to such intensive bioturbation and resulting dewatering during deposition (Fig. 9.21). Taylor and Lapre´ (1987) contend that this porosity contrast was retained into the subsurface. Mechanical and chemical compaction reduced the porosity of both chalk types equally until compaction was shut down by hydrocarbon migration into the reservoirs at about 1000m burial depth and the concurrent development of elevated pore pressures during deeper burial. Figure 9.22 illustrates a typical section through Cretaceous chalks in the Ekofisk Field illustrating the close relationship between high porosity/permeability and interpreted allochthonous chalk facies.

Maliva and Dickson (1992) conducted a detailed study on the microfacies and diagenetic controls over porosity distribution in the Eldfisk Field in the southern part of the Greater Ekofisk area (Fig. 9.17). While the authors generally agreed with previous workers concerning the importance of overpressuring and early oil migration in preservation of chalk porosity, they documented the importance of microfacies in determining the detailed vertical distribution of porosity characteristics. Figure 9.23 shows the distribution of the major microfacies, insoluble residues, and porosity of the Chalk Group in Eldfisk Field. The straight line H represents a hypothetical porosity/depth compaction curve that one would observe if the chalk were of a uniform composition. Departure from this curve represents the effects of microfacies and insoluble residue on porosity.

Maliva and Dickson (1992) concluded: 1) consolidation (mechanical compaction) is the main mechanism for porosity reduction at Eldfisk and by extension at Ekofisk to the north; 2) while some chalks have undergone extensive

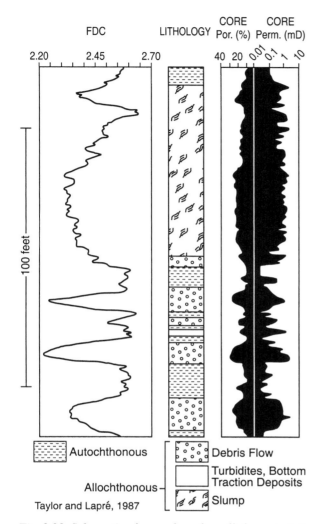

Fig. 9.22. Schematic of a core log of a well characteristic of the Ekofisk field in the Central Graben showing relationship between lithology and reservoir characteristics. Note that higher porosity and permeability coincides with allochthonous sequences. Used with permission of Elsevier.

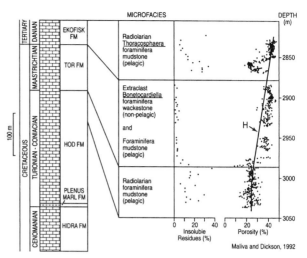

Figure 9.23. Lithostratigraphy of the Chalk Group in Eldfisk field. Ekofisk and Tor Formation thicknesses, microfacies distribution, insoluble residue concentrations, and core-plug porosity data are for well 2/7 B11. Used with permission of AAPG.

recrystallization, chemical diagenesis doesn't seem to affect porosity; 3) mode of deposition (allochthonous vs. autochthonous) by itself has little effect on porosity preservation in chalks. Rather, differential compaction related to insoluble residue leads to porosity layering; 4) insoluble residue is related directly to microfacies; and 5) the retardation of compaction by overpressuring seems to be the most important factor for the preservation of high porosities in North Sea chalks.

Ekofisk reservoirs have exceptional production characteristics. The discovery well had an IP of over 10,000 barrels of oil per day (Feazel et al, 1985). These high production rates seem to be the result of high porosity, thick productive intervals (over 150m), low oil viscosity (36° API gravity), high GOR (1547:1), and intensive fracturing that enhances their rather low permeabilities (Feazel et al., 1985).

In the following section we will consider the role of the rapid tectonically driven migration of hot basinal brines in subsurface diagenetic processes such as recrystallization, dolomitization, dissolution, and hydrothermal mineralization.

THE ACTIVE OR COLLISION MARGIN BURIAL REGIMEN

Introduction

Diagenesis in the active (collison) margin burial regimen in carbonate sequences has long elicited a great deal of interest because of its role in the development of stratified ore deposits (Mississippi Valley Type-MVT) and hydrocarbon migration and accumulation.

However, the determination of the diagenetic history of carbonate rocks in this setting is particularly difficult because late, hot, tectonically driven waters may recrystallize and chemically overprint earlier-formed diagenetic phases often leading to confusion and controversy (Moore, 1989).

Perhaps the best approach to this complex problem is to present two case histories of classic collision margins where the diagenetic history and paleohydrology of the carbonate sequence have been sufficiently documented to allow us to draw some broad conclusions and to identify the problems common to the setting.

The first case history will be the Ordovician Knox Dolomite of the southern Appalachian Basin, USA, that hosts isolated MVT and hydrocarbon deposits. The second will consist of the Devonian of the Alberta Sedimentary Basin, Canada, that hosts both major MVT and important hydrocarbon deposits.

Lower Ordovician Upper Knox Dolomite of the southern Appalachians, USA

The Lower Ordovician, Upper Knox Group and its temporal equivalents, the Arbuckle Group in Oklahoma and Arkansas and the Ellenburger Group in Texas, were deposited on a gently sloping ramp, or platform that was several hundred kilometers wide and over 5000km long, extending from west Texas to Newfoundland. The shallow water carbonates of this platform were extensively dolomitized during and shortly after deposition by marine and modified marine waters controlled by high frequency sea level changes (Montañez and Read, 1992).

This huge platform was broken up during the Alleghenian/Ouchita orogeny during Pennsylvanian-Permian time. Today, the Upper Knox group is exposed in the fold and thrust belt of the Valley and Ridge Province of the central and southern Appalachians (Fig. 9.24), while the Arbuckle and Ellenburger remain buried in the subsurface and produce prodigious quantities of gas in the subsurface of Texas and Oklahoma.

The Upper Knox Group in the Appalachians ranges in thickness from 200 to 1200m. Thick Knox sequences were deposited in several depocenters separated by positive features such as the Virginia Arch (Fig. 9.25) (Montañez and Read, 1992). Today the Knox is found both in the outcrop and buried to depths in excess of 1700m. Figure 9.26 shows a general burial curve for the Knox Group in Tennessee and Virginia. It is estimated that the Knox was buried to depths in excess of 5km during Alleghenian tectonism and suffered temperatures near 150°C. Average tectonic horizontal shortening of Knox carbonates ranges from 25-60% (Montañez, 1994).

Figure 9.27 illustrates the paragenetic sequence of the Upper Knox Group for Tennessee. The most striking aspects of this sequence are: 1) that regional burial

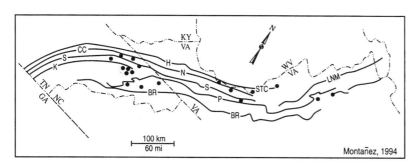

Fig. 9.24. Location map showing distribution of measured sections and cores within the Valley and Ridge Province, central and southern Appalachians. Major thrust faults are shown as heavy black lines: K = Knoxville, S = Saltville, CC = Copper Creek, BR = Blue Ridge, P = Pulaski, N = Narrows, H = Clinchport-Honaker, STC = St. Clair, LNM = Little North Mountain. Shaded area marks location of Mascot-Jefferson Mississippi Valley type mining district, east Tennessee. Used with permission of GSA.

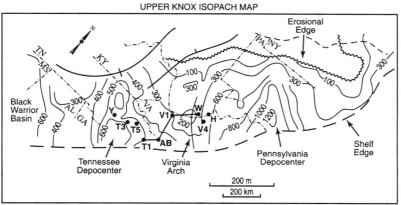

Fig. 9.25. Upper Knox isopach map on palinspastic base (Read, 1990) showing distribution of selected measured sections and cores: Y (Young Mine core, Jefferson-Mascot District, TN), T1 (ARCO core TNC-1, Green Co., TN), T3 (ARCO core TNC-3, Hamblen Co., TN), T5 (ARCO core TNC-5, Hawkins Co., TN), AB (Avens Bridge), V1 (ARCO core VAC-1, Tazewell Co., VA), V4 (ARCO core VAC-4, Alleghany Co., VA), W (ARCO core WVC-1 core, Monroe Co., WV), H (Hot Springs core, Alleghany Co., VA). Used with permission of GSA.

replacement dolomitization and dolomite cementation events seem to span most of the Paleozoic burial history of the Knox, and 2) that burial diagenesis is punctuated by regional dissolution events. Figure 9.28 shows a schematic of the relationship of the burial replacement dolomite to later dolomite cements, dissolution events, and hydrocarbon migration.

Figure 9.29 shows a cross-plot of $\partial^{18}O$ versus $^{87}Sr/^{86}Sr$ for Knox dolomites. The early syndepositional dolomites show elevated $^{87}Sr/^{86}Sr$ ratios relative to Ordovician marine waters suggesting that they have been recrystallized, probably by the fluids responsible for the zone 2 and 3 replacement dolomites. All other dolomite zones show elevated $^{87}Sr/^{86}Sr$ ratios and progressively lighter $\partial^{18}O$ values suggesting formation from waters at elevated temperatures that have interacted extensively with siliciclastics (basinal fluids) (Montañez, 1994).

While there is little doubt that the elevated temperatures and pressures encountered in the subsurface tend to favor dolomitization (Hardie, 1987; Morse and Mackenzie, 1990),

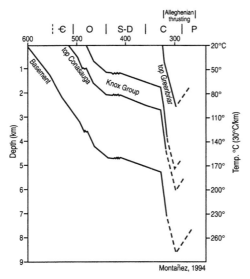

Figure 9.26 Generalized burial history plot constructed for the stratigraphic section on the Saltville thrust sheet (Fig. 9.24). Dashed lines indicate inferred burial depth by depositional and tectonic overburden, and subsequent uplift. Adapted from Mussman et al. (1988). Used with permission of AAPG.

the volume of water required for these deep burial (1-2.5km) dolomitization events is enormous. A number of studies have concluded that compactional de-watering cannot provide the amount of water or deliver the solutes necessary for regional dolomitization (Anderson and Machel, 1988; Montañez, 1994). Montañez concludes that tectonic compressional loading associated with the Alleghanian/Ouchita orogeny in the Carboniferous was responsible for the extensive and relatively rapid movement of dolomitizing and dissolutional fluids into and through the Knox carbonates (Fig. 9.30). This interpretation is supported by the thermal history of the fluids responsible for various diagenetic phases as estimated from isotopic compositions and two-phase fluid inclusions (Fig. 9.31). These data suggest a prograde (increasing temperature) then retrograde

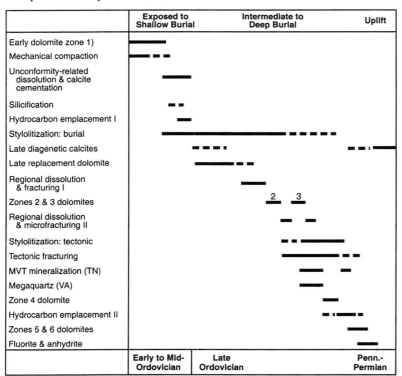

Fig. 9.27. Diagenetic sequence inferred for Upper Knox carbonates in the study area. Overlapping phases are listed according to their first appearance. Bars indicate relative depth/time ranges of the different diagenetic events. The transition between shallow burial and intermediate-to-deep burial is interpreted as the onset of stylolitization; the transition from burial to uplift is defined by the initiation of the retrograde thermal history. No temporal significance is implied for the time span between Ordovician and Pennsylvanian/Permian periods. Used with permission of AAPG.

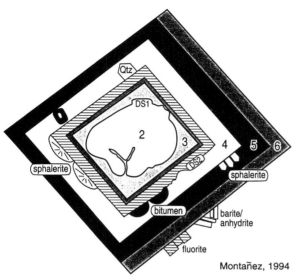

Fig. 9.28. Schematic dolomite crystal showing paragenetic relationship of dolomite zones (2 to 6) to non-carbonate diagenetic minerals and dissolution surfaces (DS). Used with permission of AAPG.

Fig. 9.29. Crossplot of dolomite $\partial^{18}O$ and $^{87}Sr/^{86}Sr$ values. Within zone 2/R field, crosses = late diagenetic replacement dolomites; filled squares = zone 2 dolomite cement. Marine field from Lohmann and Walker (1989); Popp et al. (1986); Burke et al. (1982); Gao and Land (1991b). Used with permission of AAPG.

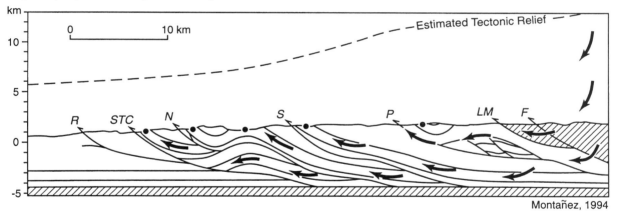

Fig. 9.30. Schematic diagram illustrating fluid flow pathways from Alleghenian highlands through the thrusted Paleozoic section. Stipple pattern = Knox carbonates; cross-hatched pattern = decollement sediments. Structure cross section for southwestern Virginia after Woodward (1987). R = Richlands thrust fault; STC = St. Clair thrust fault; N = Narrows thrust fault; S = Saltville thrust fault; P = Pulaski thrust fault; LM = Little North Mountain; F = Fries thrust fault. Magnitude of estimated tectonic relief based on estimated overburden thickness, and fluid inclusion homogenization temperatures. Arrows denote interpreted movement of fluids through Knox carbonate and along thrust planes. Flow is cross formational at the elevated recharge area of the basin and strongly focused and channeled along the deeply buried Knox carbonates and thrust planes cratonward. Used with permission of AAPG.

(decreasing temperature) thermal history that would seem to mirror the tectonic loading (prograde) then subsequent uplift (retrograde) of the sequence expected during an orogenic event. MVT mineralization and hydrocarbon migration occurred at the end of the loading

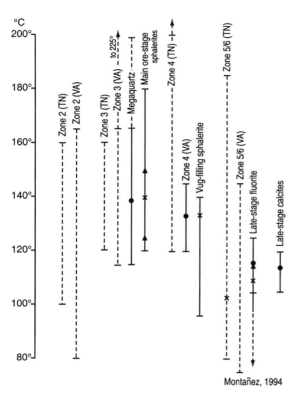

Fig. 9.31. Summary diagram of the estimated precipitation temperatures for late diagenetic minerals in Knox carbonates. Dashed lines represent temperature ranges derived indirectly from the oxygen isotopic values of zoned dolomites. Solid lines represent temperature ranges directly estimated from fluid inclusion homogenization temperatures. Calculated means indicated by symbols: dots - this study; triangles - from Kesler et al. (1989) and Haynes et al. (1989); X from Caless (1983). Used with permission of AAPG.

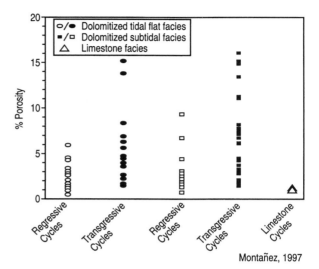

Fig. 9.32. Measured porosity of dolomitized subtidal and tidal-flat samples and limestone samples plotted by their occurrence in stacks of regressive or transgressive cycles. Plot shows that dolomitized tidal-flat and subtidal facies in transgressive cycles have porosity values that overlap to significantly exceed the range of porosity values of dolomitized tidal-flat and subtidal facies in regressive cycles. Limestone facies have the lowest measured porosity and smallest range in values. Used with permission of SEPM.

phase when temperatures seem to have been the highest.

How did these fluids move through the Knox and what impact did they have on porosity and permeability of the Knox carbonates? Montañez (1997) recognized that higher final porosities in Knox carbonates seemed to be related to the transgressive phases (TST) of low-frequency sea level cycles (Fig. 9.32). Low rates of accommodation during the regressive phases (HST) resulted in complete pervasive dolomitization, small crystal sizes and resulting low permeabilities. Dolomitization in the TST was not complete and consisted of larger isolated crystals in a matrix of pellet or ooid grainstones, packstones, and wackestones. Upon burial when the first pulse of warm dolomitizing fluids began to move through the Knox, the TST had higher porosities and permeabilities and formed a preferential fluid conduit. In the HST, small crystal sizes and low permeabilities caused rapid nucleation of dolomite cements, rapid recrystallization, and degradation of remaining porosity. Large and isolated nucleation sites resulted in coarse replacement dolomites in the TST while residual carbonates were dissolved

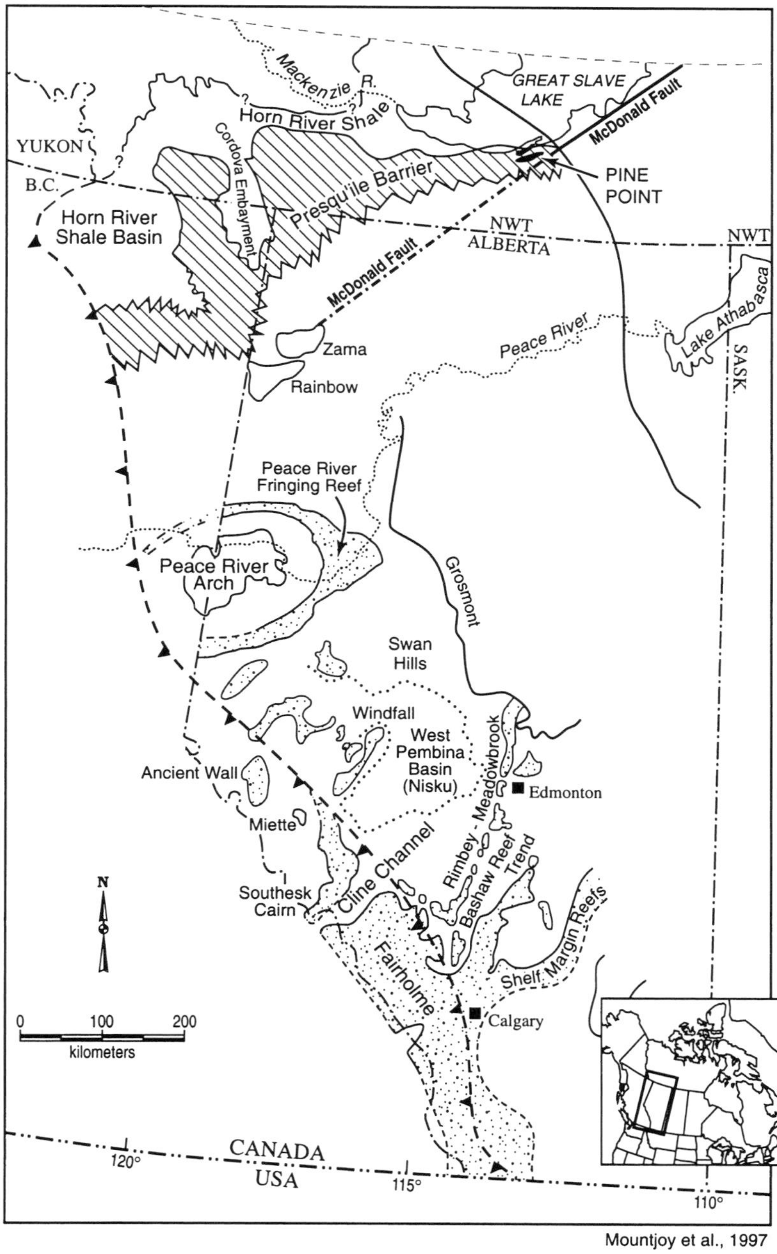

Fig. 9.33. Simplified regional geological map of the Western Canada Sedimentary Basin during Middle Devonian north of Peace River Arch and Upper Devonian south of the Arch. The Presquille barrier crops out in the Pine Point area west of the Canadian Shield. It extends westward into the subsurface of the Northwest Territories and northeastern British Columbia where its present burial depth is between 3500 and 4000 m. The McDonald-Hay River fault bounds two major geological provinces of the Canadian Shield and appears to have been active during parts of the Phanerozoic. Rainbow and Zama sub-basins formed south of the Presquille Barrier. Larger Upper Devonian reefs and reef trends (stippled pattern) are shown south of the Peace River Arch. Used with permission of SEPM.

Mountjoy et al., 1997

during the dolomitizing process leading to enhanced porosities. Montañez (1997) suggests that these conduits remained as the major pathways for later diagenetic events such as MVT mineralization and hydrocarbon migration.

The tectonically driven late diagenetic history of Knox carbonates strongly parallels the models of late dolomitization documented in the Ordovician Ellenburger in Texas (Kupecz and Land, 1991) and the Ordovician Arbuckle in Oklahoma (Land et al., 1991).

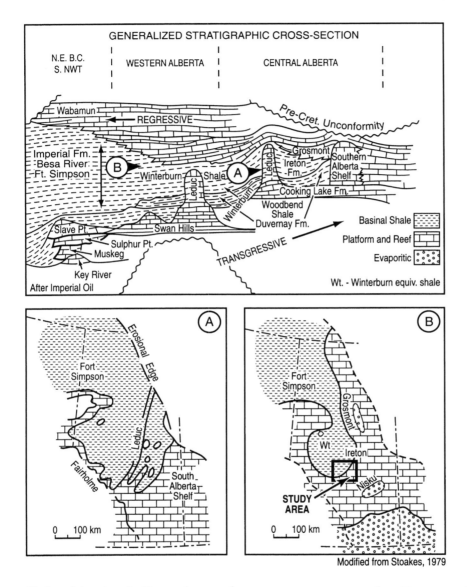

Fig. 9.34. Stratigraphic (upper diagram) and paleogeographic setting (lower diagram) of the Upper Devonian Leduc and Nisku formations. Used with permission of Elsevier.

Dolomitization in Devonian carbonates, Western Canada Sedimentary Basin, Alberta, Canada

The Western Canada Sedimentary Basin (Fig. 9.33) is a northerly trending wedge of sedimentary rocks more than 6km thick extending from the Canadian Shield to the Cordilleran foreland thrust belt. The tectonic evolution of the basin is a continuum including an early rift to passive margin phase (which includes the Middle-Upper Devonian carbonates) to foreland basin in Middle Jurassic and again from Late Cretaceous to Paleocene. Tectonic loading during these phases resulted in the deep burial of the Devonian along the western margin of the basin, uplift and exposure of the sequence along the thrust belt and development of tectonically-induced deep basinal hydrologic flow systems in the

Devonian (Stoakes et al., 1992; Bachu, 1997).

Figure 9.34 suggests that the Upper Devonian carbonates in the Western Canada Sedimentary Basin consist of one large transgressive-regressive cycle (2nd order?). This cycle may reflect the shift from a rift to a drift phase in the Devonian.

Many Frasnian age sequences in the basin, including some of Canada's most important oil reservoirs such as the Leduc reefs, are commonly pervasively dolomitized. Illing (1959) suggested that the bulk of these coarse-crystalline dolomites originated in the subsurface as a result of basinal compaction and de-watering. Subsequent workers (Mattes and Mountjoy, 1980; Machel and Mountjoy, 1987; Qing et al., 1995; Drivet and Mountjoy, 1997; Mountjoy et al., 1997; Qing, 1998) have used isotope, trace element, and fluid inclusion data to support a burial origin for these economically important dolomites. Mass balance considerations (Anderson and Machel, 1988; Amthor et al., 1993a) have led to the abandonment of basinal compaction and de-watering in favor of tectonically-induced subsurface hydrothermal flow regimes as a source for dolomitizing fluids (Qing et al., 1995; Mountjoy et al., 1997; Qing, 1998).

Figure 9.35 illustrates the paragenetic sequence for the Cooking Lake and Leduc

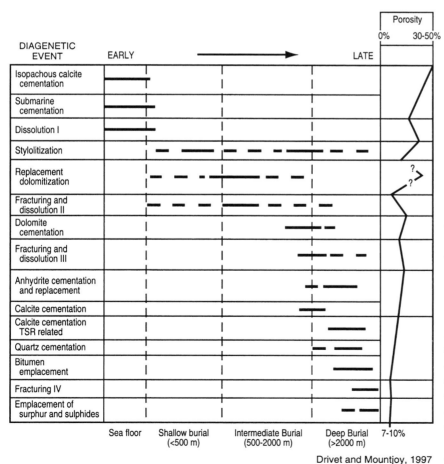

Fig. 9.35. Paragenetic sequence of Leduc Formation in the Homeglen-Rimbey and Garrington fields. The terms shallow intermediate, and deep burial are defined relative to stylolite formation. Intermediate burial represents the onset of stylolitization (>500 m from Dunnington 1967; Lind 1993), and shallow and deep burial define pre- and post-incipient stylolitization corresponding to depths of < 500m and > 1500m, respectively. Used with permission of SEPM.

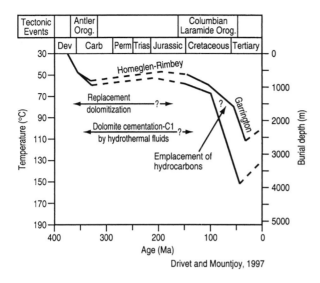

Fig. 9.36. Burial-temperature-time plot for Leduc buildups in the Homeglen-Rimbey and Garrington fields, constructed assuming a geothermal gradient of 30°C/km (Walls el al. 1979) and a surface temperature of 30°C (Hudson and Anderson 1989). The inferred burial depths, temperatures, timing of replacement dolomitization, dolomite cementation, and hydrocarbon generation are also plotted. Used with permission of SEPM.

formations in the Rimbey-Meadowbrook trend just west of Edmonton (Fig. 9.33). This relative sequence is fairly consistent with many diagenetic studies of Upper Devonian carbonates carried out by Mountjoy and his students (see last paragraph) throughout the Western Canada Sedimentary Basin. The major features are burial replacement dolomitization, burial dolomite cementation and burial dissolution. MVT mineralization such as at Pine Point coincides with hydrocarbon emplacement under deep burial conditions (Qing, 1998).

Figure 9.36 shows burial history plots for Leduc and Bearbury areas along the Rimbey-Meadowbrook trend in southern Alberta and the Pine Point and Rainbow areas in northern Alberta. These plots suggest that there was relatively rapid burial in response to the Antler Orogeny in the Carboniferous. Drivet and Mountjoy (1997) feel that Antler-induced regional flow to the northeast at this time was responsible for most of the coarse replacement dolomitization and subsurface dissolution events seen in Upper Devonian carbonates in the Western Canada Sedimentary Basin. This interpretation seems to be supported primarily by calculated temperatures of replacement near 50°C based on oxygen isotopic data. As Drivet and Mountjoy (1997) note, this interpretation is tenuous at best because the composition of the dolomitizing fluid is unknown (Chapter 4). If a more saline fluid were used in their calculations instead of Devonian seawater, the calculated temperature would be considerably higher and replacement dolomitization could be the result of fluid flow during the Columbian-Laramide orogeny in the Cretaceous.

Burial dolomites are preferentially distributed along shelf margins such as the Nisku shelf, Cooking Lake platform, Swan Hills platform and the Presquile barrier (Figs. 9.33 and 9.34) (Amthor et al., 1993a). Pinnacle reef buildups isolated in shale related to rapid transgression seem to be dolomitized only when sitting directly on antecedent dolomitized shelf margins. For example, the dolomitized Leduc buildups formed on the dolomitized Cooking Lake platform margin in the Rimbey-Meadowbrook trend (Figs. 9.33 and 9.34). This relationship suggests that these platform margins have acted as conduits for regional fluid flow and hydrocarbon migration over long periods of time (Amthor et al., 1994; Drivet and Mountjoy, 1997; Mountjoy et al., 1997).

Mountjoy et al. (1997) noted that these dolomitized Devonian subsurface conduits seem to be strikingly similar to Ordovician conduits documented by Montañez (1994; 1997) as discussed earlier in this chapter. Montañez postulated (1994) that the Ordovician conduits were initially established by partial syndepositional dolomitization of subtidal carbonate sequences and later totally dolomitized by neomorphism of the precursor

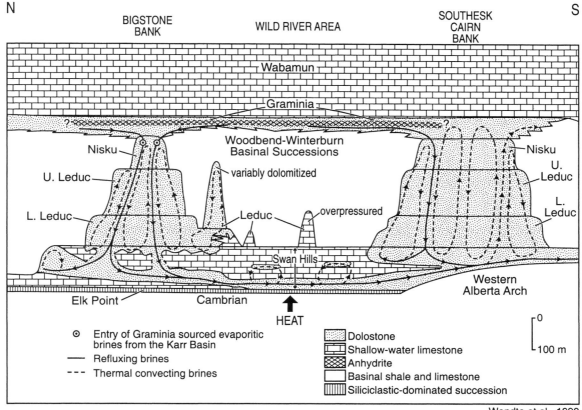

Fig. 9.37. *Conceptual model accounting for fluid flow and dolomite formation in the Swan Hills Formation. Residual brines from Graminia Formation evaporitic successions sink, as a result of their high density, through porous Nisku and Leduc bank carbonates and are focused into permeable limestones along the Swan Hills bank margin and along a belt at the base of the Swan Hills Formation in the Wild River area. As a result of elevated temperatures encountered within the first few hundred meters of burial, thermal convective cells develop in strata with vertically continuous porosity, such as along Swan Hills bank margins and possibly within large Leduc-Nisku bank complexes and isolated Leduc reefs. Convective flow is also preferentially channelled up through possible fault zones and then laterally out into permeable zones within the interior of the Swan Hills Formation in the Wild River area. Discharge of convecting brines is postulated to have occurred at Graminia Formation-equivalent bank margins in the Southesk-Cairn Complex. Most dolomite formation is concurrent with Late Frasnian evaporite formation, but thermal convection continued on after evaporite deposition throughout the duration of this thermal phase. This resulted in the additional formation of a small but significant amount of dolomite. In limestones of the Swan Hills and Leduc formations away from the dolomitizing conduits, saddle dolomite precipitated as a result of magnesium derived from cannibalization of precursor magnesium calcite during pressure solution. Used with permission of the Bulletin of Canadian Petroleum Geology.*

dolomites and replacement dolomitization, dolomite cementation and dissolution of undolomitized limestone components. This begs the question: did the Devonian dolomites follow a similar burial history?

The distribution of late dolomites along shelf margins certainly seems to suggest that subsurface fluid movements were focused along porosity-permeability pathways established during or shortly after deposition. Are the favorable porosity-permeabilities present because of more porous shelf margin limestones, or are they present because of preferential early marine dolomitization of platform margins that were later neomorphosed as suggested by Vahrenkamp and Swart (1994)?

This issue has been addressed eloquently a number of times by Mountjoy and his co-workers (in particular, Amthor et al., 1993), who have steadfastly maintained that most of the burial dolomites of the Western Canada Sedimentary Basin had no precursor early marine dolomites. However, Machel (1996) and Qing (1998) suggest a scenario involving neomorphism of earlier-formed marine dolomites for the burial dolomites of the Obed platform (Swan Hills) and the Presquile barrier respectively in the northern reaches of the basin (Fig. 9.33).

A major paper by Wendte (Wendte et al., 1998), centered on the Swan Hills platform, presented an entirely different model for subsurface dolomitization of the Leduc and Swan Hills formations. Based on an enormous petrographic, fluid inclusion and geochemical data base, these authors called upon thermal convection of evaporative brines refluxed from the overlying Grammina Formation for most of the subsurface dolomitization (Fig. 9.37). The difficult aspect of this model is the necessity for a localized high heat source to drive the brine convection. The isotopic composition of the Swan Hills and Leduc dolomites strongly suggests neomorphism of earlier marine dolomites (Fig. 9.38). Perhaps a more reasonable model would have early reflux dolomitization of the Leduc and Swan Hills shelf margin below the Grammina, then subsequent neomorphism of the original

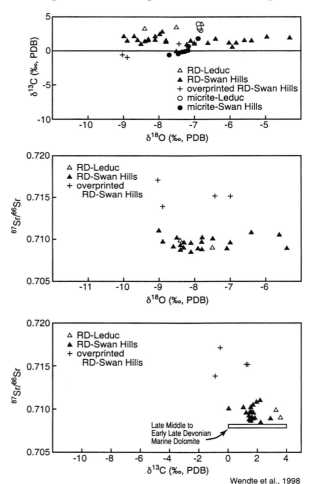

Fig. 9.38. Plots of stable isotopes versus radiogenic isotopes for replacement dolomites (RD), overprinted replacement dolomites and micrite matrix for the Leduc and Swan Hills buildups. Used with permission of the Bulletin of Canadian Petroleum Geology.

dolomites by hotter tectonically-driven fluids derived from the orogenic belt to the west.

The impact of dolomitization on porosity and permeability in the Devonian in Canada has been significant. In general terms, dolomitization (including dolomite cementation) has decreased the porosity of Devonian carbonates, particularly in those sequences encountered at depths less than 2500m. For more deeply buried sequences, however, dolomites are considerably more porous and permeable than their limestone counterparts as might be expected (Chapter 4) (Amthor et al., 1994).

In summary, there is little doubt that dolomitization of the Devonian in western Canada started early and continued over a long span of time in the burial regimen, fueled by a tectonically-driven subsurface flow system focused into favorable shelf margin conduits. While porosity/permeability may have been generally decreased, significant secondary porosity was generated by dolomitizing events. These long-lived conduits were ultimately the favored hydrocarbon migration pathways as well as pathways for the fluids responsible for MVT mineral deposits. The dolomitization model for these deposits remains a work in progress, and no doubt future studies will determine that no single model will suffice for such a complex geologic setting.

THE POST TECTONIC DIAGENETIC REGIMEN

Introduction

The post tectonic diagenetic regimen commences when active tectonism ceases and topographically driven flow, from uplifted marginal outcrops into the adjacent basin, begins to dominate (Fig. 9.5C). We will use two well-documented case histories to develop some broad conclusions concerning the diagenesis and porosity evolution in this important setting. The first concerns the Mississippian Madison carbonates in the mid-continent of the USA and the second will look at Paleozoic carbonates of the greater Permian Basin.

Mississippian Madison Aquifer of the mid-continent, USA

The Mississippian Madison is a thick carbonate sequence covering most of Montana, Wyoming, Nebraska, and North Dakota. It extends into Canada to the north as well as Colorado to the south. The Madison Group thickens to the north into the South Montana Trough and to the east-northeast into the Williston Basin in eastern Montana, North Dakota, South Dakota, and southern Canada. Shallow water shelf sequences around the margins of the Williston Basin and onto the Wyoming shelf range from ooid shoals to tidal flat evaporites (Lodgepole, Mission Canyon and Madison formations). Relatively thick marine evaporites generally cap the Madison Group (Charles Formation). While missing on the Wyoming shelf by erosion, these evaporites are still present in the subsurface of the Williston Basin and south Montana Trough.

Laramide tectonism at the end of the Cretaceous and beginning of the early Tertiary brought Madison limestones to the surface along an extended series of mountains such as the Bighorn, Laramie, Beartooth, Little Belt, Wind River and Black Hills. (Fig. 9.39). These uplifts created a series of smaller basins such

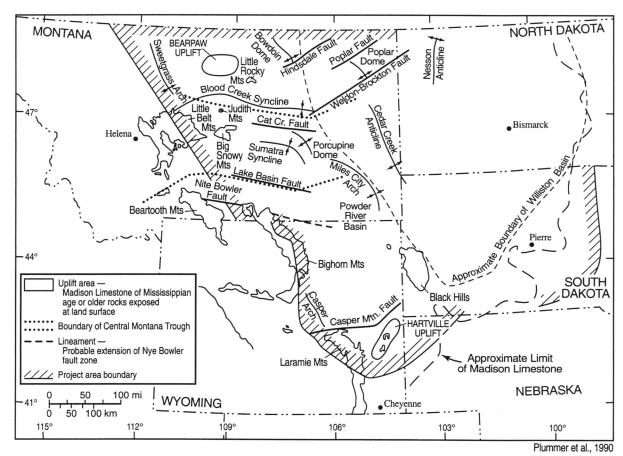

Fig. 9.39. Map showing location of major structural and physiographic features in the study area. Used with permission of Water Resources Research.

as the Powder River, Big Horn and Wind River. Meteoric recharge into the Madison Group at the outcrop set the stage for the development of the present important regional Madison Aquifer system in the Powder River, Bighorn and Williston basins. Figure 9.40 shows the aquifer and it's confining units.

Plummer et al. (1990) reported a geochemical modeling study of the Madison Aquifer in which they isolated the major diagenetic events occurring in the aquifer during the present post-orogenic hydrogeological regimen. Figure 9.41 shows the modeled vector flow paths for the Madison aquifer suggesting dominant flow to

Lower Cretaceous Rocks	Upper Boundary
Charles Formation and Big Snowy Group of Mississippian Age, and Pennsylvanian, Permian, Triassic, and Jurrasic Rocks	Confining Unit
Lodgepole and Mission Canyon Limestones of the Madison Group, and Madison Limestone	Madison Aquifer
Silurian and Devonian Rocks and Bakken Formation of Devonian and Mississippian Age	Confining Unit
Upper Part of Deadwood, and Winnipeg, Red River, and Stony Mountain Formations of Cambrian to Ordovician Age	Cambrian-Ordovician Aquifer
Precambrian Rocks	Lower Boundary

Fig. 9.40. Relationship between geologic units and units in digital simulation of flow system used by Downey in 1984. Used with permission of Water Resources Research.

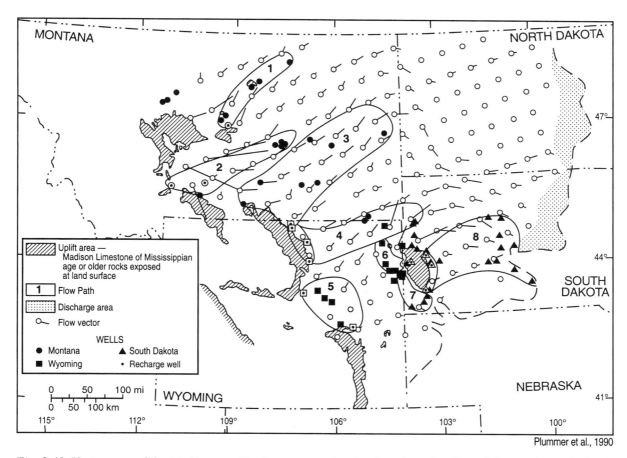

Fig. 9.41. Vector map of the Madison aquifer flow system showing location of wells and flow paths studied. Solid shapes denote location of all wells sampled in the Madison aquifer: squares (Wyoming), circles (Montana), and triangles (South Dakota). Numbers 1-8 identify flow paths. Used with permission of Water Resources Research.

the east northeast. Discharge is along the subcrop and in saline springs and lakes in the Dakotas and Canada. Groundwater ages varied from virtually modern to 23,000 years. Calculated flow velocities range from 7-87ft/yr.

One of the major features of the system is the increase in sulfate in the water as a function of distance down the flow path. A plot of the gypsum saturation index versus sulfate concentration in the water suggests that the water is strongly undersaturated with respect to gypsum (Fig. 9.42C). Plots of dolomite and calcite saturation index versus sulfate concentration (Fig. 9.42A and B) suggest that dolomite is dissolving and calcite is precipitating in the system. Plots of Ca and Mg versus sulfate (Plummer et al., 1990) support this.

The dedolomitization reaction (dissolution of dolomite, precipitation of calcite) in a groundwater system in equilibrium with calcite, dolomite and anhydrite is as follows:

$$CaMg(CO_3)_2 + CaSO_4 \iff 2CaCO_2 + MgSO_4 \text{ (aq)}$$

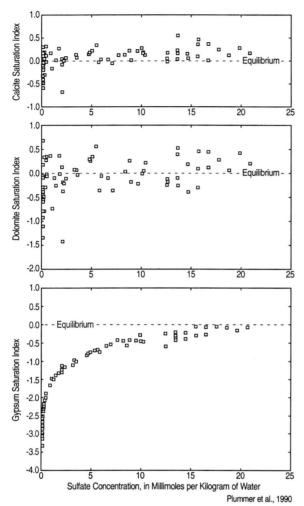

Fig. 9.42. *Comparison of calcite, dolomite, and gypsum saturation indices as a function of total dissolved sulfate content for wells and springs in the Madison aquifer. Used with permission of Water Resources Research.*

Conceptually then, 2 moles of calcite are precipitated and 1 mole of dolomite dissolved for every mole of $CaSO_4$ added to the solution by dissolution. Plummer et al. (1990) using water compositions and a mass balance model, calculated apparent rates of calcite precipitation, dolomite dissolution, and anhydrite dissolution to be 0.59, 0.24 and 0.95 μmole/L/yr, respectively.

Calcite precipitation would certainly be porosity destructive, while dolomite and anhydrite dissolution would lead to higher porosities in the Madison aquifer. The balance would seem to favor the development of significant secondary porosity along the present water flow paths in the aquifer.

Banner et al. (1989) studied the origin of saline groundwaters in springs and artesian wells developed in Paleozoic carbonates and clastics in central Missouri. These springs and artesian wells form the discharge area for the Western Interior Plains aquifer system recharged by meteoric waters along the Front Range of the Rocky Mountains in Colorado (Fig. 9.43A and B). The Mississippian Burlington-Keokuk limestones and dolomites of Missouri, the principal subject of the Banner et al. (1989) study, are the general equivalents of the Madison Group described above and crop out along the Colorado Front Range as the Leadville Formation. The Banner et al (1989) geochemical model of this system suggests that the high salinities of the ground water were the result of dissolution of Permian salts from the overlying confining unit in Kansas. There is little evidence of rock-water interaction in the Mississippian carbonates and the waters seem to be in equilibrium with the limestones and dolomites of the Mississippian Burlington-Keokuk. In other words, in the absence of significant anhydrites in the Leadville-Burlington-Keokuk carbonates, no significant porosity modification can be attributed to the present post-orogenic diagenetic regimen.

Post orogenic aquifer system in Paleozoic carbonates of the Greater Permian Basin

The Greater Permian Basin developed

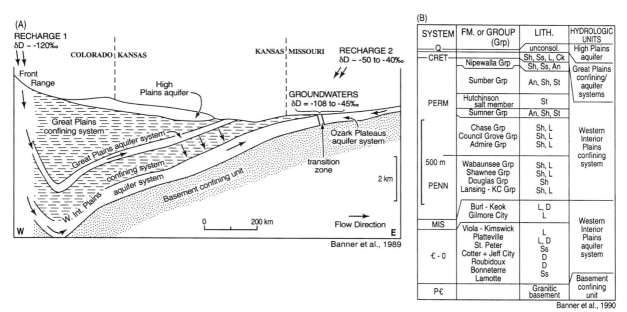

Fig. 9.43. (a) Generalized hydrologic model for the Central Great Plains area of the United States, modified from Jorgenson, et al (1989). The gravity-flow system is sustained by meteoric recharge in the Front Range. Structural relationships are simplified. The Western Interior Plains aquifer system principally comprises Cambrian-Ordovician and Mississippian carbonates and sandstones. The Western Interior Plains confining system comprises Pennsylvanian shales and limestones and Permian evaporites. Transition zone marks region where eastward-migrating saline groundwater in Western Interior Plains aquifer system encounters fresh groundwater in Ozark Plateaus aquifer system. The stable isotope data for the central Missouri ground waters range from values measured for meteoric waters of local origin (Recharge 2) to values for meteoric waters found in the Front Range (Recharge 1). Significant cross-formational flow shown through Pennsylvanian shales in the Western Interior Plains confining system is a requirement for models that invoke a Permian evaporite source of high groundwater salinities. (b) Lithostratigraphic sequence and principal geologic and hydrologic units in central Kansas (Rice and Ellsworth County), from Walters (1978), Bayne et al. (1971) and Jorgenson et al. (1989). D = dolostone, L = limestone, Sh = shale, An = anhydrite, St = halite, Ss = sandstone, Ck = chalk. Vertical scale Pertains to thickness of systems and Hutchinson Salt Member only. Used with permission of Geochemica et Cosmochimica Acta.

from an asymmetric structural depression in Precambrian basement at the southern margin of the North American plate. In the Pennsylvanian and Permian, the Greater Permian Basin was broken into several intracratonic basins (such as the Delaware and Midland basins) by fault-bounded uplifts of the basement. It was filled with shelf carbonates, marine shales, and sands of the Pennsylvanian and early Permian. Middle and upper Permian sequences consist of carbonates, evaporites, sands, and silts. Mesozoic and Tertiary deposits consist of fluvial and lacustrine clastics with some minor Cretaceous limestones. Regional uplift and tilting to the east began some 5-10 million years ago bringing the Paleozoic section to the surface in the Sacramento, Guadalupe, Apache, and Glass mountains along the western and southwestern margins of the basin (Fig. 9.44). Meteoric recharge of the Paleozoic section began at that time and the present hydrogeologic system was established. There are four hydrogeological units in the Greater Permian Basin (Fig. 9.45). The Upper Aquifer

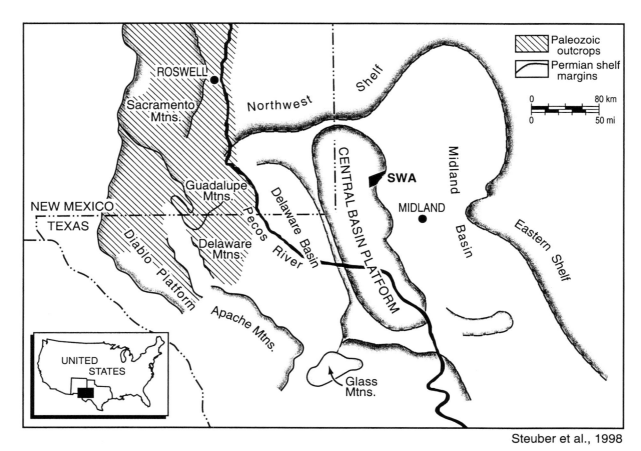

Fig. 9.44. Location of the Southwest Andrews area (SWA) within the Permian basin of west Texas and southeastern New Mexico. Used with permission of the AAPG.

system consists of Tertiary to Cretaceous clastics. There is an Evaporite Confining system consisting of Permian evaporites dominated by Ochoan salts and a Deep-Basin Brine Aquifer system consisting of Lowermost Permian to Ordovician limestones and dolomites. The lowermost unit is the Precambrian aquaclude. The Upper Aquifer is primarily fresh water and provides most of the agricultural and drinking water of west Texas and New Mexico. The Deep Basin Brine Aquifer system is filled with highly concentrated brines and form the formation water associated with the extensive hydrocarbon production from the Paleozoic sequences of the Midland and Delaware basins.

Bein and Dutton (1993) suggested that the deep basin brines were mixtures of two end-member brines. A Na-Cl brine as young as 5-10my that is derived derived from meteoric water by dissolution of halite is mixing with and is displacing an ancient Ca-Cl brine. The Ca-Cl brine is derived from evaporation of seawater modified by dolomitization of Paleozoic carbonates in the late Permian as the evaporative brines refluxed down into and displaced marine connate waters Final brine composition is a simple mixing phenomenon of the two end member brines with little or no rock-water interaction and little ultimate impact

System	Series	Group/Formation	General Lithology	Hydrogeological Unit*
Tertiary		Ogallala	fluvial and lacustrine clastics	Upper Aquifer System
Cretaceous		Fredericksburg	limestone	
		Paluxy	sandstone	
Triassic		Dockum	fluvial-deltaic and lacustrine clastics	
Permian	Ochoan	Dewey Lake	sandstone	Evaporite Confining System
		Rustler	salt, anhydrite	
		Salado	salt	
	Guadalupian	Tonsill	anhydrite	
		Yates	sandstone	
		Seven Rivers	anhydrite	
		Queen	sandstone	
		San Andres-Grayburg	dolomite-sandstone	
	Leonardian	Clear Fork	limestone-dolomite	Deep-Basin Brine Aquifer System
		Wichita		
	Wolfcampian	Wolfcamp		
Pennsylvanian		Cisco	shelf limestones, minor shale	
		Canyon		
		Strawn		
		Atokan	shale	
		Chester		
Mississippian		Mississippian Lime	limestone	
Devonian		Woodford	shale	
		Devonian	limestone	
Silurian		Silurian	shale, limestone	
Ordovician		Montoya	limestone	
		Simpson	shale, limestone	
		Ellenburger	dolomite	
PRECAMBRIAN			igneous, metamorphic	Basement Aquiclude

Steuber et al., 1998

Fig. 9.45. Stratigraphy, lithology, and hydrogeological units in the Southwest Andrews area. Stratigraphic units from which formation water samples were produced are shaded. Used with permission of the AAPG.

on porosity.

Steuber et al. (1998) reached similar conclusions in their study of brines from Paleozoic reservoirs along the eastern margin of the Central Basin Platform (Figs. 9.44, 9.45 and 9.46). They did find, however, that the Permian San Andres-Greyburg and Devonian limestones of the platform showed considerably lower salinity values than the Pennsylvanian and Wolfcamp reservoirs. They suggested that the San Andres-Greyburg and Devonian reservoirs have been almost completely flushed by meteoric fluids related to the present hydrogeologic regimen while the Pennsylvanian and Wolfcamp reservoirs have been only partially flushed. This situation suggests a complex post- orogenic plumbing system leading to a wide range of formation fluid compositions and results in significant production and logging difficulties.

PREDICTING CHANGES IN POROSITY WITH DEPTH

In 1983, Scholle published a series of curves illustrating the porosity-depth trends for chalks in North America and Europe (Fig. 9.19) which illustrated clearly the progressive loss of porosity with depth occasioned by closed-system chemical compaction in these fine-grained calcite sediments. It was obvious that these curves would be very helpful in future

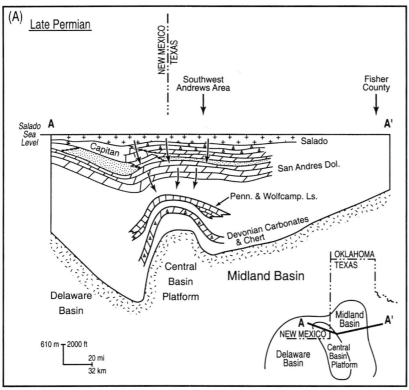
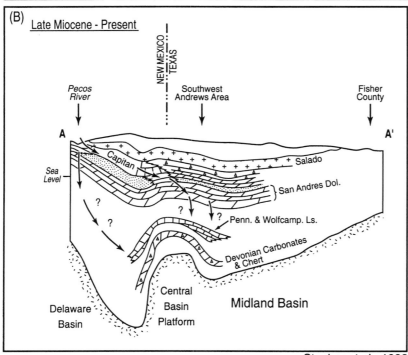

Fig. 9.46. Highly schematic west-east cross sections through the Delaware and Midland basins. (A) Cross section during Salado deposition (Late Permian) showing descent of halite-saturated brine that entered Pennsylvanian and Wolfcampian shelf limestones on the Central Basin Platform. (B) Cross section showing suggested migration of brines of meteoric origin that began 5-10 m.y. ago and has reached Permian-Devonian carbonate strata on the Central Basin platform. Used with permission of AAPG.

PREDICTING CHANGES IN POROSITY WITH DEPTH

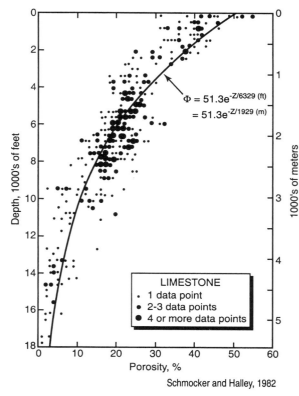

Fig. 9.47. Porosity-depth values and exponential representation of composite data for limestones of the South Florida Basin. Used with permission of Elsevier.

ultimate porosity evolution.

When one looks at the spread of the actual data swarms for depth-porosity relationships in shallow marine limestones (Fig. 9.47), it is obvious that this view is an oversimplification. As we have discussed in the preceding chapters, however, there are other, important diagenetic processes that may actively destroy, as well as generate, porosity throughout the burial history of a sequence (see Figure 9.48 for a theoretical burial curve that incorporates many of these

chalk exploration efforts (Lockridge and Scholle, 1978). In 1982 Schmoker and Halley presented a depth-porosity curve for south Florida subsurface limestones derived from log analysis of selected wells in the south Florida basin (Fig. 9.47). The smooth depth versus porosity curve was derived from a least squares exponential fit of a rather scattered data swarm.

Halley and Schmoker (1983), and later Scholle and Halley (1985), concluded that these porosity trends were basically the result of chemical compaction during burial in a semiclosed system, paralleling the processes controlling porosity changes with depth in chalks. They concluded that near surface diagenetic processes did not significantly affect

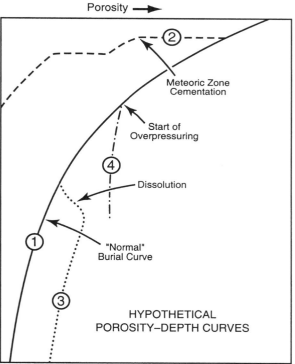

Fig. 9.48. Hypothetical porosity-depth curves in response to early diagenetic overprint, geopressured situations, and organic-related late diagenesis. (1) Normal burial curve based on chalk curves (2) Porosity destruction and development related to early meteoric zone diagenesis.(3)Porosity development related to late stage dissolution, as a result of hydrocarbon maturation and destruction. (4) Porosity preservation as a result of overpressuring. Used with permission of Elsevier.

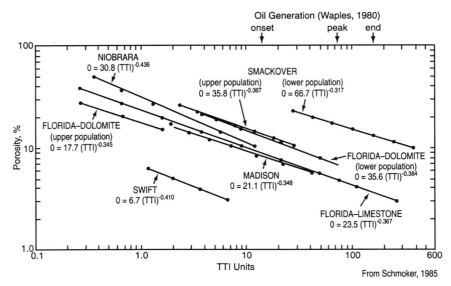

Fig. 9.49. Porosity versus Lopatin's time-temperature index of thermal maturity (TTI) for various carbonate units. Used with permission of Elsevier.

processes).

Schmoker (1984) increased the utility of these types of plots as well as our understanding of the role of early diagenesis in porosity evolution when he convincingly demonstrated the influence of thermal maturity, that is the temperature-time history of a rock, over the ultimate paths of carbonate porosity evolution.

Taking the porosity-depth relationships of a number of carbonate sequences from separate basins with different thermal histories, Schmoker plotted porosity versus calculated thermal maturity based on Lopatin models of the individual basins on a logarithmic base (Fig. 9.49).

The results were a series of distinct, but subparallel straight line plots related by a power function in the form $\phi = a(TTI)^b$. This relationship strongly supports the concept that subsurface processes of porosity loss are functions of time-temperature history. Coefficients of the power functions relating porosity and thermal maturity for Schmoker's data sets are compiled in Table 9.2. Schmoker indicated that the multiplying coefficient (a), represents the net effect on porosity of all petrologic parameters, and obviously includes early diagenetic effects such as cementation, secondary porosity and dolomitization. This exponent is widely variable in response to

Coefficients of Power Function [$\theta = a(TTI)^b$] Relating Porosity and Thermal Maturity

	Multiplying coefficient a	Exponent b
Florida		
limestone	23.5	-0.367
dolomite (upper)	17.7	-0.345
dolomite (lower)	35.6	-0.384
Niobrara	30.8	-0.436
Swift	6.7	-0.410
Madison	21.1	-0.348
Smackover		
(upper)	35.8	-0.367
(lower)	66.7	-0.317
	Avg. = 0.372 ± 0.038	

Schmoker, 1985

Table 9.2. Coefficients of power function relating porosity and thermal maturity. Used with permission of Elsevier.

variations in original depositional fabrics and subsequent diagenetic overprint (Table 9.2). The exponent (b), however, is nearly constant (Table 9.2) and results in the subparallel plots of porosity populations shown in Figure 9.49. Schmoker feels that exponent (b) represents the rate limiting step in chemical compaction, that is, diffusive solution film transport from the site of solution to site of precipitation, and hence is essentially independent of the rock matrix. It is the positive temperature influence over rates of diffusion that links thermal history to porosity destruction in the subsurface.

The Schmoker approach nicely accommodates the dramatic initial fabric and textural differences within a carbonate sequence that may result from depositional processes and local or subregional intensive early diagenesis. These diagenetic overprints would be reflected as separate, but parallel lines on a porosity-TTI plot, such as the two Smackover populations shown on Figure 9.49.

What then, is the real significance of depth-porosity curves, and how should they be viewed? If constructed carefully (Brown, 1984), TTI-porosity plots can help define, and most importantly, predict the average or typical porosity of a given potential reservoir within a basin's depth-thermal maturity framework. This average then becomes a standard or reference porosity against which subsurface porosity-diagenetic studies can be compared in order to isolate those extraordinary conditions that may lead to new drilling frontiers and discoveries. See for example, the earlier discussion in this chapter of the North Sea Chalks.

SUMMARY

There are three major tectonic hydrogeologic settings under which burial diagenesis occurs. The passive margin diagenetic regimen is marked by relatively rapid burial with steadily increasing temperatures and lithostatic pressures. Tectonic stress and uplift causing the dislocation and rapid deployment of relatively hot complex fluids through porous and permeable conduits mark the active margin diagenetic regimen. The post-orogenic setting is one of topography controlled meteoric recharge into deeply buried sequences

In the passive margin setting, mechanical and chemical compaction are the dominant porosity modifying agents. Environmental factors such as, sediment texture (presence-absence of mud and grain size), the relative compactability of grain types, and the presence or absence of bound water may affect the early mechanical compaction history of carbonate sediments. Early cementation, organic framework development, reservoir overpressuring, pores filling with oil, and pervasive dolomitization all tend to retard the onset and efficiency of chemical compaction. Aggressive pore fluids combined with active hydrologic systems, the presence of metastable mineral phases, and admixtures of siliciclastics all tend to accelerate the chemical compaction process.

In comparison to compaction, cementation has a relatively minor impact on porosity evolution in the passive margin setting. The major source of $CaCO_3$ for cementation is thought to be pressure solution. The volume of cement that can be generated from pressure solution, however, is limited; and porosity in reservoirs is generally filled with oil before chemical compaction is complete.

Additional $CaCO_3$ for cementation may be gained from carbonate dissolution by aggressive fluids associated with thermally driven organic diagenesis. While carbonate reservoir enhancement by subsurface dissolution has been documented in the passive margin setting, its actual economic importance has not to date been fully evaluated.

Geologic setting, petrography and geochemistry must be used in concert to determine the burial origin of passive cements and replacement dolomites.

The primary feature of the active margin diagenetic realm is the relatively rapid movement of significant volumes of warm-to-hot basinal fluids mobilized by tectonism through complex subsurface conduit systems. It seems clear that these conduit systems were pre-conditioned by depositional setting and early diagenetic overprint such as environment-related dolomitization. These conduits are important migration pathways for both hydrocarbons and mineralizing fluids.

Tectonically related fluids interact extensively with the rocks forming the conduits. Reactions include recrystallization of earlier dolomites and calcites, replacement dolomitization, and associated dissolution of calcite, dolomite and anhydrite cementation, and sulfide emplacement.

Calcite dissolution associated with dolomitization and possible dissolution of evaporites will impact porosity positively in the subsurface conduits. Porosity will be impacted negatively by dolomite, calcite and anhydrite cementation enhanced by thermochemical sulfate reduction.

Post-orogenic meteoric recharge will have minimal impact on carbonate porosity unless these waters encounter and dissolve anhydrite/gypsum. If no anhydrite/gypsum is involved, the meteoric water will quickly equilibrate with the carbonate rock pile and minimal rock-water interaction and porosity modification will ensue. Production problems associated with tilted oil/water contacts, reservoir edge water drives, and logging problems related to variation in brine chemistry are common.

If post-orogenic meteoric recharge encounters carbonates with significant anhydrite/gypsum the dissolution of gypsum will drive the dissolution of dolomite and precipitation of calcite. Porosity modification can be significant since 1 mole of dolomite will be dissolved and 2 moles of $CaCO_3$ will be precipitated for each mole of anhydrite/gypsum dissolved. On balance, porosity enhancement should result.

While the introduction of meteoric waters into deep reservoirs can have a significant impact on hydrocarbon production by fueling hydrocarbon biodegradation and water washing, the rapid movement of waters through the aquifer can provide significant hydrodynamic trapping opportunities.

Finally, porosity-thermal maturity plots may be important porosity predictive tools in the future.

Chapter 10

POROSITY EVOLUTION FROM SEDIMENT TO RESERVOIR: CASE HISTORIES

INTRODUCTION

In this chapter, three economically important case histories will be considered. The principles and concepts developed in the earlier chapters will be used to outline the evolution of these reservoir pore systems from deposition until they are encountered by the drill. These three examples were chosen because their sequence stratigraphic framework and total diagenetic history has been well documented. The author has done extensive stratigraphic and diagenetic research on the first two, the Paleozoic Madison Formation of central Wyoming, and the Mesozoic upper Jurassic Smackover of the central Gulf of Mexico and a significant amount of original work is included in their discussion. The Madison was deposited during a transition from greenhouse to icehouse conditions while the Smackover was deposited under greenhouse conditions. The third case history details an integrated 3-D seismic-based reservoir modeling study of the Tertiary Malampaya and Camago buildups, off shore Philippines. These buildups were deposited during the Oligocene-Miocene greenhouse-icehouse transition.

THE MISSISSIPPIAN MADISON AT MADDEN FIELD WIND RIVER BASIN, WYOMING, USA

Introduction

Dolomite of the Mississippian Madison Formation is the reservoir for a giant, super deep gas field (23,000 feet) at Madden Field on the northern flank of the Wind River Basin in central Wyoming. This field has enormous potential (probably in excess of a trillion cubic feet of gas) reservoired under exceedingly hostile conditions of high temperatures, high pressures and high H_2S. The situation is unique, however, because the dolomite reservoir facies crops out on the hanging wall of a Laramide thrust fault some 6 miles north of the gas producing deep wells of the Madden Field. The comparative sequence stratigraphic framework, burial history, porosity characteristics and diagenetic history of outcrop versus subsurface were determined to establish the viability of reservoir characterization by using the adjacent outcrops as a surrogate for the Madison reservoir in the subsurface.

General setting

A compressional structural regime during the Late Devonian Antler Orogeny was responsible for the formation of the Antler trough and rise of the Antler Highlands in the area west of Montana, Wyoming and Colorado. The Antler Orogeny culminated in Early Mississippian overthrusting along the western margin of the ancestral North American continent resulting in a widespread post-Devonian erosion surface fronting the Antler Trough. The Early Mississippian sea transgressed across this surface onto the North American Craton in response to the increased accommodation resulting from thrust-related tectonic loading, ultimately forming an enormous carbonate shelf extending the length of the present Rocky Mountains (Sonnenfeld, 1996). Warm tropical seas covered the Wyoming shelf because the Mississippian paleoequator passed just to the south of Madden Field in central Wyoming. Mississippian paleogeography of the central Rockies is shown in Figure 10.1. Major features affecting Mississippian sedimentation included the positive Transcontinental Arch to the south, the Mississippian carbonate shelf margins facing the Antler Trough to the west and the Central Montana Trough and Williston Basin to the north.

Laramide tectonism at the end of the Cretaceous, extending into the Early Tertiary in Wyoming brought Pre-Cambrian and Paleozoic sequences, including the Mississippian Madison, to the surface by faulting along an extended series of mountains such as the Bighorns, Wind Rivers, and Owl Creeks. These positive features formed a series of isolated structural basins including the Big Horn and Wind River basins in Wyoming (Fig. 10.2). The Madden Field is located near the northern margin of the Wind River Basin near the Owl Creek Mountains. Outcrops of the Madison and a stratigraphic core test occur on the hanging wall of the Owl Creek/Cedar Ridge

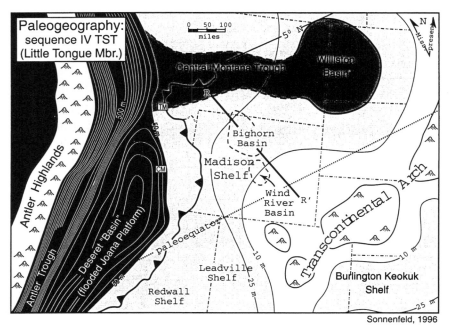

Fig. 10.1 Regional paleogeographic reconstruction for Mississippian late Osagean Little Tongue Member showing location of regional cross-section R-R' (Figure 10.8). Used with permission of the Rocky Mtn. Section, SEPM.

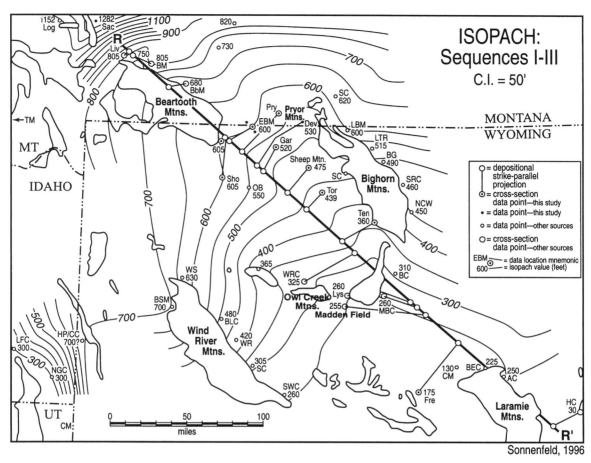

Fig. 10.2 Data basemap and isopach of sequences I-III used to guide depositional strike-parallel projections onto the dip-oriented cross-section line shown in Figure 10.8. Precambrian-cored Laramide uplifts surrounded by outcrops of Paleozoic sedimentary rocks are shown in dark gray. The Bighorn and Wind River basins are Laramide structural basins superimposed on the extensive Mississippian Madison platform. Mnemonic abbreviations for data points projected onto the cross-section are as follows: BM=Baker Mtn.; BbM=Benbow Mine; CF=Clarks Fork; CM=Crawford Mtns. EBM=Elk Basin Madison Unit; Fre=Fremont Can.; Gar=Garland Field; HC=Hartville Can.; Liv=Livingston; Lys=Lysite Mtn.; Mad=Madden Deep Field; MBC=Middle Buffalo Creek;Pry=Pryor Mtns. (Bear Can. and King Can.); SM=Sheep Mtn.; Sho=Shoshone Can.; TM=Tendoy Mtns.; Ten=Tensleep Can.; Tor=Torchlight Field; WRC=Wind River Can. Used with permission of the Rocky Mtn. Section, SEPM.

thrust from Lysite Mountain just to the north of the field to the Wind River Canyon some 30km northwest of the field (Fig. 10.2). Figure 10.3 illustrates the relative structural setting between these outcrops, the core test and Madden Field.

Stratigraphic and depositional setting

A time stratigraphic chart for the Mississippian in Montana and Wyoming is shown in Figure 10.4. Early workers such as Rose (1976) recognized two major transgressive-regressive cycles in the Mississippian bounded by regional

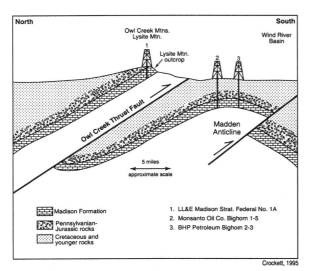

Fig. 10.3 Schematic diagram of the structural relationship between Lysite Mountain, Madden Anticline and the Cedar Ridge Thrust Fault, showing the relative location of wells in the study area. Displacement along the Cedar Ridge Thrust Fault is more than seven miles. Used with permission of author.

unconformities. The Lower Mississippian cycle was termed the Madison Group, and the Upper Mississippian cycle, the Big Snowy Group. The Madison is the subject of this study.

In central Wyoming the Madison is termed a formation whereas in Montana it is generally recognized as a group. A series of members of the Madison can be recognized in northern Wyoming and Montana, but are not differentiated in the Madden area (Crockett, 1994; Sonnenfeld, 1996).

The regional sequence stratigraphic framework of north and central Wyoming was summarized and refined by Sonnenfeld (1996) following the earlier work of Elrick (1991) Dorobeck (1991) and Crockett (1994). Sonnenfeld recognized two second order supersequences coinciding with Rose's (1976) "Lower and Upper Mississippian depositional complexes" (Fig. 10.4). Only a truncated lower 2nd order sequence, equivalent to the Madison, is present in the Madden area.

In northern Wyoming, Sonnenfeld (1996) divided the Madison into six 3rd order depositional sequences each representing some 2+million years (Fig. 10.4). In addition he recognized two 3rd order "composite sequences", a lower (sequences I and II) and an upper (sequences III-VI) "composite sequence" (Fig. 10. 4). In outcrop at Lysite Mountain along the northern flanks of the Wind River Basin Crockett (1994) recognized four distinct 3rd order depositional sequences (Figs. 10.4, 10.5 and 10.6). Subsequently, these sequences were correlated with Sonnenfeld's sequences I-IV (Moore et al., 1995; Sonnenfeld, 1996) (Fig. 10.4). Sequences V and VI were removed at Madden by mid-Mississippian and post-Mississippian erosion with sequence IV overlain unconformably by the Pennsylvanian Amsden Sandstone. The top Mississippian unconformity at Lysite Mountain near Madden represents a hiatus in excess of 15 million years. At Lysite, Madison depositional sequence I rests unconformably on the Cambrian Gallatin Formation.

Figure 10.6 is a schematic measured section of the Madison Formation at Lysite Mountain taken approximately 1/2 mile south of the Federal #1 core. The sequence boundaries are shown and are generally marked by a clay or silt-rich recessive ledge (Fig. 10.6). Sequences I-III are totally dolomitized while sequence IV is limestone. At Lysite, sequence I is dominated by dolomitized fossiliferous wackestones to packstones and represents middle to upper ramp depositional conditions (Crockett, 1994; Moore et al., 1995, Sonnenfeld, 1996) (Fig. 10.7). Sequence II is a shoaling upward sequence characterized by dolomitized cross-bedded ooid and bioclastic

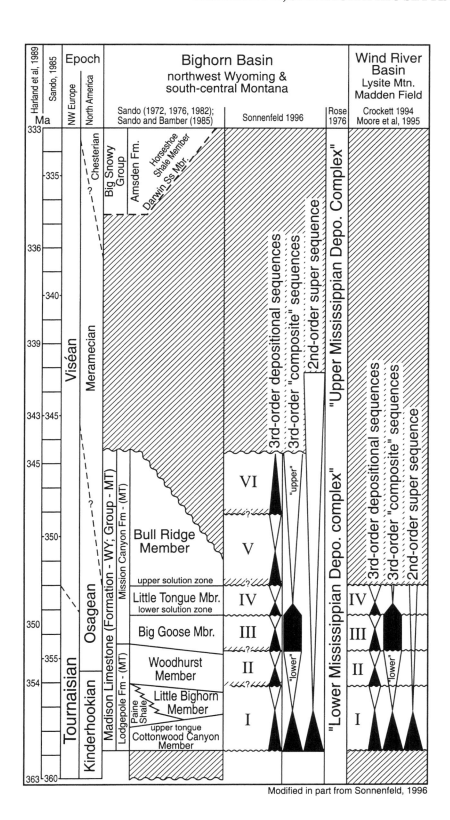

Fig. 10.4. Time stratigraphic chart comparing the Bighorn Basin with the Wind River Basin. Triangles represent increasing and decreasing inferred accommodation/sediment supply ratios at three different sequence scales; black=increasing A/S; white=decreasing A/S. Diagonal patterns represent hiatuses. (hiatus duration schematic and unconstrained for intra-Madison unconformities). Used with permission of the Rocky Mtn. Section, SEPM.

Fig. 10.5. Madison outcrops on Lysite Mountain located on the hanging wall of the Owl Creek Thrust just north of Madden Field. Photo courtesy Connie Hawkins, Denver CO. Contacts of the Madison with the Cambrian Galletin Formation and sequences I-IV of the Madison are shown. The approximate location of the Federal #1 Stratigraphic core is also shown.

grainstones with some thin evaporite solution collapse breccias just below the sequence boundary. Sequence II seems to represent inner ramp shoreface depositional conditions (Crockett, 1994; Moore et al., 1995, Sonnenfeld, 1996) (Fig. 10.7B and C). Sequence III is generally covered, but based on observations of the Federal Stratigraphic Core #1 it generally consists of cherty dolomicrites that are often intensely brecciated (Fig. 10.7D and E). These breccias occur as widespread stratigraphic breccias (Sonnenfeld, 1996). Moore et al. (1995) and Sonnenfeld (1996) felt that sequence III represented a highly restricted lagoon originally dominated by subaqueous evaporites, now represented by the breccias marking sequence III. Sequence IV is a cliff forming limestone composed of bioclastic chert-bearing wackestones to grainstones. It is often highly brecciated, shows evidence of intense karsting (Fig. 10.7F) which may occur as remnant karst towers. Sequence IV is thought to represent open marine shelf conditions.

Using the framework of his two composite sequences Sonnenfeld (1996) postulated that the Madison initially started as a ramp (sequences I and II, his lower composite sequence) that ultimately evolved into a flat topped rimmed shelf (sequences III-VI, his upper composite sequence, Fig. 10.8). His upper composite sequence developed a restricted lagoon behind shelf margin grainstone barriers in western and northern Wyoming fronting the Antler and Montana troughs. Widespread, possibly subaqueous evaporites were developed in the lagoon in response to the development of significant new accommodation space in response to continental margin tectonic loading during the TST of the upper composite sequence (Fig. 10.8).

Correlation of surface exposures into the deep subsurface at Madden

In order to use surface outcrops as surrogates for subsurface reservoirs one must be certain that the stratigraphic/sedimentologic architecture of outcrop and subsurface reservoir is similar. If outcrop and reservoir are located reasonably close and if the depositional sequences recognized on the surface can be correlated into the subsurface then our assumption of equivalency can be supported.

Correlation into the subsurface at Lysite Mountain is facilitated by access to the Madison core immediately behind the outcrop. Figure 10.9 shows a schematic core log of the Federal Stratigraphic Core #1 compared to the measured section at Lysite Mountain. A solid correlation of general lithologies and sequence boundaries between core and outcrop is evi-

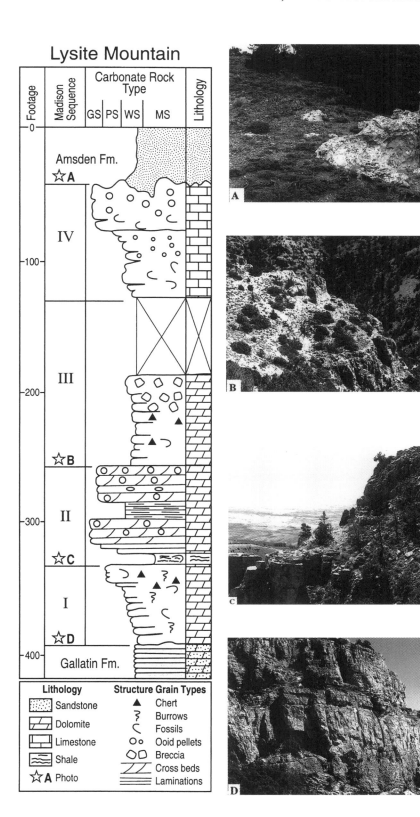

Fig. 10.6. Generalized measured section of the Madison Formation at Lysite Mountain with field photographs characteristic of the four stratigraphic sequences present at Lysite. A. Corroded remnant of Mississippian Madison tower karst surrounded by Pennsylvanian Amsden Sand. B. The cliff with the sharp upper contact is sequence II, the slope to the left is sequence III. Sequence III consists mainly of breccias, while sequence II is bedded dolomitized grainstones and laminated tidal flats. C. Bench formed by sequence II contact at left, with sequence III dolomites and breccias to the right. D. Sequence I contact with the Cambrian Gallatin Formation is at the bottom of the photograph. Contact of sequences I and II is marked by a recessive bench two thirds of the way up the cliff face.

Fig. 10.7. Field and core photographs of the Madison Limestone at Lysite Mountain. Access to core, core photos and field area courtesy Connie Hawkins, Denver, CO. A. Cambrian Gallatin-Mississippian Madison unconformable contact marked by sunglasses. Gallatin is sandy laminated dolomite. Madison sequence I at this site is massive dolomitized bioclastic wackestone. B. Crossbedded dolomitized grainstones of Madison sequence II. C. Dolomitized evaporite solution collapse breccia in the upper 5m of Madison sequence II. D. Core sample at 259 feet in the Federal #1 Stratigraphic core. Dolomitized evaporite solution collapse breccia of sequence III. E. Core sample taken at 227 feet in Federal #1 Stratigraphic core. Evaporite solution collapse breccia with brecciated chert. F. Bioclastic wackestone to packstone with silicified shell fragments and large silicified brachiopod occurring some 10m from the top of the Madison Formation in sequence IV at Lysite.

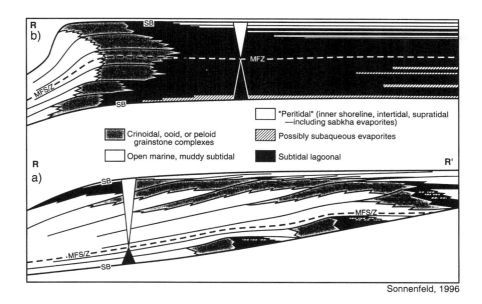

Fig. 10.8. Schematic diagram of long-term evolution of the Madison shelf. a) Ramp model for Sequences I and II. b) Flat topped shelf model for Sequences III-VI (but figure is based on sequence III). Note that increased lagoonal volumes in b) reflect a condition of balanced to increasing A/S in the early portion of the upper composite sequence. Also note the similar, but higher frequency, stratigraphic context for enhanced lagoonal accumulations within the 3rd-order TST and earliest HST of a). Used with permission of the Rocky Mtn. Section, SEPM.

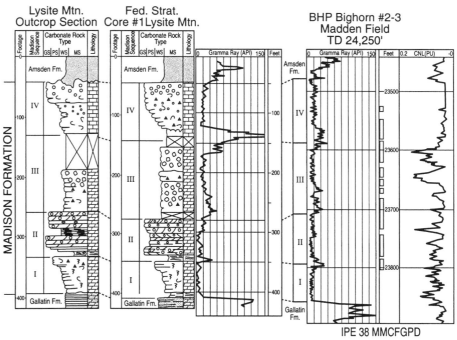

Fig. 10.9. Correlation of the Lysite Mtn. section into the deep subsurface at Madden Field through the Federal Stratigraphic Core #1 into the BHP Big Horn #2-3. Logs furnished by Connie Hawkins and LL and E, Denver, CO. Section and core description in part from Crockett, 1994. Perforations shown on the Big Horn #2-3 in the footage column.

dent. Sequence boundaries in the core hole show a gamma ray excursion that is no doubt a result of the argillaceous zones noted by Crockett (1994) in outcrop. Subsequent Madison outcrop work in the vicinity by Grammer and others using hand-held detectors confirms this observation (Grammer personal communication). Since the volume of core recovered from the adjacent deep wells at Madden is small and fragmentary correlation of depositional sequences from the core hole to the Madden Field is based on these gamma ray excursions at the

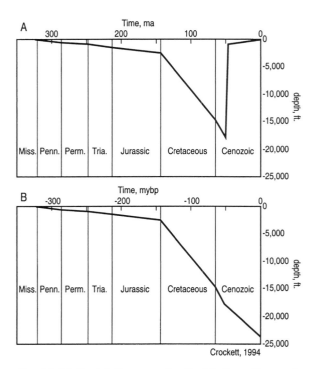

Fig. 10.10. Burial diagrams for the Madison Formation on Lysite Mountain and in Madden Field. A. Burial history for Madison Formation outcrop samples at Lysite Mountain. B. Buria] history of Madison subsurface samples in the Bighorn#1-5 well. Difference in burial history is due to Larmide thrust faulting during the early Eocene.

sequence boundaries. The correlations shown in Figure 10. 9 certainly seem to be reasonable and can be carried regionally through the available deep Madison tests along the northern rim of the Wind River Basin (Connie Hawkins, personal communication). A valid question at this point concerning equivalency of outcrop to subsurface is the relative burial history of outcrop versus reservoir.

Burial history of the Madison

Figure 10.10 illustrates the burial history of the of the Madison in the Big Horn #2 well at Madden Field. The plot is approximate and was constructed by using the thickness of younger stratigraphic units occurring above the Madison in the Madden Field and was uncorrected for compaction. Since Wyoming structural basins are notoriously asymmetric it cannot be assumed that the Lysite Mountain outcrops shared an identical burial history until dislocated by Laramide thrust faulting in the Eocene because of their location on the flanks of the basin. At the time of thrusting, the Madison at Madden Field, however, was buried to approximately 17,000'.

Porosity/permeability of Madison reservoirs and outcrop dolomites

The major Madison porosity in the wells at Madden occurs in the dolomites of depositional sequences I, II and III (note the position of the perforated zones and the porosity log for the Big Horn #2-3 as shown on figure 10. 9). The pore systems of the

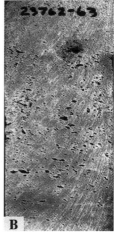

Fig. 10.11. Comparison of pore types, surface versus subsurface. A. Outcrop Lysite Mtn., top sequence II in dolomitized grainstone. Porosity is vuggy moldic dolomite. Note large vugs. B. Core sample from 23762-63 feet in BHP Big Horn #2-3. Vuggy moldic dolomite in sequence II. Core diameter is 3 inches.

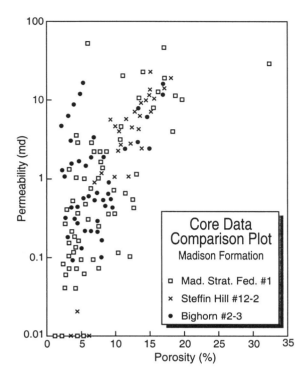

Fig. 10.12. Porosity-permeability cross-plot of the Madison Formation comparing the pore systems found in the Federal Stratigraphic Core #1 on Lysite Mtn. with two deep Madison wells at the Madden Field. Data courtesy Connie Hawkins and LLandE, Denver, Colorado.

subsurface reservoirs at Madden consist of coarse vuggy dolomite (Fig. 10. 11A). The pore systems of Lysite Mountain dolomites are similar both in outcrop and in the Federal Stratigraphic #1 core (Fig. 10.11B). When porosity/permeability values of outcrop dolomites and deep subsurface reservoir dolomites are compared, the similarity is striking (Fig. 10.12). Were the pore systems of these outcrop and deeply buried dolomites fixed by early pre-burial diagenesis, or did their porosity develop along different, but converging pathways? A look at their diagenetic histories should shed some light on the problem.

Diagenetic history of the Madison, Wind River Basin

Figure 10.13 is a paragenetic sequence developed by Sonnenfeld (1996) for the Madison in the Big Horn Basin, just north of Madden. The sequence is very similar to that observed in the Wind River Basin at Lysite and Madden. The two major differences are: the late anhydrite cements seen in the Big Horn are not generally observed in the Wind River; and there seems to be just a single period of post-Laramide hydrocarbon migration at Madden (Crockett, 1994).

One of the most important questions concerning the Madison reservoir is the origin of the dolomites. Sonnenfeld (1996) and Moore et al (1995) both developed an early evaporite reflux model for these dolomites. The interpretation is based on the tie between the distribution of restricted lagoonal sediments and regional stratigraphic collapse breccias in sequence III to the pervasive dolomitization of the open ramp facies of sequences I and II below. The rimmed shelf developed in sequence III combined with the increased accommodation generated during the TST of Sonnenfeld's upper composite sequence (Fig. 10.8) allowed the deposition of widespread lagoonal evaporites including the area of Madden. This setting is ideal for the reflux of Mg-rich, post-gypsum fluids down into and through previously deposited limestones and has the potential for supporting pervasive regional dolomitization (Chapter 6, and discussion of the Smackover which follows). It is felt, therefore, that most of the Madison dolomites at Madden formed relatively early from evaporated marine waters.

The limestones of sequence IV are

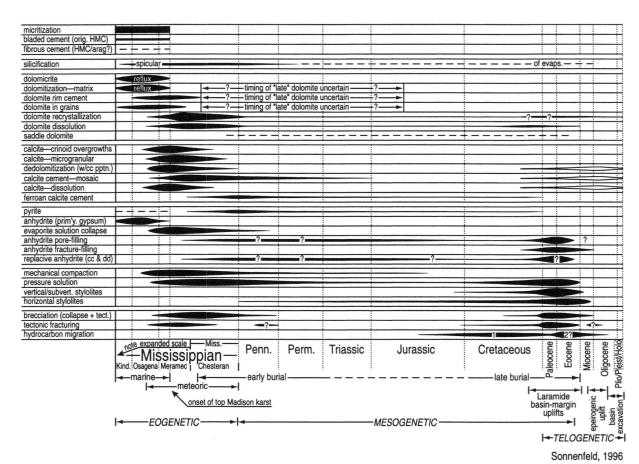

Fig. 10.13. Paragenetic sequence in relation to key stages in the Madison's post-depositional history. Developed by Sonnenfeld (1996) for northern Wyoming. The sequence is basically the same at Madden Field and at Lysite Mtn., except for the absence of the late anhydrite that occurs in northern Wyoming. Used with permission of the Rocky Mtn. Section, SEPM.

relatively non-porous and may actually form the reservoir top seal. Petrographically, porosity destruction is dominantly by chemical compaction (Crockett, 1994). What has happened to these limestones and early Madison dolomites and their pore systems after their formation in the upper Mississippian until encountered on outcrop at Lysite Mountain and by the drill in the deep subsurface at Madden Field? Perhaps isotopic geochemistry can shed some light. Figure 10.14 shows an oxygen/carbon isotopic cross plot of surface and subsurface Madison dolomites and limestones. Most of the dolomites plot in a relatively compact swarm along the zero oxygen axis with just a couple outlying light oxygen values. The carbon isotopic composition, however, shows a wide spread, ranging from +8 to 0. The limestones show a compact grouping centered around a $\partial^{18}O$ value of –5 and a $\partial^{13}C$ value of 0.

One would expect a dolomite originating from evaporative marine water to have a strongly positive oxygen isotopic (+4) composition and a carbon composition near +2 (Chapter 6). If the dolomite originated in the

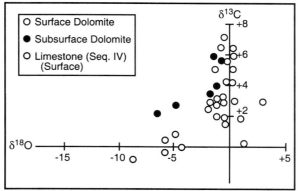

Fig. 10.14. Comparison of stable isotopes in surface limestones and dolomites of the Madison Formation (Lysite Mtn.) versus subsurface dolomites at Madden Field.

deep subsurface one would expect very light oxygen isotopic values (like the two outliers) because of high temperatures, and one would expect light carbon values because of the thermochemical degradation of hydrocarbons in the subsurface (Chapter 9). The pattern that we see on Figure 10.14, however, strongly suggests pervasive meteoric water influence (Chapter 7). If the dolomites originated as we think, from evaporative marine waters, then these early dolomites have interacted intensely with meteoric water, been recrystallized and their isotopic signature has been shifted toward meteoric values (Chapters 4, 5 and 6). This is a very plausible scenario given the fact that these rocks may have been exposed to subaerial conditions for up to 15 million years after deposition. Long exposure to meteoric water would dissolve all relict calcite matrix as well as undolomitized bioclasts leaving molds and vugs in a background of porous, coarse, recrystallized dolomite.

The isotopic composition of the limestones also suggests some meteoric influence while the petrography is typical of burial conditions.

It seems likely that the chemical compaction suffered by these limestones occurred during shallow burial under the influence of a strong meteoric water system much like the Mississippian limestones of New Mexico discussed in Chapter 9.

What has happened to the Madison dolomites and their pore systems during their subsequent burial into the deep subsurface? Strontium isotopes, as discussed in Chapter 4, may give us a clue. Figure 10.15 is a cross plot of $^{87}Sr/^{86}Sr$ versus Sr for Madison dolomites at Lysite Mountain and Madison reservoir dolomites from the deep cores at Madden Field. The single limestone value rests within the within the expected $^{87}Sr/^{86}Sr$ range for Mississippian/Pennsylvanian seawater. All values for dolomites with one exception lie above this range. There is some overlap between subsurface dolomites and outcrop dolomites, but subsurface dolomites generally have higher $^{87}Sr/^{86}Sr$ values. There is a tendency for the subsurface Madison Dolomites to have slightly higher Sr values than their surface counterparts.

The elevated $^{87}Sr/^{86}Sr$ values for the Lysite

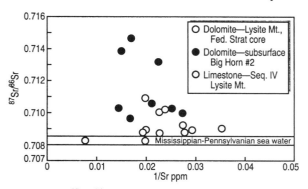

Fig. 10.15. $^{87}Sr/^{86}Sr$ versus 1/Sr for surface dolomites and limestones at Lysite Mtn. and subsurface dolomites at Madden Field. Nine data points from Crockett, 1994, the rest from the author. Mississippian sea water line from Burke et al. 1982.

Mountain dolomites suggest that they have interacted with waters with elevated $^{87}Sr/^{86}Sr$ compositions, though certainly not to same extent as their subsurface counterparts. The very low strontium values of all the Madison dolomites is probably the result of the recrystallization of the originally marine reflux waters in contact with Sr-poor meteoric waters subsurface Madison reservoirs consisted primarily of microscale dissolution and re-precipitation of the cores of dolomite rhombs. The reprecipitated dolomite exhibits bright luminescence and probably is the source for the bulk of the elevated $^{87}Sr/^{86}Sr$ and Sr shown by these dolomites (Fig. 10.16).

 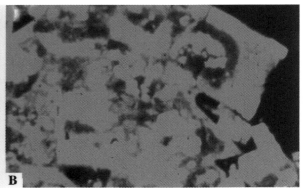

Fig. 10.16. Cathodoluminescent photomicrographs of Madison dolomites from the surface at Lysite Mtn. and the deep subsurface at 23,600 feet from the BHP Big Horn #2-3. Both samples are from sequence II. Both samples exhibit bright red strong luminescence. Largest rhomb in each sample is approximately 200 microns. A. Surface sample. Note that the rhombs have a distinctly mottled appearance, perhaps from recrystallization. B. Subsurface. Note that several rhombs show evidence of dissolution with re-precipitation of non-luminescent dolomite in the void. This phase is not present in the surface samples.

(Chapter 4). There is a tendency for subsurface basinal fluids to exhibit elevated Sr values, hence those dolomites interacting longer with subsurface fluids may contain higher Sr values. These data strongly suggest that surface and subsurface dolomites in the Madden area shared some common burial diagenetic history until dislocated by Laramide faulting. How deep were the Lysite Mountain dolomites buried? As indicated earlier, we don't know, but they were buried deeply enough to come under the influence of basinal fluids with elevated $^{87}Sr/^{86}Sr$, probably to depths in excess of 10,000ft.

Cathodoluminescence suggests that deep burial diagenetic processes affecting the

Conclusions and lessons learned from the Madison at Madden Field, Wyoming

1. The most striking feature of the Madison study at Madden Field is the fact that the reservoir pore system was established early and preserved into the deep subsurface with minimal diagenetic and pore system changes.
2. The interaction of the Madison limestones and dolomites with the meteoric water system, established during the long hiatus at the 2nd order sequence boundary at the top of the Madison, is the single most important diagenetic event suffered by these rocks. The basic pore system was

established by dissolution of residual calcite grains and matrix leaving a highly porous, vuggy dolomite. Meteoric waters recrystallized and geochemically reset the early marine protodolomites ensuring that the most stable mineral phase was present at the time of burial. These stable dolomites helped preserve the pore system in the face of deep burial (Chapter 9 discussion on mineralogy and pressure solution). Long exposure to meteoric water during the hiatus ensured that all evaporites were removed from the Madison prior to burial. The subsurface anhydrite and dolomite dissolution coupled with calcite precipitation seen in the Williston Basin, therefore, is not a factor in the Wind River Basin (Chapter 9 discussion of the Madison aquifer).

3. The use of sequence stratigraphic concepts was central to the understanding of the depositional evolution of the Madison Shelf in Wyoming. In turn, understanding the evolution of the Madison Shelf from ramp to rimmed shelf allowed the construction of a reasonable early dolomitization model for the Madison at Madden that satisfies the geological constraints of the system.

4. The extension of a viable regional sequence stratigraphic framework into the Lysite Mountain outcrops and its correlation into the subsurface at Madden allows the contiguous outcrop to be used as a surrogate for the deep reservoir dolomites during future reservoir simulation and exploitation phases.

THE UPPER JURASSIC SMACKOVER AND RELATED FORMATIONS, CENTRAL U.S. GULF COAST: A MATURE PETROLEUM FAIRWAY

Introduction

While the previous case history centered on a single deep dolomite gas reservoir and its surface surrogates, this discussion will be a bit more regional in scope involving a mature petroleum fairway, the Upper Jurassic shallow carbonates of the U.S. Gulf Coast.

Hydrocarbon exploration for oil and gas in the Upper Jurassic Smackover Formation along the U.S. Gulf Coast from east Texas to Florida commenced some 80 years ago with drilling, albeit subdued, still occurring along the trend today. The following discussion of the Smackover and related formations in the central Gulf Coast will apply the concepts of sequence stratigraphy and diagenetically driven

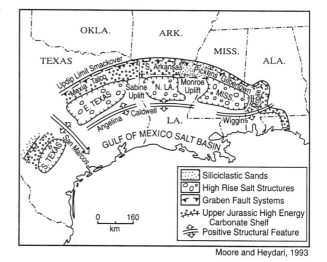

Fig. 10.17. Regional structural setting of the Upper Jurassic in the central Gulf of Mexico. Used with permission of AAPG.

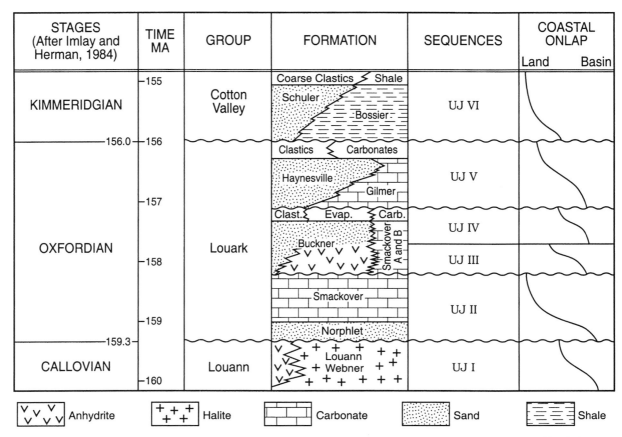

Fig. 10.18. Stratigraphic setting, Upper Jurassic, central Gulf of Mexico. Used with permission of AAPG.

porosity evolution developed in the previous chapters of this book to outline a reasonable exploration-exploitation model of this important petroleum system. This discussion will not include the so-called "Cotton Valley" pinnacle reef play in east Texas (Montgomery, 1996). Reliable data concerning the nature of these off-shelf sequences is so scarce, that their stratigraphic relationships to the units under discussion here are uncertain.

General setting

Rifting which marked the opening of the Gulf of Mexico commenced in the Triassic resulting in the development of a series of continental rift basins filled with coarse continental siliciclastics accompanied by some igneous activity. This rifting resulted in the formation of a number of closed interior basins (South Texas, East Texas, North Louisiana and Mississippi Basins) separated from the main Gulf Basin by a series of buoyant continental blocks to the south (Fig. 10.17).

The initial Mesozoic marine incursion of the region occurred in the Jurassic Callovian, flooding the main Gulf Basin and the rift related interior basins. Thick evaporites (Louann/ Werner) were subsequently deposited in the Gulf Basin and in the marginal interior basins

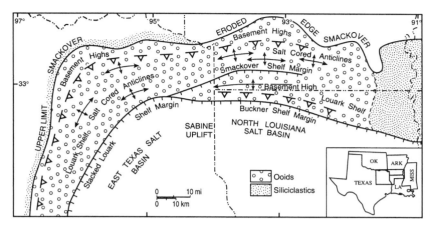

Fig. 10.19. Structural setting, Louark Shelf, central Gulf of Mexico

to the north. Thin to no salt was deposited over the marginal highs. During subsequent sediment loading and basin-ward tilting, these evaporite depocenters became the sites of intense economically important salt-related structural activity (Fig. 10.17).

There followed a major sea level low stand during which most of the interior salt basins were exposed to continental conditions, leading to some evaporite solution and the deposition of thin continental red beds to the west and extensive aeolian dune fields to the east (Norphlet Formation, Fig. 10.19). The next marine incursion occurred in the Jurassic Oxfordian (Smackover Formation, Fig. 10.19). Marine waters flooded all interior basins and the sea ultimately transgressed across the more stable northern margins of the interior salt basins to form a regionally extensive carbonate shelf (Fig. 10.17). This shelf extended with some interruptions from east Texas to Alabama (interrupted by siliciclastics in eastern Arkansas, western Mississippi) (Figs. 10.17 and 10.18).

The Smackover Formation, the target of most upper Jurassic hydrocarbon exploration in the Gulf Coast, was deposited on this shelf as a carbonate wedge ranging in thickness from 0 to 1000 feet (Fig. 10.20A). During subsequent burial and tilting toward the Gulf Basin, the wedge began to slide to the south, lubricated by the evaporites below. A major graben fault system formed at the up-dip limits of the salt (Figs. 10.17 and 10.18). In Arkansas, a major E-W trending basement structure, which seems to have been active during the Jurassic, occurs along the Arkansas-Louisiana state line. This structure acted as a fulcrum during Jurassic deposition resulting in the formation of the state line graben fault system along the northern margin of the North Louisiana Salt Basin (Fig. 10.18). As the Smackover wedge (and the underlying salt) began to slide basin-ward during subsequent burial, the state line structural high acted as a buttress resulting in the development a series of E-W trending salt cored anticlines in south Arkansas (Fig. 10.18). Most of the Jurassic production in south Arkansas is from structural traps developed over these anticlines. The following discussion will center on the east Texas, Arkansas and north Louisiana shelf (termed the Louark Shelf) fronting the east Texas and north Louisiana salt basins (Figs. 10.18 and 10.19).

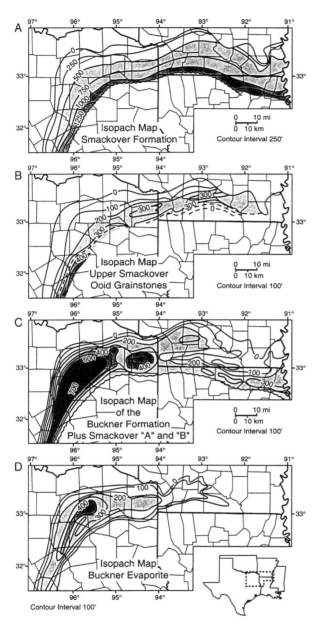

Fig. 10.20. Isopach maps of units within the Louark Group. Note that the isopach of the Buckner Formation (C) includes Smackover "A" and "B" units in northern Louisiana. Also note that (D) is an isopach of evaporites in the Buckner Formation. Bedded halite (up to 100 feet thick) occurs in the area of thickest evaporites in the westernmost area.

Sequence stratigraphic and depositional framework of the Upper Jurassic

The Upper Jurassic of the U. S. Gulf Rim is only encountered in the subsurface. The sequence has traditionally been divided into three rock stratigraphic groups: the Louann Group, consisting of the Louann and Werner evaporites at the base, followed by the mixed carbonate-siliciclastic-evaporite Louark Group, which is capped by the siliciclastic Cotton Valley Group (Fig. 10.19) (Moore and Heydari, 1993). Moore and Druckman (1991) recognized six 3rd order depositional sequences in the Upper Jurassic. They are designated UJ I-VI (Fig. 10.19).

Figure 10.20 details the isopachs of three of the major lithostratigraphic units in the Louark Group, the Smackover, Buckner and Buckner evaporites.

Figure 10.21 shows the log, lithology, interpreted depositional environments and stratigraphic interpretation of a continuous core from the base of the Buckner through the Smackover-Norphlet formations into the Louann Salt. This well, the Humble #12 McKean located in Columbia County Arkansas may be considered typical of the Smackover-Norphlet trend of the central Gulf of Mexico.

The terrestrial red desert sands and shales of the Norphlet lie unconformably on the Louann Salt and represents the lowstand systems tract of sequence II. The upper 2 meters of the Norphlet are gray, re-worked and represent the transgressive surface of the Smackover transgressive systems tract (Fig. 10.22A). A 50m sequence of organic-rich, varved, marine lime mudstones occurs immediately above this surface (Fig. 10.22B). These mudstones are thought to be relatively deep-water anoxic ba-

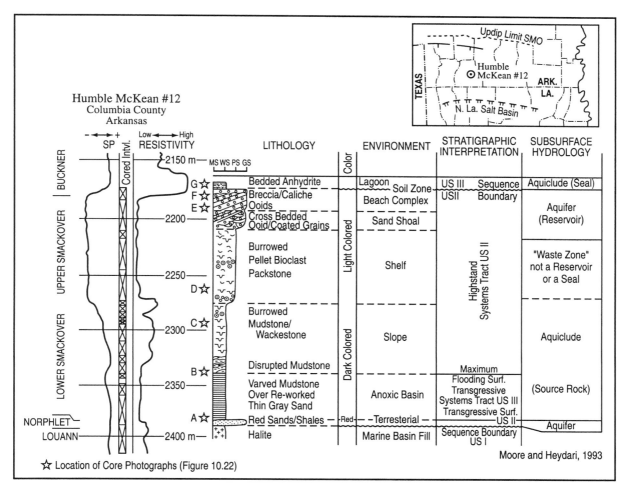

Fig. 10.21. Core description, environmental and stratigraphic interpretation of the Louann to Buckner formations in the Humble McKean #12 well located in the Haynesville Field, central Columbia County, Arkansas. Locations of core photographs (Fig. 10.22) are shown next to the graphic core description. Core courtesy Shell Oil Company, Houston, Texas. Used with permission of AAPG.

sin deposits associated with salinity stratification established soon after the Smackover flooding event and represent the transgressive systems tract of UJ II. This varved sequence occurs across the entire trend of the Smackover from Texas to Florida and acts as the source rock for Jurassic hydrocarbons (Sassen and Moore, 1988).

At the top of the varved sequence, gray lime muds showing soft sediment slump structures are thought to represent the base of the slope and the beginning of the UJ II highstand systems tract (Figs. 10.21 and 10.22B). The top of the varved sequence then, would be a downlap surface, and represents the maximum flooding surface for the UJ II sequence (Fig. 10.21).

The UJ II highstand systems tract, in the center of the Louark shelf, is a classic continuous shoaling upward sequence. Environments represented include deeper water burrowed slope (fossiliferous wackestones) and

Fig. 10.22. Core photographs from the Humble McKean #12 well, Columbia Co., Arkansas (see Fig. 10.21). A. 2391-2388m. Varved lower Smackover lime mudstones to wackestones (TST) lying with a sharp contact on Norphlet Formation red anhydritic sands and shales (LST). Upper .5m of Norphlet reworked by marine burrows. Varved mudstones are the main hydrocarbon source rock for the Jurassic. B. 2325m. Disrupted bioclastic lime mudstone exhibiting soft sediment slump structures marking the base of the slope and hence the MFS. C. 2291m. Slope to shelf transisition, from dark burrowed bioclastic lime wackestone to burrowed light colored pellet lime packstone (HST). D. 2267m. Burrowed pellet lime wackestone. Subtidal middle shelf (HST). E. 2195m. Cross-bedded dolomitized ooid lime grainstone. Upper shoreface, beach (HST). Main reservoir facies. Note high amplitude stylolite. Black stain is hydrocarbon. F. 2194m Laminated, dolomitized ooid grainstone, uppermost beach shoreface (HST). Reservoir facies. Top of slab exhibits typical beach intertidal keystone vugs as developed in modern beach rock. G. 2190m. Sequence bounding unconformity between sequences UJII and UJIII. Bedded marine anhydrite of sequence UJIII lying on quartzose sandy soil breccia developed on upper surface of sequence UJII.

burrowed outer shelf (pellet, oncolite wackestones-packstones) sequences (Figs. 10.22C and 10.22D). The succeeding shallow-water high-energy shelf sequences is dominated by sand shoal environments (ooid, pellet and oncolitic packstones and grainstones) (Fig. 10.22 E), culminating in an ooid, pisoid-dominated shoreline beach complex (Figs. 10.21 and 10.22F) at the top of sequence UJ II.

The upper sequence boundary of sequence UJ II in the Humble McKean #12 well in Arkansas is marked by a distinct unconformity. The UJ II-III sequence boundary exhibits a soil profile, soil breccias, thin sands and silts and

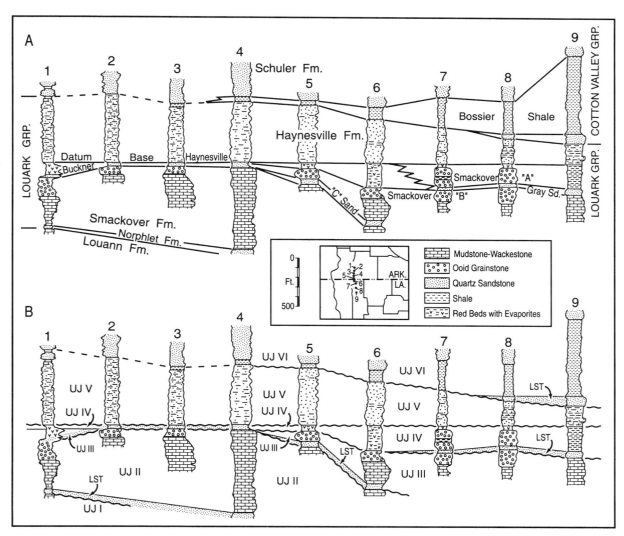

Fig. 10.23. North-south cross sections of the Upper Jurassic, from southwest Arkansas into northwest Louisiana across the state line basement structure. Well #1 is the Humble McKean #12 well, Columbia Co., Arkansas (Figures 10.21 and 10.22). Lithologic data from cores and sample logs. Sample data furnished by David Stoudt and Mosbacher Petroleum, Houston, Texas. A. Lithostratigraphic cross-section. B. Sequence stratigraphic overlay of A.

caliches strongly suggesting exposure during the relative sea level lowstand separating the sequences (Figs. 10.21F and 10.22G). The transgressive systems tract of sequence UJ III is represented by bedded anhydrites of the Buckner Formation (Figs. 10.21F and 10.22G).

Figure 10.23A is a north-south subsurface stratigraphic cross section located in western southern Arkansas and north Louisiana. It is drawn across the state line basement high into north Louisiana. It starts with the Humble McKean #12 well and is datumed on the base of the Bossier shale. It would seem that the Buckner evaporites onlap onto the sequence II bounding unconformity and that most of the upper HST shelf margin grainstones of

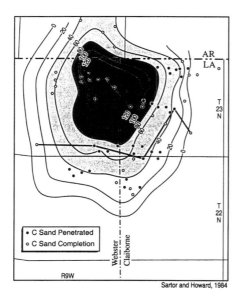

Fig. 10.24. Isopach map of the Upper Jurassic "C" sand, northern Louisiana, just south of the Louisiana-Arkansas boundary. Used with permission of GCAGS.

sequence II are missing (presumably removed by erosion) across the crest of the state line structure (an area measuring some 5 miles wide). South of the structural crest, into Louisiana, two ooid grainstone bodies separated by deeper water lime wackestones are encountered. Traditionally, these grainstones have been termed the Smackover A and Smackover B and occur as offset elongate E-W trending hydrocarbon productive carbonate sandbodies (Fig. 10. 23A). This pattern suggests prograding shoreline complexes.

The Smackover B shoaling upward sequence is separated from deep-water lime wackestones below by a gray, fine grained quartz sandstone, locally termed the "C" sand. This sand can be traced up-dip, across the crest of the state line basement high into the unconformity marking the top of the Smackover in the Humble #12 McKean well (Fig. 10.23A). Conceptually, then, the "C" sand south of the state line structure in Louisiana marks the top of the Smackover Formation and the Smackover B and A units are younger than the Smackover proper. The "C" sand is productive in north Louisiana and an isopach in a field near wells 6 and 7 (Fig. 10.23A) suggests a fan shape for the sandbody (Fig. 10.24).

Using this stratigraphic data we can now overlay the sequence stratigraphic framework of the Oxfordian in the Arkansas-Louisiana section of the trend established in the Humble McKean #12 well above (Fig. 10.23B). The Werner-Louann represents Upper Jurassic sequence I. The clastics of the Norphlet represents the low stand systems tract of Upper Jurassic sequence II. The unconformity at the top of the Smackover ooid grainstones represents Upper Jurassic sequence boundary II. This sequence boundary is traced seaward under the Smackover B ooid grainstones into the base of the "C" sands. The "C" sand slope fan represents the low stand systems tract of Upper Jurassic sequence III. Upper Jurassic sequence IV is represented by the shoaling upward sequence of the Smackover A unit. Upper Jurassic sequence V is represented by the limestones, shales and sands of the Haynesville/Gilmer. The Bossier Shale represents a major marine floodback and the sequence boundary of Upper Jurassic sequence VI (Fig. 10.11B). Is this framework regionally viable?

Figure 10.25A is an E-W stratigraphic cross section across the shelf margin in Henderson County Texas. The shelf margin in this area trends NE-SW (Fig. 10.20) and the section is approximately normal to the margin. This cross section is hung on the base of the Bossier Formation. Well #4 seems to be very

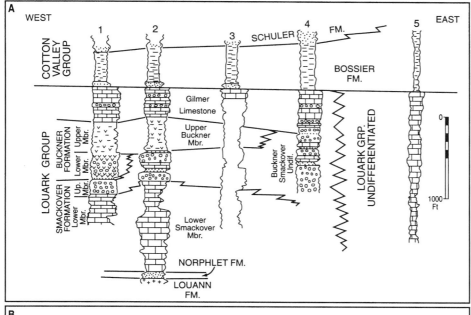

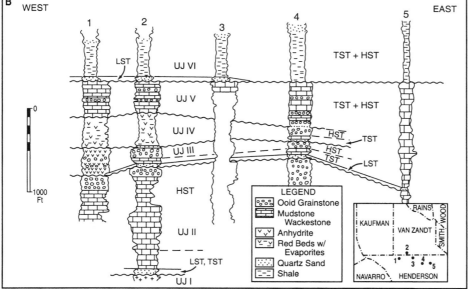

Fig. 10.25. East-west stratigraphic cross-sections of the Upper Jurassic, east Texas. Cross-section drawn normal to the general Upper Jurassic shelf margin. Lithologic data from cores and sample logs. Sample data furnished by David Stoudt and Mossbacher Petroleum, Houston, Texas. A. Lithostratigraphic cross-section. B. Sequence stratigraphic overlay of A.

close to the shelf margin because of the presence of thick ooid grainstones in the Smackover and Smackover/Buckner sequences. Buckner evaporites are not present in this well, but are present to the west in well #1. It is difficult to differentiate the Buckner Formation from the Smackover in well #4 because most of the sequence consists of ooid grainstone packages separated by thin quartzose sands. Seaward, the Buckner/Smackover interval is represented by a continuous section of deepwater lime wackestones to lime mudstones termed the Louark Group undifferentiated. Isolated quartz sandstones occur in these deep water limestones seaward of the shelf margin. Finer grained Gilmer limestones above are separated from the Buckner/Smackover below by quartz sandtone in well #4 and sit directly on upper Buckner evaporitic red beds in wells 1-3 (Fig. 10.25A). Landward, the thick ooid grainstones of the upper part of the Buckner/Smackover undifferentiated sequence in well #4 change

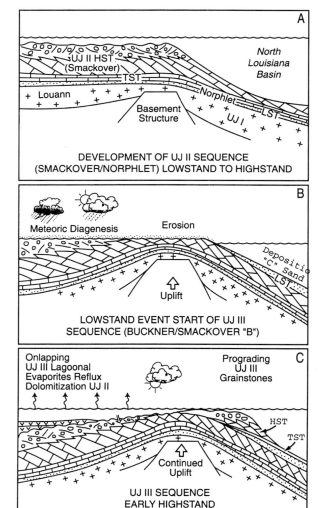

Fig. 10.26. Cartoons illustrating the development of Upper Jurassic sequences II and III in southern Arkansas and northern Louisiana.

facies into the anhydrites characteristic of the lower member of the Buckner Formation. West of well #1, the standard formational boundaries between Buckner and Smackover apply.

Figure 10.25B is the sequence stratigraphic overlay of the west to east lithostratigraphic cross section of Figure 10.25A. The six Upper Jurassic sequences seen in Arkansas and Louisiana are recognized in the Texas section some 90-100 miles to the west of the Humble McKean #12 well. The UJII, III and IV sequence boundaries are marked by the presence of low-stand quartz sands between shoaling upward sequences in the Buckner/Smackover. The abrupt marine flood back marked by the transgression of marine limestones of the Gilmer/Haynesville over the red beds of the upper Buckner mark the UJ IV-UJV sequence boundary. As in northern Louisiana, the marine flood back indicated by the Bossier marine shale marks the UJV-UJVI sequence boundary. The difference between the Upper Jurassic lithostratigraphy and chronostratigraphy is even more pronounced in this east Texas section and explains much of the stratigraphic confusion these rocks have caused workers in the Gulf Coast in the past.

Figure 10. 26 are a series of summary cartoons that represent the development of the sequence architecture in the Arkansas-Louisiana area. The development of UJII (Smackover/Norphlet) is the same in east Texas. However, there is no basement structural movement at the shelf break in east Texas, so the UJ III shelf margin is coincident with the UJII margin and UJIII ooid shoals form the barrier for the development of the lagoonal evaporites characteristic of sequence UJIII (Fig. 10.25B).

At this point it is unknown whether this

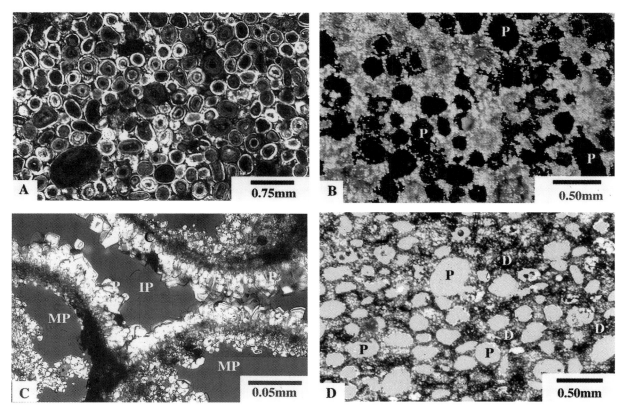

Fig. 10.27. Thin section photomicrographs illustrating major reservoir porosity types found in Upper Jurassic reservoirs in the central Gulf Coast. A. Primary intergranular porosity (8%) in sequence UJIII, Walker Creek Field, Columbia Co. Arkansas. Ooids were originally calcite. Plain light. B. Oomoldic porosity (P) in UJII. Non-reservoir rock, Mount Vernon Field, Columbia County, Arkansas. Porosity 30%, permeability 0.1md. Ooids were originally aragonite. Crossed polars. C. Cement reduced (C) interparticle (IP) and oomoldic porosity (MP), totals 40%, permeability >100md. Reservoir rock in UJII at Buckner Field, Columbia Co., Arkansas. Ooids originally aragonite. Plain light. D. Oomoldic porosity (P) developed in early reflux dolomite (D) in sequence UJII, wildcat, east Texas. Porosity and permeability not reported, but porosity should exceed 30% and rock should be permeable. Ooids were originally aragonite. Plain light.

sequence stratigraphic framework can be extended to the east into Mississippi, Alabama and Florida. Different subsidence rates and lack of biostratigraphic data make correlations uncertain.

Early diagenesis and porosity modification related to depositional sequences and sea level

Most Upper Jurassic reservoirs are developed in ooid grainstones that originally had porosities of nearly 50%. Burial depths of these reservoirs in the central Gulf Coast range from 6000-14,000 feet. Present reservoir porosity types include primary preserved intergranular; oomoldic secondary porosity and dolomitized oomoldic secondary porosity (Fig. 10.27) (see Moore, 1984 for more details).

In general terms, these pore systems are geographically and stratigraphically isolated. Primary preserved intergranular porosity occurs almost exclusively in Upper Jurassic sequences III, IV and V. The one exception is found in the ooid grainstones of UJ sequence II at the shelf margin in east Texas. Oomoldic porosity and dolomitized oomoldic porosity occurs only in Upper Jurassic sequence II. There is a direct relationship between the presence of sequence III evaporites (Buckner Formation) and dolomitized oomoldic porosity (Chapter 6) (Moore et al., 1988). The development of oomoldic porosity and dolomitization are both early diagenetic overprints that have dramatically modified the original pore system.

The original mineralogy of Oxfordian ooids in the central Gulf of Mexico is thought to have been both aragonite and calcite (Moore et al., 1986; Heydari and Moore, 1994). While the Jurassic sea was generally a calcite sea (Chapter 2), in areas of elevated salinities, such as the interior of a broad shallow shelf, ooids may well have been composed of aragonite (Moore et al., 1986). Upper Jurassic sequence II (Norphlet/Smackover) was such a shelf and ooids generated across the majority of this shelf in east Texas and southern Arkansas were originally composed of aragonite. Ooids at the shelf margin of sequence UJII, however, seem to have been calcite because of their accessibility to unmodified Jurassic seawater. While calcite ooids are present at the shelf margin in east Texas, the UJII shelf margin, and its calcite ooids have been removed by erosion across the crest of the state line basement structure in southern Arkansas.

Oomoldic porosity developed in the aragonite ooids of sequence II during exposure to a regional meteoric water system (Moore and Druckman, 1981) (Chapter 7). This meteoric water system developed during the hiatus and exposure associated with the sequence II sequence bounding unconformity (Fig. 10.26B).

During meteoric water diagenesis, aragonite was dissolved, creating oomoldic porosity (Fig. 10.27B), and down hydrologic flow calcite cements were precipitated, destroying primary porosity (Fig. 10.27C). During mineral stabilization, this porosity modification was intense and a huge range of reservoir quality resulted. Up hydrologic flow, porosity was high but calcite cements destroyed permeability. Down hydrologic flow, porosities tended to be lower, because ooids were not totally dissolved, while permeabilities were higher because calcite cements were less abundant (see Chapter 7 for a discussion of regional meteoric water systems and their effect on reservoir quality). At the time of burial porosity ranged from 5-50%, while permeability ranged from .1 to 40md.

The ooids of sequences III-V (Buckner, Hanynesville, and Gilmer), however, are thought to have been calcite because these shelves were relatively narrow and presumably covered with normal Upper Jurassic marine water (Moore et al., 1986). Even though exposed to meteoric waters at each of the 3 sequence boundaries, the main porosity type remained primary intergranular (Fig. 10.27A) because of the lack of a mineralogical diagenetic drive (Chapter 7) (Moore and Heydari, 1993). At the time of burial this porosity averaged over 30% with relatively high permeabilites.

The dolomites found in sequence II developed in association with the transgressive systems tract of sequence III. During the sequence III transgression, a widespread lagoon was formed by an ooid grainstone barrier in east Texas and by basement structural activity in southern Arkansas (Fig. 10.26C) leading to the deposition of lagoonal evaporites over much of east Texas and southwestern Arkansas (Fig. 10.20). Magnesium-rich post-gypsum brines refluxed down into the oomoldic limestones of sequence II resulting in the regional dolomitization that extends across east Texas into southwestern Arkansas. This reflux dolomitization event was described in detail in Chapter 6 (Figs. 6.33-6.36).

Reservoir quality seems to have been enhanced by this reflux dolomitization. Since the dolomitizing fluids were Mg rich, but $CaCO_3$ poor, the existing $CaCO_3$ of the oomoldic limestones of sequence II had to be dissolved to provide the $CaCO_3$ needed for dolomitization, hence enhancing both porosity and permeability (Chapters 3 and 6). At the time of burial these dolomites had a wide range of porosity (15-40%) (Fig. 10.27D).

Burial diagenesis and subsurface porosity evolution, Oxfordian reservoirs, central Gulf of Mexico

Figure 10.28 is the burial curve for the sequence II reservoir in the Humble McKean #12. This burial curve and timing of oil migration should be characteristic of the Upper Jurassic for the central Gulf of Mexico. The Humble McKean reservoir is presently at

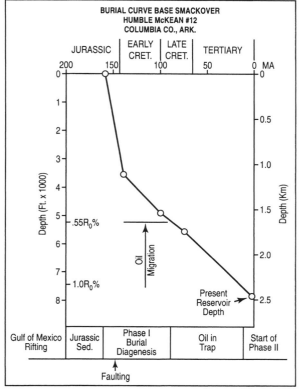

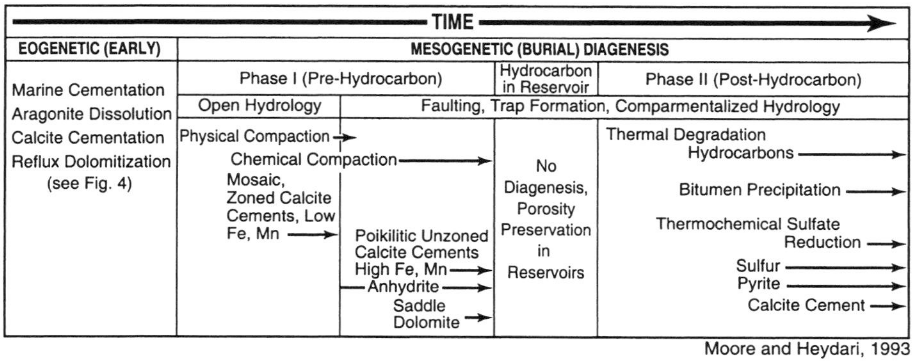

Fig. 10.28. Burial curve of the Humble Mckean#12 well. The curve was constructed from log-derived chronostratigraphic unit thicknesses. The curve was not corrected for compaction or for unconformity section loss. R_o values calculated. Used with permission of AAPG.

Fig. 10.29. Relative paragenetic sequence of upper Smackover. Phase 1 burial diagenesis and early diagenesis is based on Arkansas-Louisiana studies (Moore and Druckman, 1981). Phase II burial diagenesis is from the Mississippi Salt Basin (Heydari and Moore, 1989).

8,000' (2450m). To the west in east Texas sequence II reservoirs are presently at 10,000' to 14,000' (3049-4268m) while in north Louisiana the sequence III and IV reservoirs are presently at 10,000' (3049m). In each case, however, oil migration took place during the Cretaceous somewhere between 5000 and 6500' (1500-2000m). Since burial diagenesis essentially ceases when the reservoir is filled with hydrocarbons, burial diagenetic processes and associated porosity modification has operated on these reservoirs from the surface to a depth of some 6500' (2000m) and should be similar throughout the section of the trend under discussion (Chapter 9).

Figure 10.29 is the relative paragenetic sequence suffered by ooid grainstone reservoirs of Upper Jurassic sequences II-V. The most important burial diagenetic porosity modification process is chemical compaction where porosity is lost both by pressure solution and its related calcite cementation (Chapter 9). Chemical compaction effects are dominant in sequences III-V. These reservoirs have no early cement and are not dolomitized. Porosity loss under these conditions can be greater than 50% and final porosities in the reservoirs range from 8-15% and average 10% (Fig. 10.27A). Chemical compaction in the oomoldic limestone reservoirs of sequence II is somewhat muted by the presence of early meteoric water cements (Fig. 10.27B and C) but will still decrease porosity in the subsurface by greater than 30%, particularly where stylolites are involved (Fig. 9.9) (Chapter 9). Final porosities in these reservoirs have a much wider range of values than the intergranular porosities discussed above with values ranging from 5% to 30%. In the dolomite reservoirs of sequence II, chemical compaction is greatly reduced by the chemical stability of the dolomites (Chapter 9). The porosity seen at depth in the reservoir is nearly the porosity present at the time of burial.

Saddle dolomite cements (Fig. 9.13C) occur in most reservoirs of all sequences and porosities, but represent less than 1% total volume and hence have little effect on total reservoir porosity. Late stage replacement anhydrite (Fig. 9.13A), while mostly replacing calcite ooids and cements does tend to destroy a bit of porosity, but like the saddle dolomite has little effect on total reservoir porosity.

As was noted in Chapter 9, an important, but often overlooked late diagenetic process is the thermal degradation of hydrocarbons in the reservoir after hydrocarbon migration. One of the by-products of this degradation is bitumen precipitation that effectively destroys porosity in the reservoir. Bitumen porosity destruction is more important in the deeper reservoirs of east Texas and to the east in southern Mississippi and Alabama (Fig. 9.13A). Calcite cements associated with thermochemical sulfate reduction are only important in the deepest reservoirs (16,000', 4878m, or greater) where most of the hydrocarbons have been destroyed and most of the porosity plugged by bitumen, as described by Heydari and Moore (1989) for reservoirs in southern Mississippi (Fig. 9.13D).

Upper Jurassic exploration and production constraints and strategies, central Gulf of Mexico

The development of a viable diagenetic history and porosity evolution within a reasonable sequence stratigraphic framework has had a profound effect on our understanding of the Upper Jurassic hydrocarbon fairway in the central Gulf.

The recognition of the five Oxfordian 3rd order sequences, their distribution and facies architecture, has for the first time allowed the explorationist in this region to view the entire petroleum system without the complications of lithostratigraphic nomenclature.

For example, it is now obvious that the evaporites of sequence III are isolated in time and space by a sequence bounding unconformity from the ooid grainstones of sequence II. This negates the ramp model for sequence II and allows the development of a shelf model and the appropriate placement of related facies in the model. The shelf model also explains the distribution of contrasting ooid mineralogy seen in the central Gulf and mentioned in the discussion above. In the shelf interior, higher salinities prevail and the favored mineralogy is aragonite. Along the shelf margins facing the open marine, the normal Upper Jurassic calcite mineralogy prevails.

In east Texas and southwest Arkansas, the recognition of sequence III and its thick shelf margin calcite ooid grainstones allows the explorationist to postulate the development of a barred shelf lagoon filled with subaqueous evaporites. The long-standing battle of sabkha versus lagoon origin for the Buckner evaporites is resolved and the occurrence of thick, laminated anhydrites and halites in the Buckner are no longer enigmas. The evaporite lagoon developed in sequence III sets the stage for regional reflux dolomitization of sequence II during the sequence III marine transgression. This explains nicely the relationship between dolomitization in sequence II and evaporites in sequence III.

For east Texas then, the major reservoirs are developed in the dolomitized oomoldic grainstones of the late HST of sequence II. There are no viable stratigraphic traps because of pervasive dolomitization, so the reservoirs are dominantly structural with a sequence III evaporite seal. There is little production in sequences II and III along the shelf margin in east Texas because the stacked calcite ooid grainstones apparently cannot provide a seal. Grainstones of sequence IV can be sealed against fine-grained lagoonal limestones of sequence V if structure is available. However,

Table 10.1

	EAST TEXAS	EAST TEXAS SHELF MARGIN	ARKANSAS	LOUISIANA
Trap type	Structural	Structural	Structural	Stratigraphic
Seal	Evaporites	Transgressive marine carbonate mudstones	Evaporites coastal plain muds	Lateral-lagoonal Top-TST mudstones
Porosity type	oomoldic dolomite secondary	intergranular primary	oomoldic oomoldic dolomite secondary	intergranular primary
Reservoir in sequence	sequence UJII	sequence UJIV	sequence UJII	sequence UJIII/IV
Source rock	TST UJII	TST UJII	TST UJII	TST UJII

Table 10.1. Upper Jurassic petroleum system, central Gulf of Mexico.

most major reservoirs in east Texas show secondary oomoldic dolomite porosity (Table 10.1).

In south Arkansas and northern Louisiana the importance of understanding the sequence stratigraphic framework and its impact on porosity evolution and exploration/production strategy in the contiguous fairway is particularly compelling.

The rejuvenation of the state line basement structure during and subsequent to the top Smackover unconformity separates the Upper Jurassic fairway into northern (Arkansas) and southern (southernmost Arkansas and northern Louisiana) provinces. Correlations between these provinces were always difficult and led to much misunderstanding of the true Upper Jurassic stratigraphic equivalencies. Application of sequence stratigraphic concepts in this region, as outlined above, led us to realize that the so-called Smackover of north Louisiana was in actual fact younger, and separated spatially from the Smackover of Arkansas. The diagenetic models, exploration strategies and the scheme for porosity evolution developed for Arkansas reservoirs therefore are not applicable to north Louisiana.

For example, the calcite ooid grainstones of the sequence II shelf margin were removed by erosion along the crest of the rejuvenated basement structure. This left the shelf interior aragonite ooids as the dominant ooid mineralogy for sequence II in the area. During the sequence bounding unconformity and exposure, these aragonite ooids underwent dissolution, calcite cements were precipitated and cement reduced oomoldic porosity became the dominant porosity developed in the sequence II high stand systems tract. Subsequent flooding during the sequence III transgression led to the formation of a structurally barred lagoon, evaporite precipitation and reflux dolomitization of the sequence II HST. At this point, southwestern Arkansas is identical to east Texas with reservoirs characterized by dolomitized oomoldic secondary porosity and evaporite seals (Table 10.1).

Because of the positive basement structure along the Arkansas-Louisiana state line, the shelves developed during the HST of sequences III and IV were relatively narrow, separated from the sequence II. shelf by the state line structure and were dominated by calcite ooids. The ooid grainstone reservoirs of north Louisiana therefore show no oomoldic porosity and consist dominantly of primary intergranular porosity. Lower rates of subsidence that are related to the positive state line feature forced these sequences into an offset progradational geometry rather than the aggradational stacked architecture seen in east Texas. This geometry lead to the development of a series of strike oriented classic stratigraphic traps with shale dominated lagoonal lateral seals and deeper water lime mudstone top seals developed during the transgression of the subsequent sequence (Table 10.1).

Southern Arkansas exploration therefore, is dominated by structural traps developed over low relief salt cored anticlines with anhydrite or coastal plain shale seals. Porosity is secondary oomoldic and may be dolomite when associated with anhydrites.

In contrast, north Louisiana exploration is keyed to stratigraphic traps developed at progradational shorelines, parallel to depositional strike and sealed laterally by lagoonal shales and top sealed by transgressive marine mudstones. Reservoir porosity is dominantly primary intergranular. Siliciclastic low stand slope fans at each sequence boundary are an added attractive exploration target (Table 10.1).

Conclusions and lessons learned from the Smackover of the central Gulf of Mexico

1. The depositional model applied to a carbonate rock sequence must satisfy all geological constraints including lithofacies, chronostratigraphy, well and seismic-derived stratigraphy and diagenesis. The depositional model applied to the Smackover and related formations in the central Gulf has long been misinterpreted due to a lack of integrated regionally-based studies.
2. Local structure can dramatically influence sequence geometries. Upper Jurassic depositional sequence sets are aggradational in east Texas. They are however, strongly progradational in south Arkansas and northern Louisiana due to the influence of the state line basement structure.
3. Local oceanographic and climatic conditions can affect the mineralogy, and hence the diagenetic potential of abiotic carbonate components such as ooids and marine cement. Smackover ooids and marine cements are both aragonite and calcite even though the Jurassic is a period dominated by calcite. High salinity in the shelf interior favors aragonite ooids, while calcite ooids and marine cements were formed at the platform margins. Reservoir rocks in areas dominated by aragonite grains exhibit moldic porosity and extensive early meteoric cements. Reservoirs in areas dominated by calcite are characterized by no meteoric cement and primary intergranular porosity.
4. Early porosity modification is centered on sequence boundaries. In the Smackover this includes generation of secondary porosity, meteoric cementation and dolomitization.
5. Porosity modification during burial in a passive margin setting such as the Gulf of Mexico is dominated by porosity destruction driven by pressure solution. This

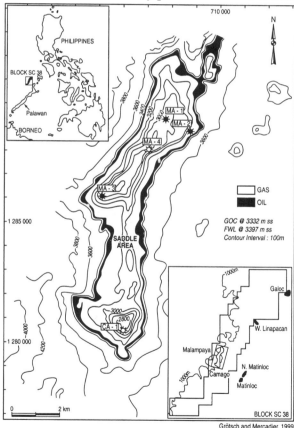

Fig. 10.30. Topographic map of the Malampaya/Camago carbonate platform located within Block SC 38, offshore western Palawan in the Philippines. The five exploration and appraisal wells discussed are indicated, as are the gas-oil contact (GOC) and the free-water level (FWL). Used with permission of AAPG.

process is stopped by arrival of hydrocarbons. Most central Gulf Jurassic reservoirs show higher than expected porosities for their present depth so represent "frozen reservoirs".

6. The sequence stratigraphic framework controls the nature of the petroleum system in the Smackover and related formations. Reasonable success and viable exploration strategies depend on an understanding of this framework.

THE TERTIARY MALAMPAYA AND CAMAGO BUILDUPS, OFFSHORE PALAWAN, PHILIPPINES: 3-D RESERVOIR MODELING

Introduction

This study was initially presented by Jurgen Grotsch and Christophe Mercadier in 1996 at the Pau AAPG Hedberg Research conference entitled "Carbonate Reservoirs:Strategies for the Future" It was subsequently published in a 1999 special theme issue of the AAPG Bulletin based on the Pau conference (Grotsch and Mercadier, 1999). It was chosen as the last case history of this book because it probably represents the future direction of carbonate reservoir studies; deterministic 3-D integrated reservoir modeling. During the early evaluation of the economics of the hydrocarbon accumulation at Malampaya and Camago limited well and rock data were integrated with 3-D seismic, isotope geochemistry, paleontology and a nearby outcrop analogue to build a viable 3-D static reservoir model. The following discussion outlines the geologic history of the buildups, the basic data components, problems faced, and the concepts utilized to build the model.

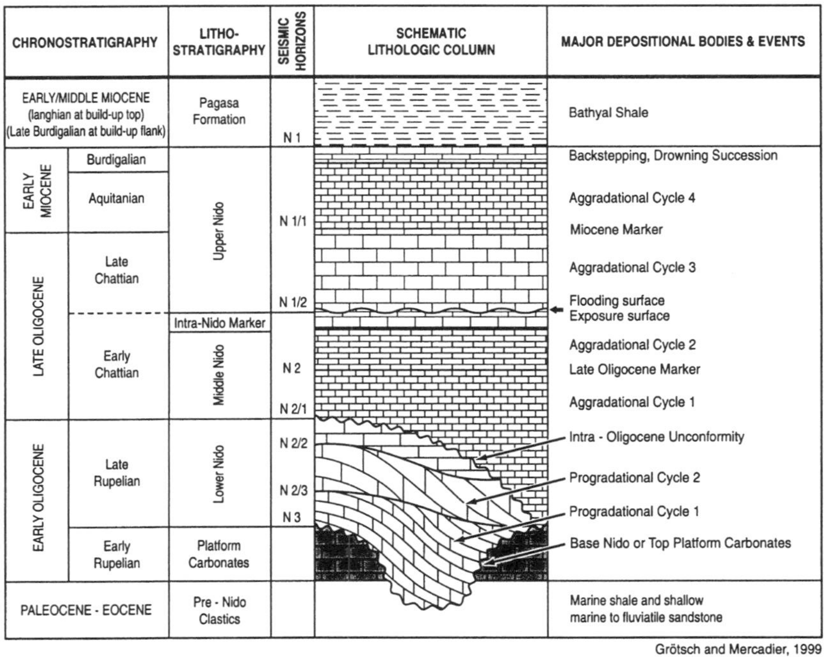

Fig. 10.31. Chronostratigraphy and lithostratigraphy of the Malampaya area. Horizons mapped from 3-D seismic. Lithology, and major depositional cycles and events during the buildup growth, are schematic. Chronostratigraphic ages are based mainly on Sr isotope stratigraphy from rotary sidewall samples and cuttings taken in all wells. Two major unconformities were identified in the early Oligocene based on seismic and well data. Note that both the drowning succession at the top of the Nido Limestone and the bathyal shale of the overlying Pagasa Formation date as Early Miocene Burdigalian. Used with permission of AAPG.

Geologic setting

The Malampaya-Camago Field is located in 850-1200m of water northwest of Palawan Island, Philippines (Fig. 10.30). The field is situated on a continental fragment in an island arc setting that marks the southeastern margin of the South China Sea. The reservoir consists of Oligocene/Miocene reefal and reef-related Nido carbonates (Fig. 10.31). After the initial opening of the South China Sea in the Paleocene-Eocene (rifting phase), the Nido Limestone was deposited during the drifting stage in the Early Oligocene-Early Miocene.

The distribution of the reefal Nido Limestone is controlled by an underlying NE-SW extensional rift-related fault system (Figs. 10.30 and 10.32). The Nido at Malampaya-Camago was deposited as an isolated reef-rimmed carbonate platform on a horst block formed by eroded Paleocene-Eocene clastics (Fig. 10.32). There is ample evidence that reef

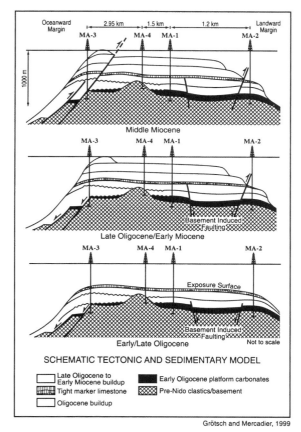

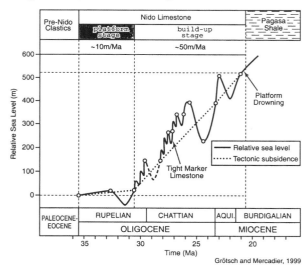

Fig. 10.32. Schematic drawing showing a composite section that illustrates the tectonic and sedimentary history of the Malampaya/Camago platform. The Oligocene part of the Nido Limestone overgrows a preexisting rugged morphology formed by top-eroded clastics deposited prior to the Nido Limestone. Note that the Eocene normal faults along the western margin of the Malampaya buildup were reactivated during the middle Miocene by transpressional movements, causing tectonic inversion. Some reverse fault movements also are observed along the eastern margin of the platform. The figure is not to scale; distances between the wells are given for reference. Used with permission of AAPG.

Fig. 10.33. Subsidence and relative sea level history of the Malampaya/Camago platform as derived from Sr isotope stratigraphy (absolute ages) and carbon isotope measurements (suggesting repeated exposure). High resolution of Sr isotope dating in the Oligocene—early Miocene allows recognition of time gaps of 0.5—2.0 m.y. during the growth history of the platform. Constrained data points are indicated as circles on the curve. Note the change in tectonic subsidence during the late Rupelian with rates changing from about 10 m/m.y. to 50 m/m.y. (not compensated for absolute eustatic sea level change). Used with permission of AAPG.

environments dominated the northwestern platform margin facing the open high energy South China Sea, while the lower energy southwestern margin was characterized by grain and mud supported carbonates.

The subsidence and relative sea level history of the Malampaya-Camago platform was derived from an extensive Sr and carbon isotope study of the available conventional and sidewall core samples (Fig. 10.33). The rate of subsidence of the platform changed abruptly during the Oligocene Rupelian from 10m/m.y.

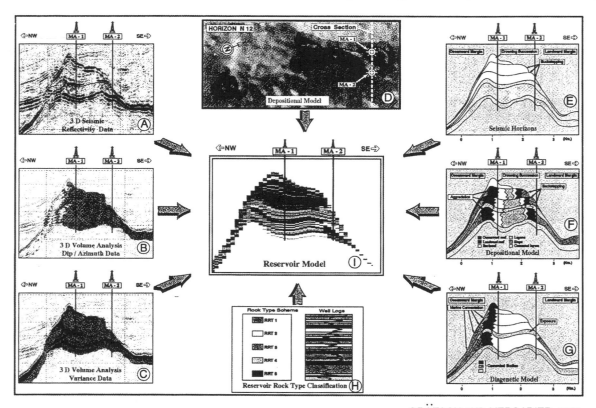

GRÖTSCH AND MERCADIER, 1999

Fig. 10.34. Schematic overview of the modeling steps illustrating the integration of different data sources into the rock type based 3-D reservoir model. For this example, the two-dimensional cross section in the northern part of Malampaya with wells MA-1 and MA-2 is used. (A) Cross-section of the 3-D seismic reflectivity data. (B and C) Body rendering using 3-D volume analysis allows identification of large-scale reservoir bodies of similar properties such as the pervasively cemented oceanward western margin. Integration of seismic and well data allows establishment of a 3-D diagenetic model. (D) Depositional areas as outlined for the seismic horizon Nl/2 (most likely scenario). The location of the cross section used is indicated as a dashed line. (E) Growth stages as outlined by seismic horizons are mapped from the 3-D reflectivity data set (A),(F) Depositional model derived from well and 3-D seismic data showing the large-scale depositional areas that are laterally constrained by 3-D seismic volume analysis and well data. (G) Diagenetic model indicating the pervasive early marine cementation along thc western margin and the cemented horizons at the buildup tops. (H) The reservoir rock type (RRT) subdivision derived from petrophysical data and neural network analysis provides the high-resolution vertical variability of RRTs in the 3-D reservoir model. (I) Cross section through the RRT-based reservoir model which is used for saturation modeling and as input for dynamic simulation. Used with permission of AAPG.

to 50m/m.y. leading to a long-term rapid relative sea level rise from the Upper Rupelian onwards (Fig. 10.33). This rapid relative sea level rise led to the vertical aggradation of the seaward reef bound platform margin (NW) and the progressive backstepping of the lower energy southeastern margin. The carbonate platform history culminated in drowning during the Miocene Burdigalian marked by deposition of the deep marine shale of the Pagasa Formation (Figs. 10.32 and 10.34) (Chapter 1).

The world climatic setting seems to have changed from green house to ice house conditions toward the end of the Oligocene Chattian as indicated by the apparent increase in the amplitude of sea level change postulated for the Malampaya-Camago platform (Fig. 10.33) (Chapter 8).

The Middle-Late Miocene collision with the Sulu Island arc, followed by Pliocene collision with the Philippines archipelago resulted in the uplift of the northwestern margin of Palawan Island. This uplift exposed age equivalent Nido limestones on Palawan and provided insight into the depositional framework of the Nido at Malampaya-Camago. The offshore margin (block SC 38) was downwarped and buried more deeply under basinal deposits (in excess of 3000m). These collision events caused fault reactivation of the basement faults beneath the platform at Malampaya-Camago and were responsible for tectonic inversion and reservoir fracturing (Fig. 10. 32).

Lithofacies and depositional model

Lithofacies of the Nido were described from limited conventional and sidewall cores from the 5 appraisal/exploration wells at Malampaya-Camago (Fig. 10.30). Lithofacies encountered include coral boundstone, bioclas-

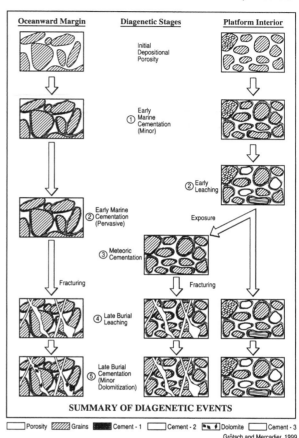

Fig. 10.35. Summary of the diagenetic history as observed in well data. Cement 1, early marine-phreatic isopachous rim cement. Cements 2 and 3, blocky calcite cement. Used with permission of AAPG.

tic rudstone, pisolitic rudstone, bioclastic packstone and grainstone, gastropod wackestone, planktonic foraminiferal wackestone-floatstone, and red algal floatstone. The nature of these lithofacies confirmed that Malampaya-Camago was indeed a reef-bound isolated platform.

The rock-based data set, however, is so sparse and spatially isolated that a viable three-dimensional depositional model could not be constructed. A 3-D seismic data set (Fig. 34A-C) was integrated with the rock data and nearby outcrop analogues (northwest Palawan Island) to develop a reasonable three- dimensional depositional model. Seismic reflectivity data gave the gross stratigraphic framework (Fig. 10.34A and E) while 3-D volume analysis was used to identify large-scale reservoir bodies of similar properties. Integration of the core rock data and the outcrop analogue with seismic allowed Grotsch and Mercadier (1999)to establish a gross depositional model (Fig. 10.34F).

The platform is bound on the northwest (oceanward) by a pervasively cemented aggradational framework reef complex (Chapter 5). Cemented debris flow slope deposits are transitional into fine-grained pelagic basinal deposits. Bioclastic packstones, wackestones and grainstones characterize the platform interior. The landward (SE) margin of the platform is dominated by bioclastic grainstones while the southeast (landward) slope is characterized by pelagic wackestones to mudstones (Fig. 10.34).

Diagenetic history

Figure 10.35 summarizes the diagenetic history of the reservoir rocks of the Malampaya-Camago platform. Diagenetic pathways, and resulting reservoir rock quality were strongly influenced by location on the platform relative to open marine conditions. On the oceanward margin, porosity destructive pervasive marine cementation was dominant. In the platform interior, dissolution and cementation associated with meteoric diagenesis strongly modified original porosity distribution. Stratiform zones tightly cemented

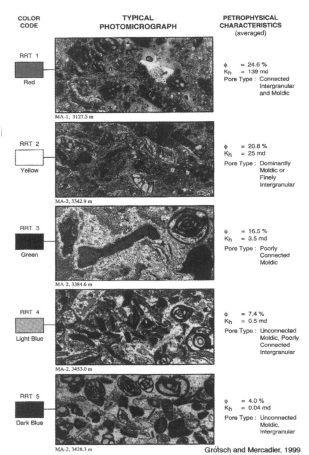

Fig. 10.36. Thin-section photomicrographs illustrating the reservoir rock types (RRTs) for the Nido Limestone in the Malampaya/Camago area. Five RRTs have been defined based on thin section analysis, porosity, permeability, and saturation-height functions. Note that RRTs do not necessarily follow the classical texture-based carbonate classification schemes, but rather combine the effects of depositional texture and diagenetic overprinting into classes of common petrophysical properties. This approach was necessary because the diagenetic overprinting is not controlled by fabric or lithofacies. Average properties for porosity and permeability and main pore types are listed for each RRT. Well name and sample depth are given below each photomicrograph. Used with permission of AAPG.

with meteoric calcite cements vertically compartmentalize the platform interior. The major zones are thick enough to be imaged

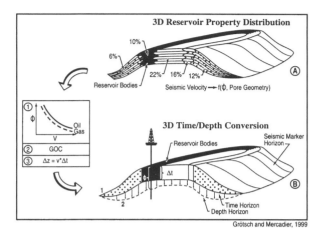

Fig. 10.37. Time-to-depth conversion in carbonates is predominantly a function of the 3-D porosity distribution within the carbonate platform. Reservoir modeling was first performed in the time domain (A). This approach allowed construction of a 3-D Nido limestone velocity model that honors the actual 3-D reservoir architecture. Based on this velocity model, the porosity dependent 3-D depth conversion of horizons was performed, matching in all scenarios the well data (B). For this approach, porosity-velocity functions for gas- and oil-bearing zones and the gas-oil contact (GOC) were provided to the model (steps 1 and 2 in the box). Using this information, depth (z) was calculated on a cell-by-cell basis (step 3 in the box). Used with permission of AAPG.

seismically (Fig. 10.34). These cemented zones can be related to long-lived unconformities associated with major relative sea level lowstands. The zones exhibit light carbon isotopic compositions, pisolites and alveolar structures suggesting extensive exposure and meteoric diagenesis (Chapter 7). Between these tightly cemented zones, grain dissolution and minor meteoric cementation driven by repeated rapid relative sea level fluctuations across the platform interior enhanced porosity. This early diagenetic history and porosity evolution directly parallels the history of the Devonian reservoirs of Alberta discussed in Chapter 5.

After burial, the tightly cemented zones are preferentially fractured during the two collision events and tectonic inversion of the late Miocene and Pliocene. These fractures, although enlarged by aggressive subsurface fluids, apparently do little to enhance reservoir quality because they are filled with subsurface calcite and dolomite cements (Fig. 10.34). Primary and moldic porosity in the platform interior is preserved during burial until filled by hydrocarbons because of the rigid framework formed by early meteoric cements (Chapters 7 and 9). These porous sequences show no fracturing.

Reservoir characteristics at Malampaya-Camago, therefore, are predominantly controlled by the platform's early diagenetic history as driven by sea level and climate and are not directly influenced by lithofacies (Chapter 8). This history integrated with 3-D seismic volume analysis (Fig. 10.34B and C) forms the basis for the 3-D diagenetic model (Fig. 10.34G)

Reservoir rock types as model input parameters

Grotsch and Mercadier (1999) used a petrophysically-based definition of reservoir rock types (RRT) (Chapter 3; Lucia, 1999) for their reservoir model input because of the non-linear relationship between porosity and permeability in the Malampaya-Camago reservoir. Each RRT was characterized by its pore geometry, an average porosity, an average permeability, an average gas saturation, and a saturation-height function, but not necessarily by its lithofacies. Emphasis was on pore geometry, pore throat connectivity and core-derived permeability cutoffs. Five characteristic RRTs were defined for the Nido

Limestone at Malampaya-Camago (Fig. 10.36). A neural network was trained to recognize the RRT classes from a set of wireline logs obtained from the 5 evaluation wells and used to generate a vertical distribution of these classes for each well (Fig. 10.34H). An average porosity was then reassigned to each RRT based on an average of its respective log porosity (Fig. 10.36).

The distribution of RRTs was entered in the model by filling each volume of a depositional area (Fig. 10.34D and F) with a vertical RRT succession that honors the cyclic architecture of the sequences by lateral extrapolation conformable to the contiguous layers. The well-derived RRT succession is the link between seismic-scale observations (layers and depositional areas) (Fig. 10.34D, E and F) and subseismic scale well data (Fig. 10.34H). This provides the means to define high-resolution flow units in the 3-D geological model in the appraisal stage, despite the complexity of the reservoir and the sparse data (Fig. 10.34I).

The model

Grotsch and Mercadier's (1999) prime objective for the static geologically-based deterministic reservoir model for Malampaya-Camago was to define the range of expected hydrocarbon reserves during the early assessment stage of the reservoir. Secondarily, they hoped to provide input and bounding conditions for further stochastic hydrocarbon volume modeling.

The cell size they used in the model was 80m long by 80m wide by 2m high. Traditionally, 3-D reservoir modeling is performed in the depth domain (Tinker, 1996). Seismic velocity in pure carbonates is dominantly a function of porosity. However, in a complex reservoir like Malampaya-Camago where porosity varies laterally, it is

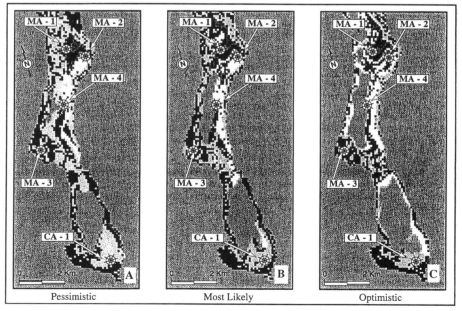

Fig. 10.38. (A-C) Reservoir rock type (RRT) distribution at the FWL for the "pessimistic," "most likely," and "optimistic" scenarios (depth models) showing their significantly different RRT distribution at the intersection with the FWL. See Figure 10.36 for RRT shade code. Used with permission of AAPG.

Grötsch and Mercadier, 1999

difficult to predict interval velocities accurately and hence make an appropriate time to depth conversion. Grotsch and Mercadier (1999) approached this problem by computing a differential 3-D time to depth conversion on a cell-by-cell basis using their time based 3-D property models (Fig. 10.34) and by integrating different porosity-velocity functions for the oil and gas-bearing parts of the reservoir (Fig. 10.37).

They reached their prime objective by modeling a series of scenarios based on a most likely base case. Figure 10.38 illustrates output from their pessimistic, most likely and optimistic scenarios. The scenarios were developed by varying the extent of the depositional areas, which directly modifies the 3-D distribution of reservoir quality.

Conclusions and lessons learned from the 3-D modeling of the Tertiary Malampaya and Camago buildups

1. Geologically sound depositional and stratigraphic frameworks are essential components in any 3-D modeling effort. Outcrop analogs and core data combined with seismic architecture were critical in building these frameworks.
2. The stratigraphic architecture of the buildups consist of a long term relative sea level rise (2nd order?) resulting in vertical reef aggradation on the windward side of the platform, retrogradation of the leeward lower energy facies, and ultimate drowning of the platform.
3. Application of the depositional and stratigraphic frameworks are essential to the determination of the controls on porosity modifying diagenetic events. The windward, platform margin reefs facing the open South China Sea had their porosity destroyed by marine cementation while the lower energy leeward deposits suffered little marine diagenesis (Chapter 5). Higher frequency sea level lowstands (3rd and 4th order) repeatedly exposed the entire upper surface of the platform to subaerial meteoric conditions resulting in unconformity-related dense non-porous platform-wide vertical baffles.
4. Because of the complexity of the diagenetic overprint suffered by the carbonates of the platform, reservoir rock type definition was based primarily on petrophysical characteristics. These parameters included; average porosity, permeability, gas saturation, and a gas height function (see discussion of Lucia's classification in Chapter 3). A neural network was trained to recognize these reservoir rock types from a suite of wireline logs taken from well on the platform.
5. The 3-D seismic data set was used in conjunction with the depositional and diagenetic models and the reservoir rock type distribution (based on the well log neural network) to build a time-based, static, reservoir model.
6. Time to depth conversion was complicated by intense porosity heterogeneity across the buildup. The time-based reservoir model was used as a platform for porosity corrected time/depth conversion on a cell by cell basis in order to arrive at a viable depth-based reservoir model suitable for hydrocarbon volume calculations.
7. The integration of modern geophysical and computer technology with solid geologic observations and concepts allows viable

economic hydrocarbon volume evaluations to be made with minimal well data during the early evaluation stage of reservoir development.

EPILOGUE

INTRODUCTION

 I finish this book by borrowing heavily from two old friends, John Morse and Fred Mackenzie. They concluded their 1990 book on the geochemistry of carbonates with an epilogue that presented their impressions, as sedimentary geochemists, of the history of carbonate geology, the state of the art and the future. I will attempt to do much the same in the year 2001, only from the vantage point of a sedimentary petrologist, sedimentologist/stratigrapher whose focus is the evolution and predictability of porosity in carbonate reservoirs.

Our common point of overlap is diagenesis.

THE LEGACY

The 1950's and 60's laid the foundation for the renaissance in the study and understanding of carbonate rocks and sediments. *Bob Ginsburg's* seminal papers on carbonate sedimentation in south Florida and the Bahamas during the '50s (Ginsburg, 1956; Ginsburg and Lowenstam, 1958) fired the imagination of sedimentologists around the world and led to the later development of sophisticated conceptual models of carbonate depositional environments. *Robin Bathurst's* landmark paper (Bathurst, 1958) on the diagenesis of Dinatian limestones set an entire generation of sedimentary petrologists on a course leading to a better understanding of carbonate rock fabrics and textures. When *Bob Folk* published his carbonate rock classification in 1959 (Folk, 1959), he showed the world's geological community that texture, fabric and composition could be used to decipher the depositional environments of ancient limestones. These three individuals provided the fuel for the exciting research explosion of the next three decades as academic and commercial research interests overlapped in the search for oil and gas.

It's at this point that the sedimentary geochemists began to contribute significantly to the problems being faced by the sedimentologists, petrographers, and stratigraphers as they wrestled with the effects of diagenesis on the nature of reservoir rock pore systems and their heterogeneity. *Hal Helgeson, Keith Chave, Bob Berner, and Heinz Lowenstam* provided us with a generation of sedimentary geochemists who aggressively attacked the fundamental diagenetic problems affecting carbonate porosity plagued the modern carbonate geologist. Rob Matthews, Lynton Land, Fred Mackenzie, John Morse, Bob Schmalz and John Milliman all played pivotal roles in bringing geochemical tools and concepts to the diagenetic-porosity table.

The next three decades (1960's, 70's and 80's) were *the golden age of carbonates*. Some fundamental understanding of the geochemical and geological constraints on complex diagenetic processes were gained by numerous intense studies of modern marine and fresh water systems. These efforts fueled the modern integrated carbonate reservoir studies combining environmental reconstruction, diagenetic history, and porosity evolution supported by basic rock geochemistry such as trace element, isotope, and fluid inclusion studies.

It was during the late 1980's and early 90's that the application of sequence stratigraphy to carbonate rocks began to make an impact. The seminal works of *Rick Sarg* (1988), *Wolfgang Schlager* (1989), *and Maurice Tucker* (1993) were critical to this effort. At this point the predictive power of sequence stratigraphy began to be focussed on early carbonate diagenesis and porosity evolution.

The 1990's saw an upsurge in the use of outcrop analogs and a full integration of sequence stratigraphy in detailed computer modeling of carbonate reservoirs. See the works of *Tinker* (1996) and *Kerans et al.* (1994). These reservoir modeling efforts necessitated the development of an engineering/geologically-based objective carbonate porosity classification scheme spearheaded by *Jerry Lucia* (1995).

The evangelical efforts of *Bob Folk* (1993) and *Hank Chavitz* (Chafetz and Buczynski, 1992) re-focussed the attention of the geological community on the importance of biological intervention in diagenetic processes and by extension carbonate porosity modification.

Finally, *Noel James* and his students (James et al., 1992) convinced us of the importance of temperate water carbonates and early porosity-modifying diagenesis in cold marine waters.

THE STATE OF THE ART

Where do we stand in 2001 relative to our understanding of the setting, geometry and predictability of carbonate reservoirs and their pore system evolution?

Sequence stratigraphy has given us a viable framework that allows a moderate level of predictability of regional facies patterns and reservoir scale heterogeneity and compartmentalization. The recognition of sediment response to changes in relative sea level has been exceptionally useful in predicting early diagenetic environments and porosity-modifying diagenetic processes. We have become increasingly aware that the role of short and long-term climate change is a fundamental linchpin in our efforts to reconstruct the diagenetic history and porosity evolution of ancient carbonate reservoirs. Our present level of understanding of paleoclimate evolution has resulted in the formulation of some preliminary but useful diagenetic models that are allowing us to do a better job of anticipating reservoir heterogeneity and compartmentalization. These models have benefited from some fundamental geohydrologic and hydrogeochemical studies of modern environments such as the Bahamas, Bermuda and the earlier Barbados work.

An exciting development of the 1990's is 3-D carbonate reservoir modeling utilizing 3-D seismic volume and horizon analysis integrated with subseismic well data. The use of neural networks may allow the model to be built with minimal well data during the early evaluation stage of reservoir development as exemplified by our last case history in Chapter 10.

Our understanding of late burial diagenesis and porosity evolution in the subsurface has increased significantly through the integration of tectonic history, hydrogeology, basic petrography, sequence stratigraphy, and the intelligent use of carbonate geochemistry (see the work of *Isabel Montenez* (1994) in Chapter 9).

The geochemical tools and concepts available to the modern carbonate researcher continue to be refined as new generations of instrumentation come on line and become more widely available to the geological community. All too often, however, basic constraints on the application of these geochemical techniques (particularly isotopic and two-phase fluid inclusion data) are still ignored, giving rise to misleading or even spurious conclusions. Ignoring the recrystallization potential of early-formed dolomite and using its present isotopic composition to discern mode of origin, is still the single most common mistake made by carbonate researchers dealing with ancient carbonate rock sequences (Chapter 9).

The active microbial intervention in marine diagenesis, and in particular, cementation under shallow marine conditions, has probably been overstated (the 90's microbe bandwagon). While microbes undoubtedly have a role to play, either passive as nucleation sites, or active as related to life processes, the evidence commonly used to determine that role is largely circumstantial and is often not compelling. The marine chemical system is perfectly capable of providing the volumes of cement observed, and the geochemistry and distributional patterns of the cement support a dominantly chemical precipitational origin. That said, there can be little doubt that microbes probably play a major role in the diagenetic processes (both cementation as well as dissolution) acting under deeper marine conditions within deep water mud mounds.

In 2001 we are still floundering over dolomite. Every month seems to bring yet another model. It has become crystal clear that there is no single all encompassing model that fits all circumstances, and that choice of model for ancient dolomite origin should be driven by geologic and hydrologic setting rather than current dolomite chemistry. Marine fluids, either normal or evaporated, seem to have become the dolomitizing fluid of choice. Evaporite reflux dolomitization has made a comeback (Chapter 8), while the Dorag mixing model continues to fade despite *John Humphrey's* (Canter et al., 1994) ongoing research in Barbados (Chapter 7). Platform-wide dolomitization by marine waters driven by thermal convection still seems to be a viable model. During relative sea level lowstands, however, marine waters entrained along the active base of deep floating meteoric lenses may provide the necessary fluid flux for large-scale marine water platform dolomitization as visualized by *Vahrenkamp and Swart* (1994) for the Little Bahama Platform (Chapter 5).

Finally, controversy over the role of dolomitization in porosity development is still festering 41 years after Weyl's 1960 paper entitled *Porosity through dolomitization-conservation of mass requirements*. Some workers have recently taken the position that all dolomitization is basically a cementation phenomenon with all necessary dolomite components being imported into the dolomitization site. This model makes mole-for-mole replacement unnecessary and results in porosity destruction rather than creation (Chapter 3). In this scenario porosity seen in dolomites is inherited rather than produced by the dolomitization event. This may certainly be true under conditions of marine water dolomitization or recycled dolomites, where all dolomite components are present in the dolomitizing fluids. Where dolomitization is associated with evaporite formation (sabkha and evaporite lagoon and basin), however, the water furnished to the dolomitization site will be Mg rich and Ca and CO_3 poor. These components will have to be furnished by dissolution of pre-existing calcites, Weyl's conservation of mass constraints will apply, and porosity will be enhanced. The dolomite beat goes on.

WHERE DO WE GO FROM HERE?

The future of carbonate reservoir studies is exemplified by the 3-D integrated model of an offshore Miocene platform detailed in the last case history of Chapter 10. The operative word for future carbonate reservoir studies will be *integration*. In the future, successful reservoir modeling will necessitate full integration of seismic geophysical data, subseismic geologic and geophysical well data, engineering rock data, rock petrography, geochemistry and biostratigraphy. The common thread that binds these data sets into a coherent model is a viable sequence stratigraphic framework based on seismic geometries, subseismic sedimentary facies analysis, biostratigraphy, and isotope geochronology. The successful carbonate reservoir geologist of the future will have to be a broadly cross-trained geoscientist with a solid understanding of carbonate sedimentology, carbonate diagenesis, petrophysics and sequence stratigraphy. A challenge, and certainly not one for the faint of heart!

Some of the more important carbonate research directions that will directly impact our ability to explore for and exploit carbonate reservoirs in the future are outlined below.

1. Refinement of seismic imaging of carbonate rock attributes at depth, particularly porosity.
2. Development of outcrop reservoir analogs in a variety of settings (depositional as well as climatic) to model the nature and predictability of depositional and diagenetic reservoir compartmentalization.
3. Refinement of basic carbonate sequence stratigraphic models through regional outcrop and modeling studies so that we can do a better job of facies prediction in the subsurface.
4. Parallel development of more refined sequence stratigraphic-based diagenetic models that will allow better prediction of favorable porosity distribution in the subsurface at the exploration as well as the reservoir scale.
5. Fundamental research on paleoclimatology stressing root-causes of cyclicity, and the impact that cyclic climate changes have on carbonate production, carbonate mineralogy and diagenesis.
6. Further research on controls over dolomite and calcite recrystallization and its recognition in ancient carbonate rock sequences.

This is critical if isotopes and two-phase fluid inclusions are to be truly useful diagenetic tools in the future.
7. A better understanding of dolomitizating processes and their impact on porosity evolution through carefully integrated geological, geochemical and geohydrological studies.
8. Assessment of the diagenetic potential of CO_2 in carbonate reservoirs during CO_2 floods and CO_2 assisted reservoir fracturing. These techniques may damage the reservoir as well as enhance recoveribility.

Finally, in the future we must not forget to look at the rocks in core, outcrop, and thin section. The basic tools of the geologist, the hammer, hand lens, microscope and an educated eye, when applied critically to carbonate reservoir rocks, can answer many of our most critical questions.

REFERENCES

Achauer, C. A., 1977, Contrasts in cementation, dissolution, and porosity development between two Lower Cretaceous reefs of Texas, in D. G. Bebout, and R. G. Loucks, eds., Cretaceous Carbonates of Texas and Mexico: Applications to Subsurface Exploration, Austin, TX, University of Texas, Bureau of Economic Geology Report of Investigations No. 89, p. 127-137.

Achauer, C. W., 1969, Origin of Capitan formation, Guadalupe Mountains, New Mexico and Texas: American Association of Petroleum Geologists Bulletin, v. 53, p. 2314-2323.

Adams, J. E., and M. L. Rhodes, 1961, Dolomitization by seepage refluxion: Bulletin American Association of Petroleum Geologists, v. 44, p. 1912-1920.

Aharon, P., R. A. Socki, and L. Chan, 1987, Dolomitization of atolls by sea water convective flow: test of a hypothesis at Niue, South Pacific: Journal of Geology, v. 95, p. 187-203.

Ahr, W. M., 1973, The carbonate ramp: an alternative to the shelf model: Transactions of the Gulf Coast Association of Geological Societies, v. 23, p. 221-225.

Aïssaoui, D. M., D. Buigues, and B. H. Purser, 1986, Model of reef diagenesis: Mururoa Atoll, French Polynesia, in J. H. Schroeder, and B. H. Purser, eds., Reef Diagenesis, New York, Springer-Verlag, p. 27-52.

Allan, J. R., and R. K. Matthews, 1982, Isotope signatures associated with early meteoric diagenesis: Sedimentology, v. 29, p. 797-817.

Allan, J. R., and W. D. Wiggins, 1993, Dolomite reservoirs — geochemical techniques for evaluating origin and distribution: Tulsa, OK, American Association of Petroleum Geologists Short Course Note Series No. 36, 129p, with appendices.

Alsharhan, A. S., and C. G. S. C. Kendall, 1986, Precambrian to Jurassic rocks of Arabian Gulf and adjacent areas: their facies, depositional setting, and hydrocarbon habitat: American Association of Petroleum Geologists Bulletin, v. 70, p. 977-1002.

Amthor, J. E., E. W. Mountjoy, and H. G. Machel, 1993a, Subsurface dolomites in Upper Devonian Leduc Formation buildups, central part of Rimbey-Meadowbrook reef trend, Alberta, Canada: Bulletin of Canadian Society of Petroleum Geology, v. 41, p. 164-185.

Amthor, J. E., E. W. Mountjoy, and H. G. Machel, 1994, Regional-Scale Porosity and Permiability variations in Upper Devonian Leduc Buildups: Implications for reservoir development and Prediction: American Association of Petroleum Geologists Bulletin, v. 78, p. 1541-1559.

Anderson, J. H., and H.-G. Machel, 1988, The Upper Devonian Nisku reef trend in central Alberta, in H. H. J. Geldsetzer, N. P. James, and G. E. Tebbutt, eds., Reefs, Canada and Adjacent Areas, Calgary, Alberta,

Canadian Society of Petroleum Geologists Memoir 13, p. 391-398.

Anderson, T. F., and M. A. Arthur, 1983, Stable isotopes of oxygen and carbon and their application to sedimentologic and paleoenvironmental problems, *in* M. A. Arthur, ed., Stable Isotopes in Sedimentary Geology, Tulsa, OK, SEPM Short Course No. 10, p. 1-1 to 1-151.

Archie, G. E., 1952, Classification of carbonate reservoir rocks and petrophysical considerations: American Association of Petroleum Geologists Bulletin, v. 36, p. 278-298.

Bachu, S., 1997, Flow of formation waters, aquifer characteristics, and their relation to hydrocarbon accumulations, northern Alberta Basin: AAPG, v. 81, p. 712-733.

Back, W., B. B. Hanshaw, T. E. Pyle, L. N. Plummer, and A. E. Weide, 1979, Geochemical significance of groundwater discharge and carbonate solution to the formation of Caleta Xel Ha, Quintana Roo, Mexico: Water Resources Research, v. 15, p. 1521-1535.

Back, W., B. B. Hanshaw, and J. N. van Driel, 1984, Role of groundwater in shaping the eastern coastline of the Yucatan Peninsula, Mexico, *in* R. G. LaFleur, ed., Groundwater as a Geomorphic Agent, Boston, Allen & Unwin, p. 157-172.

Badiozamani, K., 1973, The Dorag dolomitization model — application to the Middle Ordovician of Wisconsin: Journal of Sedimentary Petrology, v. 43, p. 965-984.

Balog, A., J. F. Read, and J. Haas, 1999, Climate-controlled early dolomite, Late Triassic cyclic platform carbonates, Hungary: Journal Sedimentary Research-Section A, v. 69, p. 267-282.

Banner, J. L., 1995, Application of the trace element and isotope geochemistry of strontium to studies of carbonate diagenesis: Sedimentology, v. 42, p. 805-824.

Banner, J. L., and G. N. Hanson, 1990, Calculation of simultaneous isotopic and trace element variations during water-rock interaction with applications to carbonate diagenesis: Geochimica et Cosmochimica Acta, v. 54, p. 3123-3137.

Banner, J. L., G. N. Hanson, and W. J. Meyers, 1988, Water-rock interaction history of regionally extensive dolomites of the Burlington-Keokuk Formation (Mississippian): isotopic evidence, *in* V. Shukla, and P. A. Baker, eds., Sedimentology and Geochemistry of Dolostones, Tulsa, OK, SEPM Special Publication No. 43, p. 97-113.

Banner, J. L., M. Musgrove, and R. C. Capo, 1994, Tracing ground-water evolution in a limestone aquifer using Sr isotopes: Effects of multiple sources of dissolved ions and mineral-solution reactions: Geology, v. 22, p. 687-690.

Banner, J. L., G. J. Wasserburg, P. F. Dobson, A. B. Carpenter, and C. H. Moore, 1989, Isotopic and trace element constraints on the origin and evolution of saline groundwaters from central Missouri: Geochimica et Cosmochimica Acta, v. 53, p. 383-398.

Barnes, V. E., P. E. Cloud, Jr., L. P. Dixon, R. L. Folk, E. C. Jonas, A. R. Palmer, and E. J. Tynan, 1959, Stratigraphy of the pre-Simpson Paleozoic subsurface rocks of Texas and southeast New Mexico, v. 1, 2: Austin, Texas, Bureau of Economic Geology, 836 p.

Bates, N. R., and U. Brand, 1991, Environmental and physiological influences on isotopic and elemental compositions of brachiopod shell calcite: implications for the isotopic evolution of Paleozoic oceans: Chemical Geology (Isotope Geosciences Section), v. 94, p. 67-78.

Bates, R. L., 1942, Lateral gradation in the Seven Rivers Formation, Rocky Arroyo, Eddy County, New Mexico: American Association of Petroleum Geologists Bulletin, v. 26, p. 80-99.

Bathurst, R. G. C., 1958, Diagenetic fabrics in some British Dinantian limestones: Liverpool and Manchester Geological Journal, v. 2, p. 11-36.

Bathurst, R. G. C., 1975, Carbonate Sediments and their Diagenesis: New York, Elsevier Science Publ. Co., 658 p.

Bathurst, R. G. C., 1980, Deep crustal diagenesis in limestones: Revista del Instituto de Investigaciones Geologicas de la Diputación de Barcelona (Universidad de Barcelona), v. 34, p. 89-100.

Bathurst, R. G. C., 1982, Genesis of stromatactis cavities between submarine crusts in Paleozoic carbonate mud buildups: Journal of the Geological Society of London, v. 139, p. 165-181.

Bathurst, R. G. C., 1983, The integration of pressure-solution with mechanical compaction and cementation, Stylolites and Associated Phenomena: Relevence to Hydrocarbon Reservoirs, Abu Dhabi, U.A.E., Abu Dhabi Reservoir Research Foundation Special Publication, p. 41-56.

Bathurst, R. G. C., 1987, Diagenetically enhanced bedding in argillaceous platform limestones: stratified cementation and selective compaction: Sedimentology, v. 34, p. 749-778.

Bathurst, R. G. C., 1991, Pressure-dissolution and limestone bedding: the influence of stratified cementation, in G. Einsele, W. Ricken, and A. Seilacher, eds., Cycles and Events in Stratigraphy, Berlin, Springer-Verlag, p. 450-463.

Bathurst, R. G. C., and L. S. Land, 1986, Carbonate depositional environments, modern and ancient. Part 5 - Diagenesis 1: Colorado School of Mines Quarterly, v. 81, p. 41.

Beach, D. K., 1982, Depositional and diagenetic history of Pliocene-Pleistocene carbonates of northwestern Great Bahama Bank; evolution of a carbonate platform: Ph.D. thesis, University of Miami, Miami, FL, 425 p.

Beach, D. K., 1995, Controls and Effects of Subaerial Exposure on Cementation and Development of secondary Porosity in the subsurface of Great Bahama Bank, in D. A. Budd, A. H. Saller, and P. M. Harris, eds., Unconformities and Porosity in Carbonate Strata, AAPG Memoir 63, p. 1-33.

Beales, F. W., 1956, Conditions of deposition of Palliser (Devonian) limestone of southwestern Alberta: American Association of Petroleum Geologists Bulletin, v. 40, p. 848-870.

Bebout, D. G., and R. G. Loucks, 1974, Stuart City Trend, Lower Cretaceous, south Texas - a carbonate shelf-margin model for hydrocarbon exploration, University of Texas, Bureau of Economic Geology Report of Investigations No., p. 80.

Bebout, D. G., and R. G. Loucks, 1983, Lower Cretaceous reefs, south Texas, in P. A. Scholle, D. G. Bebout, and C. H. Moore, eds., Carbonate Depositional Environments, Tulsa, OK, American Association of Petroleum Geologists Memoir 33, p. 441-444.

Behrens, E. W., and L. S. Land, 1972, Subtidal Holocene dolomite, Baffin Bay, Texas: Journal of Sedimentary Petrology, v. 42, p. 155-161.

Bein, A., and A. Dutton, 1993, Origin, distribution, and movement of brine in the Permian Basin (USA): A model for displacement of connate brine: Geological Society of America Bulletin, v. 105, p. 695-707.

Belka, Z., 1998, Early Devonian Kess-Kess carbonate mud mounds of the eastern Anti-Atlas (Morocco), and their relation to submarine hydrothermal venting: Journal of Sedimentary Research, v. 68, p. 368-367.

Benson, L. V., C. W. Achauer, and R. K. Matthews, 1972, Electron microprobe analyses of magnesium and iron distribution in carbonate cements and recrystallized sediment grains from ancient carbonate rocks: Journal of Sedimentary Petrology, v. 42, p. 803-811.

Benson, L. V., and R. K. Matthews, 1971, Electron microprobe studies of magnesium distribution in carbonate cements and recrystallized skeletal grainstones from the Pleistocene of Barbados, West Indies: Journal of Sedimentary Petrology, v. 41, p. 1018-1025.

Berner, R. A., 1971, Bacterial processes effecting the precipitation of calcium carbonate in sediments, *in* O. P. Bricker, ed., Carbonate Cements: Studies in Geology No. 19, Baltimore, MD, The Johns Hopkins Press, p. 247-251.

Bhattacharyya, A., and G. M. Friedman, 1979, Experimental compaction of ooids and lime mud and its implication for lithification during burial: Journal of Sedimentary Petrology, v. 49, p. 1279-1286.

Bhattacharyya, A., and G. M. Friedman, 1984, Experimental compaction of ooids under deep-burial diagenetic temperatures and pressures: Journal of Sedimentary Petrology, v. 54, p. 362-372.

Blatt, H., 1982, Sedimentary Petrology: San Francisco, CA, Freeman and Sons, 564 p.

Blatt, H., G. V. Middleton, and R. C. Murray, 1980, Origin of Sedimentary Rocks: 2nd edition: Englewood Cliffs, NJ, Prentice Hall Inc., 782 p.

Blount, D., and C. H. Moore, 1969, Depositional and non-depositional carbonate breccias, Chiantla Quadrangle, Guatemala: Geological Society of America Bulletin, v. 80, p. 429-442.

Boles, J. R., and G. Franks, 1979, Clay diagenesis in Wilcox sandstone of seoutheast Texas: implications of smectite diageneis on sandstone cementation: JSP, v. 49, p. 55-70.

Borre, M., and F. I., 1998, Chemical and mechanical processes during burial diagenesis of chalk: an interpretation based on specific surface data of deep-sea cores: Sedimentology, v. 45, p. 755-769.

Bosence, D. W. J., and P. H. Bridges, 1995, A review of the origin and evolution of carbonate mud-mounds, *in* C. L. V. Monty, D. W. J. Bosence, P. H. Bridges, and B. R. Pratt, eds., Carbonate Mud-Mounds: Their Origin and Evolution, Oxford, England, International Association of Sedimentologists, Special Publication, p. 3-9.

Boyd, D. R., 1963, Geology of the Golden Lane trend and related fields of the Tampico Embayment, Geology of Peregrina Canyon and Sierra de El Abra, Mexico [Annual Field Trip Guidebook], Corpus Chris-

REFERENCES

ti, TX, Corpus Christi Geological Society, p. 49-56.

Brand, U., and J. Veizer, 1980, Chemical diagenesis of a multicomponent carbonate system-1: Trace elements: Journal of Sedimentary Petrology, v. 50, p. 1219-1236.

Brock, F. C., and C. H. Moore, 1981, Walker Creek revisited: a reinterpretation of the diagenesis of the Smackover Formation of the Walker Creek field, Arkansas: Transactions of the Gulf Coast Association of Geological Societies, v. 31, p. 49-58.

Buczynski, C., and H. S. Chafetz, 1993, Habit of bacterially induced precipitates of calcium carbonate: examples from laboratory experiments and recent sediments, in R. Rezak, and D. L. Lavoie, eds., Carbonate Microfabrics, New York, Springer-Verlag, p. 105-116.

Budd, D. A., 1984, Freshwater diagenesis of Holocene ooid sands, Schooner Cays, Bahamas: Ph.D. thesis, University of Texas at Austin, Austin, TX, 491 p.

Budd, D. A., 1988, Aragonite-to-calcite transformation during fresh-water diagenesis of carbonates: insights from pore-water chemistry: Geological Society of America Bulletin, v. 100, p. 1260-1270.

Budd, D. A., 1992, Dissolution of high-Mg calcite fossils and the formation of biomolds during mineralogical stabilization: Carbonates and Evaporites, v. 7, p. 74-81.

Budd, D. A., U. Hammes, and H. L. Vacher, 1993, Calcite cementation in upper Floridan aquifer: a modern example for confined-aquifer cementation models?: Geology, v. 21, p. 33-36.

Budd, D. A., and E. E. Hiatt, 1993, Mineralogical stabilization of high-magnesium calcite: geochemical evidence for intracrystal recrystallization within Holocene porcellaneous foraminifera: Journal of Sedimentary Petrology, v. 63, p. 261-274.

Budd, D. A., and L. S. Land, 1990, Geochemical imprint of meteoric diagenesis in Holocene ooid sands, Schooner Cays, Bahamas: correlation of calcite cement geochemistry with extant groundwaters: Journal of Sedimentary Petrology, v. 60, p. 361-378.

Budd, D. A., and R. G. Loucks, 1981, Smackover and lower Buckner Formations, south Texas: depositional systems on a Jurassic carbonate ramp, University of Texas, Bureau of Economic Geology Report of Investigations No., p. 38.

Budd, D. A., and H. L. Vacher, 1991, Predicting the thickness of fresh-water lenses in carbonate paleo-islands: Journal of Sedimentary Petrology, v. 61, p. 43-53.

Buffler, R. T., and C. D. Winker, 1988, Paleogeographic evolution of early deep-water Gulf of Mexico and margins, Jurassic to Middle Cretaceous (Comanchean): American Association of Petroleum Geologists Bulletin, v. 72, p. 318-346.

Burke, W. H., R. E. Denison, E. A. Hetherington, R. B. Koepnick, H. F. Nelson, and J. B. Otto, 1982, Variation of seawater $^{87}Sr/^{86}Sr$ throughout Phanerozoic time: Geology, v. 10, p. 516-519.

Burri, P., R. DuDresnay, and C. W. Wagner, 1973, Teepee structures and associated diagenetic features in intertidal carbonate sands (Lower Jurassic, Morocco): Sedimentary Geology, v. 9, p. 221-228.

Bush, P. R., 1973, Some aspects of the diagenetic history of the sabkha in Abu Dhabi, Persian Gulf, in B. H. Purser, ed., The Persian Gulf, New York, Springer-Verlag, p.

395-407.

Butler, G. P., 1969, Modern evaporite deposition and geochemistry of coexisting brines, the sabkha, Trucial Coast, Arabian Gulf: Journal of Sedimentary Petrology, v. 39, p. 70-89.

Buxton, T. M., and D. F. Sibley, 1981, Pressure solution features in a shallow buried limestone: Journal of Sedimentary Petrology, v. 51, p. 19-26.

Caillet, G., N. P. Bramwell, L. Meciani, N. C. Judge, M. Green, and P. Adam, 1996, Porosity distribution and field reservoir extension in the chalk of the Greater Ekofisk Area (Norway): 1996 AAPG Hedberg Conference, Pau, France.

Cander, H. S., 1995, Interplay of water-rock interaction efficiency, unconformities and fluid flow in a carbonate aquifer, in D. A. Budd, A. H. Saller, and P. M. Harris, eds., Unconformities in Carbonate Strata: Their Recognition and Associated Porosity, AAPG, p. 103-124.

Canter, K. L., J. D. Humphrey, and T. N. Kimbell, 1994, Early Dolomitization of coralline red Algae and high-Mg Calcite-rich micrite, southeastern Barbados: Geological Society of America Annual Meeting, p. A-66.

Carballo, J. D., L. S. Land, and D. E. Miser, 1987, Holocene dolomitization of supratidal sediments by active tidal pumping, Sugarloaf Key, Florida: Journal of Sedimentary Petrology, v. 57, p. 153-165.

Carew, J. L., and J. E. Mylroie, 1997, Geology of the Bahamas, in H. L. Vacher, and T. M. Quinn, eds., Geology and Hydrogeology of Carbonate Islands: Developments in Sedimentology, Amsterdam, Elsevier Science B. V., p. 91-139.

Carozzi, A. V., and D. A. Textoris, 1967, Paleozoic Carbonate Microfacies of the Eastern Stable Interior (U.S.A.): Leiden, Neth., E.J. Brill, 41 p.

Carpenter, A. B., 1976, Dorag dolomitization model by K. Badiozamani — a discussion: Journal of Sedimentary Petrology, v. 46, p. 258-261.

Carpenter, A. B., 1978, Origin and evolution of brines in sedimentary basins, Society of Petroleum Engineers AIME, Paper No., p. 8.

Carpenter, S. J., and K. C. Lohmann, 1989, $d^{18}O$ and $d^{13}C$ variations in Late Devonian marine cements from the Golden Spike and Nevis reefs, Alberta, Canada: Journal of Sedimentary Petrology, v. 59, p. 792-814.

Chafetz, H. S., 1986, Marine peloids: a product of bacterially induced precipitation of calcite: Journal of Sedimentary Petrology, v. 56, p. 812-817.

Chafetz, H. S., and C. Buczynski, 1992, Bacterially induced lithification of microbial mats: Palaios, v. 7, p. 277-293.

Choquette, P. W., and N. P. James, 1987, Diagenesis #12. Diagenesis in limestones - 3. The deep burial environment: Geoscience Canada, v. 14, p. 3-35.

Choquette, P. W., and N. P. James, 1988, Introduction, in N. P. James, and P. W. Choquette, eds., Paleokarst, New York, Springer-Verlag, p. 1-24.

Choquette, P. W., and N. P. James, 1990, Limestones-the burial diagenetic environment, in I. A. McIlreath, and D. W. Morrow, eds., Diagenesis, Ottowa, Ontario, Canada, Geological Association of Canada, p. 75-111.

Choquette, P. W., and L. C. Pray, 1970, Geologic nomenclature and classification of porosity in sedimentary carbonates: American Association of Petroleum Geologists Bulletin, v. 54, p. 207-250.

Choquette, P. W., and R. P. Steinen, 1980, Mississippian non-supratidal dolomite, Ste. Genevieve Limestone, Illinois Basin: evidence for mixed water dolomitization, in D. H. Zenger, J. B. Dunham, and R. L. Ethington, eds., Concepts and Models of Dolomitization, Tulsa, OK, SEPM Special Publication No. 28, p. 163-196.

Choquette, P. W., and R. P. Steinen, 1985, Mississippian oolite and non-supratidal dolomite reservoirs in the Ste. Genevieve Formation, North Bridgeport Field, Illinois Basin, in P. O. Roehl, and P. W. Choquette, eds., Carbonate Petroleum Reservoirs, New York, Springer–Verlag, p. 207-225.

Collins, A. G., 1975, The Geochemistry of Oil Field Waters: New York, Elsevier Science Publ. Co., 496 p.

Coogan, A. H., 1970, Measurements of compaction in oolitic grainstone: Journal of Sedimentary Petrology, v. 40, p. 921-929.

Coogan, A. H., D. G. Bebout, and C. Maggio, 1972, Depositional environments and geologic history of Golden Lane and Poza Rica trend, Mexico, an alternative view: American Association of Petroleum Geologists Bulletin, v. 56, p. 1419-1447.

Coogan, A. H., and R. W. Manus, 1975, Compaction and diagenesis of carbonate sands, in G. V. Chilingarian, and K. H. Wolf, eds., Compaction of Coarse-Grained Sediments: Developments in Sedimentology 18A, New York, Elsevier, p. 76-166.

Craig, D. H., 1988, Caves and other features of Permian karst in San Andres Dolomite, Yates field reservoir, west Texas, in N. P. James, and P. W. Choquette, eds., Paleokarst, New York, Springer-Verlag, p. 342-363.

Crevello, P. D., P. M. Harris, D. L. Stoudt, and L. R. Baria, 1985, Porosity evolution and burial diagenesis in a Jurassic reef–debris reservoir, Smackover Formation, Hico Knowles field, Louisiana, in P. O. Roehl, and P. W. Choquette, eds., Carbonate Petroleum Reservoirs, New York, Springer–Verlag, p. 385-406.

Crockett, J., 1994, Porosity evolution of the Madison Limestone, Mississippian: Wind River Basin, Wyoming: MS thesis, Louisiana State University, Baton Rouge, Louisiana, 103 p.

Curtis, C. D., G. Oertel, and C. Petrowski, 1972, Stable carbon isotope ratios within carbonate concretions: a clue to place and time of formation: Nature, v. 235, p. 98-100.

Davies, G. R., 1975, Upper Paleozoic carbonates and evaporites in the Sverdrup Basin, Canadian Arctic Archipelago, Geological Survey of Canada Paper, 75-1B, p. 209-214.

Dickson, J. A. D., 1990, Carbonate mineralogy and chemistry, in M. E. Tucker, and V. P. Wright, eds., Carbonate Sedimentology, Oxford, Blackwell Scientific Publications, p. 284-313.

Dickson, J. A. D., 1993, Crystal growth diagrams as an aid to interpreting the fabrics of calcite aggregates: Journal of Sedimentary Petrology, v. 63, p. 1-17.

Dickson, J. A. D., L. A. González, S. J. Carpenter, and K. C. Lohmann, 1993, Inorganic Calcite Morphology: roles of fluid chemistry and fluid flow: discussion and reply: Journal of Sedimentary Petrology, v. 63, p. 560-563.

Dill, R. F., G. S. C. C. Kendall, and R. P. Steinen, 1989, Stromatolites, ooid dunes, hardgrounds, and crusted mud beds, all products of marine cementation and microbial mats in the subtidal mixing zone on the eastern margin of the Great Bahama Bank [abs.]: American Association of Petroleum Geologists Bulletin, v. 73, p. 350.

Dorobek, S. L., S. K. Reid, M. Elrick, G. C. Bond, and M. A. Kominz, 1991, Subsidence across the Antler foreland of Montana and Idaho: tectonic versus eustatic effects, in E. K. Franseen, W. L. Watney, C. G. S. C. Kendall, and W. C. Ross, eds., Sedimentary Modeling: Computer Simulations and Methods for Improved Parameter Definition, Lawrence, KS, Kansas Geological Survey Bulletin 233, p. 231-252.

Dravis, J. J., 1979, Rapid and widespread generation of Recent oolitic hardgrounds on a high energy Bahamian platform, Eleuthera Bank, Bahamas: Journal of Sedimentary Petrology, v. 49, p. 195-209.

Drew, G. H., 1914, On the precipitation of calcium carbonate in the sea by marine bacteria and on the action of denitrifying bacteria in tropical and temperate seas: Carnegie Institute of Washington, Publication 182, Papers from the Tortugas Laboratory, v. 5, p. 7-45.

Drivet, E., and E. Mountjoy, 1997, Dolomitization of the Leduc Formation (Upper Devonian) southern Rimbey-Meadowbrook reef trend, Alberta: Journal SedimentaryResearch, v. 67, p. 411-423.

Droste, H., 1990, Depositional cycles and source rock development in an epeiric intra-platform basin: the Hanifa Formation of the Arabian peninsula: Sedimentary Geology, v. 69, p. 281-296.

Druckman, Y., and C. H. Moore, Jr., 1985, Late subsurface secondary porosity in a Jurassic grainstone reservoir, Smackover Formation, Mt. Vernon field, southern Arkansas, in P. O. Roehl, and P. W. Choquette, eds., Carbonate petroleum reservoirs, New York, Springer-Verlag, p. 369-384.

Dullo, W.-C., 1986, Variations in diagenetic sequences: and example from Pleistocene coral reefs, Red Sea, Saudi Arabia, in J. H. Schroeder, and B. H. Purser, eds., Reef Diagenesis, New York, Springer-Verlag, p. 77-90.

Dunham, R. J., 1962, Classification of carbonate rocks according to their depositional texture, in W. E. Ham, ed., Classification of Carbonate Rocks, Tulsa, OK, American Association of Petroleum Geologists Memoir 1, p. 108-121.

Dunham, R. J., 1971, Meniscus cement, in O. P. Bricker, ed., Carbonate Cements, Baltimore, MD, The Johns Hopkins Press, p. 297-300.

Dunham, R. J., 1972, Capitan Reef, New Mexico and Texas: Facts and questions to aid interpretation and group discussion: Midland, TX, Permian Basin Section-SEPM Publication 72-141, 272 p.

Dunnington, H. V., 1967, Aspects of diagenesis and shape change in stylolitic limestone reservoirs, Proceedings of the Seventh World Petroleum Congress (Mexico City, Mexico), New York, Elsevier, p. 339-352.

Durlet, C., and J. P. Loreau, 1996, Sequence diagenetique intrinseque des surfaces durcies: mise en evidence de surface de'emersion et de leur ablation marine. Exemple de la plateforme bourguignonne, Bajocien (France):: Comptes Rendus de

L'Academie des Sciences, v. 323, p. 389-396.

Duval, B., C. Cramez, P. R. Vail, 1992, Types and hiearchy of stratigraphic cycles, Abstract: Mesozoic and Cenozoic Sequence Stratigraphy of European Basins International Symposium, p. 44-45.

Eberli, G. P., and R. N. Ginsburg, 1987, Segmentation and coalescence of Cenozoic carbonate platforms, northwestern Great Bahama Bank: Geology, v. 15, p. 75-79.

Eichenseer, H. T., F. R. Walgenwitz, and P. J. Biondi, 1999, Stratigraphic control on facies and diagenesis of dolomitized sequences (Pinda Group, Albian, offshore Angola): AAPG, v. 83, p. 1729-1758.

Elrick, M., and J. F. Read, 1991, Cyclic ramp-to-basin carbonate deposits, Lower Mississippian, Wyoming and Montana: a combined field and computer modeling study: Journal of Sedimentary Petrology, v. 61, p. 1194-1224.

Emery, D., and K. Myers, 1996, Sequence Stratigraphy, Oxford, Blackwell Science, p. 297.

Enos, P., 1977, Tamabra Limestone of the Poza Rica trend, Cretaceous, Mexico, in H. E. Cook, and Enos, P., ed., Deep-Water Carbonate Environments, Tulsa, OK, SEPM Special Publication No. 25, p. 273-314.

Enos, P., 1986, Diagenesis of Mid-Cretaceous rudist reefs, Valles platform, Mexico, in J. H. Schroeder, and B. H. Purser, eds., Reef Diagenesis, New York, Springer-Verlag, p. 160-185.

Enos, P., 1988, Evolution of pore space in the Poza Rica trend (mid-Cretaceous), Mexico: Sedimentology, v. 35, p. 287-326.

Enos, P., C. J. Minero, and J. E. Aguayo, 1983, Sedimentation and Diagenesis of mid-Cretaceous Platform Margin East-Central Mexico with Accompanying Field Guide: Dallas, TX, Dallas Geological Society, 168 p.

Enos, P., and L. H. Sawatsky, 1981, Pore networks in Holocene carbonate sediments: Journal of Sedimentary Petrology, v. 51, p. 961-986.

Erwin, C. R., D. E. Eby, and V. S. Whitesides, Jr., 1979, Clasticity index: a key to correlating depositional and diagenetic environments of Smackover reservoirs, Oaks field, Claiborne Parish, Louisiana: Transactions of the Gulf Coast Association of Geological Societies, v. 29, p. 52-59.

Esteban, M., 1991, Paleokarst: case histories, in V. P. Wright, M. Esteban, and P. L. Smart, eds., Paleokarsts and Paleokarstic Reservoirs: P.R.I.S. Occasional Publication Series No. 2, Reading, University of Reading, p. 89-119.

Esteban, M., and C. F. Klappa, 1983, Subaerial exposure environment, in P. A. Scholle, D. G. Bebout, and C. H. Moore, eds., Carbonate Depositional Environments, Tulsa, OK, American Association of Petroleum Geologists Memoir 33, p. 1-54.

Fairbridge, R. W., 1957, The dolomite question, in R. J. LeBlanc, and J. G. Breeding, eds., Regional Aspects of Carbonate Deposition, Tulsa, OK, SEPM Special Publication No. 5, p. 124-178.

Fairchild, I. J., A. H. Knoll, and K. Swett, 1991, Coastal lithologies and biofacies associated with syndepositional dolomitization (Draken Formation, Upper Riphean, Svalbard): Precambrian Research, v. 53, p. 165-198.

Feazel, C. T., J. Keany, and R. M. Peterson, 1985, Cretaceous and Tertiary chalk of the

Ekofisk Field area, central North Sea, *in* P. O. Roehl, and P. W. Choquette, eds., Carbonate Petroleum Reservoirs, New York, Springer–Verlag, p. 495-507.

Feazel, C. T., and R. A. Schatzinger, 1985, Prevention of carbonate cementation in petroleum reservoirs, *in* N. Schneidermann, and P. M. Harris, eds., Carbonate Cements, Tulsa, OK, SEPM Special Publication No. 36, p. 97-106.

Folk, R. L., 1954, The distinction between grainsize and mineral composition in sedimentary rock nomenclature: Journal of Geology, v. 62, p. 344-359.

Folk, R. L., 1959, Practical petrographic classification of limestones: American Association of Petroleum Geologists Bulletin, v. 43, p. 1-38.

Folk, R. L., 1968, Petrology of sedimentary rocks: Austin, TX, Hemphill's Book Store, 170 p.

Folk, R. L., 1974, The natural history of crystalline calcium carbonate: effect of magnesium content and salinity: Journal of Sedimentary Petrology, v. 44, p. 40-53.

Folk, R. L., 1993, SEM imaging of bacteria and nannobacteria in carbonate sediments and rocks: Journal of Sedimentary Petrology, v. 63, p. 990-999.

Folk, R. L., and L. S. Land, 1975, Mg/Ca ratio and salinity: two controls over crystallization of dolomite: American Association of Petroleum Geologists Bulletin, v. 59, p. 61-68.

Folk, R. L., and R. Robles, 1964, Carbonate sands of Isla Perez, Alacran reef complex, Yucatan: Journal of Geology, v. 72, p. 255-292.

Fouke, B. W., C. J. Beets, W. J. Meyers, G. N. Hanson, and A. J. Melillo, 1996, $^{87}Sr/^{86}Sr$ chronostratigraphy and dolomitization history of the Seroe Domi Formation, Curacao (Netherlands Antilles): Facies, v. 35, p. 293-320.

Friedman, I., and J. R. O'Niel, 1977, Compilation of stable isotope fractionation factors of geochemical interest [Data of Geochemistry, Sixth Edition], U. S. Geological Survey Professional Paper, p. 12.

Fruth, L. S., Jr., G. R. Orme, and F. A. Donath, 1966, Experimental compaction effects in carbonate sediments: Journal of Sedimentary Petrology, v. 36, p. 747-754.

Gaillard, C., M. Rio, Y. Rolin, and M. Roux, 1992, Fossil chemosynthetic communities related to vents or seeps in sedimentary basins: the psudobioherms of southeastern France compared to other world examples: Palaios, v. 7, p. 451-465.

Gaines, A. M., 1977, Protodolomite redefined: Journal of Sedimentary Petrology, v. 47, p. 543-546.

Gao, G., L. S. Land, and R. L. Folk, 1992, Meteoric modification of early dolomite and late dolomitization by basinal fluids, upper Arbuckle Group, Slick Hills, southwestern Oklahoma: American Association of Petroleum Geologists Bulletin, v. 76, p. 1649-1664.

Garber, R. A., G. A. Grover, and P. M. Harris, 1989, Geology of the Capitan shelf margin — subsurface data from the northern Delaware Basin, *in* P. M. Harris, and G. A. Grover, eds., Subsurface and Outcrop Examination of the Capitan Shelf Margin, Northern Delaware Basin, Tulsa, OK, SEPM Core Workshop No. 13, p. 3-269.

Garrison, R. E., 1981, Diagenesis of oceanic carbonate sediments: a review of the DSDP perspective, *in* J. E. Warme, R. G.

Douglas, and E. L. Winterer, eds., The Deep Sea Drilling Project: A Decade of Progress, Tulsa, OK, SEPM Special Publication No. 32, p. 181-207.

Garven, G., M. A. Person, and D. A. Sverjensky, 1993, Genesis of stratiform ore deposits in the midcontinent basins of North America, 1. quantitative results: American Journal of Science, v. 293, p. 497-568.

Gebelein, C. D., 1977, Dynamics of Recent Carbonate Sedimentation and Ecology, Cape Sable, Florida: Leiden, Neth., E. J. Brill, 120 p.

Gebelein, C. D., R. P. Steinen, P. Garrett, E. J. Hoffman, J. M. Queen, and L. N. Plummer, 1980, Subsurface dolomitization beneath the tidal flats of central west Andros Island, Bahamas, in D. H. Zenger, J. B. Dunham, and R. L. Ethington, eds., Concepts and Models of Dolomitization, Tulsa, OK, SEPM Special Publication No. 28, p. 31-49.

Gill, I. P., C. H. Moore, and P. Aharon, 1995, Evaporitic mixed-water dolomitization on St. Croix, U.S.V.I.: Journal of Sedimentary Research, v. A65, p. 591-604.

Ginsburg, R. N., 1956, Environmental relationships of grain size and constituent particles in some south Florida carbonate sediments: American Association of Petroleum Geologists Bulletin, v. 40, p. 2384-2427.

Ginsburg, R. N., and D. R. Choi, 1984, Late Cenozoic geology and reefs of the Belize shelf, Central America, University of Miami, Comparative Sedimentology Laboratory, Field Trip Guidebook (March 20-24, 1984), p. 17.

Ginsburg, R. N., and H. A. Lowenstam, 1958, The influence of marine bottom communities on the depositional environments of sediments: Journal of Geology, v. 66, p. 310-318.

Ginsburg, R. N., and J. H. Schroeder, 1973, Growth and submarine fossilization of algal cup reefs, Bermuda: Sedimentology, v. 20, p. 575-614.

Ginsburg, R. N., J. H. Schroeder, and E. A. Shinn, 1971, Recent synsedimentary cementation in subtidal Bermuda reefs, in O. P. Bricker, ed., Carbonate Cements, Baltimore, MD, John Hopkins University Press, p. 54-59.

Given, R. K., and B. H. Wilkinson, 1985, Kinetic control of morphology, composition, and mineralogy of abiotic sedimentary carbonates: Journal of Sedimentary Petrology, v. 55, p. 109-119.

Goldhammer, R. K., 1997, Compaction and Decompaction Algorithms for Sedimentary Carbonates: Journal Sedimentary Research, Section A, v. 67, p. 26-35.

Goldstein, R. H., and T. J. Reynolds, 1994, Systematics of Fluid Inclusions in Diagenetic Minerals: SEPM Short Course Note Series, v. 31: Tulsa, OK, SEPM, 199 p.

Gonzalez, L., H. M. Ruiz, B. E. Taggart, and A. F. Budd, 1997, Geology of Isla de Mona, Puerto Rico, in H. L. Vacher, and T. M. Quinn, eds., Geology and Hydrology of Carbonate Islands: Developments in Sedimentology, Amsterdam, Elsevier Science B.V, p. 327-358.

González, L. A., S. J. Carpenter, and K. C. Lohmann, 1992, Inorganic calcite morphology: roles of fluid chemistry and fluid flow: Journal of Sedimentary Petrology, v. 62, p. 382-399.

González, L. A., and K. C. Lohmann, 1988, Controls on mineralogy and composition of spelean carbonates: Carlsbad Caverns, New Mexico, *in* N. P. James, and P. W. Choquette, eds., Paleokarst, New York, Springer-Verlag, p. 81-101.

Grammer, G. M., C. M. Crescine, D. F. McNeill, and L. H. Taylor, 1999, Quantifying rates of syndepositional marine cementation in deeper plafform environments-new insight into a fundamental process: Journal of Sedimentary Research, v. 69, p. 202-207.

Grammer, G. M., G. R. Eberli, F. S. P. Van Buchem, G. M. Stevenson, and P. Homewood, 1996, Ajpplication of high-resolution sequence stratigraphy to evaluate lateral variability in outcrop and subsurface-Desert Creek and Ismay intervals, Paradox Basin, *in* M. W. Longman, and M. D. Sonnenfeld, eds., Paleozoic Systems of the Rocky Mountain Region, Denver CO, Rocky Mountain Section of the SEPM, p. 235-266.

Graton, L. C. a. F., H. J., 1935, Systematic packing of spheres-with particular relation to porosity and permeability: Journal of Geology, v. 43, p. 785-909.

Grotsch, J., and C. Mercadier, 1999, Integrated 3-D reservoir modeling based on 3-D seismic: The Tertiary Malampaya and Camago buildups, offshore Palawan, Philippines: AAPG Bulletin, v. 83, p. 1703-1728.

Halley, R. B., 1987, Burial diagenesis of carbonate rocks: Colorado School of Mines Quarterly, v. 82, p. 1-15.

Halley, R. B., and P. M. Harris, 1979, Fresh water cementation of a 1,000 year-old oolite: Journal of Sedimentary Petrology, v. 49, p. 969-988.

Halley, R. B., P. M. Harris, and A. C. Hine, 1983, Bank margin environments, *in* P. A. Scholle, D. G. Bebout, and C. H. Moore, eds., Carbonate Depositional Environments, Tulsa, OK, American Association of Petroleum Geologists Memoir 33, p. 464-506.

Hanor, J., 1987, Origin and Migration of Subsurface Sedimentary Brines: Tulsa, OK, SEPM Short Course No. 21, 247 p.

Hanshaw, B. B., W. Back, and R. G. Deike, 1971, A geochemical hypothesis for dolomitization by ground water: Economic Geology, v. 66, p. 710-724.

Hardie, L. A., 1987, Dolomitization: a critical view of some current views: Journal of Sedimentary Petrology, v. 57, p. 166-183.

Harper, M. L., 1971, Approximate geothermal gradients in the North Sea Basin: Nature, v. 230, p. 235-236.

Harper, M. L., and B. B. Shaw, 1974, Cretaceous-Tertiary carbonate reservoirs in the North Sea, Offshore North Sea Technology Conference (Stavanger, Norway), Paper, p. 20.

Harris, P. M., and S. H. Frost, 1984, Middle Cretaceous carbonate reservoirs, Fahud field and northwestern Oman: American Association of Petroleum Geologists Bulletin, v. 68, p. 649-658.

Harris, P. M., A. H. Saller, and J. A. Simo, 1999, Introduction, *in* P. M. Harris, A. H. Saller, and J. A. Simo, eds., Advances in Carbonate Sequence Stratigraphy: Applications to Reservoirs, Outcrops and Models, Tulsa, OK, SEPM, p. 1-10.

Harris, W. H., and R. K. Matthews, 1968, Subaerial diagenesis of carbonate sediments: efficiency of the solution-reprecipitation process: Science, v. 160, p. 77-79.

Harrison, R. S., 1975, Porosity in Pleistocene grainstones from Barbados: some preliminary observations: Bulletin of Canadian Petroleum Geology, v. 23, p. 383-392.

Harrison, W. J., and L. L. Summa, 1991, Paleohydrology of the Gulf of Mexico Basin: American Journal of Science, v. 291, p. 109-176.

Hemming, N. G., W. J. Meyers, and J. C. Grams, 1989, Cathodoluminescence in diagenetic calcites: the roles of Fe and Mn as deduced from electron probe and spectrophotometric measurements: Journal of Sedimentary Petrology, v. 59, p. 404-411.

Herrington, P. M., K. Pederstad, and J. A. D. Dickson, 1991, Sedimentology and diagenesis of resedimented and rhythmically bedded chalks from the Eldfisk Field, North Sea Central Graben: American Association of Petroleum Geologists Bulletin, v. 75, p. 1661-1674.

Heydari, E., 1997a, Hydrotectonic models of burial diagenesis in platform carbonates based on formation water geochemistry in north american sedimentary basins, in I. P. Montanez, J. M. Gregg, and K. L. Shelton, eds., Basin-wide Digenetic Patterns: Integrated Petrologic, Geochemical, and Hydrologic Considerations, Tulsa OK, SEPM, p. 53-79.

Heydari, E., 1997b, The role of burial diagenesis in hydrocarbon destruction and H2S accumulation, Upper Jurassic Smackover Formation, Black Creek Field, Mississippi: AAPG, v. 81, p. 26-45.

Heydari, E., 2000, Porosity loss, fluid flow and mass transfer in limestone reservoirs: Application to the Upper Jurassic Smackover Formation: AAPG, v. 84, p. 100-118.

Heydari, E., and C. H. Moore, 1989, Burial diagenesis and thermochemical sulfate reduction, Smackover Formation, southeastern Mississippi salt basin: Geology, v. 17, p. 1080-1084.

Heydari, E., and C. H. Moore, 1993, Zonation and geochemical patterns of burial calcite cements: Upper Smackover Formation, Clarke County, Mississippi: Journal of Sedimentary Petrology, v. 63, p. 44-60.

Heydari, E., and C. H. Moore, 1994, Paleoceanographic and paleoclimatic controls on ooid mineralogy of the Smackover Formation, Mississippi salt basin: implications for Late Jurassic seawater composition: Journal of Sedimentary Research, Section A: v. 64, p. 101-114.

Hills, J. M., 1972, Late Paleozoic sedimentation in west Texas Permian basin: American Association of Petroleum Geologists Bulletin, v. 56, p. 2303-2322.

Hine, A. C., R. J. Wilber, J. M. Bane, A. C. Neumann, and K. R. Lorenson, 1981, Offbank transport of carbonate sands along leeward bank margins, northern Bahamas: Marine Geology, v. 42, p. 327-348.

Hird, K., and M. E. Tucker, 1988, Contrasting diagenesis of two Carboniferous oolites from South Wales: a tale of climatic influence: Sedimentology, v. 35, p. 587-602.

Horowitz, A. S., and P. E. Potter, 1971, Introductory Petrography of Fossils: New York, Springer-Verlag, 302 p.

Hovorka, S. D., H. S. Nance and C. Kerans, 1993, Parasequence geometry as a control on permeability evolution: examples from the San Andres and Grayburg Formations in the Guadalupe Mountains, New Mexico, in R. L. a. J. F. Sarg, ed., Carbonate Sequence Stratigraphy: Recent Developments and Applications: AAPG Memoir, Tulsa, Oklahoma, AAPG, p. 493-514.

Hudson, J. D., 1977, Stable isotopes and limestone lithification: Journal of the Geological Society of London, v. 133, p. 637-660.

Humphrey, J. D., 1988, Late Pleistocene mixing zone dolomitization, southeastern Barbados, West Indies: Sedimentology, v. 35, p. 327-348.

Humphrey, J. D., 1997, Geology and hydrology of Barbados, in H. L. Vacher, and T. M. Quinn, eds., Geology and Hydrology of Carbonate Islands: Developments in Sedimentology, Amsterdam, Elsevier Science B.V, p. 381-406.

Humphrey, J. D., and R. P. Howell, 1999, Effect of differential stress on strontium partitioning in calcite: Journal of Sedimentary Research Section A, v. 69, p. 208-215.

Humphrey, J. D., and T. N. Kimbell, 1990, Sedimentology and sequence stratigraphy of upper Pleistocene carbonates of southeastern Barbados, West Indies: American Association of Petroleum Geologists Bulletin, v. 74, p. 1671-1684.

Humphrey, J. D., and R. K. Matthews, 1986, Deposition and diagenesis of the Pleistocene Coral Cap of Barbados, Eleventh Caribbean Geological Congress (Bridgetown, Barbados), Field Guide for Trip "C", p. 86-105.

Humphrey, J. D., and E. M. Radjef, 1991, Dolomite stoichiometric variability resulting from changing aquifer conditions, Barbados, West Indies: Sedimentary Geology, v. 71, p. 129-136.

Humphrey, J. D., K. L. Ransom, and R. K. Matthews, 1986, Early meteoric diagenetic control of Upper Smackover production, Oaks Field, Louisiana: American Association of Petroleum Geologists Bulletin, v. 70, p. 70-85.

Hurley, N. F., and K. C. Lohmann, 1989, Diagenesis of Devonian reefal carbonates in the Oscar Range, Canning Basin, Western Australia: Journal of Sedimentary Petrology, v. 59, p. 127-146.

Illing, L. V., 1959, Deposition and diagenesis of some Upper Paleozoic carbonate sediments in western Canada, Proceedings of the Fifth World Petroleum Congress (New York), Geology and Geophysics, Section I, p. 23-52.

Inden, R. F., and C. H. Moore, 1983, Beach environments, in P. A. Scholle, D. G. Bebout, and C. H. Moore, eds., Carbonate depositional environments, Tulsa, OK, American Association of Petroleum Geologists Memoir 33, p. 211-265.

James, N. P., 1979, Facies models 9. Introduction to carbonate facies models, in R. G. Walker, ed., Facies Models: Geoscience Canada Reprint Series 1, Toronto, Geological Association of Canada, p. 105-107.

James, N. P., 1983, Reef environment, in P. A. Scholle, D. G. Bebout, and C. H. Moore, eds., Carbonate Depositional Environments, Tulsa, OK, American Association of Petroleum Geologists Memoir 33, p. 345-440.

James, N. P., 1984, Shallowing upward sequences in carbonates, in R. G. Walker, ed., Facies Models, Geoscience Canada, p. 213-228.

James, N. P., Y. Bone, C. C. von der Borch, and V. A. Gostin, 1992, Modern carbonate and terrigenous clastic sediments on a cool water, high energy, mid-latitude shelf: Lacepede, southern Australia: Sedimentology, v. 39, p. 877-903.

James, N. P., and P. W. Choquette, 1983, Diagenesis 6. Limestones — The sea floor diagenetic environment: Geoscience Canada, v. 10, p. 162-179.

James, N. P., and P. W. Choquette, 1984, Diagenesis 9. Limestones — the meteoric diagenetic environment: Geoscience Canada, v. 11, p. 161-194.

James, N. P., and P. W. Choquette, 1990, Limestones-the sea-floor diagenetic environment, in I. A. McIlreath, and D. W. Morrow, eds., Diagenesis, Ottowa, Ontario, Canada, Geological Association of Canada, p. 13-34.

James, N. P., and J. D. A. Clark, 1997, Cool Water Carbonates: SEPM Special Publication, v. 56: Tulsa Oklahoma, SEPM, 400 p.

James, N. P., and R. N. Ginsburg, 1979, The seaward margin of Belize barrier and atoll reefs: Oxford, International Association of Sedimentologists Special Pubication No. 3, 191 p.

James, N. P., and C. F. Klappa, 1983, Petrogenesis of Early Cambrian reef limestones, Labrador, Canada: Journal of Sedimentary Petrology, v. 53, p. 1051-1096.

Jordan, C. F., Jr., T. C. Connally, Jr., and H. A. Vest, 1985, Middle Cretaceous carbonates of the Mishrif Formation, Fateh Field, offshore Dubai, U. A. E., in P. O. Roehl, and P. W. Choquette, eds., Carbonate Petroleum Reservoirs, New York, Springer–Verlag, p. 425-442.

Kaufman, E. G., M. A. Arthur, B. Howe, and P. A. Scholle, 1996, Widespread venting of methane-rich fluids in Late Cretaceous (Campanian) submarine springs (Teepee Buttes), Western Interior seaway, U.S.A.: Geology, v. 24, p. 799-802.

Kaufman, J., 1994, Numerical models of fluid flow in carbonate platforms: implications for dolomitization: Journal of Sedimentary Research, Section A: Sedimentary Petrology and Processes, v. A64, p. 128-139.

Kaufman, J., G. N. Hanson, and W. J. Meyers, 1991, Dolomitization of the Devonian Swan Hills Formation, Rosevear field, Alberta, Canada: Sedimentology, v. 38, p. 41-66.

Kaufmann, B., 1997, Diagenesis of Middle Devonian carbonate mounds of the Mader Basin (eastern Anti-Atlas, Morocco): Journal of Sedimentary Research, v. 67, p. 945-956.

Kendall, A. C., 1977a, Fascicular-optic calcite: a replacement of bundled acicular carbonate cements: Journal of Sedimentary Petrology, v. 47, p. 1056-1062.

Kendall, A. C., 1977b, Origin of dolomite mottling in Ordovician limestones from Saskatchewan and Manitoba: Bulletin of Canadian Petroleum Geology, v. 25, p. 480-504.

Kendall, A. C., 1984, Evaporites, in R. G. Walker, ed., Facies Models (2nd edition), Geoscience Canada Reprint Series, p. 259-296.

Kendall, A. C., 1985, Radiaxial fibrous calcite: a reappraisal, in N. Schneidermann, and P. M. Harris, eds., Carbonate Cements, Tulsa, OK, SEPM Special Publication No. 36, p. 59-77.

Kendall, A. C., 1988, Aspects of evaporite basin stratigraphy, in B. C. Schreiber, ed., Evaporites and Hydrocarbons, New York, Colombia University Press, p. 11-65.

Kendall, C. G. S. C., 1969, An environmental re-interpretation of the Permian evaporite-carbonate shelf sediments of the Guadalupe Mountains: Geological Society of America Bulletin, v. 80, p. 2503-2526.

Kendall, C. G. S. C., and J. K. Warren, 1988, Peritidal evaporites and their sedimentary assemblages, *in* B. C. Schreiber, ed., Evaporites and Hydrocarbons, New York, Colombia University Press, p. 66-138.

Kennedy, W. J., 1987, Sedimentology of Late Cretaceous-Paleocene chalk reservoirs, North Sea Central Graben, *in* J. Brooks, and K. Glennie, eds., Petroleum Geology of North West Europe, London, Graham and Trotman, p. 469-481.

Kerans, C., 1989, Karst-controlled reservoir heterogeneity and an example from the Ellenburger (Lower Ordovician) of west Texas, University of Texas, Bureau of Economic Geology Report of Investigations No. 186, 40p.

Kerans, C., 1993, Description and interpretation of karst-related breccia fabrics, Ellenburger Group, west Texas, *in* R. D. Fritz, J. L. Wilson, and D. A. Yurewicz, eds., Paleokarst Related Hydrocarbon Reservoirs, New orleans, SEPM, Society of Sedimentary Geology, p. 181-200.

Kerans, C., F. J. Lucia, and R. K. Senger, 1994, Integrated characterization of carbonate ramp reservoirs using Permian San Andreas Formation outcrop analogs: AAPG Bulletin, v. 78, p. 181-216.

Kimbell, T. N., J. D. Humphrey, and R. K. Stoessell, 1990, Quaternary mixing zone dolomite in a cored borehole, southeastern Barbados, West Indies [abs.]: Geological Society of America, Abstracts with Programs, v. 22, p. 179.

Kindler, P., and P. J. Hearty, 1997, Geology of the Bahamas:architecture of Bahamian islands, *in* H. L. Vacher, and T. M. Quinn, eds., Geology and Hydrogeology of Carbonate Islands: Developments in Sedimentology, Amsterdam, Elsevier Science B. V., p. 141-160.

Kinsman, D. J. J., 1969, Interpretation of Sr^{2+} concentration in carbonate minerals and rocks: Journal of Sedimentary Petrology, v. 39, p. 486-508.

Kinsman, D. J. J., and H. D. Holland, 1969, The co-precipitation of cations with $CaCO_3$ - IV. The co-precipitation of Sr^{+2} with aragonite between 16 and 96° C: Geochimica et Cosmochimica Acta, v. 33, p. 1-17.

Kirkland, B. L., J. A. D. Dickson, R. A. Wood, and L. S. Land, 1998, Microbialite and microstratigraphy: the origin of encrustations in the middle and upper Capitan Formation, Guadalupe Mountains, Texas and New Mexico: Journal Sedimentary Research, v. 68, p. 956-969.

Kirkland, B. L., and C. H. Moore, 1996, Microfacies analysis of the Tansell outer shelf, Permian Capitan reef complex, *in* M. R. L., ed., Permian Basin Oil and Gas Fields:Keys to Success that Unlock Future Reserves, Midland Texas, West Texas Geological Society, p. 99-106.

Kitano, Y., and D. W. Hood, 1965, The influence of organic material on polymorphic crystallization of calcium carbonate: Geochimica et Cosmochimica Acta, v. 29, p. 29-41.

Klosterman, M. J., 1981, Applications of fluid inclusion techniques to burial diagenesis in carbonate rock sequences, Louisiana State University, Applied Carbonate Research Program, Technical Series Contribution No.7, 101p.

Klovan, J. E., 1974, Development of western Canadian Devonian reefs and comparison with Holocene analogues: American Association of Petroleum Geologists Bulletin, v. 58, p. 787-799.

REFERENCES

Koepnick, R. B., 1983, Distribution and vertical permeability of stylolites within a Lower Cretaceous carbonate reservoir, Abu Dhabi, United Arab Emirates, Stylolites and Associated Phenomena: Relevence to Hydrocarbon Reservoirs, Abu Dhabi, U.A.E., Abu Dhabi Reservoir Research Foundation Special Publication, p. 261-278.

Kohout, F. A., 1965, A hypothesis concerning cyclic flow of salt water related to geothermal heating in the Floridian aquifer: Transactions New York Academy fo Science Series 2, v. 28, p. 249-271.

Krynine, P. D., 1941, Petrographic studies of variations in cementing material in the Oriskany Sand: Pa. State. College. Mineral Industry Experiment Station, Bulletin., v. 33, p. 108-116.

Kupecz, J. A., and L. S. Land, 1991, Late-stage dolomitization of the Lower Ordovician Ellenburger Group, west Texas: Journal of Sedimentary Petrology, v. 61, p. 551-574.

Kupecz, J. A., and L. S. Land, 1994, Progressive recrystallization and stabilization of early-stage dolomite:Lower Ordovician Ellenburger Group, west Texas, in B.Purser, M. Tucker, and D. Zenger, eds., Dolomites, International Association of Sedimentologists Special Publication 21, p. 255-279.

Kupecz, J. A., I. P. Montanez, and G. Gao, 1992, Recrystallization of dolomite with time, in R. Rezak, and D. Lavoie, eds., Carbonate Microfabrics: Frontiers in Sedimentology, New York, Springer Verlag.

Lafon, G. M., and H. L. Vacher, 1975, Diagenetic reactions as stochastic processes:application to Bermudian eolianites: GSA Memoir, v. 142, p. 187-204.

Lahann, R. W., 1978, A chemical model for calcite crystal growth and morphology control: Journal of Sedimentary Petrology, v. 48, p. 337-344.

Lahann, R. W., and R. M. Siebert, 1982, A kinetic model for distribution coefficients and application to Mg-calcite: Geochimica et Cosmochimica Acta, v. 46, p. 229-237.

Land, L. S., 1973, Contemporaneous dolomitization of middle Pleistocene reefs by meteoric water, north Jamaica: Bulletin of Marine Science, v. 23, p. 64-92.

Land, L. S., 1980, The isotopic and trace element geochemistry of dolomite: the state of the art, in D. H. Zenger, J. B. Dunham, and R. L. Ethington, eds., Concepts and Models of Dolomitization, Tulsa, OK, SEPM Special Publication No. 28, p. 87-110.

Land, L. S., 1985, The origin of massive dolomite: Journal of Geological Education, v. 33, p. 112-125.

Land, L. S., 1986, Limestone diagenesis — some geochemical considerations, in F. A. Mumpton, ed., Studies in Diagenesis, Washington, D.C., U. S. Geological Survey Bulletin 1578, p. 129-137.

Land, L. S., G. Gao, and J. A. Kupecz, 1991, Diagenetic history of the Arbuckle Group, Slick Hills, southwestern Oklahoma: a petrographic and geochemical summary and comparison with the Ellenburger Group, Texas, in K. S. Johnson, ed., Arbuckle Group Core Workshop and Field Trip, Norman, Oklahoma Geological Survey Special Publication 91-3, p. 103-110.

Land, L. S., and T. F. Goreau, 1970, Submarine lithification of Jamaican reefs: Journal of Sedimentary Petrology, v. 40, p. 457-462.

Land, L. S., F. T. Mackenzie, and S. J. Gould, 1967, Pleistocene history of Bermuda: Geological Society of America Bulletin, v. 78, p. 993-1006.

Land, L. S., and C. H. Moore, 1980, Lithification, micritization and syndepositional diagenesis of biolithites on the Jamaican island slope: Journal of Sedimentary Petrology, v. 50, p. 357-370.

Land, L. S., and D. R. Prezbindowski, 1981, The origin and evolution of saline formation waters, Lower Cretaceous carbonates, south-central Texas: Journal of Hydrology, v. 54, p. 51-74.

Landa, G. H., W. H. Holm, and E. V. Hough, 1998, Ekofisk area reservoir management and redevelopment: OshoreTechnology Conference, Houston, TX, paper #8654.

Lees, A., 1975, Possible influence of salinity and temperature on modern shelf carbonate sedimentation: Marine Geology, v. 19, p. 159-198.

Lees, A., and J. Miller, 1995, Waulsortian banks, in C. L. V. Monty, D. W. J. Bosence, P. H. Bridges, and B. R. Pratt, eds., Carbonate Mud-Mounds: Their Origin and Evolution, Oxford, England, International Association of Sedimentologists, Special Publication.

Lighty, R. G., 1985, Preservation of internal reef porosity and diagenetic sealing of submerged early Holocene barrier reef, southeast Florida shelf, in N. Schneidermann, and P. M. Harris, eds., Carbonate Cements, Tulsa, OK, SEPM Special Publication No. 36, p. 123-151.

Lindsay, R. F., and C. G. S. C. Kendall, 1985, Depositional facies, diagenesis, and reservoir character of Mississippian cyclic carbonates in the Mission Canyon Formation, Little Knife Field, Williston Basin, North Dakota, in P. O. Roehl, and P. W. Choquette, eds., Carbonate Petroleum Reservoirs, New York, Springer–Verlag, p. 175-190.

Lindsay, R. F., and M. S. Roth, 1982, Carbonate and evaporite facies, dolomitization and reservoir distribution of the Mission Canyon Formation, Little Knife field, North Dakota, in J. E. Christopher, and J. Kaldi, eds., Fourth International Williston Basin Symposium, Regina, Saskatchewan Geological Society Special Publication 6, p. 153-179.

Lippmann, F., 1973, Sedimentary Carbonate Minerals: New York, Springer-Verlag, 228 p.

Lloyd, A. R., 1977, The basement beneath the Queensland Continental Shelf, in O. A. Jones, and others, eds., Biology and Geology of Coral Reefs IV, New York, Academic Press, p. 261-266.

Lloyd, E. R., 1929, Capitan Limestone and associated formations of New Mexico and Texas: American Association of Petroleum Geologists Bulletin, v. 13, p. 645-658.

Lockridge, J. P., and P. A. Scholle, 1978, Niobrara gas in eastern Colorado and northwestern Kansas, in J. D. Pruit, and P. E. Coffin, eds., Energy Resources of the Denver Basin: 1978 Symposium Guidebook, Denver, CO, Rocky Mountain Association of Geologists, p. 35-49.

Logan, B. W., 1987, The MacLeod Evaporite Basin, Western Australia: Tulsa, OK, American Association of Petroleum Geologists Memoir 44, 140 p.

Lohmann, K. C., 1988, Geochemical patterns

of meteoric diagenetic systems and their application to studies of paleokarst, in N. P. James, and P. W. Choquette, eds., Paleokarst, New York, Springer-Verlag, p. 58-80.

Lohmann, K. C., and W. J. Meyers, 1977, Microdolomite inclusions in cloudy prismatic calcites: a proposed criterion for former high magnesium calcites: Journal of Sedimentary Petrology, v. 47, p. 1075-1088.

Longman, M. W., 1980, Carbonate diagenetic textures from nearsurface diagenetic environments: American Association of Petroleum Geologists Bulletin, v. 64, p. 461-487.

Longman, M. W., 1985, Fracture porosity in reef talus of a Miocene pinnacle–reef reservoir, Nido B Field, The Philippines, in P. O. Roehl, and P. W. Choquette, eds., Carbonate Petroleum Reservoirs, New York, Springer–Verlag, p. 547-560.

Loreau, J.-P., and B. H. Purser, 1973, Distribution and ultrastructure of Holocene ooids in the Persian Gulf, in B. H. Purser, ed., The Persian Gulf, New York, Springer-Verlag, p. 279-328.

Loucks, R. G., 1999, Paleocave carbonate reservoirs: origins, burial-depth modifications, spatial complexity, and reservoir implications: Journal of Sedimentary Research, v. 83, p. 1795-12834.

Loucks, R. G., and J. H. Anderson, 1985, Depositional facies, diagenetic terranes, and porosity development in Lower Ordovician Ellenburger Dolomite, Puckett Field, west Texas, in P. O. Roehl, and P. W. Choquette, eds., Carbonate Petroleum Reservoirs, New York, Springer–Verlag, p. 19-37.

Loucks, R. G., D. G. Bebout, and W. E. Galloway, 1977, Relationship of porosity formation and preservation to sandstone consolidation history — Gulf Coast Lower Tertiary Frio Formation: Transactions of the Gulf Coast Association of Geological Societies, v. 27, p. 109-120.

Loucks, R. G., and C. R. Handford, 1992, Origin and recognition of fractures, breccias, and sediment fills in paleocave-reservoir networks, in M. P. Candelaria, and C. L. Reed, eds., Paleokarst, Karst-Related Diagenesis and Reservoir Development: Examples from Ordovician-Devonian Age Strata of West Texas and the Mid-Continent, Midland, TX, Permian Basin Section-SEPM Publication 92-33, p. 31-44.

Loucks, R. G., and M. Longman, 1982, Lower Cretaceous Ferry Lake Anhydrite, Fairway field, east Texas, in C. R. Handford, R. G. Loucks, and G. R. Davies, eds., Depositional and Diagenetic Spectra of Evaporites - a core workshop (Calgary, 1982), Tulsa, OK, SEPM Core Workshop No. 3, p. 130-173.

Loucks, R. G., and J. F. Sarg, 1993, Carbonate Sequence Stratigraphy: recent developments and applications: AAPG Memoir, v. 57, AAPG, 545p. p.

Lucia, F. J., 1972, Recognition of evaporite - carbonate shoreline sedimentation, in J. K. Rigby, and W. K. Hamblin, eds., Recognition of Ancient Sedimentary Environments, Tulsa, OK, SEPM Special Publication No. 16, p. 160-191.

Lucia, F. J., 1983, Petrophysical parameters estimated from visual descriptions of carbonate rocks: a field classification of carbonate pore space: Journal of Petroleum Technology, v. 35, p. 629-637.

Lucia, F. J., 1995a, Lower Paleozoic Cavern

Development, Collapse, and Dolomitization, Franklin Mountains, El Paso, Texas, *in* D. A. Budd, A. H. Saller, and P. M. Harris, eds., Unconformities and Porosity in Carbonate Strata, AAPG Memoir 63, p. 279-300.

Lucia, F. J., 1995b, Rock-fabric/Petrophysical Classification of carbonate pore space for reservoir characterization: American Association of Petroleum Geologists Bulletin, v. 79, p. 1275-1300.

Lucia, F. J., 1999, Carbonate Reservoir Characterization: Berlin Heidelberg, Springer-Verlag, 226 p.

Lucia, F. J., and R. P. Major, 1994, Porosity evolution through hypersaline reflux dolomitization, *in* B. Purser, M.Tucker, and D. Zenger, eds., Dolomites, International Association of Sedimentologists Special Publication 21, p. 325-341.

M'Rabet, A., M. H. Negra, B. H. Purser, S. Sassi, and N. Ben Ayed, 1986, Micrite diagenesis in Senonian rudist build-ups in central Tunisia, *in* J. H. Schroeder, and B. H. Purser, eds., Reef Diagenesis, New York, Springer-Verlag, p. 291-310.

Machel, H. G., 1985, Cathodoluminescence in calcite and dolomite and its chemical interpretation: Geoscience Canada, v. 12, p. 139-147.

Machel, H. G., 1987, Some aspects of diagenetic sulphate-hydrocarbon redox reactions, *in* J. D. Marshall, ed., Diagenesis of Sedimentary Sequences, London, Geological Society (London) Special Publication No. 36, p. 15-28.

Machel, H. G., P. A. Cavell, and K. S. Patey, 1996, Isotopic evidence for carbonate cementation and recrystallization, and for tectonic expulsion of fluids into the Western Canada Sedimentary Basin: GSA, v. 108, p. 1108-1119.

Machel, H. G., and E. W. Mountjoy, 1986, Chemistry and environments of dolomitization — a reappraisal: Earth-Science Reviews, v. 23, p. 175-222.

Machel, H. G., and E. W. Mountjoy, 1987, General constraints on extensive pervasive dolomitization — and their application to the Devonian carbonates of western Canada: Bulletin of Canadian Petroleum Geology, v. 35, p. 143-158.

Macintyre, I. G., 1977, Distribution of submarine cements in a modern Caribbean fringing reef, Galeta Point, Panama: Journal of Sedimentary Petrology, v. 47, p. 503-516.

Macintyre, I. G., and R. P. Reid, 1992, Comment on the origin of aragonite needle mud: a picture is worth a thousand words: Journal of Sedimentary Petrology, v. 62, p. 1095-1097.

Mackenzie, F. T., and J. D. Pigott, 1981, Tectonic controls of Phanerozoic rock cycling: Journal of the Geological Society of London, v. 138, p. 183-196.

Maiklem, W. R., 1971, Evaporative drawdown — a mechanism for water-level lowering and diagenesis in the Elk Point basin: Bulletin of Canadian Petroleum Geology, v. 19, p. 487-503.

Maiklem, W. R., and D. G. Bebout, 1973, Ancient anhydrite facies and environments, Middle Devonian Elk Point basin, Alberta: Bulletin of Canadian Petroleum Geology, v. 21, p. 287-343.

Majewske, O. P., 1969, Recognition of Invertebrate Fossil Fragments in Rocks and Thin Sections: Leiden, Neth., E. J. Brill, 101 p.

Malek-Aslani, M., 1977, Plate tectonics and

sedimentary cycles in carbonates: Trans. Gulf Coast Assoc. Geol. Socs., v. 27, p. 125-133.

Maliva, R. G., and J. A. D. Dickson, 1992, Microfacies and diagenetic controls of porosity in Cretaceous/Tertary chalks, Eldfisk Field, Norwegian North Sea: American Association of Petroleum Geologists Bulletin, v. 76, p. 1825-1838.

Malone, M. J., P. A. Baker, and S. J. Burns, 1996, Hydrothermal dolomitization and recrystallization of dolomite breccias from the Miocene Monterey Formation, Tepusquet area, California: Journal of Sedimentary Research, v. 66, p. 976-990.

Marshall, J. F., 1986, Regional distribution of submarine cements within an epicontinental reef system: central Great Barrier Reef, Australia, in J. H. Schroeder, and B. H. Purser, eds., Reef Diagenesis, New York, Springer-Verlag, p. 8-26.

Matter, A., R. G. Douglas, and K. Perch-Nielsen, 1975, Fossil preservation, geochemistry and diagenesis of pelagic carbonates from the Shatsky Rise, northwest Pacific: Initial Reports of the Deep Sea Drilling Project, v. 32, p. 891-921.

Mattes, B. W., and E. W. Mountjoy, 1980, Burial dolomitization of the Upper Devonian Miette buildup, Jasper National Park, Alberta, in D. H. Zenger, J. B. Dunham, and R. L. Ethington, eds., Concepts and Models of Dolomitization, Tulsa, OK, SEPM Special Publication No. 28, p. 259-297.

Matthews, R. K., 1967, The Coral Cap of Barbados: Pleistocene studies of possible significance to petroleum geology, in J. G. Elam, and S. Chuber, eds., Cyclic Sedimentation in the Permian Basin, Midland, TX, West Texas Geological Society, p. 28-40.

Matthews, R. K., 1968, Carbonate diagenesis: equilibration of sedimentary mineralogy to the subaerial environment; Coral Cap of Barbados, West Indies: Journal of Sedimentary Petrology, v. 38, p. 1110-1119.

Matthews, R. K., 1974, A process approach to diagenesis of reefs and reef-associated limestones, in L. F. Laporte, ed., Reefs in Time and Space, Tulsa, OK, SEPM Special Publication No. 18, p. 234-256.

Matthews, R. K., 1984, Dynamic Stratigraphy: Englewood Cliffs, NJ, Prentice-Hall, 489 p.

Mazzullo, L. J., and P. M. Harris, 1991, An overview of dissolution porosity development in the deep-burial environment, with examples from carbonate reservoirs in the Permian Basin, in M. P. Candellaria, ed., Permian Basin Plays — Tomorrow's Technology Today, Midland, TX, West Texas Geological Society Publication No. 91-89, p. 125-138.

Mazzullo, S. J., 1980, Calcite pseudospar replacive of marine acicular aragonite, and implications for aragonite cement diagenesis: Journal of Sedimentary Petrology, v. 50, p. 409-422.

Mazzullo, S. J., W. D. Bischoff, and C. S. Teal, 1995, Holocene shallow-subtidal dolomitization by near-normal seawater, northern Belize: Geology, v. 23, p. 341-344.

Mazzullo, S. J., and A. M. Reid, 1988, Sedimentary textures of Recent Belizean peritidal dolomite: Journal of Sedimentary Petrology, v. 58, p. 479-488.

Mazzullo, S. J., A. M. Reid, and J. M. Gregg, 1987, Dolomitization of Holocene Mg-calcite supratidal deposits, Ambergris Cay, Belize: Geological Society of America

Bulletin, v. 98, p. 224-231.

McKenzie, J. A., 1981, Holocene dolomitization of calcium carbonate sediments from the coastal sabkhas of Abu Dhabi, U.A.E.: a stable isotope study: Journal of Geology, v. 89, p. 185-198.

McKenzie, J. A., K. J. Hsu, and J. F. Schneider, 1980, Movement of subsurface waters under the sabkha, Abu Dhabi, UAE, and its relation to evaporative dolomite genesis, *in* D. H. Zenger, J. B. Dunham, and R. L. Ethington, eds., Concepts and Models of Dolomitization, Tulsa, OK, SEPM Special Publication No. 28, p. 11-30.

Mckenzie, J. A., C. Vasconcalos, S. Bernasconi, D. Grujie, and A. Tien, 1995, Microbial mediation as a possible mechanism for natural dolomite formation under low-temperature conditions: Nature..

McQuillan, H., 1985, Fracture–controlled production from the Oligo–Miocene Asmari Formation in Gachsaran and Bibi Hakimeh Fields, southwest Iran, *in* P. O. Roehl, and P. W. Choquette, eds., Carbonate Petroleum Reservoirs, New York, Springer–Verlag, p. 511-523.

Meissner, F. F., 1972, Cyclic sedimentation in middle Permian strata of the Permian basin, *in* J. G. Elam, and S. Chuber, eds., Cyclic Sedimentation in the Permian Basin, second edition, Midland, TX, West Texas Geological Society Publication 72-60, p. 203-232.

Melim, L. A., and J. L. Masaferro, 1997, Subsurface geology of the Bahamas Banks, *in* H. L. Vacher, and T. M. Quinn, eds., Geology and Hydrology of Carbonate Islands, Amsterdam, Elsevier Science B.V., p. 161-182.

Meyers, W. J., 1974, Carbonate cement stratigraphy of the Lake Valley Formation (Mississippian), Sacramento Mountains, New Mexico: Journal of Sedimentary Petrology, v. 44, p. 837-861.

Meyers, W. J., 1988, Paleokarstic features on Mississippian limestones, New Mexico, *in* N. P. James, and P. W. Choquette, eds., Paleokarst, New York, Springer-Verlag, p. 306-328.

Meyers, W. J., and B. E. Hill, 1983, Quantitative studies of compaction in Mississippian skeletal limestones, New Mexico: Journal of Sedimentary Petrology, v. 53, p. 231-242.

Meyers, W. J., and K. C. Lohmann, 1985, Isotope geochemistry of regionally extensive calcite cement zones and marine components in Mississippian limestones, New Mexico, *in* P. M. Harris, and N. Schneidermann, eds., Carbonate Cements, Tulsa, OK, SEPM Special Publication No. 36, p. 223-239.

Miller, J. A., 1985, Depositional and reservoir facies of the Mississippian Leadville Formation, Northwest Lisbon Field, Utah, *in* P. O. Roehl, and P. W. Choquette, eds., Carbonate Petroleum Reservoirs, New York, Springer–Verlag, p. 161-173.

Milliman, J. D., 1974, Marine Carbonates. Part 1, Recent Sedimentary Carbonates: New York, Springer-Verlag, 375 p.

Miran, Y., 1977, Chalk deformation and large scale migration of calcium carbonate: Sedimentology, v. 24, p. 333-360.

Mitchum, R. M., and M. A. Uliana, 1985, Seismic stratigraphy of carbonate depositional sequences, Upper Jurassic-Lower Cretaceous, Neuquén Basin, Argentina, *in* O. R. Berg, and D. G. Woolverton, eds., Seismic Stratigraphy, II: An Integrated Approach to Hydrocarbon Exploration, Tulsa, OK, American Association of Petroleum Geologists Memoir 39, p. 255-274.

Mitra, S., and W. C. Beard, 1980, Theoretical models of porosity reduction by pressure solution for well-sorted sandstones: Journal of Sedimentary Petrology, v. 50, p. 1347-1360.

Montañez, I., 1997, Secondary porosity and late diagenetic cements of the Upper Knox Group, central Tennessee region: A temporal and spatial history of fluid flow conduit development within the Knox regional aquifer, in I. P. Montanez, J. M. Gregg, and K. L. Shelton, eds., Basin-wide Digenetic Patterns: Integrated Petrologic, Geochemical, and Hydrologic Considerations, Tulsa, Oklahoma, SEPM Special Publication 57, p. 101-117.

Montañez, I. P., 1994, Late diagenetic dolomitization of Lower Ordovician, Upper Knox Carbonates: a record of the hydrodynamic evolution of the southern Appalachian Basin: American Association of Petroleum Geologists Bulletin, v. 78, p. 1210-1239.

Montanez, I. P., J. M. Gregg, and K. L. Shelton, 1997, Basin-wide Digenetic Patterns: Integrated Petrologic, Geochemical, and Hydrologic Considerations, SEPM Special Publication 57, Tulsa, OK, SEPM, p. 300.

Montañez, I. P., and D. A. Osleger, 1993, Parasequence Stacking patterns, third-order accomodation events, and sequence stratigraphy of middle to upper Cambrian Platform Carbonates, Bonanza King Formation, southern Great Basin, in R. G. Loucks, and J. F. Sarg, eds., Carbonate Sequence Stratigraphy: recent developments and applications, AAPG Memoir 57, p. 305-326.

Montañez, I. P., and J. F. Read, 1992, Eustatic control on early dolomitization of cyclic peritidal carbonates: evidence from the Early Ordovician Upper Knox Group, Appalachians: Geological Society of America Bulletin, v. 104, p. 872-886.

Montgomery, S. L., 1996, Cotton Valley Lime Pinnacle Reef Play: Branton Field: American Association of Petroleum Geologists Bulletin, v. 80, p. 617-629.

Monty, C. L. V., 1995, The rise and nature of carbonate mud-mounds: and introductory actualistic approach, in C. L. V. Monty, D. W. J. Bosence, P. H. Bridges, and B. R. Pratt, eds., Carbonate Mud-Mounds: Their Origin and Evolution, Oxford, England, Blackwell Scientific, p. 11-48.

Moore, C., J. Crockett, and C. Hawkins, 1995, The Deep Madden Field, a super deep Madison gas reservoir, Wind River Basin, Wyoming (Abstract): AAPG.

Moore, C. H., Jr., 1977, Beach rock origin: some geochemical, mineralogical, and petrographic considerations: Geoscience and Man, v. 18, p. 155-163.

Moore, C. H., 1984, Regional patterns of diagenesis, porosity evolution and hydrocarbon production, upper Smackover, Gulf Rim, Jurassic of the Gulf Rim, Austin, Texas, SEPM Gulf Coast Section, p. 283-308.

Moore, C. H., 1985, Upper Jurassic subsurface cements: a case history, in N. Schneidermann, and P. M. Harris, eds., Carbonate Cements, Tulsa, OK, SEPM Special Publication No. 36, p. 291-308.

Moore, C. H., 1989, Carbonate Diagenesis and Porosity: New York, Elsevier, 338 p.

Moore, C. H., 1996, Anatomy of Sequence Boundary - Lower Cretaceous Glenrose/

Fredericksburg, Central Texas Platform: Gulf Coast Association of Geological Societies Transaction, p. 313-320.

Moore, C. H., Jr., and F. C. Brock, Jr., 1982, Porosity preservation in the upper Smackover (Jurassic) carbonate grainstone, Walker Creek Field, Arkansas: response to paleophreatic lenses to burial processes — discussion: Journal of Sedimentary Petrology, v. 52, p. 19-23.

Moore, C. H., A. Chowdhury, and L. Chan, 1988, Upper Jurassic platform dolomitization, northwestern Gulf of Mexico: a tale of two waters, in V. Shukla, and P. A. Baker, eds., Sedimentology and Geochemistry of Dolostones, Tulsa, OK, SEPM Special Publication No. 43, p. 175-190.

Moore, C. H., A. Chowdhury, and E. Heydari, 1986, Variation of ooid mineralogy in Jurassic Smackover Limestone as control of ultimate diagenetic potential: American Association of Petroleum Geologists Bulletin, v. 70, p. 622-628.

Moore, C. H., and Y. Druckman, 1981, Burial diagenesis and porosity evolution, Upper Jurassic Smackover, Arkansas and Louisiana: American Association of Petroleum Geologists Bulletin, v. 65, p. 597-628.

Moore, C. H., and Y. Druckman, 1991, Sequence stratigraphic framework fo the Upper Jurassic Smackover and related units, western Gulf of Mexico (abstract): AAPG, v. 75, p. 639.

Moore, C. H., E. A. Graham, and L. S. Land, 1976, Sediment transport and dispersal across the deep fore-reef and island slope (-55 m to -305 m), Discovery Bay, Jamaica: Journal of Sedimentary Petrology, v. 46, p. 174-187.

Moore, C. H., and E. Heydari, 1993, Burial diagenesis and hydrocarbon migration in platform limestones: a conceptual model based on the Upper Jurassic of Gulf Coast of USA, in A. D. Horbury, and A. G. Robinson, eds., Diagenesis and Basin Development: Studies in Geology 36, Tulsa, Oklahoma, AAPG, p. 213-229.

Moore, C. H., Jr., and W. W. Shedd, 1977, Effective rates of sponge bioerosion as a function of carbonate production, in D. L. Taylor, ed., Proceedings of the Third International Coral Reef Symposium, Miami, FL, University of Miami, Rosenstiel School of Marine and Atmospheric Science, p. 499-505.

Moore, C. H., Jr., J. M. Smitherman, and S. H. Allen, 1972, Pore system evolution in a Cretaceous carbonate beach sequence, 24th International Geological Congress, Proceedings, Section 6, p. 124-136.

Morrow, D. W., 1982a, Diagenesis 1. Dolomite - Part 1: The chemistry of dolomitization and dolomite precipitation: Geoscience Canada, v. 9, p. 5-13.

Morrow, D. W., 1982b, Diagenesis 2. Dolomite - Part 2: Dolomitization models and ancient dolostones: Geoscience Canada, v. 9, p. 95-107.

Morse, J., J. Hanor, and S. He, 1997, The role of mixing and migration of basinal waters in carbonate mineral mass transport, in I. P. Montanez, J. M. Gregg, and K. L. Shelton, eds., Basin-wide Digenetic Patterns: Integrated Petrologic, Geochemical, and Hydrologic Considerations, Tulsa, Oklahoma, SEPM, p. 41-50.

Morse, J. W., and F. T. Mackenzie, 1990, Geochemistry of Sedimentary Carbonates: New York, Elsevier Scientific Publ. Co., 696 p.

Moshier, S. O., 1989a, Development of microporosity in a micritic limestone reservoir, Lower Cretaceous, Middle East: Sedimentary Geology, v. 63, p. 217-240.

Moshier, S. O., 1989b, Microporosity in micritic limestones: a review: Sedimentary Geology, v. 63, p. 191-216.

Moshier, S. O., C. R. Handford, R. W. Scott, and R. D. Boutell, 1988, Giant gas accumulation in "chalky"-textured micritic limestone, Lower Cretaceous Shuaiba Formation, eastern United Arab Emirates, in A. J. Lomondo, and P. M. Harris, eds., Giant Oil and Gas Fields — A Core Workshop, Tulsa, OK, SEPM Core Workshop No. 12, p. 229-272.

Mountjoy, E., S. Whittaker, and A. Williams-Jones, 1997, Variable fluid and heat flow regimes in three Devonian Dolomite conduit systems, Western Canada Sedimentary Basin: isotopic and fluid inclusion evidence/constraints, in I. P. Montanez, J. M. Gregg, and K. L. Shelton, eds., Basinwide Digenetic Patterns: Integrated Petrologic, Geochemical, and Hydrologic Considerations, Tulsa, Oklahoma, SEPM Special Publication 57, p. 119-137.

Mucci, A., and J. W. Morse, 1983, The incorporation of Mg^{2+} and Sr^{2+} into calcite overgrowths: influences of growth rate and solution composition: Geochimica et Cosmochimica Acta, v. 47, p. 217-233.

Mullins, H. T., and H. E. Cook, 1986, Carbonate apron models: alternatives to the submarine fan model for paleoenvironmental analysis and hydrocarbon exploration: Sedimentary Geology, v. 48, p. 37-79.

Murray, G. E., 1961, Geology of the Atlantic and Gulf Coast Province of North America: New York, Harper & Bros., 692 p.

Murray, R. C., 1960, Origin of porosity in carbonate rocks: Journal of Sedimentary Petrology, v. 30, p. 59-84.

Murray, R. C., 1969, Hydrology of south Bonaire, N.A. — a rock selective dolomitization model: Journal of Sedimentary Petrology, v. 39, p. 1007-1013.

Mylroie, J. E., and J. L. Carew, 1995, Karst development on carbonate islands, in D. A. Budd, A. H. Saller, and P. M. Harris, eds., Unconformities and Porosity in Carbonate Strata, p. 55-76.

Neugebauer, J., 1973, The diagenetic problem of chalk — the role of pressure solution and pore fluid: Neues Jahrbuch für Geologie und Paläontologie, Abhandlungen, v. 143, p. 223-245.

Neugebauer, J., 1974, Some aspects of cementation in chalk, in K. J. Hsu, and H. C. Jenkyns, eds., Pelagic Sediments: on Land and Under the Sea, Oxford, International Association of Sedimentologists Special Publication No. 1, p. 149-176.

Neumann, A. C., and L. S. Land, 1975, Lime mud deposition and calcareous algae in the Bight of Abaco, Bahamas: A budget: Journal of Sedimentary Petrology, v. 45, p. 763-786.

Neumann, A. C., and I. Macintyre, 1985, Reef response to sea level rise: keep–up, catch–up or give–up: Proceedings of the Fifth International Coral Reef Congress, Tahiti, v. 3, p. 105-110.

Nicolaides, S., and M. W. Wallace, 1997, Submarine cementation and subaerial exposure in Oligo-Miocene temperate carbonates, Torquay Basin, Australia: Journal of Sedimentary Research, v. 67, p. 397-410.

Niemann, J. C., and J. F. Read, 1988, Regional cementation from unconformity-recharged aquifer and burial fluids, Mississippian Newman Limestone, Kentucky: Journal of Sedimentary Petrology, v. 58, p. 688-705.

O'Hearn, T. C., 1985, A fluid inclusion study of diagenetic mineral phases, Upper Jurassic Smackover Formation, southwest Arkansas and northeast Texas, Louisiana State University, Applied Carbonate Research Program, Technical Series Contribution No.24, 189p.

Oldershaw, A. E., and T. P. Scoffin, 1967, The source of ferroan and non-ferroan calcite cements in the Halkin and Wenlock Limestones: Geological Journal, v. 5, p. 309-320.

Oliver, J., 1986, Fluids expelled tectonically from orogenic belts; their role in hydrocarbon migration and other geologic phenomena: Geology, v. 14, p. 99-102.

Palmer, A. N., 1995, Geochemical Models for the Origin of Macroscopic Solution Porosity in Carbonate Rocks, in D. A. Budd, A. H. Saller, and P. M. Harris, eds., Unconformities and Porosity in Carbonate Strata, AAPG Memoir 63, p. 77-101.

Palmer, T. J., J. D. Hudson, and M. A. Wilson, 1988, Palaeoecological evidence for early aragonite dissolution in ancient calcite seas: Nature, v. 335, p. 809-810.

Patterson, R. J., and D. J. J. Kinsman, 1977, Marine and continental groundwater sources in a Persian Gulf coastal sabkha, in S. H. Frost, M. P. Weiss, and J. B. Saunders, eds., Reefs and Related Carbonates — Ecology and Sedimentology, Tulsa, OK, American Association of Petroleum Geologists Studies in Geology No. 4, p. 381-399.

Perkins, R. D., G. S. Dwyer, D. B. Rosoff, J. Fuller, P. A. Baker, and R. M. Lloyd, 1994, Salina sedimentation and diagenesis: West Caicos Island, British West Indies, in B. Purser, M. Tucker, and D. Zenger, eds., Dolomites: A Volume in Honour of Dolomieu, Oxford, Special Publication Number 21 of the International Association of Sedimentologists, by Blackwell Scientific Publications, p. 37-54.

Perkins, R. D., and S. D. Halsey, 1971, Geologic significance of microboring fungi and algae in Carolina shelf sediments: Journal of Sedimentary Petrology, v. 41, p. 843-853.

Perkins, R. D., M. D. McKenzie, and P. L. Blackwelder, 1971, Aragonite crystals within Codiacean algae: distinctive morphology and sedimentary implications: Science, v. 175, p. 624-626.

Petta, T. J., 1977, Diagenesis and geochemistry of a Glen Rose patch reef complex, Bandera Co., Texas, in D. G. Bebout, and R. G. Loucks, eds., Cretaceous Carbonates of Texas and Mexico: Applications to Subsurface Exploration, Austin, TX, University of Texas, Bureau of Economic Geology Report of Investigations No. 89, p. 138-167.

Pettijohn, F. J., 1957, Sedimentary Rocks (Second Edition): New York, Harper Brothers, 718 p.

Pingitore, N. E., Jr., 1976, Vadose and phreatic diagenesis: processes, products and their recognition in corals: Journal of Sedimentary Petrology, v. 46, p. 985-1006.

Playford, P. E., 1980, Devonian "Great Barrier Reef" of Canning Basin, Western Australia: American Association of Petroleum Geologists Bulletin, v. 64, p. 814-840.

Playford, P. E., and A. E. Cockbain, 1989, Devonian reef complexes, Canning Basin,

Western Australia: a review, Memoirs of the Association of Australasian Paleontology Nr., p. 401-412.

Plummer, L. N., 1975, Mixing of sea water with calcium carbonate ground water, *in* E. H. T. Whitten, ed., Quantitative Studies in the Geological Sciences, Boulder, CO, Geological Society of America Memoir 142, p. 219-236.

Plummer, L. N., H. L. Vacher, F. T. Mackenzie, O. P. Bricker, and L. S. Land, 1976, Hydrogeochemistry of Bermuda: A case history of ground-water diagenesis of biocalcarenites: Geological Society of America Bulletin, v. 87, p. 1301-1316.

Plummer, N., J. Busby, R. Lee, and B. Hanshaw, 1990, Geochemical modeling of the Madison Aquifer in parts of Montana, Wyoming, and South Dakota: Water Resources Research, v. 26, p. 1981-2014.

Posamentier, H. W., M.T.Jervey, and P.R.Vail, 1988, Eustatic controls on clastic deposition I-conceptual framework:, *in* e. a. C. K. Wilgus, ed., Sea Level Changes: an integrated approach, Tulsa, Oklahoma, S.E.P.M. Special Publication 42, p. 109-124.

Powers, R. W., 1962, Arabian Upper Jurassic carbonate reservoir rocks, *in* W. E. Ham, ed., Classification of Carbonate Rocks — a symposium, Tulsa, OK, American Association of Petroleum Geologists Memoir 1, p. 122-192.

Prezbindowski, D. R., 1985, Burial cementation — is it important? A case study, Stuart City trend, south central Texas, *in* N. Schneidermann, and P. M. Harris, eds., Carbonate Cements, Tulsa, OK, SEPM Special Publication No. 36, p. 241-264.

Prezbindowski, D. R., and R. E. Larese, 1987, Experimental stretching of fluid inclusions in calcite — implications for diagenetic studies: Geology, v. 15, p. 333-336.

Purdy, E. G., W. C. Pusey, III, and K. F. Wantland, 1975, Continental shelf of Belize — regional shelf attributes, *in* K. F. Wantland, and W. C. Pusey, III, eds., Belize Shelf — Carbonate Sediments, Clastic Sediments, and Ecology, Tulsa, OK, American Association of Petroleum Geologists Studies in Geology No. 2, p. 1-39.

Purdy, E. G., and D. Waltham, 1999, Reservoir implications of modern karst topography: AAPG, v. 83, p. 1774-1794.

Purser, B. H., and D. M. Aïssaoui, 1985, Reef diagenesis: dolomitization and dedolomitization at Mururoa Atoll (French Polynesia): Proceedings of the Fifth International Coral Reef Congress (Tahiti), v. 3, p. 263-269.

Purser, B. H., A. Brown, and Aissaoui, 1994, Nature, origins and eveolution of porosity in dolomites, *in* B.Purser, M. Tucker, and D. Zenger, eds., Dolomites, International Association of Sedimentologists Special Publication 21, p. 283-308.

Purser, B. H., and J. H. Schroeder, 1986, The diagenesis of reefs: a brief review of our present understanding, *in* J. H. Schroeder, and B. H. Purser, eds., Reef Diagenesis, New York, Springer-Verlag, p. 424-446.

Purvis, K., and V. P. Wright, 1991, Calcretes related to phreatophytic vegetation from the Middle Triassic Otter Sandstone of south west England: Sedimentology, v. 38, p. 539-551.

Qing, H., 1998, Petrography and geochemistry of early-stage, fine-and medium-crystalline dolomites in the Middle Devonian Presquile Barrier at Pine Point, Canada:

Sedimentology, v. 45, p. 433-446.

Qing, H., E. W. Mountjoy, B. Nesbitt, E., and K. Muehlenbachs, 1995, Paleohydrology of the Canadian Rocies and origins of brines, Pb-Zn deposits and dolomitization in the Western Canada Sedimentary Basin, Comment and Reply: Geology, v. 23, p. 189-190.

Radjef, A. M., 1992, Geochemical and stoichiometric variability of dolomite as a result of changing aquifer conditions, Barbados, West Indies, MS Thesis, University of Texas at Dallas, Richardson, Texas, 82 p.

Railsback, L. B., 1993, Contrasting styles of chemical compaction in the Upper Pennsylvanian Dennis Limestone in the Midcontinent region: Journal of Sedimentary Petrology, v. 63, p. 61-72.

Railsback, L. B., and E. C. Hood, 1993, Vertical sutured contacts caused by intergranular pressure disslution during tectonic compression in Jurassic limestones, high Atlas Mountains, Morocco: Geological Society of America Annual Meeting, p. A-162.

Rassmann-McLaurin, 1983, Holocene and ancient hardgrounds: petrographic comparison: AAPG Bulletin (abst), v. 65, p. 975-976.

Read, J. F., 1985, Carbonate platform facies models: American Association of Petroleum Geologists Bulletin, v. 69, p. 1-21.

Reeder, R. J., 1981, Electron optical investigation of sedimentary dolomites: Contributions to Mineralogy and Petrology, v. 76, p. 148-157.

Reeder, R. J., 1983, Crystal chemistry of the rhombohedral carbonates, in R. J. Reeder, ed., Carbonates: Mineralogy and Chemistry: Reviews in Mineralogy, Vol. 11, Washington, D.C., Mineralogical Society of America, p. 1-47.

Riding, R., 1991, Classification of microbial carbonates, in R. Riding, ed., Calcareous Algae and Stromatolites, Berlin, Springer-Verlag, p. 21-51.

Rittenhouse, G., 1971, Pore-space reduction by solution and cementation: American Association of Petroleum Geologists Bulletin, v. 55, p. 80-91.

Roberts, H. H., R. Sassen, R. Carney, and P. Aharon, 1989, ^{13}C-depleted authigenic carbonate buildups from hydrocarbon seeps, Louisiana continental slope: Transactions of the Gulf Coast Association of Geological Societies, v. 39, p. 523-530.

Roberts, H. H., and T. Whalen, 1977, Methane-derived carbonate cements in barrier and beach sands of a subtropical delta complex: Geochim. cosmochim. Acata, v. 39, p. 1085-1089.

Robinson, R. B., 1967, Diagenesis and porosity development in Recent and Pleistocene oolites from southern Florida and the Bahamas: Journal of Sedimentary Petrology, v. 37, p. 355-364.

Roedder, E., 1979, Fluid inclusion evidence on the environment of sedimentary diagenesis, a review, in P. A. Scholle, and P. R. Schluger, eds., Aspects of Diagenesis, Tulsa, OK, SEPM Special Publication No. 25, p. 89-108.

Roedder, E., and R. J. Bodnar, 1980, Geologic pressure determinations from fluid-inclusion studies: Annual Review of Earth and Planetary Sciences, v. 8, p. 263-301.

Roehl, P. O., and P. W. Choquette, 1985, Carbonate Petroleum Reservoirs, New York, Springer-Verlag, p. 622.

Roehl, P. O., and R. M. Weinbrandt, 1985, Geology and production characteristics of fractured reservoirs in the Miocene Monterey Formation, West Cat Canyon Oilfield, Santa Maria Valley, California, *in* P. O. Roehl, and P. W. Choquette, eds., Carbonate Petroleum Reservoirs, New York, Springer–Verlag, p. 525-545.

Rose, P. R., 1976, Mississippian carbonate shelf margins, western United States: U. S. Geological Survey Journal of Research, v. 4, p. 449-466.

Runnells, D. D., 1969, Diagenesis, chemical sediments, and the mixing of natural waters: Journal of Sedimentary Petrology, v. 39, p. 1188-1201.

Ruzyla, K., and G. M. Friedman, 1985, Factors controlling porosity in dolomite reservoirs of the Ordovician Red River Formation, Cabin Creek Field, Montana, *in* P. O. Roehl, and P. W. Choquette, eds., Carbonate Petroleum Reservoirs, New York, Springer–Verlag, p. 39-58.

Saller, A. H., 1984a, Diagenesis of Cenozoic Limestones on Enewetak Atoll, Louisiana State University, Baton Rouge, Louisiana, 363 p.

Saller, A. H., 1984b, Petrologic and geochemical constraints on the origin of subsurface dolomite, Enewetak Atoll: an example of dolomitization by normal sea water: Geology, v. 12, p. 217-220.

Saller, A. H., 1986, Radixial calcite in Lower Miocene strata, subsurface Enewetak Atoll: Journal of Sedimentary Petrology, v. 56, p. 743-762.

Saller, A. H., 1992, Calcitization of aragonite in Pleistocene limestones of Enewetak Atoll, Bahamas, and Yucatan — an alternative to thin-film neomorphism: Carbonates and Evaporites, v. 7, p. 56-73.

Saller, A. H., D. A. Budd, and P. M. Harris, 1994a, Unconformities and porosity development in carbonate strata: ideas from a Hedberg Conference: American Association of Petroleum Geologists Bulletin, v. 78, p. 857-872.

Saller, A. H., J. A. D. Dickson, and S. A. Boyd, 1994b, Cycle Stratigraphy and Porosity in Pennsylvanian and Lower Permian Shelf Limestones, Eastern Central Basin Platform: American Association of Petroleum Geologists Bulletin, v. 78, p. 1820-1842.

Saller, A. H., J. A. D. Dickson, and F. Matsuda, 1999, Evolution and distribution of porosity associated with subaerial exposure in Upper Paleozoic platform limestones, west Texas: AAPG, v. 83, p. 1835-1854.

Saller, A. H., and N. Henderson, 1998, Distribution of porosity and permeability in platform dolomites: insight from the Permian of west Texas: AAPG, v. 82, p. 1528-1550.

Saller, A. H., and R. B. Koepnick, 1990, Eocene to early Miocene growth of Enewetak Atoll: insight from strontium isotope data: Geological Society of America Bulletin, v. 102, p. 381-390.

Sandberg, P. A., 1983, An oscillating trend in Phanerozoic non-skeletal carbonate mineralogy: Nature, v. 305, p. 19-22.

Sandberg, P. A., 1985, Aragonite cements and their occurrence in ancient limestones, *in* N. Schneidermann, and P. M. Harris, eds., Carbonate Cements, Tulsa, OK, SEPM Special Publication No. 36, p. 33-57.

Sandberg, P. A., and J. D. Hudson, 1983, Aragonite relic preservation in Jurassic calcite-replaced bivalves: Sedimentology, v. 30, p. 879-892.

Sandberg, P. A., N. Schneidermann, and S. J. Wunder, 1973, Aragonitic ultrastructural relics in calcite-replaced Pleistocene skeltons: Nature Physical Science, v. 245, p. 133-134.

Saner, S., and W. M. Abdulghani, 1995, Lithostratigraphy and Depositional Environments of the Upper Jurassic Arab-C Carbonate and Assiciated Evaporites in the Abqaiq Field, Eastern Saudi Arabia: American Association of Eptroleum Geologists Bulletin, v. 79, p. 394-409.

Sarg, J. F., 1976, Sedimentology of the carbonate-evaporite facies transition of the Seven Rivers Formation (Guadalupian, Permian) in southeast New Mexico: Ph.D. thesis, University of Wisconsin - Madison, Madison, WI, 313 p.

Sarg, J. F., 1977, Sedimentology of the carbonate-evaporite facies transition of the Seven Rivers Formation (Guadalupian, Permian) in southeast New Mexico, in M. E. Hileman, and S. J. Mazzullo, eds., Upper Guadalupian Facies, Permian Reef Complex, Guadalupe Mountains, New Mexico and West Texas (1977 Field Conference Guidebook), Midland, TX, Permian Basin Section-SEPM Publication 77-16, p. 451-478.

Sarg, J. F., 1981, Petrology of the carbonate-evaporite facies transition of the Seven Rivers Formation (Guadalupian, Permian), southeast New Mexico: Journal of Sedimentary Petrology, v. 51, p. 73-95.

Sarg, J. F., 1988, Carbonate sequence stratigraphy, in C. K. Wilgus, B. S. Hastings, C. G. S. C. Kendall, H. W. Posamentier, C. A. Ross, and J. C. Van Wagoner, eds., Sea-Level Changes: An Integrated Approach, Tulsa, OK, SEPM Special Publication No. 42, p. 155-182.

Sassen, R., and C. H. Moore, 1988, Framework of hydrocarbon generation and destruction in eastern Smackover trend: American Association of Petroleum Geologists Bulletin, v. 72, p. 649-663.

Sassen, R., C. H. Moore, and F. C. Meendsen, 1987, Distribution of hydrocarbon source potential in the Jurassic Smackover Formation: Organic Geochemistry, v. 11, p. 379-383.

Schlager, W., 1989, Drowning unconformities on carbonate platforms, in P. D. Crevello, J. L. Wilson, J. F. Sarg, and J. F. Read, eds., Controls on Carbonate Platform and Basin Development, Tulsa, OK, SEPM Special Publication No. 44, p. 15-26.

Schlager, W., 1992, Sedimentology and sequence stratigraphy of reefs and carbonate platforms: Continuing Education Course Note Series, v. 34: Tulsa, OK, American Association of Petroleum Geologists, 71 p.

Schlager, W., and N. P. James, 1978, Low-magnesian calcite limestones forming at the deep-sea floor, Tongue of the Ocean, Bahamas: Sedimentology, v. 25, p. 675-702.

Schlager, W., J. J. Reijmer, G,, and A. Droxler, 1994, Highstand shedding of carbonate platforms: Journal of Sedimentary Research, v. B64, p. 270-281.

Schlanger, S. O., 1981, Shallow-water limestones in oceanic basins as tectonic and paleoceanographic indicators, in J. E. Warme, R. G. Douglas, and E. L. Winterer, eds., The Deep Sea Drilling Project: A decade of progress, Tulsa, OK, SEPM Special Publication No. 32, p. 209-226.

Schlanger, S. O., and R. G. Douglas, 1974, Pelagic ooze-chalk-limestone transition and its implications for marine stratigraphy, in K. J. Hsu, and H. C. Jenkyns, eds.,

Pelagic Sediments: on Land and under the Sea, Oxford, International Association of Sedimentologists Special Publication No. 1, p. 117-148.

Schmidt, V., and D. A. McDonald, 1979, The role of secondary porosity in the course of sandstone diagenesis, in P. A. Scholle, and P. Schluger, eds., Aspects of Diagenesis, Tulsa, OK, SEPM Special Publication No. 26, p. 175-207.

Schmidt, V., I. A. McIlreath, and A. E. Budwill, 1985, Origin and diagenesis of Middle Devonian pinnacle reefs encased in evaporites, "A" and "E" Pools, Rainbow Field, Alberta, in P. O. Roehl, and P. W. Choquette, eds., Carbonate Petroleum Reservoirs, New York, Springer–Verlag, p. 141-160.

Schmoker, J. W., 1984, Empirical relation between carbonate porosity and thermal maturity: an approach to regional porosity prediction: American Association of Petroleum Geologists Bulletin, v. 68, p. 1697-1703.

Scholle, P. A., 1971, Diagenesis of deep-water carbonate turbidites, Upper Cretaceous Monte Antola Flysch, Northern Apennines, Italy: Journal of Sedimentary Petrology, v. 41, p. 233-250.

Scholle, P. A., 1977a, Chalk diagenesis and its relation to petroleum exploration: oil from chalks, a modern miracle?: American Association of Petroleum Geologists Bulletin, v. 61, p. 982-1009.

Scholle, P. A., 1977b, Deposition, diagenesis, and hydrocarbon potential of "deeper-water" limestones: Tulsa, OK, American Association of Petroleum Geologists Short Course in Exploration Geology Notes No. 7, 27 p.

Scholle, P. A., M. A. Arthur, and A. A. Ekdale, 1983a, Pelagic environment, in P. A. Scholle, D. G. Bebout, and C. H. Moore, eds., Carbonate Depositional Environments, Tulsa, OK, American Association of Petroleum Geologists Memoir 33, p. 619-691.

Scholle, P. A., D. G. Bebout, and C. H. Moore, 1983b, Carbonate Depositional Environments, Tulsa, OK, American Association of Petroleum Geologists Memoir 33, p. 708.

Scholle, P. A., and R. B. Halley, 1985, Burial diagenesis: out of sight, out of mind!, in N. Schneidermann, and P. M. Harris, eds., Carbonate Cements, Tulsa, OK, SEPM Special Publication No. 36, p. 309-334.

Schreiber, B. C., 1988, Evaporites and Hydrocarbons, New York, Columbia University Press, p. 475.

Schreiber, B. C., M. S. Roth, and M. L. Helman, 1982, Recognition of primary facies characteristics of evaporites and differentiation of the forms from diagenetic overprints, in C. R. Handford, R. G. Loucks, and G. R. Davies, eds., Depositional and Diagenetic Spectra of Evaporites - a core workshop (Calgary, 1982), Tulsa, OK, SEPM Core Workshop No. 3, p. 1-32.

Schroeder, J. H., 1969, Experimental dissolution of calcium, magnesium, and strontium from Recent biogenic carbonates: a model of diagenesis: Journal of Sedimentary Petrology, v. 39, p. 1057-1073.

Schroeder, J. H., and B. H. Purser, 1986, Reef Diagenesis, New York, Springer-Verlag, p. 448.

Sears, S. O., and F. J. Lucia, 1980, Dolomitization of northern Michigan Niagara reefs by brine refluxion and freshwater/seawater mixing, *in* D. H. Zenger, J. B. Dunham, and R. L. Ethington, eds., Concepts and Models of Dolomitization, Tulsa, OK, SEPM Special Publication No. 28, p. 215-235.

Sellwood, B. W., 1978, Shallow water carbonate environments, *in* H. G. Reading, ed., Sedimentary Environments and Facies, Oxford, Blackwell Scientific Publications, p. 259-313.

Sellwood, B. W., J. Scott, B. James, R. Evans, and J. D. Marshall, 1987, Regional significance of 'dedolomitization' in Great Oolite reservoir facies of southern England, *in* J. Brooks, and K. W. Glennie, eds., Petroleum Geology of North West Europe, London, Graham and Trotman, p. 129-137.

Shearman, D. J., and J. G. Fuller, 1969, Anhydrite diagenesis, calcitization, and organic laminites, Winnipegosis Formation, Middle Devonian, Saskatchewan: Bulletin of Canadian Petroleum Geology, v. 17, p. 496-525.

Shinn, E. A., 1968, Practical significance of birdseye structures in carbonate rocks: Journal of Sedimentary Petrology, v. 38, p. 215-223.

Shinn, E. A., 1969, Submarine lithification of Holocene carbonate sediments in the Persian Gulf: Sedimentology, v. 12, p. 109-144.

Shinn, E. A., 1983, Tidal flat environment, *in* P. A. Scholle, D. G. Bebout, and C. H. Moore, eds., Carbonate Depositional Environments, Tulsa, OK, American Association of Petroleum Geologists Memoir 33, p. 171-210.

Shinn, E. A., W. E. Bloxsom, and R. M. Lloyd, 1974, Recognition of submarine cements in Cretaceous reef limestones from Texas and Mexico (Abstract): Am. Assoc. Petrol. Geol. Abstract Volume, v. 1, p. 82-83.

Shinn, E. A., R. N. Ginsburg, and R. M. Lloyd, 1965, Recent supratidal dolomite from Andros Island, Bahamas, *in* L. C. Pray, and R. C. Murray, eds., Dolomitization and Limestone Diagenesis, Tulsa, OK, SEPM Special Publication No. 13, p. 112-123.

Shinn, E. A., H. J. Hudson, R. B. Halley, B. Lidz, D. M. Robbin, and I. G. Macintyre, 1982, Geology and sediment accumulation rates at Carrie Bow Cay, Belize, *in* K. Rutzler, and I. G. Macintyre, eds., The Atlantic Barrier Reef Ecosystem at Carrie Bow Cay, Belize, I. Structure and Communities, Washington, DC, Smithsonian Contrib. to the Mar. Sci, p. 63-75.

Shinn, E. A., R. M. Lloyd, and R. N. Ginsburg, 1969, Anatomy of a modern carbonate tidal-flat, Andros Island, Bahamas: Journal of Sedimentary Petrology, v. 39, p. 1202-1228.

Shinn, E. A., and D. M. Robbin, 1983, Mechanical and chemical compaction in fine-grained shallow-water limestones: Journal of Sedimentary Petrology, v. 53, p. 595-618.

Shinn, E. A., D. M. Robbin, B. H. Lidz, and J. H. Hudson, 1983, Influence of deposition and early diagenesis on porosity and chemical compaction in two Paleozoic buildups: Mississippian and Permian age rocks in the Sacramento Mountains, New Mexico, *in* P. M. Harris, ed., Carbonate Buildups — a core workshop (Dallas, April 16-17, 1983), Tulsa, OK, SEPM Core Workshop No. 4, p. 182-222.

Shinn, E. A., R. P. Steinen, B. H. Lidz, and P. K. Swart, 1989, Whitings, a sedimentologic dilemma: Journal of Sedimentary Petrology, v. 59, p. 147-161.

Sibley, D. F., 1980, Climatic control of dolomitization, Seroe Domi Formation (Pliocene), Bonaire, N.A., *in* D. H. Zenger, J. B. Dunham, and R. L. Ethington, eds., Concepts and Models of Dolomitization, Tulsa, OK, SEPM Special Publication No. 28, p. 247-258.

Sibley, D. F., 1982, The origin of common dolomite fabrics: clues from the Pliocene: Journal of Sedimentary Petrology, v. 52, p. 1087-1100.

Silver, B. A., and R. G. Todd, 1969, Permian cyclic strata, northern Midland and Delaware Basins, west Texas and southeastern New Mexico: American Association of Petroleum Geologists Bulletin, v. 53, p. 2223-2251.

Simms, M. A., 1984, Dolomitization by groundwater-flow systems in carbonate platforms: Transactions of the Gulf Coast Association of Geological Societies, v. 34, p. 411-420.

Skall, H., 1975, The paleoenvironment of the Pine Point lead-zinc district: Economic Geology, v. 70, p. 22-47.

Smalley, P. C., P. K. Bishop, J. A. D. Dickson, and D. Emery, 1994, Water-rock interaction during meteoric flushing of a limestone: implications for porosity development in karstified petroleum reservoirs: Journal of Sedimentary Research, v. A64, p. 180-189.

Smart, P. L., and F. F. Whitaker, 1991, Karst processes, hydrology and porosity evolution, *in* V. P. Wright, M. Esteban, and P. L. Smart, eds., Paleokarsts and Paleokarstic Reservoirs: P.R.I.S. Occasional Publication Series No. 2, Reading, University of Reading, p. 1-55.

Sonnenfeld, M., 1996, Sequence evolution and hierarchy within the Lower Mississippian Madison Limestone of Wyoming, *in* M. Longman, and M. Sonnenfeld, eds., Paleozoic Systems of the Rocky Mountain Region, Denver, Colorado, Rocky Mountain Section, SEPM, p. 165-192.

Steinen, R. P., and R. K. Matthews, 1973, Phreatic vs. vadose diagenesis: stratigraphy and mineralogy of a cored borehole on Barbados, W.I.: Journal of Sedimentary Petrology, v. 43, p. 1012-1020.

Stoakes, F., J. Wendte, and C. Campbell, 1992, Devonian-Early Mississippian Carbonates of the Western Canada Sedimentary Basin: A Sequence Stratigraphic Framework: Tulsa, OK, SEPM Short Course Notes No. 28, 268 p.

Stoessell, R. K., and C. H. Moore, 1983, Chemical constraints and origins of four groups of Gulf Coast reservoir fluids: American Association of Petroleum Geologists Bulletin, v. 67, p. 896-906.

Stueber, A., A. Saller, and H. Ishida, 1998, Origin, migration and mixing of brines in the Permian Basin: Geochemical evidence from the eastern Central Basin Platform: AAPG, v. 82, p. 1652-1672.

Stueber, A. M., P. Pushkar, and E. A. Hetherington, 1984, A strontium isotopic study of Smackover brines and associated solids, southern Arkansas: Geochimica et Cosmochimica Acta, v. 48, p. 1637-1649.

Sumner, D. Y., and J. P. Grotzinger, 1993, Numerical modeling of ooid size and the problem of Neoproterozoic giant ooids: Journal of Sedimentary Petrology, v. 63, p. 974-982.

Sun, S. Q., 1994, A reappraisal of dolomite abundance and occurrence in the Phanerozoic: Journal of Sedimentary research, v. A64, p. 396-404.

Surdam, R. C., S. W. Boese, and L. J. Crossey, 1984, The chemistry of secondary porosity, *in* D. A. McDonald, and R. C. Surdam, eds., Clastic Diagenesis, Tulsa, OK, American Association of Petroleum Geologists Memoir 37, p. 127-150.

Swartz, J. H., 1958, Geothermal measurements on Enisetok and Bikini atolls, Washington D.C., U.S.Geological Survey Professional Paper, p. 127-149.

Swartz, J. H., 1962, Some physical constraints for the Marshall Island area, Washington D.C., U.S.Geological Survey Professional Paper, p. 953-989.

Swinchatt, J. P., 1965, Significance of constituent composition, texture, and skeletal breakdown in some Recent carbonate sediments: Journal of Sedimentary Petrology, v. 35, p. 71-90.

Taylor, S. R., and J. F. Lapré, 1987, North Sea chalk diagenesis: its effects on reservoir location and properties, *in* J. Brooks, and K. Glennie, eds., Petroleum Geology of North West Europe, London, Graham and Trotman, p. 483-495.

Thibodaux, B. L., 1972, Sedimentological Comparison Between Sombrero Key and Looe Key: Masters thesis, Louisiana State University, Baton Rouge, 83 p.

Tinker, S. W., 1996, Building the 3-D jigsaw pizzle: applications of sequence stratigraphy to 3-D reservoir characterization, Permian basin: AAPG Bulletin, v. 80, p. 460-485.

Tinker, S. W., and D. H. Mruk, 1995, Reservoir characterization of a Permian giant: Yates field, West Texas, *in* E. L. Stoudt, and P. M. Harris, eds., Hydrocarbon reservoir characterization: Geologic framework and flow unit modeling, SEPM Short Course Notes, p. 51-128.

Tissot, B. P., and D. H. Welte, 1978, Petroleum Formation and Occurrence: a new approach to oil and gas exploration: New York, Springer-Verlag, 538 p.

Tucker, M. E., 1993, Carbonate Diagenesis and Sequence Stratigraphy, *in* V. P. Wright, ed., Sedimentology Review, Oxford, Blackwell, p. 51-72.

Tucker, M. E., and V. P. Wright, 1990, Carbonate Sedimentology: Oxford, Blackwell Scientific Publications, 482 p.

Usiglio, J., 1849, Analyse de L'eau de la Mediterranee sur le Cotes de France: Chem. Phys, v. 27, p. 92-107.

Vacher, H. L., 1997, Introduction: varieties of carbonate islands and a historical perspective, *in* H. L. Vacher, and T. M. Quinn, eds., Geology and Hydrology of Bermuda, Amsterdam, Elsevier Science B.V., p. 1-33.

Vacher, H. L., T. O. Bengtsson, and L. N. Plummer, 1990, Hydrology of meteoric diagenesis: residence time of meteoric ground water in island fresh-water lenses with application to aragonite-calcite stabilization rate in Bermuda: Geological Society of America Bulletin, v. 102, p. 223-232.

Vacher, H. L., and M. P. Rowe, 1997, Geology and Hydrology of Bermuda, *in* H. L. Vacher, and T. M. Quinn, eds., Geology and Hydrology of Carbonate Islands: Developments in Sedimentology, Amsterdam, Elsevier Science B.V., p. 35-89.

Vahrenkamp, V. C., and P. Swart, 1987, Stable isotopes as tracers of fluid/rock interactions during massive platform dolomitization, Little Bahama Bank: Abstract Volume SEPM Midyear meeting, v. 4, p. 85-86.

Vahrenkamp, V. C., and P. K. Swart, 1990, New distribution coefficient for the incorporation of strontium into dolomite and its implications for the formation of ancient dolomites: Geology, v. 18, p. 387-391.

Vahrenkamp, V. C., and P. K. Swart, 1991, Episodic dolomitization of late Cenozoic carbonates in the Bahamas: evidence from strontium isotopes: Journal of Sedimentary Petrology, v. 61, p. 1002-1014.

Vahrenkamp, V. C., and P. K. Swart, 1994, Late Cenozoic dolomites of the Bahamas: metastable analogues for the genesis of ancient platform dolomites, in B.Purser, M.Tucker, and D. Zenger, eds., Dolomites, International Association of Sedimentologists Special Publication 21, p. 133-153.

Vahrenkamp, V. C., P. K. Swart, and J. Ruiz, 1988, Constraints and interpretations of $^{87}Sr/^{86}Sr$ ratios in Cenozoic dolomites: Geophysical Research Letters, v. 15, p. 385-388.

Van den Bark, E., and O. D. Thomas, 1981, Ekofisk: first of the giant oil fields in western Europe: American Association of Petroleum Geologists Bulletin, v. 65, p. 2341-2363.

Van Wagoner, J. C., R. M. Mitchum Jr., R. G. Todd, K. M. Campion and V. D. Rahmanian, 1991, Siliciclastic sequence stratigraphy in well logs, cores, and outcrops: Concepts for high-resolution correlation of time and facies: AAPG Methods in Exploration Series, v. 7: Tulsa, Oklahoma, AAPG, 55 p.

Veizer, J., 1983a, Chemical diagenesis of carbonates: theory and application of trace element technique, in M. A. Arthur, ed., Stable Isotopes in Sedimentary Geology, Tulsa, OK, SEPM Short Course No. 10, p. 3-1 to 3-100.

Veizer, J., 1983b, Trace elements and isotopes in sedimentary carbonates, in R. J. Reeder, ed., Carbonates: Mineralogy and Chemistry: Reviews in Mineralogy, Vol. 11, Washington, D.C., Mineralogical Society of America, p. 265-299.

Veizer, J., and J. Hoefs, 1976, The nature of O^{18}/O^{16} and C^{13}/C^{12} secular trends in sedimentary carbonate rocks: Geochimica et Cosmochimica Acta, v. 40, p. 1387-1395.

von der Borch, C. C., D. E. Lock, and D. Schwebel, 1975, Ground-water formation of dolomite in the Coorong region of South Australia: Geology, v. 3, p. 283-285.

Walls, R. A., and G. Burrowes, 1985, The role of cementation in the diagenetic history of Devonian reefs, western Canada, in N. Schneidermann, and P. M. Harris, eds., Carbonate Cements, Tulsa, OK, SEPM Special Publication No. 36, p. 185-220.

Wanless, H. R., 1979, Limestone response to stress: pressure solution and dolomitization: Journal of Sedimentary Petrology, v. 49, p. 437-462.

Ward, W. C., 1997, Geology of coastal islands, northeastern Yucatan peninsula, in H. L. Vacher, and T. M. Quinn, eds., Geology and Hydrogeology of Carbonate Islands: Developments in Sedimentology, Amsterdam, Elsevier Science B. V., p. 275-298.

Ward, W. C., and M. J. Brady, 1979, Strandline sedimentation of carbonate grainstones, Upper Pleistocene, Yucatan Peninsula, Mexico: American Association of Petroleum Geologists Bulletin, v. 63, p. 362-369.

Ward, W. C., and R. B. Halley, 1985, Dolomitization in a mixing zone of near-seawater composition, Late Pleistocene, northeastern Yucatan Peninsula: Journal of Sedimentary Petrology, v. 55, p. 407-420.

Ward, W. C., A. E. Weidie, and W. Back, 1985, Geology and hydrogeology of the Yucatan and Quaternary geology of northeastern Yucatan Peninsula: New Orleans, LA, New Orleans Geological Society Guidebook, 160 p.

Wendte, J., 1974, Sedimentation and diagenesis of the Cooking Lake platform and lower Leduc reef facies, Upper Devonian, Redwater, Alberta: Ph.D. thesis, University of California - Santa Cruz, Santa Cruz, CA, 221 p.

Wendte, J., J. Dravis, L. Stasiuk, H. Qing, S. Moore, and G. Ward, 1998, High-temperature saline (thermoflux) dolomitization of Devonian Swan Hills platform and bank carbonates, Wild River area, west-central Alberta: Bulletin of Canadian Petroleum Geology, v. 46, p. 210-265.

Weyl, P. K., 1959, Pressure solution and the force of crystallization — a phenomenological theory: Journal of Geophysical Research, v. 64, p. 2001-2025.

Weyl, P. K., 1960, Porosity through dolomitization: conservation-of-mass requirements: Journal of Sedimentary Petrology, v. 30, p. 85-90.

Wheeler, C. W., and P. Aharon, 1993, It ISn't the thermal convection after all: the Dolomite record at Niue revisited: Geological Society of America Annual Meeting, p. A-398.

Wheeler, C. W., and P. Aharon, 1997, The geology and hydrogeology of Niue, South Pacific, in H. L. Vacher, and T. M. Quinn, eds., Geology and Hydrogeology of Carbonate Islands, Amsterdam, Elsevier, p. 537-564.

Whitaker, F. F., and P. L. Smart, 1997, Hydrogeology of the Bahamian Archipelago, in H. L. Vacher, and T. M. Quinn, eds., Geology and Hydrology of Carbonate Islands: Developments in Sedimentology, Amsterdam, Elsevier Science B.V., p. 183-216.

Wilderer, P. A., and W. G. Characklis, 1989, Structure and function of biofilms, in W. G. Characklis, and P. A. Wilderer, eds., Biofilms, Chichester, U.K., Wily, p. 5-18.

Wilkinson, B. H., 1979, Biomineralization, paleoceanography, and the evolution of calcareous marine organisms: Geology, v. 7, p. 524-527.

Wilkinson, B. H., and R. K. Given, 1986, Secular variation in abiotic marine carbonates: constraints on Paleozoic atmospheric carbon dioxide contents and oceanic Mg/Ca ratios: Journal of Geology, v. 94, p. 321-333.

Wilkinson, B. H., S. U. Janecke, and C. E. Brett, 1982, Low-magnesian calcite marine cement in Middle Ordovician hardgrounds from Kirkfield, Ontario: Journal of Sedimentary Petrology, v. 52, p. 47-57.

Wilkinson, B. H., R. M. Owen, and A. R. Carroll, 1985, Submarine hydrothermal weathering, global eustasy, and carbonate polymorphism in Phanerozoic marine oolites: Journal of Sedimentary Petrology, v. 55, p. 171-183.

Wilson, A. O., 1985, Depositional and diagenetic facies in the Jurassic Arab–C and –D reservoirs, Qatif Field, Saudi Arabia, in P. O. Roehl, and P. W. Choquette, eds., Carbonate Petroleum Reservoirs, New York, Springer–Verlag, p. 319-340.

REFERENCES

Wilson, J. L., 1975, Carbonate Facies in Geologic History: New York, Springer Verlag, 471 p.

Wilson, J. L., 1980, A review of carbonate reservoirs, *in* A. D. Miall, ed., Facts and Principles of World Petroleum Occurrence, Calgary, Alberta, Canadian Society of Petroleum Geologists Memoir 6, p. 95-117.

Wong, P. K., and A. Oldershaw, 1981, Burial cementation in the Devonian Kaybob reef complex, Alberta, Canada: Journal of Sedimentary Petrology, v. 51, p. 507-520.

Wood, R., J. A. D. Dickson, and B. L. Kirkland, 1996, New observations on the ecology of the Permian Reef complex, Texas and New Mexico: Paleontology, v. 39, p. 733-762.

Wood, R., J. A. D. Dickson, and B. Kirkland-George, 1994, Turning the Capitan Reef upside down: a new appraisal of the ecology of the Permian Capitan Reef, Guadalupe Mountains, Texas and New Mexico: PALAIOS, v. 9, p. 422-427.

Woronick, R. E., and L. S. Land, 1985, Late burial diagenesis, Lower Cretaceous Pearsall and lower Glen Rose Formations, south Texas, *in* N. Schneidermann, and P. M. Harris, eds., Carbonate Cements, Tulsa, OK, SEPM Special Publication No. 36, p. 265-275.

Wright, V. P., 1991, Paleokarst: types, recognition, controls and associations, *in* V. P. Wright, M. Esteban, and P. L. Smart, eds., Paleokarsts and Paleokarstic Reservoirs: P.R.I.S. Occasional Publication Series No. 2, Reading, University of Reading, p. 56-88.

Zankl, H., 1993, The origin of high-Mg-calcite microbialites in cryptic habitats of Caribbean coral reefs-their dependence on light and turbulence: Facies, v. 29, p. 55-60.

Zenger, D. H., and J. B. Dunham, 1980, Concepts and models of dolomitization—an introduction, *in* D. H. Zenger, J. B. Dunham, and R. L. Ethington, eds., Concepts and Models of Dolomitization, Tulsa, OK, SEPM Special Publication No. 28, p. 1-10.

INDEX

Symbols

∂ notation 80
3-D diagenetic model 377
3-D distribution of reservoir quality.
　Tertiary, Philippines 379
87Rb 85
87Sr 85

A

abiotic 95
　marine cementation 95, 126
　　characteristics 97
　　distribution 98
　　mineralogy 97
　　porosity destruction 376
　　rates 97
　　recognition 98
　　trace elements 97
Abo/Clear Fork shelves
　Central Basin Platform 276
Abu Dhabi 150
abundance modifiers
　Choquette and Pray 42
accommodation space 22
　HST 260
　in a retrogradatinal sequence set 270
accretionary 116
accumulation rates of ancient carbonate platforms 12
active margin 296
　basin hydrology 296
　　porosity evolution/diagenesis 318
adsorbed cations 71
aeolian processes 29
aggradational platform margins 10
aggradational sequence sets 269
aggressive subsurface fluids 54
Alacran Reef, Mexico 5
Alberta, Canada 177
Albian/Aptian 15
algal mud mounds 14
Alleghenian/Ouchita Orogeny 318
allochthonous
　chalk

　　effect on porosity 316
alluvial base level 21
alluvial processes 27
amplitudes of glacio-eustatic sea level changes 273
anaerobic bottom environments 26
Andrews Field, west Texas, USA 276
Andros Bank 33
Andros Island 151
Angola offshore Cretaceous reservoirs 283
　geologic setting 283
anhydrite 150, 154, 156, 171, 177
　dissolution 47
　evaporative marine 150, 156
　sabkha 150
anoxic basin deposits
　source rock 358
Anti-Atlas Mountains, Morocco and Algeria 133
antiform "turtle-back" structures 283
Antler Orogeny 326, 342
Aptian Cow Creek Limestone, Texas, USA
　ancient ramp 16
aquaclude
　gypsum under evaporative conditions 146
Arab Formation, Middle East 180
Arabian Gulf 150
aragonite 53, 63, 93
　dissolution 137, 206, 271, 366
　　microscale 192
　　moldic porosity 192
　dissolution in the meteoric environment
　　fueling calcite precipitation 190, 215
　evaporative marine water 146
　marine cement
　　recognition in ancient rocks 100
　undersaturated
　　thermocline 93
aragonite botryoidal cements 114
aragonite cements
　fibrous crust or mesh
　　massive botryoids 97
aragonite compensation depth 112
aragonite lysocline 137, 251, 257, 259, 263
aragonite needles 3
　effect on porosity 50

electrical bi-polarity 50
felted framework 49
aragonite relicts in calcite cements 100
aragonite speleothems 186
aragonitic seas
  icehouse periods 62
Archie 43
  porosity classification 38
architecture of sedimentary sequences
  influence of climate 272
arid climate 283
asymmetrical cycles 10
asymmetrical parasequences 34
Atascosa Formation, Texas, USA 118
Atlantic rifting 281
atoll 16
atoll central lagoon 16
atomic number 79
auto cyclicity 26, 266

B

backreef 136
backstepping 29
bacteria 3
  role in diagenesis 3
bafflers 114
Bahama Platform 56, 139, 151, 195, 382
  geologic history 210
  marine hardgrounds 104
  regional meteoric water system
    lowstand 209
Barbados 188, 196, 214, 382
  geology 214
  hydrologic setting 214
  mixing zone dolomitization 235
    dolomite cements 237
barrier beach/tidal channel complexes 16
barrier reef 13
basic accommodation model 20
basic porosity types
  Choquette and Pray 42
basin 14
basin floor fans 14, 27
Bathurst, Robin 381
bathymetric highs 16
beach progradation
  as a factor in marine cementation 103
beach rock 67, 96, 102, 255, 265

cement mineralogy 104
cements 104
beach 14
Belize 108
Belize. 151
Bermuda 108, 195, 382
  floating fresh water lens 202
  geology 202
  hydrology 203
  water geochemistry relative to carbonates 203
Big Snowy Group, Wyoming, Montana, USA 344
Bimini Bank 32
binding 3
bioclastic wackestones 16
bioclasts 39
bioeroded cavities 4
bioeroders 4
  boring 4
  grazing 4
  rasping 4
bioerosion 109
biohermal mud mounds 13
biologic production 9
  excess
    transport into deep water 9
biological sediment origin 1
biologically mediated marine carbonate cementation 101
bittern salts 145
bitumen porosity destruction 368
bladed calcite 68
Bliss Fm. west Texas 159
Bonaire 170
Bonaire, Netherland West Indies 57
boring organisms 102, 109
  Clionid sponges 102, 109, 110
borings
  Clionids
    pelecypods 51
botryoidal aragonite 108, 112
boundstone 8
breccia porosity 47
brine 63
brine mixing
  Permian Basin, USA 334
bryozoans 114
Buckner Formation, Arkansas, USA 170
buoyancy effect of seawater 29

burial diagenesis 291, 383
  pressure 291
burial history 52, 122
  Devonian Leduc, western Canada 326
  Jurassic Smackover Fm. 367
  Knox Group 318
  Madison Fm., Wyoming, USA 350
  Stuart City, Texas, USA 122
by-pass escarpment margin 29

C

C-axis direction
  crystal elongation
    marine environment 65
calcareous phytoplankton 109
calcian dolomites 147
calcite 93
calcite cement 63, 65
  distribution 68
    microstalictitic cement 69
  distribution patterns
    meniscus cement 69
  meteoric 220
    seismic imaging 376
  morphology
    control 65
  morphology marine 65
  morphology-meteoric 65
  source 222
  subsurface 307
    geochemistry 310
    impact on reservoir porosity 312
    petrography 308
    source 308
  timing of cement 70
calcite cements and dolomites
  trace element geochemistry 71, 218
    distribution coefficient 72
    Fe and Mn 78
    Sr and Mg 74
calcite compensation depth. 112
calcite crystallography 67
calcite lysocline 68, 138
calcite seas 100
  greenhous periods 62
calcite solubility curve 187
calcite's orthorhombic polymorph 34
caliche 199

crusts
  impermeable 255, 281
  pisolite 199
capillarity 38, 68
carbon
  marine origin
    isotopic signature 83
carbon mass in the various natural reservoirs 83
carbonate factory 29, 34
carbonate minerals 34
  controls 35
  controls over
    carbon dioxide 36
  kinetics 146
    dolomite 146
    mineral stabilization in the meteoric environment 190
    rate steps 191
  long term changes 271
    affect on diagenesis/porosity 271
  relative stability 34
    aragonite 35
    calcite 34
    dolomite 34
  saturation state
    evaporative waters 146
    marine waters 93, 139
    meteoric waters 186
carbonate mud 3
carbonate platform 12
  types 13
    eperic platform 13
    isolated platform 13
    ramp 13
    rimmed shelf 13
carbonate porosity 37
  classification 37
    Archie 38
    Choquette and Pray 39
    fabric selective 39
    Lucia 43
    petrophysical 44
  comparison with siliciclastics 37
  eogenetic stage 42
  genetic modifiers 43
  mesogenetic stage 42
  meteoric diagenetic system 212
    evolution 220

regional meteoric aquifer 215, 366
moldic 257
pore size distribution 44
   permeability 44
porosity/permeability fields
   particle size classes 46
primary 41, 216, 270
   depositional porosity of muddy sediments 49
   framework and fenestral 50
   intergrain 48
   intragrain 49
   preserved 315
secondary 41, 52, 297, 332
   assocaited with dolomitization 52
   associated with dolomitization 52
   breccia and fracture porosity 58
vuggy 43, 257
carbonate prodction in temperate waters 2
Carbonate production in tropical water 2
carbonate rock classification 6
   Dunham 8
   Folk 6
carbonate sediments 1
   grain composition 5
   origin 1
      biological 1
      chemical 3
   production 9
      high stand shedding 10
      law of sigmoidal growth 10
   texture and fabric 5
      textural inversion 5
carbonate starved 29
carbonate-secreting organisms 3
   ability to modify environment 3
   binding 3
   encrusting 3
   framebuilding 3
Caribbean
   intertidal marine cementation
      thickness of zones 103
Carlsbad New Mexico 165
Carrie Bow Key, Belize
   Belize barrier reef
      cementation gradient 109
catch-up phase. 10
cathodoluminesence 218, 219
   dolomites

Madison Fm., Wyoming, USA 354
   marine cements 97
cation substitution 65
cavernous porosity 47
caverns
   Choquette and Pray 54
Cedar Creek Anticline, Montana, USA 153
celestite 59
cement
   internal sediment 4
cement distributional patterns 68
Central Basin Platform, west Texas, USA 167, 275
Central Montana Uplift, USA 177
chalk 14
   burial curve 315
   mass-movement
      effect on porosity 316
chalk 313
channel cutouts 14
channels
   Choquette and Pray 54
chemical compaction 269, 300, 315
   factors affecting efficiency 305
   porosity loss as a result of 301, 337
      Jurassic Smackover Fm. 368
chemical precipitate 3
chemoautotrophic bacteria 133
chemosynthetic biota 134
chert 14
chlorinity
   effect on anhydrite precipitation 150
chronostratigraphy 20
climate 29
   effect on meteoric diagenesis 192, 209, 213
      arid 193, 248
      karst 193, 222
   ice house vs. green house 270
      diagenetic trends 273
clinoforms 27
Clionid
   chips 109
closed framework reef
   coralline algae, sponges, stromatoporoids 51
closed system
   transformation of metastable to stable mineralogy 100
CO_2
   dissolution events 52

meteoric access 185
CO_2 degassing 59, 98, 102, 105
   reef-related marine cementation 106
CO_3 ion
   control over saturation 68
coalescence isolated platforms 33
coastal dune complexes 27
coastal onlap 26
coastal salina 175, 180, 181
coccoliths 109
compaction 42
   mechanical 297
      experiments 298
      pelagic oozes 298
      porosity loss 298, 316
compaction events
   timing of 70
complex polyhedral calcite cement 68
concepts unique to the carbonate regimen, 1
condensation 79
confined regional aquifer 196, 255
continent-attached platforms 30
continental crust
   Sr isotope end member source 86
continental meteoric waters 145
continental rift basins 356
continental weathering
   control over Sr isotope source 86
Cooking Lake platform, western Canada 123
coral polyp
   porosity 49
coralline algae 51
Coulommes Field, Paris Basin, France
   de-dolomite 57
covariation of trace elements and isotopes 78
Cretaceous Golden Lane, Mexico 115
crosscutting relationships 70
cryptic calcareous sponge fauna 114
crystal dislocations
   effect on fluid inclusions 88
crystal growth patterns 68
crystal lattice
   effect on distribution coefficientient 74
crystal morphology 65
cyanobacteria 101
cycle terminus 33
   high-energy shoreface 34
   low energy tidal flat 34

   stacked marine shoaling upward paarasequence 34

D

de-watering 297
   compactional 320
dedolomitization 332
deep basin brines 334
deep fore reef 108
deep marine 128
deep marine oozes
   porosity of 50
deep ramp
   algal mud mounds 16
Delaware Basin, west Texas, USA 159
density stratification 26, 164, 359
depositional environments 12
depositional particles 39
deterministic reservoir model
   used to determine expected hydrocarbon reserves
      Tertiary, Philippines 378
Devonian carbonates, western Canada
   dolomitization 325
Devonian Reef Complexes, western Canada 123
   porosity evolution 123
   Presqu'ile reef complex, Canada
      Elk Point Basin, Canada 177
diagenesis/porosity models 246
   HST 259
      isolated platform 265
      ramp 259
      ramp-arid 261
      rimmed shelf 263
      rimmed shelf-arid 263
   LST 246
      isolated platform 251
      isolated platform-arid 253
      ramp 246
      ramp-arid conditions 248
      steep-rimmed carbonate shelf 249
      steep-rimmed carbonate shelf-arid 251
   TST 253
      ramp 253
      ramp-arid 255
      rimmed isolated platform 257
      rimmed isolated platform-arid 259
      rimmed shelf 255
      rimmed shelf-arid 257
diagenetic environments 42, 61

calcite cements
   morphology 68
   marine 62
   meteoric 63
   subsurface 63
   tools for recognition in the rock record 64, 383
      petrography-cement morphology 65
diagenetic terrain 160
   interior sabkha 153
diagenetic/porosity history,
   western Canada reef complexes 125
differential compaction 59
diffusive flow 195
dilute waters 63
dissolution of limestones 52
distally steepened ramp 26
distribution coefficient 71
dolomite 93, 383
   association with evaporites 145
      crystal sizes 161
   association with reflux
      crystal sizes 175
   dissolution 47
      porosity 57
   recrystallization 147, 328
      meteoric water 153, 162, 353
      subsurface 319
   synthesis 73
   volume in sedimentary record 272
dolomite cementation 170, 319
   subsurface 326
dolomite crystal size 44
dolomite distribution
   through time 152
dolomite enriched in CaCO3, 35
dolomite ideal composition 34
dolomitization 56, 126, 138
   deep marine 138
   effect on rock fabric 46
   evaporative reflux
      140, 158, 164, 166, 169, 171, 176, 226, 261, 269, 351, 383
      criteria for recognition 174
      Jurassic Smackover 366
      porosity enhancement 257
   late subsurface 319, 325
   marine sabkha 156
   marine waters 138, 140, 178, 251, 383
   mixed meteoric-marine water 232, 235, 286
      attributes 237
      porosity 239
      recognition in the record 233
      reservoir 238
      validity of model 232, 233
      where do we stand? 242
   mole for mole 56, 173, 383
   porosity enhancement 57, 173
      Jurassic Smackover Fm. 367
   porosity impact
      Devonian carbonates, western Canada 329
   porosity loss 56, 57, 384
   sabkha
      Abu Dhabi 150
   through time 152
dorag model of dolomitization 187, 232
   validity of model 232, 233, 383
down-lap 24
downlap surface 359
drowning unconformity 12
   causes
      environmental deterioration 12
      Tertiary, Philippines 375
ductile fine-grained siliciclastics 59
Dunham rock classification 8
   boundstone 8
   energy continuum 8
   grainstone 8
   mudstone 8
   packstone 8
   wackestone 8
Duperow Formation, western Canada 123

E

Early Devonian "Kess Kess" mounds in Morocco 133
east Texas and north Louisiana salt basins 357
Eh and pH
   control over incorporation of Fe and Mn in carbonates 75
Ekofisk Field, North Sea
   chalk reservoir 313
El Abra Formation, Mexico 118
Elk Point Basin, Canada 177
encrusting 3
endolithic algae
   associated with hardgrounds 106
   in beach rock 102
energy continuum 8

Enewetak Atoll, western Pacific 112, 136, 188
  Sr isotope stratigraphy 87
environmental parameters for carbonate production 2
  temperature 2
  presence absence of siliciclastics 2
  salinity 2
  substrate 2
environmental reconstruction 6
eogenetic zone 42
eperic platform. 13
epikarst 211, 246
epilogue 381
epitaxial overgrowths
  fibrous aragonite
    on aragonitic reef frame 108
equant crystal shapes 65
erosion 23
escarpment shelf margins 136
eustatic cycles, 21
eustatic sea level 21
evaporation 79
  effect on isotopic composition 190
evaporative brines 145, 152
evaporative drawdown 173, 178, 179
evaporative lagoon 162, 170, 180, 257
  density gradients 162
  dolomite
    porosity/permeability 169
  Jurassic Buckner Fm., Arkansas, Texas, USA 366
  Madison Fm., Wyoming, USA 346
  modeling studies 164
  Permian South Cowden Field, Central Basin
    Platform, Texas, USA 166
  Permian west Texas
    reflux dolomitization 163
  porosity destruction
    seal 265
  sill or barrier 162, 164, 168, 171, 176, 177
  Upper Permian Guadalupian, west Texas 164
evaporative lagoons 156
evaporative marine 42, 145
evaporative marine environment 145
evaporative mineral series 146
evaporite basin 175, 177
evaporite basin inlet 163
evaporite solution collapse
  breccias
    reservoirs 58

evaporite solution collapse breccias 346
evaporites 145
  gypsum 146, 164
  subaqueous 178
exploration and production constraints and strategies
  Jurassic Smackover Fm. 368
extensional basins 269

F

fabric 39
fabric selectivity 37, 39
  factors to determine 39
fabric-selective
  mineralogical control 53
facies model 13
  isolated platform 16
  ramp 15
  rimmed carbonate shelf 13
facies shift 23
failed rift systems 12
fair weather wave base. 16
falling inflection point 21
falling limb 21
fascicular-optic calcite 111
Fe 75
fenestral porosity 47
fermentation 83
fibrous crystal shapes 65
fibrous-to-bladed circumgranular crusts 97
Fisher plot 276
flank margin cavernous porosity 251
floating fresh water lenses 141, 195, 255, 261
  Bahama Platform 210
  effectiveness of diagenetic processes 200
  porosity preservation by early mineral stabilizati
    207
  Schooner Cay, Bahamas 201
  size control 200, 205
    predictability 209
flood recharge
  sabkha 149
Florida 151, 170
fluid conduits
  Devonian, western Canada 328
  Knox Group 323
fluid inclusions 59, 88
  constraints 89
  freezing behavior 89

homogenization temperature 88
fluorite 59
Folk, Bob 381
Folk rock classification 6
foraminifera 5
fore-reef 136
fore-reef debris 123
foreland basins 12
fractionation factor
   dolomite 81
fracture 47
   permeability enhancement 160
      Ekofisk Field 315
   width
      permeability control 48
fracture fills 59
fracture porosity
   vs fracture permeability 48
fractures 59, 249, 297, 375, 377
   solution enhanced 249
framebuilding 3
framework organisms 4
   attached 4
   branching metazoic 4
   massive 4
framework porosity 37, 50
freezing behavior
   of fluids in inclusions 89
freshwater flushing
   porosity enhancement 158
fringing reef 13
fungi
   increase in porosity 49

G

galena 59
Gaschsaran Field, Iran
   fractured reservoir 59
generation of new accommodation space 10
genetic modifiers
   Choquette and Pray 42
genetic packages 20
genetic porosity classification
   Choquette and Pray 37
geochemical modeling study
   Madison aquifer 332
geochemical trends 217
   regional meteoric aquifer system 217, 255

geochemistry
   marine cements 97
geopetal fabrics 109
geopressure 293
geothermal gradients. 293, 307
Ghawar Field, Middle East 180
Ginsburg, Bob 381
glacial episodes 21
glacio-eustasy 26
glauconite
   associated with marine hardgrounds 106
Glorietta Formation, west Texas, USA 170
golden age of carbonates. 382
"Golden Lane", Mexico 15
graben fault system
   central Gulf Coast, USA 357
grain composition 5
grain contacts
   sutured 300
grain density 70
grain interpenetration 70
grain packing 300
   effect on porosity 301
   orthorhombic 301
grain ultrastructure 192
grain versus matrix support
   Dunham carbonate rock classification 8
grain-supported sediments
   compaction 300
grainstone 8, 168
gravity processes 29
gravity-driven unconfined aquifers 196
   lowstand 213, 246
Grayburg Fm. west Texas, USA 167
Great Barrier Reef, Pacific 14
greenhouse times 209, 270, 271
   diagenetic activity 213
   high frequency cycles 272
   reservoirs 283
growth rates of modern carbonate producers 12
Guadalupe Mountains, Texas and New Mexico USA 113
Gulf of Mexico, USA
   rifting 356
gypsum
   dissolution
      driving calcite cementation 332

H

H2S
  Madison Fm.
    Madden Field, Wyoming, USA 341
Halimeda 138
halite 156, 171, 177, 178, 179
  sabkha 150
halokinesis 283
  Gulf Coast, USA 357
  North Sea
    central graben 314
hermatypic corals 2
heterogeneity of reservoir porosity/permeability 34
hierarchy of stratigraphic cycles 26
  super sequence 182
high frequency cycles 33, 182, 283
  origin 33
    auto cycles 33
    Milankovitch cyclicity 33
  porosity control 278
high-energy grain shoals. 13
high-energy shoreline complex 15
high-resolution stratigraphy
  use of Sr isotopes 86
highstand sediment shedding 10
highstand systems tract (HST) 25, 170
homoclinal ramp 26
homogenization temperature
  two-phase fluid inclusions 88
horst blocks 14
  control of isolated platforms 30
humid supratidal 151
hydrates
  Mg hydrates effect on dolomitization 146
hydraulic fracturing
  associated with overpressured zones 59
hydrocarbon maturation
  carbonate dissolution 312
hydrocarbon migration
  early
    effect on porosity preservation 315
  Knox Group 321
  Madison Fm., Wyoming, USA 351
hydrocarbon source rock 26
hydrologic zones 194
hydrostatic pressure 293, 315
hydrothermal and cold hydrocarbon vents 133
hydrothermal flow regimes
  source for dolomitizing fluids 325
hydrothermal ore mineralization 225
hypersaline 14
hypersalinity 26

I

ice caps 26
ice-free 26
icehouse intervals 209, 270, 275
  high frequency cycles 273
  porosity evolution
    Upper Paleozoic reservoirs 275
ideal dolomite 35
immature meteoric phase 195
in-situ sediment production 26
incisement 27
inclusions 71
incongruent dissolution
  magnesian calcite 53
inner shallow ramp 16
insoluble residue 316
integration. 384
interconnected pore system 43
intercrystal porosity 43
intercrystalline porosity 52
  dolomite 55
interdependency of sediment texture and energy 6
interglacial episodes 21
intergrain porosity
  modern sediments 48
interior salt basin
  Gulf of Mexico, USA 357
internal sediment 4
internal sediments 52
  associated with marine hardgrounds 106
interparticle porosity
  Lucia carbonate porosity classification 43
interrhomb calcite
  residual calcite after dolomitization 55
intertidal
  abiotic
    marine cementation 102
intertidal zone 102
  marine cementation
    beach rock 63
intrafossil porosity
  Lucia carbonate porosity classification 47

intragrain porosity
   modern sediments 49
Ireton Shale, Devonian western Canada 123
isolated oceanic platforms 32
isolated platform 13
isotopes 78, 79
isotopic compostion of water
   rock buffering 188
isotopic fractionation 85
   temperature dependency 80
isotopic re-equilibration 82
isotopic sampling 84

J

J-shaped curve
   isotopic signature of meteoric diagenesis 83
Jamaica 108
   reef rock 97
James, Noel 382
joint systems 41
Jurassic Great Oolite, England
   reservoir
      dedolomite 57
Jurassic Hith Formation 179
   evaporite 182
Jurassic Kimmeridgian shales
   source rock
      Ekofisk Field 315
Jurassic Smackover Formation 170
Jurassic Walker Creek Field
   ancient beach rock 104

K

karst 123, 222, 263
   cavernous porosity 59, 226
   collapse breccias 346
   features 223
   mature 223, 246
   porosity 225
keep-up phase 10
kinetics
   marine cementation 98
Kohout (thermal) convection 139
Kwanza Basin, Angola 281

L

lacolith intrusion 134

lag phase 10
lagoon/sabkha complex 26
Laramide tectonism
   thrust faulting 342
Laser Raman Microprobe
   gross chemical composition of fluids in inclusions 89
lateral seal 113
latitudinal control over the distribution of carbo 2
lattice distortion
   crystal morphology control 65
lattice geometry 34
Leduc reefs, western Canada 123
limestone solution collapse
   breccia 58
      reservoirs 58
Lisbon field, Utah, USA 59
lithophagus pelycepods 110
lithosphere 83
lithostatic load 300
lithostatic pressure 315
lithostratigraphic units 358
lithostratigraphy 20
   Jurassic Smackover Fm.
      comparison with chronostratigraphy 364
Little Knife Field. Williston Basin USA 156
local datum 21
long term cycles, termed 1st and 2nd order 26
Lower Congo Basin 283
Lower Cretaceous 14
Lower Cretaceous Edwards Formation, U.S. Gulf Coast
   Sr isotopes 87
lower slope 14
lowstand
   diagenesis/porosity evolution 246
lowstand prograding complex 248
   diagenetic processes 248
      porosity evolution 248, 249
lowstand shedding 10
lowstand system tract (LST) 25
Lucia, Jerry 382
Lysite Mountain, Wyoming, USA 343

M

MacLeod salt basin, Australia 175
   similarity to Jurassic Arab sequences 182
Madden Field, Wyoming, USA

Madison Fm. 341
Madison Formation, Wyoming, USA
   giant, super deep gas field
      dolomitized reservoir 341
magnesian calcite 34, 53, 93
   conversion to calcite 206
   peloidal aggregates
      in reef rock 108
   reef cements 108
   transformation to calcite 100
magnesium substitution 34
manganese
   associated with marine hardgrounds 106
Manitoba, Canada 177
marine cement
   polygonal suture pattern 97
   reef related
      magnesian calcite 108
marine cementation 63
   factors favoring 97
   porosity destruction in shelf margin reefs 125
marine chemical precipitates 35
marine diagenetic environment 61, 93
marine flooding surface 34
marine hardgrounds 96, 123, 265
   cement mineralogy
      aragonite, magnesian calcite 104
matrix porosity 38
mature hydrologic system 213
   Bahama Platform 209
mature phase 195
maximum flooding surface 26, 255, 359
   Jurassic Smackover Fm. 359
Mayan Mountains 14
mechanical compaction 315
mega-breccia debris flows 249
megabreccia debris flows 14
megapores 43
meniscus cement
   vadose zone 69
mercury displacement pressure
   as a measure of average particle size 45
mesogenetic zone 42, 54
metastable mineral suite 53
metastable phases
   modern marine cements 98
meteoric calcite line 190
meteoric diagenetic environment 61, 63

meteoric phreatic 42
meteoric vadose 42
meteoric water 173, 185
   geochemistry 185
methane 83, 101
   oxidation 101, 130, 133
methane generation
   biochemical fermentation
      thermochemical 83
methane oxidation 134
methane seeps
   microbial diagenesis 101
Mexico 118, 170
Mg poisoning 67
Mg/Ca
   control on dolomitization 146
Mg/Ca ratio 65, 149, 164
Michigan Basin 178
   barred basin 178
micrite 101
   original mineralogy
      calcite 271
   pelleted 101
      reef 108
micro-fractures
   in fluid inclusions 88
microbial 101, 383
   cementation 101, 108
      beach rock 102
   crusts 113, 114, 129
   mud mounds 128, 261
microbial mats 114
microbial mediation 73
microbial ooze 101
microboring algae
   relationship to porosity 49
microcodium 199
microdolomite 100
microfacies 6
   chalk
      porosity control 316
micropores 43
microporosity 47, 272
microstalictitic cements 69
Midale beds of the Charles Formation, Canada
   relationship between % dolomite and porosity 55
Midland Basin, west Texas, USA 167
Milankovitch cyclicity 33

mineral equilibrium conditions 190
mineral solubility 35
mineral stabilization 36, 41
mineralogical drive in diagenesis 190
    Barbados 215
    meteoric system
        role of aragonite 190
mineralogical evolution of the major groups of car 35
minimum temperature of formation
    fluid inclusion 88
Mississippi Valley type ore deposits 297
    Knox Group 321
    Pine Point, Canada 177, 326
Mississippian Lake Valley, New Mexico, USA 219
Mississippian Madison Fm., Wyoming, USA
    giant super deep gas reservoir
        dolomite 341
Mississippian Mission Canyon Formation 154
Mississippian North Bridgeport Field, USA 238
mixed carbonate-siliciclastics 283
mixed meteoric/marine zones 190, 194
mixing zone 141
    diagenetic processes 201, 247
        reflux of marine waters 249, 253
Mn 75
mobile dune fields 27
molar ratios 34
moldic porosity 54, 138, 286, 287
    associated with dolomites 153
Monterrey Shale, at West Cat Canyon Field, California, USA
    fractured reservoir 59
morphology of calcite cement 65
mud mound 114
mud mounds 113, 128
    Cretaceous of Colorado, USA 135
    diagenesis 131
    Gulf of Mexico 134
    North Africa 133
mud-bearing sediments
    porosity 49
mudstone 8
multivalent cations
    Fe and Mn 75
Mururoa Atoll, Pacific 108
Muskeg evaporites, western Canada 177

N

nannobacteria 101
neural network
    trained to recognize reservoir rock types
        Tertiary, Philippines 378
neutrons 79
new accommodation space 29
Nisku shelf, Alberta, Canada 326
nitrate reduction 101
non-deposition 23
non-ideal dolomites 35
non-skeletal grains
    pellets
        ooids 41
normal marine diagenetic environments 93
Norphlet Formation
    Jurassic continental siliciclastics
        LST 357
North American Craton 342
North Dakota 156
North Sea Basin 313
Norwegian and Danish sectors, North Sea 313
not fabric selective porosity 41
not-fabric selective dissolution 54
Nuclear Magnetic Resonance
    gross chemical composition of fluids in inclusions 89
nuclei
    calcite/dolomite cementation 68

O

oceanic crust
    Sr isotope end member source 86
oceanic hydrothermal processes
    control over Sr isotope source 86
oceanic ridge volume 21
oceanic volcanism 16
oceanic water volume 21
offlapping coastal zone 26
oil field brines 75
    Sr isotopes 87
oil field waters 295
    chemical composition 295
oil migration into reservoir
    porosity preservation 315
Oligocene Asmari Limestone
    fractured reservoirs 59

on-lap 24
oncoids 5, 284
onlap 26, 167
ooid grainstones 170, 283
   reservoir 365
ooid shoals 364
ooids 39, 100
   compaction
      porosity loss 303
   mineralogy 100
      calcite 271
      greenhouse times 271
      Jurassic 366
      magnesian calcite 208
      through time 100, 112
   reservoir 170, 238
      Oaks Field, Louisiana, USA 208
oomoldic porosity
   Jurassic 173, 366
open framework reef
   scleractinian coral 51
Ordovician Ellenburger 158
Ordovician Knox Dolomite of the southern Appalachians, USA 318
Ordovician Red River Fm. Montana, USA 153
   Cabin Creek Field, Williston Basin 153
organic acids 294
organic carbon
   isotopic composition of 83
organic material preservation 26
organic-rich lake deposits
   Angola
      Cretaceous source rock 281
outcrop reservoir surrogate 342, 375
overpressured zones
   hydraulic fracturing 59
   porosity preservation 314
oversteepning 29
oxidation of organic material 83
oxygen isotopes
   latitudinal effect 188

P

packstone 8,
packstones 50, 136
   porosity in 50
paleoequator
   Mississippian, Wyoming 341

Pennsylvanian/Permian 276
paleogeography
   Mississippian
      central Rocky Mountains 342
paleokarsts 223, 269
   reservoirs 223, 225
paleosols 210
   assocaited with sequence bounding unconformity 360
paleowater depth 21
Paleozoic carbonates of the Greater Permian Basin, Texas, USA 332
palm hammocks
   Andros Island
      dolomitization 151
Panama 108
paragenetic sequence 64
   Jurassic Smackover Fm. 368
   Knox Group 318
   Madison Fm., Wyoming, USA 351
   Tertiary Malimpaya-Camago buildup, Philippines 376
   Upper Devonian Leduc, western Canada 325
parasequence 265
   diagenesis and porosity 265
   thickness 34
parasequence stacking pattern 34
   cumulative diagenesis/porosity evolution 266
      early highstand 267
      lowstand 267
      transgressive systems tract 267
particle shape 48
particle-size boundaries
   Lucia carbonate porosity classification 45
passive margins 269
   basin hydrology 296
      diagenesis/porosity 297
passive nucleation on bacterial cells 101
patch reef 13
PCO2
   control on dolomitization 152
   de-gassing 186
PDB 80
Peace River Arch 177
Peace River Arch, western Canada 123
pelagic foraminifera
   effect on porosity 50
pelagic oozes 50

pellet 8
pelleted 108
   micrite 128
pelleted micrite
   in beach rock 102
pelleted micrite crusts 97
pelloids 101
peloidal wackestones 16
pendant cryptobionts 114
Pennsylvanian-Permian Andrews Field, Central Basin
      Platform, Texas, USA
   geologic setting 276
peri-platform oozes 138
permeability 44, 47, 48, 63
   low 315, 322
   relationship of grain support, mud and porosity 50
Permian Capitan reef complex 113
   framework reef 114
persaturated wit 93
Persian Gulf
   marine hardgrounds 104
   sabkhas 152
petrographic relationships
   grain to cement 71
petroleum system
   Upper Jurassic Smackover Fm. 369
petrophysical and rock-fabric classes
   classes 1-3 46
petrophysical parameters
   porosity/permeability
      saturation 37
phase changes 294
phase transformations 79
Philippines
   Palawan Island
      Malampaya-Camago Field 373
photosynthesis 79
   effect on geochemical behavior of carbon 83
phreatic 63, 194
   diagenetic processes 200, 246, 251
phreatic caves 251
Pine Point, Alberta, Canada
   MVT 326
pinnacle geometry 32
pinnacle reefs 177, 179
   Devonian, western Canada 326
pisolite 199
pit caves 210

planktonic foraminifera 109
planktonic mollusks 109
platform drowning 10
platform margin progradation 10
playa 182
poorly ordered dolomite 147
population dynamics 5
pore fluid pressure
   affect on chemical compaction 306
pore fluid pressure 315
pore size 38
pore systems 37
pore volume 52
pore-type terminology
   comparison 43
porosity
   trends
      down hydrologic flow 255
porosity and permeability,
   increase as a result of dolomitization 56
porosity classifications
   comparison 43
porosity designation 43
porosity destruction
   by dolomitization 55
   marine cements 62
   pressure solution 63
porosity evolution in the subsurface 383
porosity-depth relationships 306
porosity-depth trends 335
porosity/permeability
   Jurassic Smackover Fm. 368
   Madison Fm., Wyoming, USA
      surface vs. subsurface 350
porosity/permeability fields
   Lucia carbonate porosity classification 46
post tectonic
   basin hydrology 296
   diagenetic regimen 329
post-gypsum brine reflux 57
post-orogenic uplift 52
post-rift 269
precipitation rate
   effect on trace element distribution coefficients 73
predicting changes in porosity with depth 335
predictive conceptual models 64
preferential dolomitization
   muddy matrix 153

presence or absence of organic binding
   Dunham carbonate rock classification 8
preserved primary porosity 126
pressure
   hydrostatic
      lithostatic 292
pressure correction
   two phase fluid inclusions 89
pressure solution 63, 125, 300
   associated cementation 301
   cement volume 301
   doner and receptor beds 305
primary growth cavities 4
primary inclusions 89
primary intergranular porosity or 39
primary intragranular porosity 39
primary porosity
   predepositional
      depositional 41
   preservation
      in chalk 315
production reservoir characterization 39
progradation 16
progradational 116
   shoreline
      Jurassic Oaks Field, Louisiana USA 206
      Jurassic Smackover Fm. 370
progradational parasequence stacking pattern 34
progradational sequence set
   diagenesis and porosity evolution 268
progradational shelf margin 29, 167, 276
prograding tidal flats 29
progressive mineral stabilization 36
protodolomite 82, 147, 151
Proton Microprobe
   gross chemical composition of fluids in inclusions 89
protons 79
Puckett Field, west Texas, USA 59
pycnocline 162

Q

Qatif Field, Saudi Arabia 181
Quintana Roo, Mexico 195
   strand plain 205

R

radiaxial fibrous calcite
   97, 111, 114, 122, 125, 130, 137
   below aragonite lysocline
      Enewetak Atoll 138
radioisotopes 79
Rainbow reefs, western Canada 123
ramp 113, 282, 284
rapid precipitation 35
rate of diffusion
   thermal history
      porosity destruction 339
rates of relative sea level rise 12
reciprocal sedimentation 30
recrystallization 39
   protodolomite 82
Red Sea 108
reef 3
   -related reservoir
      Malampaya-Camago, Philippines 373
   as a lagoonal sill 164
   cement distribution 108
   definition 3
   diagenetic setting 106
   exploration strategy 127
   framework
      shelter voids 109
   Golden Lane 115
   internal sediments 109
   organism-sediment mosaic 3
   porosity
      destruction by marine cementation 263
      preservation 256
   recognition in the geological record 110
   reservoir 179
   shelter voids 109
   Stuart City, Texas, USA 115
   unique depositional environment 3
reef ancillary organisms 4
   green algae
      Halimeda 4
reef crest 109, 136
reef frame
   corals and red algae 4
reef organisms
   evolution 4
   rudists 4

stromatoporoids 4
reef rock 4
reef-bound isolated platform
   Tertiary, Philippines 375
reef-forming organisms
   periods when diminished 4
reflection seismic 376
reflux dolomitization 57
regional meteoric aquifer 219
regional unconformities 223
relative sea level 21
relative sea level curve 24
relict calcite 57
relict topography 28
remobilization of dolomite 57
replacive calcite mosaic 100
reservoir
   compartmentilization 104, 125
reservoir model
   3-D 231, 384
      Malampaya-Camago Field, Philippines 373
reservoir overpressuring
   primary porosity preservation 315
reservoir rock types
   based on petrophysics 377
reservoir seal 145, 182
   gypsum 146
   shale
      Ekofisk Field 314
retrogradational parasequence stacking pattern 34, 287
retrogradational sequence set 269, 287
   hydrocarbon potential 270
retrograde diagenesis
   anhydrite to gypsum
      interior sabkha 150
rhizoids 199
rhodoliths 5, 286
rhombohedral crystal system 34
rift
   SE Asia 373
   western Africa 373
rifting tectonic setting 283
Rimbey-Meadowbrook trend, southern Alberta, Canada 326
rimmed shelf 13
rising inflection point 21
rising limb. 21

rock bulk volume 301
rock fabric/petrophysical porosity classification
   Lucia carbonate rock classification 39
rock fabrics 38
rock-buffering
   Sr isotopes in meteoric environment 87
rock-water interaction 54, 83
   control on final isotopic composition 188
Rocky Mountains 342
roundness
   controlled by shape of organisms 5
rudist
   reef complex 121

S

sabkha 14, 148
   continental clastics 150
   hydrological system 152
   meteoric water influence 153
      dolomite recrystallization 162
   pore fluids
      precipitation of gypsum 149
   recognition criteria 161
   reservoirs 153, 154, 158
      porosity 154, 156, 160–183
saddle dolomite cements 368
salina 182, 283
salinity
   control on dolomitization 146
salt cored anticlines 357
salt dome movement, 59
San Andres Formation, west Texas, USA 170
sandstone compositional classes 6
Sarawak
   Miocene isolated platforms 16
Sarg, Rick 382
Saskatchewan, Canada 177
saturated
   with respect to most carbonate minerals 63
saturation 44
Saudi Arabia 179
Schlager, Wolfgang 382
scleractinian coral 51
sea level cycle 24
sea level highstands
   platform margin progradation 10
sea level highstands 10
sea level inflection points 20

sea level lowstands 137
sea margin caves 211
secondary fluid inclusions 88
secondary porosity 37, 41, 63
   dolomitization 63
   moldic 154
secondary porosity associated with dolomitization 55
sediment accommodation space 182
sediment compaction 22
sediment fill 21
sediment supply 21
sedimentary sequence 25
seismic record 24
seismic reflector terminations 24
seismic scale cycles 33
seismic stratigraphy 24
seismic velocity
   impacted by porosity 378
seismic volume analysis
   3-D 377
separate vugs
   Lucia carbonate porosity classification 43
Sequence 20
sequence boundary 23, 154
   amalgamated 269
   characteristics 344
   Jurassic Smackover Fm. 360
sequence stratigraphic models 26
   escarpment margin 29
      margin incisement 29
   isolated platform 30
   ramp 26
   rimmed shelf 28
sequence stratigraphy 20–34, 170
   Jurassic
      Persian Gulf 179
   Jurassic Smackover Fm., central Gulf, USA 358
   major variables 20
   Mississippian Madison, Wyoming, USA 344
   Oxfordian Jurassic
      Gulf Coast USA 170
   Yates Field 226
shale dewatering 52
Shark Bay, western Australia
   marine hardgrounds 104
shelf crest model
   Capitan reef, Texas New Mexico, USA 114
shelf lagoon 13

shelf margin 13, 326
shelf margin collapse 30
shelf margin facies tract 13
shelf margin reef
   marine cementation 63
shelf margin relief 29
shelter porosity 47
shoaling-upward parasequences 34
shoreface 34
shoreline beach complex 360
short-term 3rd order cycles 26
short-term, high frequency 4th order 26
shorthand porosity designations
   MO for moldic
      BP for interparticle 43
sidewise poisoning
   calcite crystal morphology 65
siliciclastic encroachment 30
siliciclastic slope fan
   LST 362
siliciclastics 3, 283
   associated with sequence bounding unconformity
      LST 360
   rock water interaction 319
site defects 71
Size and sorting
   contrasts between carbonates and clastics 5
size and sorting control
   natural size distribution of components 5
size and sorting in carbonate sediments 5
size modifiers
   Choquette and Pray carbonate porosity classification 42
slope to basin facies tracts 14
SMOW (Standard Mean Ocean Water) 80
soil gas 189
soil zone carbon dioxide 63, 185
solid bitumen 59
solution-collapse breccia 153
   evaporite 160
   karst 223
source rock 179
   carbonate
      Jurassic Smackover Fm. 358
South Cowden Field, west Texas, USA 166
Southeast Asia 30
speleothems 68, 186
sphalerite 59

spherical fiber fascioles 5
spherulitic aragonite 108
sponges 114
   framework porosity 51
Sr 74
   in botryoidal aragonite 108
Sr residence time 85
St. Croix 108
stable isotopes 79
   carbon 80
      fermentation 83
      thermochemical degredation of organic matter 83
   characteristic of diagenetic environments 80
   dolomite 81, 147, 169, 173
      Knox Group 319
      Madison Fm., Wyoming, USA 352
   interpretation 84
   marine cements 97
      changes through time 101
      reef 108
   meteoric waters 188, 219
      regional trends 218
      vadose zone 207
   oxygen 79
      temperature dependence of fractionation 80
   sampling 84
   subsurface calcite cements 311
   trends 83
      meteoric "J" trend 189
start-up phase 10
state of the art 382
storm events 103
storm processes 16
Straits of Andros 32
stratal patterns 24
stratigraphic traps 276
   Jurassic Smackover
      progradational shoreline ooid grainstones 370
stratigraphy 20
Strawn/Wolfcamp shelf, west Texas, USA 278
stromatactis 133
stromatoporoids 4
   framework porosity 51, 123
strontianite 59
strontium isotopes 84
   basinal fluids 311
   dolomite
      Madison Fm., Wyoming, USA 353
   dolomites 173
   late subsurface dolomites
      Knox Group 319
   marine water
      marine water isotope curve 85
   meteoric waters 87
   oil field brines
      Lower Cretaceous Edwards 87
   subsurface calcite cements 311
structural traps
   Jurassic Smackover Fm. 357, 369
structural-stratigraphic trap
   Little Knife Field 156
structurally controlled platforms 32
Stuart City Fm., Texas, USA 115
subaerial 42
subaqueous 42
   evaporites 156, 165
submarine hardgronds 104
   recognition 106
submarine volcanic eruption 134
subsidence 24
subsurface calcite cement
   Jurassic Smackover Fm. 76
subsurface diagenetic environment 61
subsurface dissolution
   in a passive margin 312
      importance 313
sucrosic dolomite 153
sulfate cements; 39
sulfate reduction 101, 130
sulfur isotopes 80
super deep gas field 341
super-continent break-ups 26
supersaturated 35
supersequence (2nd order) 268
   Cretaceous, Angola 281
   Devonian carbonates, western Canada 325
   Jurassic, Saudi Arabia 179
   Madison Fm., Wyoming, USA 343
supersequence (2nd order) stacking patterns
   diagenesis and porosity evolution 268
supratidal 34, 52, 154, 159
   fenestral porosity 52
   south Florida 151
supratidal flats 35
surface potential

due to calcite crystallography 67
surface-active cations 67
surficial diagenetic processes 42
suspended sediments
   source of reef internal sediments 109
Swan Hills reef, western Canada 123
systems tracts 20, 25, 158
   genetically linked depositional systems 20
   highstand 25
      prograding shoreline Oaks Field 206
      sabkha 152
   low stand 25
      hydrologic setting 209
   transgressive 25

T

talus apron 29
tectonic compressional loading
   diagenetic fluid movement
      dolomitization and dissolution 320, 324
tectonic subsidence 22
tectonics and basin hydrology 295
teepees 199
telogenetic 42
telogenetic zone 62
temperate climate 94
   marine 94
temperature 293
   effect on chemical reactions 294
      catagenesis of organic compounds 294
temporal changes in abiotic and biotic carbonate mineraks 36
temporal variation of chemically precipitated carbonate 36
temporal variation of stable isotopes 36, 84
temporal variations in oceanic $^{87}Sr/^{86}Sr$ 86
terrestrial clays 87
textural scale 8
texturally based classification 6
thermal convection 63, 139, 251, 253, 259
   Devonian, western Canada
      dolomitization 329
thermal degradation 54
thermal degradation of hydrocarbons 63
thermal history 306
   Knox Group 320
thermal maturity
   relationship to porosity 338

thermochemical sulfate reduction 368
thermocline
   aragonite saturation within 93
thrust faulting 297, 342
   loading 342
tidal channel complexes 159
tidal creek natural levees
   Andros Island
      dolomitization 151
tidal flat organisims
   gastropod 5
tidal flats 14
tidal pumping
   south Florida
      dolomitization 151
   trigger for beach rock cementation 102
Tongue of the Ocean, Bahamas 138
tools for the recognition of diagenetic environments 64
   constraints 65
top seal
   Madison Fm., Wyoming, USA 352
top sets 29
top-lap 24
touching vugs
trace element geochemistry of calcite cements and
trace elements
   constraints on their use in carbonates 74
Transcontinental Arch 342
transgressive system tract (TST) 25
transgressive systems tract
   porosity/diagenesis evolution 253
transport of $CaCO_3$ 54
Trucial Coast, Middle East 16
TTI-porosity plots 339
Tucker, Maurice 382
turbidites 14
Tuxpan Platform, Mexico 15, 118
two-phase fluid inclusions 81, 171

U

ultrastructure of grains 49
   coral
      Halimeda 5
      spherical fiber fascioles 5
   ooids
      peloids 49
   use of in identification of grain composition 6

unconformities 42, 211
　exhumed limestones 54
　karsting 225
　　sequence boundary 344
　porosity destruction by cementation 377
　sequence bounding
　　caliche crusts 255
undersaturated 54
　with respect to most carbonate minerals 63, 185
undulose extinction 111
unit cell
　aragonite 74
uplift 22
Upper Jurassic Smackover 26
　north Louisiana, USA 206
Upper Jurassic Smackover Fm., central Gulf Coast, mature petroleum fairway
　case history 355
　late subsurface porosity development 54
upper slope 14

V

vadose 194, 246
　lower zone
　　zone of capillarity 197, 199
　　meniscus cement 212
　　porosity development 199, 261
　　　lower vadose 199
　upper zone 197
vadose zone 63
Valles-San Luis Potosi Platform, Mexico 120
vapor bubble
　in a two-phase fluid inclusion 88
varved limestones 26
visible porosity 38
vitrinite reflectence 294
vug 160
vug interconnection 46
vuggy porosity 286
　in the sense of the Lucia carbonate porosity classification 43
vugs 154
　after gypsum 153
　Choquette and Pray carbonate porosity classification 54

W

wackestone 8
wadies 27
water density
　evaporative waters 146
water flood 278
water saturation 47
water table 194
　diagenetic activity 210
　paleo-water table 228
Western Canada Sedimentary Basin 324
whisker crystal cement
　soil zone 68
whitings 3
Williston Basin, USA 330, 342
Wind River Canyon, Wyoming, USA 343
windward platform margin 257, 259
　reef development
　　Philippines 373
　reefs
　　porosity destruction 265
windward reef bound margins 32

Y

Yates Field, west Texas, USA: 225
　geologic setting 226
Yucatan Peninsula, Mexico 195
　dorag dolomitization 237
　strand plain
　　floating fresh water lenses 205

Z

Zama reefs, western Canada 178
zooplankton 109